Surface Engineering and Mechanical Properties of Building Materials

Surface Engineering and Mechanical Properties of Building Materials

Guest Editors

Jiwang Jiang
Hao Jin
Xing Cai
Hui Li
Li Ai

MDPI

Basel • Beijing • Wuhan • Barcelona • Belgrade • Novi Sad • Cluj • Manchester

Guest Editors

Jiwang Jiang
School of Transportation
Southeast University
Nanjing
China

Hao Jin
School of Transportation
Southeast University
Nanjing
China

Xing Cai
School of Transportation
Southeast University
Nanjing
China

Hui Li
School of Transportation
Southeast University
Nanjing
China

Li Ai
Department of Civil and
Environmental Engineering
University of South Carolina
Columbia, SC
USA

Editorial Office
MDPI AG
Grosspeteranlage 5
4052 Basel, Switzerland

This is a reprint of the Special Issue, published open access by the journal *Coatings* (ISSN 2079-6412), freely accessible at: https://www.mdpi.com/journal/coatings/special_issues/Building_Mechanical.

For citation purposes, cite each article independently as indicated on the article page online and as indicated below:

Lastname, A.A.; Lastname, B.B. Article Title. *Journal Name* **Year**, *Volume Number*, Page Range.

ISBN 978-3-7258-5699-2 (Hbk)
ISBN 978-3-7258-5700-5 (PDF)
https://doi.org/10.3390/books978-3-7258-5700-5

Contents

Preface . . . . **vii**

Zili Zhao, Jiwang Jiang and Fujian Ni
Characterization of the Cracking Resistance Gradient of Bitumen Emulsion-Based Cold In-Place Recycling Mixtures over Curing by Semi-Circular Bending Test
Reprinted from: *Coatings* **2024**, *14*, 46, https://doi.org/10.3390/coatings14010046 . . . . **1**

Yajin Han, Jiwang Jiang, Jiahao Tian, Zhu Zhang, Fujian Ni and Sheng Zhang
An Investigation of the Dynamic Curing Behavior and Micro-Mechanism of a Super-Tough Resin for Steel Bridge Pavements
Reprinted from: *Coatings* **2023**, *13*, 1567, https://doi.org/10.3390/coatings13091567 . . . . **17**

Zhanming Zhang, Duo Xu, Yunlong Fan, Hui Ma, Zihao Ding, Jiwang Jiang, et al.
Investigation of Fatigue Performance for In-Service Asphalt Pavement Mixtures Using Push-Pull Fatigue Test
Reprinted from: *Coatings* **2025**, *15*, 178, https://doi.org/10.3390/coatings15020178 . . . . **34**

Duo Xu, Fujian Ni, Hui Du, Zili Zhao, Jingling Wang and Sheng Chen
Investigation of Factors Affecting the Intermediate-Temperature Cracking Resistance of In-Situ Asphalt Mixtures Based on Semi-Circular Bending Test
Reprinted from: *Coatings* **2023**, *13*, 384, https://doi.org/10.3390/coatings13020384 . . . . **50**

Yueran Zhang, Baoyong Cao, Heng Zhang, Xiong Zhang and Yan He
Study on Key Mechanical Parameters of High-Strength Grouting Material
Reprinted from: *Coatings* **2023**, *13*, 440, https://doi.org/10.3390/coatings13020440 . . . . **65**

Fujian Tang, Hao Cui, Gang Li and Shangtong Yang
Effect of Strain Levels on the Corrosion Resistance of an Enamel-Coated Steel Rebar
Reprinted from: *Coatings* **2023**, *13*, 510, https://doi.org/10.3390/coatings13030510 . . . . **78**

Yajin Han, Zhu Zhang, Jiahao Tian, Fujian Ni and Xingyu Gu
Long-Term In Situ Performance Evaluation of Epoxy Asphalt Concrete for Long-Span Steel Bridge Deck Pavement
Reprinted from: *Coatings* **2023**, *13*, 545, https://doi.org/10.3390/coatings13030545 . . . . **89**

Dong Zhao, Xiang Ma, Hao Wang and Chaolin Zhang
Investigation of the Wet and Thermal Conditions Effect on the Micro-Scale Characteristics of Interfacial Transition Zone of Porous Asphalt Mixture
Reprinted from: *Coatings* **2023**, *13*, 566, https://doi.org/10.3390/coatings13030566 . . . . **107**

Jianhu Xu, Hui Wang, Wanzhen Wang and Feiting Shi
The Influence of CO_2-Cured Incinerated Waste Fly Ash on the Performance of Reactive Powder Concrete
Reprinted from: *Coatings* **2023**, *13*, 709, https://doi.org/10.3390/coatings13040709 . . . . **120**

Xianghui Kong, Gaoqiang Wang, Shu Rong, Yunpeng Liang, Mengmeng Liu and Yanhao Zhang
Utilization of Fly Ash and Red Mud in Soil-Based Controlled Low Strength Materials
Reprinted from: *Coatings* **2023**, *13*, 893, https://doi.org/10.3390/coatings13050893 . . . . **134**

Xiang Ma, Jinquan Kan, Song Liu, Mingyuan Tu and Dongjia Wang
Investigation of the Effect of Filler on Cohesive Bond Strength of Asphalt Mastic Using Binder Bond Strength (BBS) Test
Reprinted from: *Coatings* **2023**, *13*, 1001, https://doi.org/10.3390/coatings13061001 . . . . **146**

Haozhen Wang, Xin Cai, Xi Peng, Hui Wang and Pengqian Wang
Study on Soil Corrosion Resistance Reinforced with Reactive Powder Concrete in Chloride Environment
Reprinted from: *Coatings* **2023**, *13*, 1134, https://doi.org/10.3390/coatings13071134 . . . . **161**

Zili Zhao, Fujian Ni, Junqiu Zheng, Zhiqiang Cheng and Shengjia Xie
Evaluation of Curing Effects on Bitumen Emulsion-Based Cold In-Place Recycling Mixture Considering Field-Water Evaporation and Heat-Transfer Conditions
Reprinted from: *Coatings* **2023**, *13*, 1204, https://doi.org/10.3390/coatings13071204 . . . . **180**

Liheng Shu, Fujian Ni, Hui Du and Yajin Han
An Evaluation of Asphalt Mixture Crack Resistance and Identification of Influential Factors
Reprinted from: *Coatings* **2023**, *13*, 1382, https://doi.org/10.3390/coatings13081382 . . . . **197**

Yueyang Shi, Zongliang Tong, Qingbo Ren, Zhigang Li and Jingling Wang
Modification of the Crumb Rubber Asphalt by Eucommia Ulmoides Gum under a High-Temperature Mixing Process
Reprinted from: *Coatings* **2024**, *14*, 1059, https://doi.org/10.3390/coatings14081059 . . . . **218**

Hangyang Wang, Kaiwei Gong, Bingling Cheng, Xi Peng, Hui Wang and Bin Xu
Effect of Chloride Salt Erosion on the Properties of Straw Fiber Reactive Powder Concrete
Reprinted from: *Coatings* **2024**, *14*, 1069, https://doi.org/10.3390/coatings14081069 . . . . **228**

Yehan Yu, Bing Xiao, Zihao Cao, Bingling Cheng, Xi Peng and Hui Wang
The Relationship between the Fresh Sludge Ceramsite Concrete's Fluidity and the Sludge Ceramsite's Dispersion
Reprinted from: *Coatings* **2024**, *14*, 1095, https://doi.org/10.3390/coatings14091095 . . . . **245**

Mingxin Lai, Xiaoying Gao, Lin Kong, Lizong Chen, Guoan Gan, Haixing Lin, et al.
Enhancing Interface Performance Through Self-Assembly Mechanisms of APTES on Surface-Modified Tuff Aggregates
Reprinted from: *Coatings* **2024**, *14*, 1422, https://doi.org/10.3390/coatings14111422 . . . . **261**

Wojciech Owczarzak, Sławomir Sommer and Grzegorz Ronowski
Method of Forming Road Surface Replicas Using 3D Printing Technology
Reprinted from: *Coatings* **2024**, *14*, 1455, https://doi.org/10.3390/coatings14111455 . . . . **279**

Lian Li, Chunxiang Huang, Xu Zeng, Jinfeng Yang, Wei Gao, Yaping Chen and Haibo Yang
Rheological Performance and UV Aging Resistance of Buton Rock Asphalt and Styrene-Butadiene-Styrene Composite Modified Asphalt Under High Temperature
Reprinted from: *Coatings* **2024**, *14*, 1589, https://doi.org/10.3390/coatings14121589 . . . . **297**

Xiang Cheng, Qizhi Li, Peng Liu, Jingxiang Huang, Lingling Wang, Ying Chen, et al.
Experimental and Numerical Investigation on the Mechanical Properties of Concrete with High Volumes of Modified Phosphogypsum
Reprinted from: *Coatings* **2025**, *15*, 65, https://doi.org/10.3390/coatings15010065 . . . . **315**

Xudi Huang, Xi Peng, Fengjiang Qin, Qiuwei Yang and Bin Xu
Damage Detection of Beam Structures Using Displacement Differences and an Artificial Neural Network
Reprinted from: *Coatings* **2025**, *15*, 289, https://doi.org/10.3390/coatings15030289 . . . . **342**

Preface

The ever-increasing demands on our transportation infrastructure, coupled with the urgent need for sustainable development, have placed the performance and durability of building materials at the forefront of engineering research. This Reprint, stemming from the closed Special Issue "Surface Engineering and Mechanical Properties of Building Materials," was conceived to address these critical challenges. Our primary aim was to compile a body of work that not only diagnoses the failure mechanisms in conventional materials but also pioneers innovative solutions for enhancement and sustainability.

The motivation for this project was to bridge the gap between fundamental material science and practical engineering applications. We sought to bring together diverse yet interconnected studies that explore the entire lifecycle of construction materials—from their micro-scale interfacial properties to their long-term in-service performance. The scope of this Reprint is broad, encompassing advanced testing methodologies for asphalt and concrete, novel modification techniques, and the strategic incorporation of industrial by-products to create a new generation of eco-friendly construction materials.

This Collection is addressed to a wide audience, including academic researchers, civil and materials engineers, pavement technologists, and graduate students who are dedicated to advancing the field of construction materials. We believe that the insights contained within will provide valuable perspectives for both fundamental research and the development of next-generation infrastructure solutions.

As the Guest Editors, we extend our deepest gratitude to all the contributing authors for sharing their valuable research and expertise, which form the very foundation of this Reprint. We are also profoundly thankful to the numerous dedicated reviewers whose rigorous and insightful comments were instrumental in ensuring the high quality of the published works. Finally, we acknowledge the unwavering support and professionalism of the editorial team and staff at MDPI, whose efforts were vital in bringing this significant Collection to fruition.

Jiwang Jiang, Hao Jin, Xing Cai, Hui Li, and Li Ai
Guest Editors

coatings

MDPI

ticle

haracterization of the Cracking Resistance Gradient of itumen Emulsion-Based Cold In-Place Recycling Mixtures ver Curing by Semi-Circular Bending Test

i Zhao, Jiwang Jiang * and Fujian Ni

College of Transportation Engineering, Southeast University, Nanjing 211189, China; zili_zhao@seu.edu.cn (Z.Z.); 101004746@seu.edu.cn (F.N.)
* Correspondence: jiang_jiwang@seu.edu.cn

Abstract: To better reveal the performance development of bitumen emulsion-based cold in-place recycling (BE-CIR) mixture over curing, a semi-sealed laboratory curing method was proposed in this research to simulate the in situ moisture evaporation process and cracking resistance of the BE-CIR specimen at different depths during a curing time of 28 days, which was also investigated by the semi-circular bending (SCB) test. The influencing factors of cement content (1.5% to 2.5%), initial moisture content (3.5% to 4.5%), curing temperature (25 °C to 45 °C) and relative humidity were investigated, and the significance of different factors affecting the performance development was also analyzed. The results indicate significant variations in cracking performance parameters at different depths, with the top part exhibiting notably higher tensile strength and fracture energy compared to the bottom part, and a gradient index (GI) is proposed to describe the difference. Cement content affected early tensile strength and fracture energy, while the initial moisture content affected the development rate of the performance. The influence of curing temperature was extensive, and as the temperature increased beyond 40 °C, the strength of the effect decreased. High humidity during the early stage of curing inhibited the strength formation and development of fracture energy. The performance development of the BE-CIR mixture is more significantly influenced by the moisture migration process, which is governed by curing temperature and relative humidity, as opposed to the cement content and initial moisture content.

Keywords: bitumen emulsion; cold-in-place recycling; curing; gradient characteristics; crack resistance

1. Introduction

In recent years, with the continuing attention given by society to environmental issues, low-energy and low-emission technologies for asphalt pavements have also received rapid development. Bitumen emulsion-based cold in-place recycling (BE-CIR) as a cost-effective and environment-friendly maintenance technology has gained worldwide application [1–4]. The composition of the BE-CIR mixture includes emulsified bitumen, virgin aggregates, recycled asphalt pavement (RAP), additional water, and various inorganic additives (e.g., Portland cement) [5]. Due to the low viscosity of bitumen emulsion, BE-CIR mixture can be mixed, paved and compacted at an ambient temperature, as opposed to the traditional hot-mix asphalt (HMA) [6,7]. The adding of Portland cement can improve the mechanical properties of mixtures by promoting the dispersion of the bitumen, regulating the breaking of the emulsion and accelerating the curing time [8–10]. The additional water is an essential component, since it contributes to homogeneous distribution of bitumen droplets and good encapsulation of aggregates [11–13].

Because of the wide variation in the component materials, the mechanical performance of BE-CIR mixtures is a combination of the properties of the bitumen and the cement [14]. Bitumen is a viscoelastic binder with low stiffness, while cement is a hydraulic binder with brittle hydration products and high compressive strength and stiffness. Increasing the cement content in this composite leads to variations in the microstructure of the co-binder, and consequently increases the stiffness and decreases the temperature dependence of the mechanical properties [15–17]. Despite such variability, a common feature of BE-CIR mixtures' mechanical performance is their evolutionary behavior with curing time [18–20]. During the curing process, the water content is reduced to equilibrium with the surroundings due to evaporation, and the mixture evolves from the short-term "fresh" state to the long-term "hardened" state under the interaction of different physical and chemical mechanisms, including emulsion breaking and setting and cement hydration.

The BE-CIR mixtures have been investigated in a number of studies, showing good rutting resistance and relatively high modulus, while its crack resistance and fatigue durability are relatively weak compared to HMA [21–26]. Given that the fracture characteristics of the BE-CIR layer directly affect the service life of the pavement structure, it is important to study the growth behavior of the fracture properties in order to facilitate the enhancement of the cracking resistance of the mixes. Factors influencing the mechanical properties of BE-CIR include not only material composition and compaction, which are similar to those of HMA, but also the curing process [27–29]. Saadoon et al. [30,31] studied the performance of CIR mixtures with different cements and emulsified bitumen. The Marshall stability was found to have a good relation with the water loss, which is mainly influenced by the bitumen emulsion, but not with the type and content of cement. Kim [32] confirmed that the moisture content and length of the curing period have significant effects on the properties of the CIR mixtures made with emulsion asphalt. Effects of different curing conditions in terms of temperature (varied 15 °C–60 °C) and relative humidity (varied 6%–95%) were studied [33–35]. It was demonstrated that higher curing temperatures can lead to an increase in strength, and that higher humidity at an early stage can improve the hydration of cement [27,36]. However, the extent of influence between material design parameters and curing conditions has not been comprehensively compared.

In addition, from the perspective of field performance, a unique gradient distribution of the in situ moisture content during curing in BE-CIR pavement was identified by monitoring the humidity at different depths using humidity sensors [37]. The tensile strength of field cores determined using semi-circular bending tests (SCB) also exhibited gradient features which are higher in the top half and lower in the bottom. However, the reasons for the gradient features of field cores include moisture evaporation and vehicle loading pressure. It is essential to identify the reason for the gradient feature and investigate the forming process. Additionally, another work has evaluated the effects of field-water evaporation and heat-transfer conditions in the laboratory [38]. A semi-sealed laboratory curing condition was promoted as being more consistent with field performance.

To the authors' best knowledge, the BE-CIR mixtures are mostly viewed as an integrated whole when evaluating their mechanical properties, which may display a great difference from examples in the field. Unlike the hot-mix asphalt, the strength formation of BE-CIR is time-variable due to the coupling effects of cement hydration, emulsion breaking and water evaporation. However, the observed depth variation mechanism and its corresponding influencing factors have not been investigated. Therefore, it is essential to identify the reason for the gradient feature and investigate the forming process. This paper intends to reach a better understanding of the gradient developmental behavior of crack resistance in BE-CIR mixtures under different influencing factors.

2. Scope and Objectives

The crack resistance associated with different depths in BE-CIR mixtures was evaluated by an SCB test under semi-sealed laboratory curing conditions. The main objectives are listed as follows:

1. Investigate the gradient characteristics of the cracking resistance of BE-CIR mixture affected by water evaporation;
2. Characterize the influence of factors including mixture design parameters and curing conditions on the gradient developmental behavior of fracture properties;

3. Evaluate the influence of various factors on the fracture properties and gradient distribution.

3. Materials and Experimental Design

3.1. Materials

The RAP materials were collected from a highway pavement maintenance project using the BE-CIR technique in the Jiangsu Province of China. As shown in Figure 1, the gradation of RAP material from this project complied with the medium-gradation limitations specified by the Jiangsu Province agency [39]. The type of emulsified bitumen was cationic slow-setting, and the cement was Portland Ordinary Cement P.O. 42.5; the properties of these materials are listed in Table 1, and are in accordance with the Chinese standards for cold recycling [40]. In the BE-CIR project, 100% of RAP materials were recycled and no addition aggregates were added in the field. So, similarly, the mixture samples were prepared with a 100% reuse of RAP.

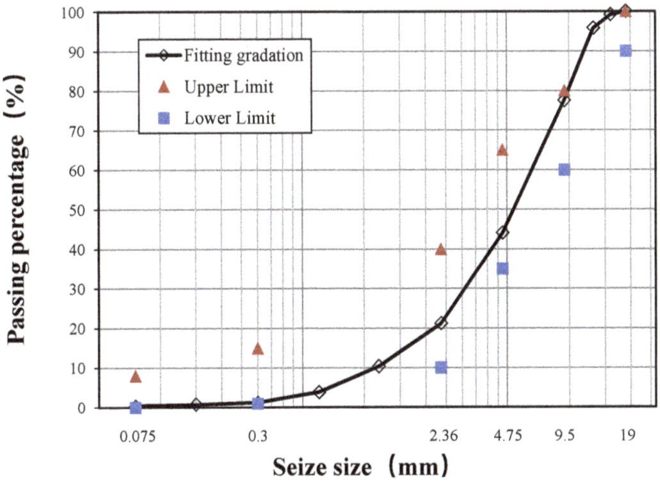

Figure 1. Aggregate gradation of BE-CIR mixtures.

Table 1. Properties of emulsified bitumen and P.O. 42.5 cement.

Properties of the Emulsified Bitumen		Value
Residue content (%)		64
Penetration (25 °C; 0.1 mm)		59.1
Soft point (°C)		59.9
Ductility (15 °C; cm)		110
Sieve residue (1.18 mm; %)		0.03
Storage stability (5 d, 25 °C; %)		1.47
Storage stability (1 d, 25 °C; %)		0.42
Properties of Cement		
Setting time (min)	Initial set	3.14
	Final set	4.62
Flexural strength (MPa)	3 d	4.9
	28 d	7.2
Compressive strength (MPa)	3 d	24.1
	28 d	51.4

3.2. Specimen Preparation and Curing Conditions

Firstly, the RAP was dried in an oven at 60 °C to a constant mass and then cooled naturally to an ambient temperature in preparation for mixing. The mixing was performed in three steps: firstly, the RAP and cement were mixed, then water was added to the mix, and finally the emulsified bitumen was added. The mixing time for each step was 60 s. This procedure enabled the aggregates and fillers to be wet before coming into contact with the binder so that the adsorption capacity of the emulsified bitumen could be reduced, which favored the homogeneity of the mix. Moreover, the Chinese specification stipulates that the mixing time should not be longer than 60 s after the addition of asphalt emulsion, in order to avoid the breaking of asphalt emulsion. After mixing, the specimens of 150 mm in diameter and 105 ± 2 mm in height were compacted by the Superpave Gyratory Compaction (SGC); approximately 3800 g of the mixture was used.

The accelerated curing condition usually employed in most studies involves placing the specimen in the oven unsealed at an ambient temperature. However, these methods ignore the boundary condition of water evaporation under field conditions, and the moisture migration may contribute to the inconsistent strength-growth behavior.

Based on a previous study, a semi-sealed laboratory curing condition was designed to simulate the in situ curing condition, and proved to be more consistent with field cores [38]. This method employed a waterproof fabric made of Teflon material wrapped around the side and bottom surfaces of the specimen to ensure that moisture within the mixture could only evaporate from the upper surface, as shown in Figure 2. The environmental chamber was then used for curing at different combinations of temperature and relative humidity.

Figure 2. Curing of selected full-depth specimens.

3.3. Testing Program

In this paper, a total of four influencing factors were investigated, including the two mixture-design factors of cement content and initial moisture content and the two curing environmental factors of temperature and relative humidity. The orthogonal experimental design method was used; a total of 10 different BE-CIR mixtures were designed for testing, and the influencing factors corresponding to each BE-CIR were as shown in Table 2. The asphalt emulsion content of all BE-CIR was set at 3.3% according to the mix design used during the field project. In addition, a relatively high cement content of 2% was adopted in the previous project to address the lack of high temperature performance and weak early strength. Therefore, in the orthogonal experimental design, a cement content of 2%, an initial moisture content of 4% and the average curing temperature of 35 °C used in the field project were centered. Three cement contents of 1.5%, 2% and 2.5%, and three initial moisture contents of 3.5%, 4% and 4.5% were studied, respectively. The initial moisture content consists of the water in the emulsified bitumen and the extra water. All BE-CIR had the same emulsified bitumen content of 3.3%, and the moisture in the emulsified bitumen was 1.188% according to the residue content. The different initial moisture content was controlled by the extra moisture. Since the BE-CIR pavement is not suitable for construction in environments with low temperature, the range of the curing temperature was set to extend from 25 °C to 45 °C, and every 5 °C was set for different groups. In terms of the relative humidity of curing, except for a constant relative humidity of 30%, this paper considers one situation in which relative humidity was higher in the early 5 days of curing (90%), which represents rain in this curing period.

Table 2. Design of different BE-CIR.

Mix ID	Cement Content (%)	Initial Moisture Content (%)	Curing Temperature (°C)	Relative Humidity of Curing (%)
BE-CIR1	1.5			
BE-CIR 2	2	4		
BE-CIR 3	2.5		35	
BE-CIR 4		3.5		
BE-CIR 5		4.5		30%
BE-CIR 6			25	
BE-CIR 7	2		30	
BE-CIR 8		4	40	
BE-CIR 9			45	
BE-CIR 10			35	90% (first 5 days)

In order to investigate the growth of crack resistance over curing time, for each BE-CIR, 12 parallel specimens were prepared for the following tests at different curing days. Four different curing durations of 3, 7, 14 and 28 days were selected, respectively, and there were three parallel specimens for test for each curing day. At each curing day set for a test, the mass of each sample was firstly measured to calculate the water-loss rate by comparing the result with the initial mass after compaction. The wrapped waterproofing was then removed and the specimens cut for an SCB test according to the process described in Figure 3. The sample size for the SCB test was a semicircle with a diameter of 150 mm and a thickness of 50 mm. Since this paper investigated the differences in performance development in the depth direction, four semicircular samples obtained from one specimen should be divided into two groups, top and bottom, respectively. The top part represented the portion containing a surface in direct contact with air during curing, and the bottom part was the remaining portion. Therefore, during each curing interval, three parallel specimens were utilized to obtain a total of 12 semicircular samples—six from the top portion and six from the bottom portion.

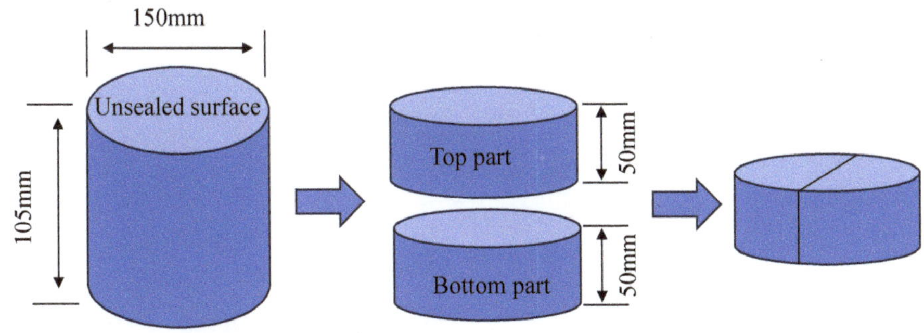

Figure 3. Specimen preparation for SCB test.

3.4. SCB Test

A UTM-25 universal testing system was used to perform the SCB tests in an environmental chamber according to AASHTO TP 124 [41]. The span between supporting rollers was set to 0.8 times the specimen diameter, which was 120 mm, as shown in Figure 4a,b. Before the test, specimens were placed into the environmental chamber for 4 h at the test temperature of 25 °C. The loading rate was set to 50 mm/min, and the applied load and displacement of the load actuator could be obtained. According to the load–displacement curve, as shown in Figure 4c, two indices, tensile strength (σ_t) and fracture energy (G_f), could be calculated by Equation (1) and Equation (2), respectively [42]. Apparently, these two indicators represent different perspectives on the cracking resistance of a mixture. The tensile strength denotes the ultimate load that the sample can withstand; the greater the strength, the stiffer the sample and the more difficult it is to crack. The fracture energy represents the flexibility of the sample, and the greater the fracture energy, the better the ability of the sample to work with cracks.

$$\sigma_t = \frac{4.8F_{max}}{BD} \tag{1}$$

$$G_f = \frac{W_f}{Area_{lig}} \tag{2}$$

where σ_t = SCB tensile strength (MPa); F_{max} = peak load (kN); B = specimen thickness (mm); D = specimen diameter (mm); G_f = fracture energy (J/m^2); W_f = fracture work (J), $\int Pdu$, where P = applied load (kN), u = vertical displacement (m); $Area_{lig}$ = ligament area (m^2), $BD/2$.

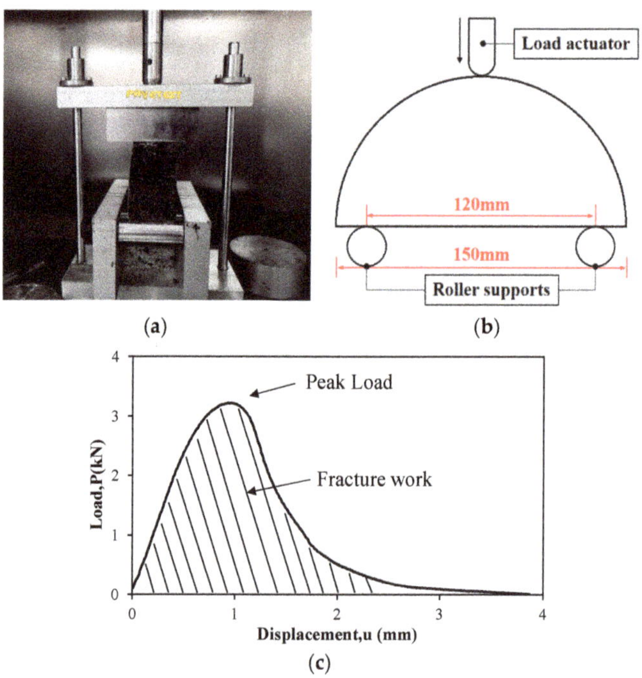

Figure 4. Scheme and curves for SCB tests. (**a**) Specimen under SCB test. (**b**) sketch of SCB test. (**c**) typical load-displacement curve.

In addition, the gradient index (GI) is defined to characterize the difference in cracking resistance in the depth direction. The gradient indices of tensile strength and fracture energy were calculated by Equation (3) and Equation (4), respectively.

$$GI - \sigma_t = \frac{\sigma_t^{top} - \sigma_t^{bottom}}{\sigma_t^{bottom}} \times 100 \quad (3)$$

$$GI - G_f = \frac{G_f^{top} - G_f^{bottom}}{G_f^{bottom}} \times 100 \quad (4)$$

where $GI - \sigma_t$ = the gradient index of tensile strength; $GI - G_f$ = the gradient index of fracture energy; σ_t^{top}, σ_t^{bottom} = tensile strength of top or bottom layer; G_f^{top}, G_f^{bottom} = fracture energy of top or bottom layer.

4. Results

4.1. Water Loss

The results for water loss for different BE-CIRs over curing time are shown in Figure 5. Overall, there was a trend of rapid increase followed by gradual stabilization of water loss for all samples except BE-CIR10. The rapid increase was mainly observed within 0–7 days of curing, and the variation was very small after a curing of 14 days.

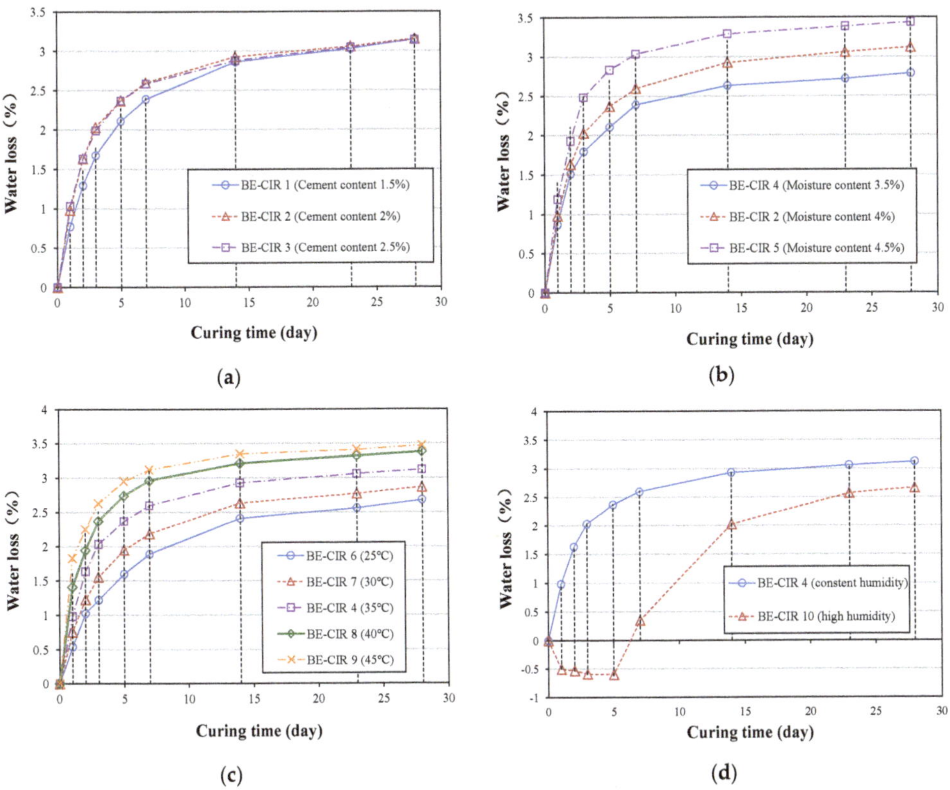

Figure 5. Water loss of different BE-CIR over curing. (**a**) cement content. (**b**) initial moisture content. (**c**) curing temperature. (**d**) relative humidity of curing.

According to Figure 5a, representing BE-CIRs with different levels of cement content, the water loss at 1.5% cement content was significantly smaller than with the other two cement contents, and the difference appeared mainly in the period of 0–7 days curing, while it became smaller after curing for 14 days. This phenomenon arises from the direct relationship between cement content and the quantity of free water combined with the cement, leading to an elevated rate of water loss. However, when the cement content was increased from 2% to 2.5%, the difference in water loss was minor and only slightly increased, from 0.78% to 0.83% for BE-CIR with 2.5% cement content at curing 1 d. The potential explanation for this occurrence could be that when the cement content surpassed 2%, an excess of cement particles caused uneven bonding with the bitumen film. As a result, some cement particles failed to make contact with the free water, thereby impeding further increases in water loss. The results for different initial water contents are shown in Figure 5b, which indicates that the water loss of the three mixtures was relatively similar to each other at curing 1 d, at approximately 1%, and with the increase of the curing, the water-loss rates of the three mixtures progressively exhibited the law of increasing according to the increased initial water content. Nevertheless, by using the initial moisture content to back-calculate the residual moisture content, it can be found that the residual moisture content was almost the same, at 0.8%, for different mixtures after curing for 28 days. This indicates that the initial moisture content could only affect the early rate of water loss but not the final residual moisture content.

Figure 5c illustrates the variation of water loss of BE-CIRs over time under different curing temperatures. It can be noticed that the water-loss rates for different days increased with the increase of the curing temperature. There was a significant increase in the water loss for every 5 °C increase in the interval from 25 °C to 40 °C, while the water loss at 40 °C and 45 °C were much closer to each other, which suggests that when the curing temperature was greater than 40 °C, the rate of water evaporation inside the BE-CIR mixtures might have reached its peak. This indicates that the curing temperature significantly affects the water migration and evaporation process of the mixture. In addition, comparing the water-loss curves at 40 °C and 25 °C, the water loss of the BE-CIR had reached 3% at 7 d when the curing temperature was 40 °C, and became stable after 14 d, while the water loss at a curing temperature of 25 °C was about 0.7% lower than that at 40 °C after curing 28 days. It can be assumed that there was still some residual water inside the BE-CIR mix at lower conditioning temperatures, and thus it might still take 2–3 months for the water to be completely evaporated.

The variation in water loss for two different humidity conditions is presented in Figure 5d. The water loss of BE-CIR10 was negative under the high-humidity condition for the first 5 curing days, which implies that the mixture could absorb a certain amount of additional water, and there exists an extreme value for this water absorption, which is about 0.5%. When the ambient relative humidity decreased to normal humidity, the water loss started to increase, and the rate of this increment was basically the same as that of BE-CIR4. This suggests that the occurrence of rainy conditions at the early stage of the curing will delay the water dissipation process accordingly.

4.2. *Performance Parameters from SCB and Gradient Distribution*

4.2.1. Effect of Cement Content

Figure 6 illustrates the SCB indices and GI over curing of different cement contents. It can be observed that both tensile strength and fracture energy continuously become larger along with the curing. However, the incremental rates of the two indices were different, and the tensile strength was constantly increasing in different curing periods, while the fracture energy increased significantly in curing 3–7 d, and the growth rate decreased in the subsequent period. This is explained by the fact that the tensile strength could more significantly be influenced by the hydration of the cement, which continues throughout the 28 days of curing. On the other hand, the fracture energy could be significantly influenced by the emulsified bitumen. During the curing period of 3–7 days, there was a rapid dissipation of water, causing the emulsified bitumen to rupture. Consequently, the adhesion between the bitumen film and the aggregate increased rapidly, resulting in a swift escalation of the fracture energy.

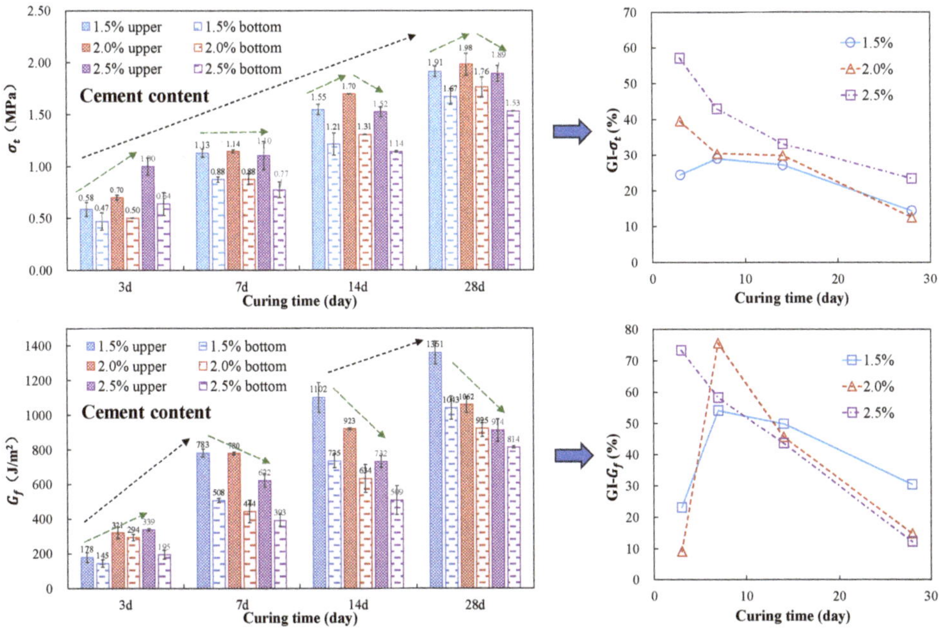

Figure 6. SCB Indices and GIs of different cement content.

Considering different cement contents, at 3 days of curing, the higher the cement content, the higher the tensile strength σ_t, especially when the cement content increased from 2% to 2.5% and the tensile strength of the top part increased from 0.7 MPa to 1.00 MPa, with an increment of 42.8%; the increment for the bottom part was slightly smaller but also went up, rising to 28%, indicating that an increase in the cement content can improve the early strength of the BE-CIR mixture. However, as the curing proceeded, the tensile strength of the BE-CIR with 2.5% cement content grew more slowly and was already smaller than those of the other two cement content levels at 7 days of curing. After curing 14 days, the tensile strength of the BE-CIR with 2.0% cement content was the highest, followed by that of 1.5%, and that of the 2.5% was the lowest. This indicates that there is an optimum cement content for the tensile strength of BE-CIR mixes in the long term. When the cement content is higher, it will lead to a decrease, in the long term, of tensile strength, although it can improve the early strength. The fracture energy was different from tensile strength to some extent. At 3 days of curing, the mixture with the lower cement content similarly had the lower fracture energy, while after 7 days, the higher the cement content the lower the fracture energy. The underlying cause is attributed to the fact that as the cement content increases, the mixture becomes stiffer, leading to reduced resistance against bending and tensile deformation. Consequently, this results in a decrease in fracture energy.

Further, comparing the performance of the top and bottom parts, the tensile strength and fracture energy of the top part was always greater than those of the bottom part, regardless of days of curing. This phenomenon was mainly due to the restriction of unidirectional evaporation of water, which can lead to an uneven distribution of residual water within the mixture; different water contents could lead to different degrees of cement hydration, forming different microscopic morphologies and adhesive capacities of the cement-emulsified asphalt mortar. As the curing proceeded, the gradient index (GI) of both

σ_t and G_f kept decreasing, which also meant that the difference in cracking performance between the top and bottom parts kept decreasing. For a GI of σ_t, the higher the cement content, the greater the GI. The GI-σ_t of BE-CIR with 2.5% cement content could even rise beyond 50% at 3 d of curing. Regarding a GI of G_f, the mixture with 2.5% cement content had the highest GI at 3 d of curing, while the mixture with 1.5% cement content was the highest after curing for 28 d. This also demonstrates that the higher the cement content, the more significant the influence on the inhomogeneous distribution of the residual water is at the early stage of curing. Meanwhile, cement plays a weaker role in the fracture-energy growth behavior, compared to its role in tensile strength.

4.2.2. Effect of Initial Moisture Content

Results for the SCB Indices and GIs of different initial moisture content are illustrated in Figure 7. The development of both σ_t and G_f for BE-CIRs with different initial water contents were almost similar. The tensile strengths of the different mixtures were almost the same at 3 and 28 days of curing. And only during 7 to 14 days of curing there was a (small) increase in σ_t with an increase in initial moisture content, which may have been due to a more rapid evaporation of water in this period. However, there were relatively large differences in G_f between three BE-CIRs. The mixture with the largest initial moisture content, of 4.5%, had the largest G_f, indicating that the rapid migration of moisture during the curing period could contribute to fracture resistance development.

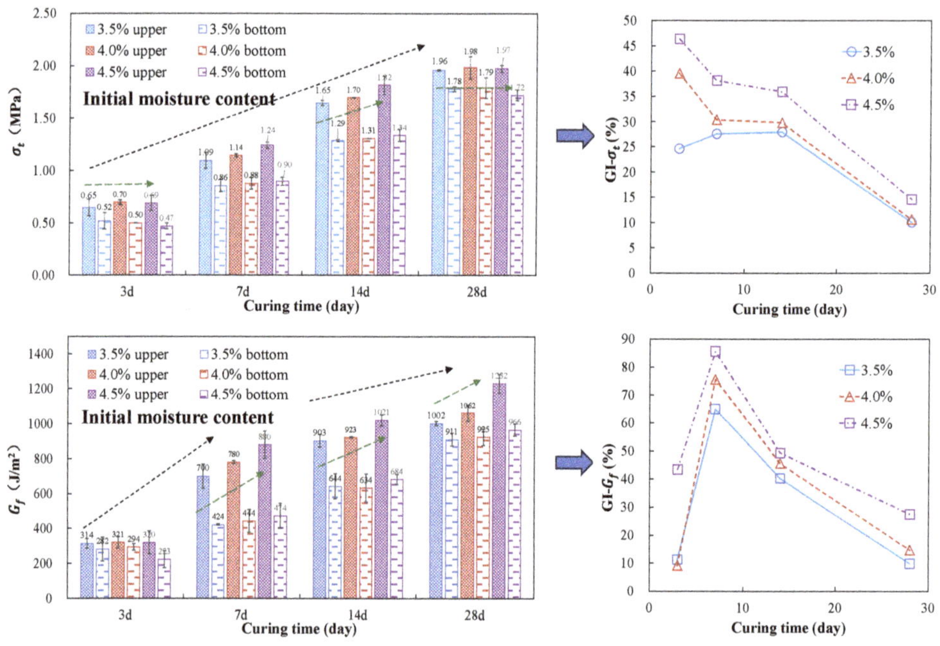

Figure 7. SCB Indices and GIs of different initial moisture content.

From the perspective of GI, the GI of σ_t decreased continuously with curing time, whereas the GI of G_f increased from 3 to 7 days, reached a maximum at 7 days of curing, and then decreased until the 28-day mark. The maximum of GI-G_f could be over 80%. And the GI-G_f could drop to 20% after curing 28 days. In addition, the greater the initial water content, the greater the GI of both σ_t and G_f. The rationale behind this is straightforward: a higher initial water content can cause a less uniform distribution of residual water in the depth direction.

4.2.3. Effects of Curing Temperature

Figure 8 shows the SCB Indices and GIs of BE-CIRs with different curing temperatures. At 3 days of curing, there was a large difference between the σ_t and G_f of BE-CIR mixtures at different curing temperatures. The σ_t of the BE-CIR at the lowest temperature of 25 °C was only 0.44 MPa for the top part and 0.40 MPa for the bottom part, while it reached 1.14 MPa and 0.92 MPa for the top and bottom parts, respectively, at a curing temperature of 45 °C, which were both more than twice as much as those at 25 °C. Similarly, the G_f of the BE-CIR at 45 °C was more than three times that of the mix at 25 °C. This demonstrates that there was a significant impact made by the curing temperature on the early development of anti-cracking properties. After curing for 7 days, the σ_t of BE-CIR with a curing temperature of 40 °C was already closer to that of the mixture at 45 °C, with a difference of only 0.13 MPa in the top layer, while the difference with the mixture cured at 35 °C was 0.42 MPa. The results for G_f presented larger differences between curing temperatures from 25 °C to 35 °C, with smaller differences between those from 35 °C to 45 °C. Combined with the results of the water loss, when the curing temperature rose to more than 40 °C, the water loss was basically the same at 7 days of curing, showing similar results with the tensile strength, indicating that the variation of the moisture inside the mixture generated quite a significant influence. Moreover, after 28 days of curing, the σ_t and G_f of BE-CIRs still increased with the increase of the curing temperature, and the fracture energy was significantly less at curing temperatures of 25 and 30 °C.

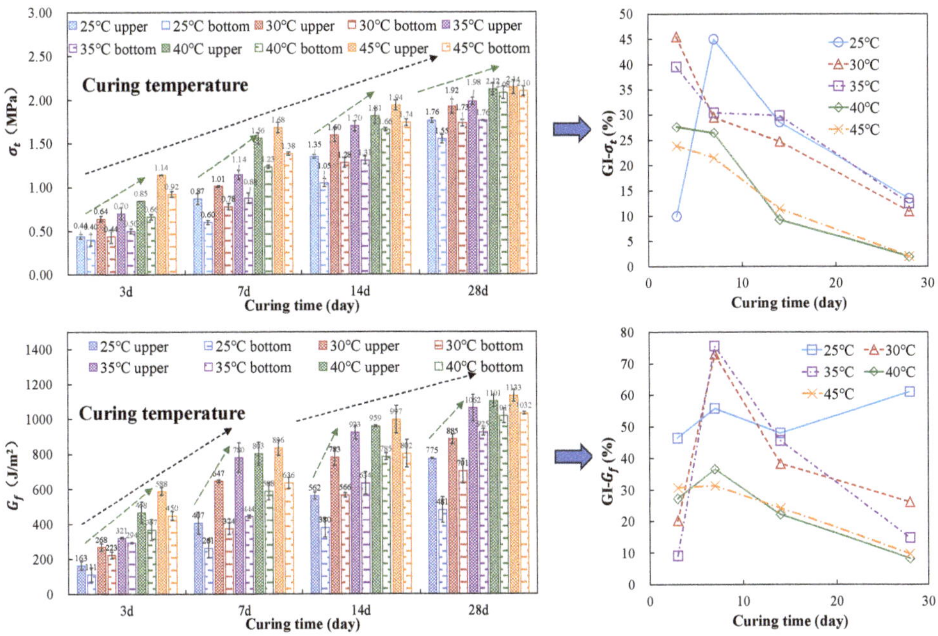

Figure 8. SCB Indices and GIs of different curing temperatures.

Considering the differences between top and bottom parts, the effects of the curing temperature were obvious. The GIs for the higher curing temperatures of 40 and 45 °C varied relatively closely over time and corresponded to smaller values, reaching only 30% during the first 7 days of the curing and dropping to less than 10% after 28 days. The other three curing temperatures had similar GI-σ_t, at 15%, after curing for 28 days, while the GI-G_f of the lowest curing temperature (25 °C) presented a much larger value, at 60%, than the other two values, which reached about 30%. It is known that different curing temperatures will result in different rates of moisture evaporation, which leads to different residual moisture contents as well as a gradient distribution, ultimately contributing to an inconsistent development of properties.

4.2.4. Effects of Relative Humidity

For easy expression, the constant relative humidity and early high relative humidity are replaced by 30% and 90% in this section; the results are presented in Figure 9. Due to the mixtures of 90% curing at a high relative humidity during the first 5 days, and the fact that water did not dissipate, or even the possible existence of water absorption, both the σ_t and the G_f of the top and bottom parts were basically less than those for the mixtures curing in a constant low relative humidity. The difference of performance between the two categories of mixtures was larger in the top parts than in bottom parts. In addition, the GI of σ_t and G_f was close to 0 or even negative at 3 and 7 days. This is further evidence that evaporation of moisture is directly responsible for the gradient in anti-cracking properties of the BE-CIR mixture. The presence of rain in the early stages of curing could lead to a delay in moisture migration, and consequently to a delay in the occurrence of performance development.

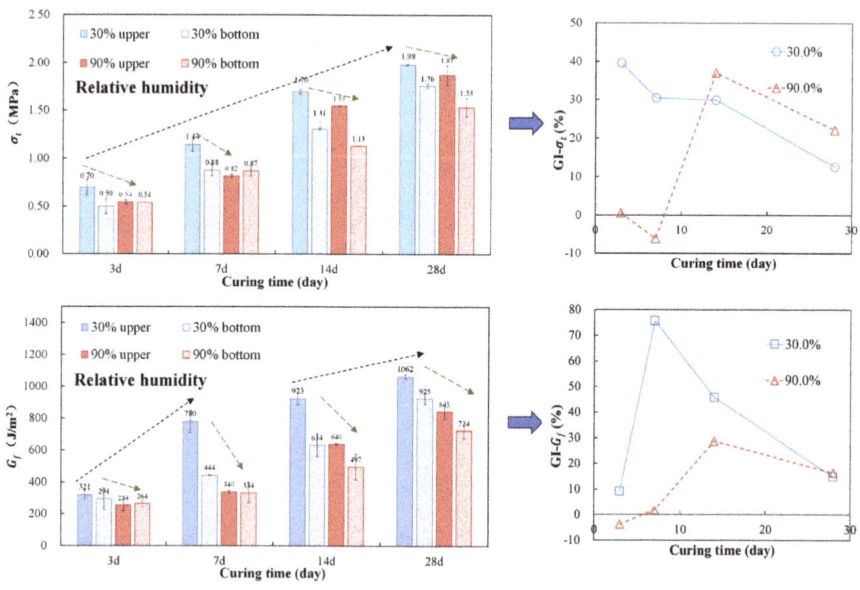

Figure 9. SCB Indices and GIs of different curing relative humidity.

4.3. Multifactor Analysis of Variance

The multi-factor analysis of variance (MANOVA) test was conducted by MATLAB to investigate the significance of the four factors. A variable is significant when the p-value is less than the level of significance (0.05). Results of MANOVA for both σ_t and G_f at different curing times are listed in Table 3. According to the p-value, the value for cement content was only significant in the σ_t of the top part at 3 days of curing, the top part at day 14, and the bottom part at 28 days of curing. This indicates that the cement content will primarily exert a significant impact on both the early strength and the long-term ability to resist deformation and function in the presence of cracks. The initial moisture content was not significant for properties at any curing time. The effect of curing temperature at each curing time was significant, with a p-value of less than 0.01, while the p-value of relative humidity was only smaller than 0.05 for day 3 or day 7, which was mainly due to the humidity mode used in this paper. As humidity conditions represent the weather during curing, the results prove that the presence of rain during the curing process has a significant influence on the anti-cracking properties. Therefore, compared to moisture content, the migration or evaporation process of water has a greater influence on the formation of the strength and long-term performance.

Table 3. MANOVA of factors for properties at different curing times.

p-Value	Top Layer				Bottom Layer			
	3 d	7 d	14 d	28 d	3 d	7 d	14 d	28 d
Factors				σ_t				
Cement content	0.0018	0.8175	0.8047	0.7591	0.1292	0.3635	0.6669	0.4106
Initial moisture content	0.6732	0.2672	0.1855	0.8778	0.6322	0.7351	0.7579	0.7195
Curing temperature	<0.0001	<0.0001	0.0007	0.0001	0.0005	<0.0001	0.001	0.0038
Relative humidity	0.0211	0.0042	0.2022	0.0726	0.9652	0.6468	0.1225	0.0787
Factors				G_f				
Cement content	0.5065	0.4467	0.0083	0.0013	0.2497	0.091	0.0544	0.0228
Initial moisture content	0.8663	0.5907	0.3563	0.1447	0.6329	0.6269	0.5379	0.6985
Curing temperature	<0.0001	0.011	0.0046	0.0014	0.0001	0.001	0.0019	0.002
Relative humidity	0.2641	0.045	0.6027	0.4465	0.0047	0.799	0.8142	0.3328

In addition, MANOVA of factors for the gradient index was also employed, as shown in Table 4. The most significant of the four factors is the curing temperature. However, comparing the different days of curing, the significance of curing temperature on each performance level is mainly evident after 7 days of curing. For the other three factors, relative humidity had a significant effect on the GI at 7 days due to the particular curing condition, while the cement content only manifested significance at σ_t after 7 days, which was due to the effects of the different cement contents on the early strength. Therefore, the more significant factors influencing the gradient characteristics presented by BE-CIR mixtures are the curing temperature and relative humidity.

Table 4. MANOVA of factors for the gradient index.

p-Value	GI			
	3 d	7 d	14 d	28 d
Factors		σ_t		
Cement content	0.1199	0.0236	0.6743	0.2873
Initial moisture content	0.2592	0.0558	0.5349	0.5541
Curing temperature	0.6127	0.0022	0.0534	0.0578
Relative humidity	0.008	<0.0001	0.2465	0.0918
Factors		G_f		
Cement content	0.7288	0.0962	0.029	0.001
Initial moisture content	0.1566	0.8608	0.6378	0.0839
Curing temperature	0.7818	0.8263	0.0210	0.0095
Relative humidity	0.071	0.0007	0.1634	0.6983

5. Discussion of the Influencing Factors

The curing process of BE-CIR mixture is usually accompanied by a continuous decrease in water content due to evaporation and an increase in strength due to emulsion breaking and cement hydration [31,34,43,44]. The depth and time variations of BE-CIR pavement are affected by both the material component and environmental conditions. In this study, the two material-component factors of cement content and initial moisture content as well as the two environmental factors of curing temperature and relative humidity were examined. The addition of cement can improve the early strength of BE-CIR, but the improvement effect is diminished when the cement content exceeds 2%. Furthermore, it was observed that the depth fluctuation of BE-CIR became more pronounced with an increase in cement content, suggesting that the hydration extent was influenced by the gradient evaporation process. As for the initial moisture content, it can influence the rate of water dissipation significantly, but it has little effect on the final residual moisture content trapped in the BE-CIR. The initial moisture content may mainly affect the time and depth variation of BE-CIR pavement during a curing process of around 2 weeks. Compared with the material component, the environmental factors showed more extensive effects on the performance variation of the BE-CIR. Especially after curing 28 days, when the difference of σ_t between 25 °C and 45 °C could be 0.4 MPa, while it was less than 0.1 MPa between different cement contents and moisture contents. As the temperature surpassed 40 °C, the acceleration impact became less pronounced. Moreover, during the initial stage of curing, high humidity hindered water dissipation from the mixture, and the mixture could even have absorbed approximately 0.5% more water from the air. Excess water also impeded cement hydration, suspending the development of crack-resistance properties [17,35]. Overall, the growth in the cracking resistance of BE-CIRs is extremely dependent on the moisture environment they are subjected to. Consequently, understanding and possibly controlling moisture levels throughout the construction and curing phases is crucial for ensuring the longevity of BE-CIR pavements.

6. Conclusions

This paper evaluated the cracking resistance of different parts of BE-CIR mixtures by SCB tests based on a semi-sealed laboratory curing condition simulating in situ moisture evaporation. The influence of four factors, comprising cement content, initial moisture content, curing temperature and relative humidity, were investigated. Based on the analysis, the following conclusions can be drawn:

(1) The performance growth characteristics of BE-CIR mixes at various depths exhibited notable differences, evident through the top part's higher tensile strength and fracture energy compared to the bottom part. This occurrence can be attributed to the non-uniform distribution of residual water within the mixture due to moisture evaporation. Consequently, this results in distinct microscopic morphologies and adhesive capabilities of the cement-emulsified asphalt mortar, ultimately leading to differences in crack resistance.

(2) Increasing cement content can only improve the early tensile strength of the BE-CIR. Meanwhile, BE-CIR with higher cement content become stiffer and consequently less deformable, resulting in lower fracture energy. The initial moisture content mainly influenced the rate of water dissipation, but not the final residual moisture content. In addition, higher initial moisture content contributed to a more significant gradient characterization of the fracture properties due to the increase in residual water.

(3) Elevating the temperature resulted in a considerable acceleration of water loss, leading to enhanced fracture properties and a reduction in the gradient index. In addition, during the initial stages of curing, high humidity prevented water dissipation from the mixture, which resulted in hindering the development of fracture resistance.

(4) The findings indicated that the moisture migration process was governed by curing temperature and relative humidity, which were more significant than cement content and initial moisture content to the formation of cracking resistance performance. As for the gradient index of the fracture properties, the effect of curing temperature was the most significant, particularly after a curing period of 7 days.

In summary, it could be concluded that BE-CIR mixture subjected to different moisture environments will exhibit different cracking resistance developmental behaviors. Therefore, future research efforts should investigate the change of moisture in the BE-CIR layer directly, and figure out the impact of moisture on the microstructure and micromechanics of cement-emulsified asphalt mortar during curing to better understand the gradient development of performance.

Author Contributions: Z.Z.: Methodology, Formal analysis, Data curation, Visualization, Writing—original draft. J.J.: Conceptualization, Supervision, Validation, Writing—review and editing. F.N.:

Conceptualization, Resources, Validation, Data curation, Funding acquisition. J.J.: Testing, Data curation. All authors have read and agreed to the published version of the manuscript.

Funding: This work was supported by the National Natural Science Foundation of China (Grant No. 52108421), and the Scientific Research Foundation of the Graduate School of Southeast University (Grant No. YBPY2155).

Institutional Review Board Statement: Not applicable.

Informed Consent Statement: Not applicable.

Data Availability Statement: The data presented in this study are available on request from the corresponding author.

Conflicts of Interest: The authors declare no conflict of interest.

eferences

Wang, J.; Li, Q.; Lu, Y.; Luo, S. Effect of Waste-Oil regenerant on diffusion and fusion behaviors of asphalt recycling using molecular dynamics simulation. *Constr. Build. Mater.* **2022**, *343*, 128043. [CrossRef]

Gu, F.; Ma, W.; West, R.C.; Taylor, A.J.; Zhang, Y. Structural performance and sustainability assessment of cold central-plant and in-place recycled asphalt pavements: A case study. *J. Clean. Prod.* **2019**, *208*, 1513–1523. [CrossRef]

Turk, J.; Pranjić, A.M.; Mladenovič, A.; Cotič, Z.; Jurjavčič, P. Environmental comparison of two alternative road pavement rehabilitation techniques: Cold-in-place-recycling versus traditional reconstruction. *J. Clean. Prod.* **2016**, *121*, 45–55. [CrossRef]

Stochino, F.; Alibeigibeni, A.; Milia, A.; Zucca, M.; Pani, L.; Simoncelli, M. Environmental and Economic Analysis of Using Recycled Concrete Aggregates in Composite Steel-Concrete Slabs. In Proceedings of the International Conference on Computational Science and Its Applications, Athens, Greece, 3–6 July 2023; pp. 277–288.

Flores, G.; Gallego, J.; Miranda, L.; Marcobal, J.R. Cold asphalt mix with emulsion and 100% rap: Compaction energy and influence of emulsion and cement content. *Constr. Build. Mater.* **2020**, *250*, 118804. [CrossRef]

Deb, P.; Lakshman Singh, K. Mix design, durability and strength enhancement of cold mix asphalt: A state-of-the-art review. *Innov. Infrastruct. Solut.* **2021**, *7*, 1–22. [CrossRef]

Ouyang, J.; Pan, B.; Xu, W.; Hu, L. Effect of water content on volumetric and mechanical properties of cement bitumen emulsion mixture. *J. Mater. Civ. Eng.* **2019**, *31*, 04019085. [CrossRef]

Golian, M.; Katibeh, H.; Singh, V.P.; Ostad-Ali-Askari, K.; Rostami, H.T. Prediction of tunnelling impact on flow rates of adjacent extraction water wells. *Q. J. Eng. Geol. Hydrogeol.* **2020**, *53*, 236–251. [CrossRef]

Yang, Y.; Wang, H.; Yang, Y.; Zhang, H. Evaluation of the evolution of the structure of cold recycled mixture subjected to wheel tracking using digital image processing. *Constr. Build. Mater.* **2021**, *304*, 124680. [CrossRef]

Lin, J.; Wei, T.; Hong, J.; Zhao, Y.; Liu, J. Research on development mechanism of early-stage strength for cold recycled asphalt mixture using emulsion asphalt. *Constr. Build. Mater.* **2015**, *99*, 137–142. [CrossRef]

Grilli, A.; Graziani, A.; Bocci, E.; Bocci, M. Volumetric properties and influence of water content on the compactability of cold recycled mixtures. *Mater. Struct.* **2016**, *49*, 4349–4362. [CrossRef]

Du, S. The optimum pre-mixing water content in asphalt emulsion mixture with cement. *J. Test. Eval.* **2021**, *49*, 4560–4575. [CrossRef]

Jiang, J.; Zhao, Z.; Jiang, X.; Leng, Z.; Yang, B.; Ni, F. Moisture migration characterization of bitumen emulsion-based cold in-place recycling mixture over curing. *Mater. Struct.* **2023**, *56*, 122. [CrossRef]

Raschia, S.; Mignini, C.; Graziani, A.; Carter, A.; Perraton, D.; Vaillancourt, M. Effect of gradation on volumetric and mechanical properties of cold recycled mixtures (CRM). *Road Mater. Pavement Des.* **2019**, *20*, S740–S754. [CrossRef]

Miljković, M.; Poulikakos, L.; Piemontese, F.; Shakoorioskooie, M.; Lura, P. Mechanical behaviour of bitumen emulsion-cement composites across the structural transition of the co-binder system. *Constr. Build. Mater.* **2019**, *215*, 217–232. [CrossRef]

Ziyani, L.; Gaudefroy, V.; Ferber, V.; Hammoum, F. A predictive and experimental method to assess bitumen emulsion wetting on mineral substrates. *Colloids Surf. A Physicochem. Eng. Asp.* **2016**, *489*, 322–335. [CrossRef]

Kong, X.-M.; Liu, Y.-L.; Zhang, Y.-R.; Zhang, Z.-L.; Yan, P.-Y.; Bai, Y. Influences of temperature on mechanical properties of cement asphalt mortars. *Mater. Struct.* **2014**, *47*, 285–292. [CrossRef]

Cardone, F.; Grilli, A.; Bocci, M.; Graziani, A. Curing and temperature sensitivity of cement–bitumen treated materials. *Int. J. Pavement Eng.* **2015**, *16*, 868–880. [CrossRef]

Graziani, A.; Iafelice, C.; Raschia, S.; Perraton, D.; Carter, A. A procedure for characterizing the curing process of cold recycled bitumen emulsion mixtures. *Constr. Build. Mater.* **2018**, *173*, 754–762. [CrossRef]

Orosa, P.; Pérez, I.; Pasandín, A. Short-term resilient behaviour and its evolution with curing in cold in-place recycled asphalt mixtures. *Constr. Build. Mater.* **2022**, *323*, 126559. [CrossRef]

Xiao, F.; Amirkhanian, S.; Juang, C.H. Rutting resistance of rubberized asphalt concrete pavements containing reclaimed asphalt pavement mixtures. *J. Mater. Civ. Eng.* **2007**, *19*, 475–483. [CrossRef]

22. Wang, D.; Guo, T.; Chang, H.; Yao, X.; Chen, Y.; Wang, T. Research on the performance of regenerant modified cold recycle mixture with asphalt emulsions. *Sustainability* **2021**, *13*, 7284. [CrossRef]
23. Deng, C.; Jiang, Y.; Lin, H.; Chen, Z.; Ji, X. Influence of gradations on performance of emulsified asphalt cold recycled mixtu produced using vertical vibration compaction method. *Road Mater. Pavement Des.* **2021**, *22*, 983–1003. [CrossRef]
24. Zhang, J.; Zheng, M.; Pei, J.; Zhang, J.; Li, R. Research on low temperature performance of emulsified asphalt cold recycle mixture and improvement measures based on fracture energy. *Materials* **2020**, *13*, 3176. [CrossRef] [PubMed]
25. Yang, Y.; Yue, L.; Cui, H.; Yang, Y. Simulation and evaluation of fatigue damage of cold recycled mixtures with bitumen emulsio *Constr. Build. Mater.* **2023**, *364*, 129976. [CrossRef]
26. Grilli, V.; Virgili, A.; Graziani, A. Evaluation of complex modulus and fatigue properties of cold recycled material mixtures usir small-scale specimens. *Road Mater. Pavement Des.* **2023**, 1–16. [CrossRef]
27. Miljković, M.; Radenberg, M. Characterising the influence of bitumen emulsion on asphalt mixture performance. *Mater. Stru* **2015**, *48*, 2195–2210. [CrossRef]
28. Kuchiishi, A.K.; Vasconcelos, K.; Bariani Bernucci, L.L. Effect of mixture composition on the mechanical behaviour of co recycled asphalt mixtures. *Int. J. Pavement Eng.* **2021**, *22*, 984–994.
29. Abbasnejad, S.Y.; Modarres, A. Effect of setting accelerator additive on short-and long-term properties of cold recycled mixtu containing bitumen emulsion–cement composites. *Road Mater. Pavement Des.* **2020**, *21*, 1932–1954. [CrossRef]
30. Saadoon, T.; Gómez-Meijide, B.; Garcia, A. Prediction of water evaporation and stability of cold asphalt mixtures containin different types of cement. *Constr. Build. Mater.* **2018**, *186*, 751–761. [CrossRef]
31. Saadoon, T.; Garcia, A.; Gómez-Meijide, B. Dynamics of water evaporation in cold asphalt mixtures. *Mater. Des.* **2017**, *13* 196–206. [CrossRef]
32. Kim, Y.; Im, S.; Lee, H.D. Impacts of curing time and moisture content on engineering properties of cold in-place recyclin mixtures using foamed or emulsified asphalt. *J. Mater. Civ. Eng.* **2011**, *23*, 542–553. [CrossRef]
33. Doyle, T.A.; McNally, C.; Gibney, A.; Tabaković, A. Developing maturity methods for the assessment of cold-mix bituminou materials. *Constr. Build. Mater.* **2013**, *38*, 524–529. [CrossRef]
34. Graziani, A.; Godenzoni, C.; Cardone, F.; Bocci, M. Effect of curing on the physical and mechanical properties of cold-recycle bituminous mixtures. *Mater. Des.* **2016**, *95*, 358–369. [CrossRef]
35. García, A.; Lura, P.; Partl, M.N.; Jerjen, I. Influence of cement content and environmental humidity on asphalt emulsion an cement composites performance. *Mater. Struct.* **2013**, *46*, 1275–1289.
36. Fang, X.; Garcia, A.; Winnefeld, F.; Partl, M.N.; Lura, P. Impact of rapid-hardening cements on mechanical properties of cemer bitumen emulsion asphalt. *Mater. Struct.* **2016**, *49*, 487–498. [CrossRef]
37. Zhao, Z.; Jiang, J.; Chen, Z.; Ni, F. Moisture migration of bitumen emulsion-based cold in-place recycling pavement afte compaction: Real-time field measurement and laboratory investigation. *J. Clean. Prod.* **2022**, *360*, 132213. [CrossRef]
38. Zhao, Z.; Ni, F.; Zheng, J.; Cheng, Z.; Xie, S. Evaluation of Curing Effects on Bitumen Emulsion-Based Cold In-Place Recyclin Mixture Considering Field-Water Evaporation and Heat-Transfer Conditions. *Coatings* **2023**, *13*, 1204. [CrossRef]
39. Dot, J.P. *Applied Manual for Asphalt Emulsion Cold In-Place Recycling Technology*; Department of Transport of Jiangsu Provinc Nanjing, China, 2010. (In Chinese)
40. *JTG E20-2011*; Standard Test Methods of Bitumen and Bituminous Mixtures for Highway Engineering. Ministry of Transport c the People's Republic of China: Beijing, China, 2011. (In Chinese)
41. *AASHTO TP 124-2018*; Standard Method of Test for Determining the Fracture Potential of Asphalt Mixtures Using the Flexibilit Index Test (FIT). American Association of State Highway and Transportation Officials: Washington, DC, USA, 2018.
42. Du, H.; Ni, F.J.; Ma, X. Crack Resistance Evaluation for In-Service Asphalt Pavements by Using SCB Tests of Layer-Core Sample *J. Mater. Civ. Eng.* **2021**, *33*, 04020418. [CrossRef]
43. Xiao, F.; Yao, S.; Wang, J.; Li, X.; Amirkhanian, S. A literature review on cold recycling technology of asphalt pavement. *Const Build. Mater.* **2018**, *180*, 579–604.
44. Li, Q.; Wang, J.; Song, S.; Wang, R.; Jiang, J.; Yan, C. Study on the Adhesion Characteristics of Asphalt-Aggregate Interface in Col Recycled Asphalt Mixtures. *J. Mater. Civ. Eng.* **2023**, *35*, 04023283. [CrossRef]

coatings

ticle

n Investigation of the Dynamic Curing Behavior and Iicro-Mechanism of a Super-Tough Resin for Steel ridge Pavements

in Han [1], Jiwang Jiang [1,*], Jiahao Tian [1], Zhu Zhang [1], Fujian Ni [1,*] and Sheng Zhang [2]

[1] School of Transportation, Southeast University, Nanjing 211189, China; yajinhan@seu.edu.cn (Y.H.); 230198289@seu.edu.cn (J.T.); 230218386@seu.edu.cn (Z.Z.)
[2] China Railway 19th Bureau Group No. 3 Engineering Co., Ltd., Shenyang 110136, China; shengzhang12@outlook.com
* Correspondence: jiang_jiwang@seu.edu.cn (J.J.); 101004746@seu.edu.cn (F.N.)

Abstract: To overcome challenging service conditions, a groundbreaking thermoset, "Super-Tough Resin" (STR), has been specifically designed for steel bridge deck paving. Currently, investigations of paving thermosets mainly focus on cured materials. Detailed investigations of the curing process and its impact on the evolving properties of STR are lacking. Therefore, this study aims to explore the curing kinetics and the performance evolution of STR. Specifically, spectroscopy test, time sweep, linear viscoelastic region, and weight loss tests were conducted. Our results show that the curing degrees increase significantly with the curing durations and temperatures at the initial stage. When cured for 10 h, the curing degrees at four temperatures all exceed 80%. Then, a kinetic model with an *n*th-order of 1.551 was established. Upon increasing the temperature from 35 to 80 °C, the gel point time decreases from 480 to 189 min but the corresponding curing degree remains constant at 75.73%. When curing time is increased from 2.5 to 4 h, the linear viscoelastic regions decrease from 20% to 3%. Finally, after 400 h, the weight losses of STR at 35 and 80 °C are about 8% and 20%, respectively. These outcomes are beneficial to understanding the dynamic curing behaviors of STR and similar thermosets.

Keywords: curing-dependent rheological properties; curing kinetics; steel bridge deck pavement; super-tough resin; weight loss monitoring

1. Introduction

Compared with concrete bridges, steel bridges demonstrate superior advantages, including low self-weight, good prefabrication, favorable mechanical performance, and excellent bearing capacity [1,2]. Consequently, steel bridges are extensively constructed to connect two separated regions with a considerably long span. Typically, to provide a flat driving surface for vehicles, a deck pavement system is constructed on the steel bridge's deck plate [3,4]. It should be noted that the service environment of a steel bridge deck pavement is significantly divergent from the concrete deck pavement and subgrade pavement [5,6]. Specifically, under the integrated influences of traffic vehicles and wind load, the steel bridge may generate relatively significant deformation and vibration. In addition, the comparatively high thermal conductivity of steel plates may increase the service temperature of the deck pavement. As a temperature-sensitive material, a traditional asphaltic pavement is prone to melting at high temperatures [7]. Hence, under vehicle loading, rutting distress may emerge on the steel deck pavement [2].

It is well acknowledged that rutting distress is primarily attributed to the viscoelastic nature of the asphalt binder [8]. Therefore, scholars are devoted to improving asphalt binder temperature stability and performance by altering the viscoelastic instinct by coupling versatile polymers [9], such as styrene butadiene rubber (SBR) latex [8] and the styrene-butadiene-styrene (SBS) copolymer [10,11]. Among these polymers, the thermosetting polymer epoxy resin was developed in 1939 [12]. Due to its great mechanical performance, superior corrosion resistance, and thermal stability, epoxy resin is extensively utilized in the fields of civil engineering, coatings, adhesives, structural composites, and electrical elements [13]. Epoxy resin was first utilized together with asphalt binder in pavement construction in the 1950s to overcome the challenge of jest blast erosion of airplanes on the pavement [14]. Afterward, epoxy resin asphalt or epoxy asphalt (EA) was used for demanding projects, such as tunnel pavements, steel bridge deck pavements, and intersection pavements [15–17].

EA is fabricated by mechanically mixing the epoxy resin's main agent, its curing agent, and base asphalt. Upon mixing, the main agent spontaneously reacts with the curing agent and forms a three-dimensional thermoset network. After the curing process, the asphalt is distributed as spheres in the epoxy resin spatial network [18]. The thermoset network endows the asphalt with great temperature stability [7]. Hence, the performance of the EA is highly dominated by the thermosetting epoxy resin. Apart from EA, currently, some thermosetting materials are also being developed as binders for steel bridge deck paving, such as the polyurethane (PU) [11] and super-tough resin (STR) [7] studied in this paper. Similar to EA, the strength of the thermosetting polymers is gradually formed along with the curing reaction process [17]. Typically, the curing process, which is also known as polymerization, includes the formation, branching, and crosslinking of polymeric chains [19], as shown in Figure 1 [20]. Initially, individual polymeric chains are generated in the form of linear or branched shapes and the molecular weight demonstrates an increasing trend. As the curing reaction continues, the linear chains may further branch and transform into branched chains. The pronounced branched chains may eventually lead to the gelation of thermosets. At the point of gelation or gel point, the thermoset converts from the original viscous liquid to a viscoelastic solid and the molecular weight tends to be infinite [12,19,21]. Afterward, the curing reaction proceeds but is restricted due to the low mobility of polymeric chains. Consequently, the polymerization may transform from a chemical-controlled reaction to a diffusion-controlled reaction after the gel point [12,22].

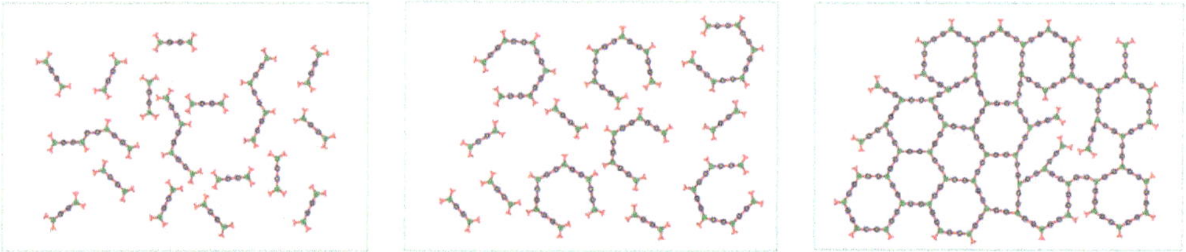

(a) (b) (c)

Figure 1. Formation process of thermoset network: (**a**) formation; (**b**) branching; (**c**) crosslinking.

As reviewed above, the polymerization of the thermoset is a considerably complex process that involves chemical component variation and mechanical performance evolution. It is well acknowledged that there is no uniform standard for thermoset curing characterization. Typically, the mechanical or chemical parameters varying with the curing process can be monitored, and then, adopted to depict the curing degree and curing reaction rate of thermosets. Consequently, the methods for characterizing the curing reaction cam be generally divided into two categories [23,24]: (1) direct methods, which monitor the variation in reactant concentration, such as spectroscopy techniques, and (2) indirect methods, which characterize the evolution of thermosets' mechanical properties, such as thermal and rheological techniques. By monitoring the curing degree-related parameters, the curing degree and curing reaction rate can be determined. Finally, suitable curing kinetic models can be established to reveal the curing process of thermosets and therefore determine the appropriate manufacturing procedures.

Currently, in road engineering, previous studies [25,26] mainly emphasize the mechanical and rheological performance of the cured polymer binders and their mixture for paving. The high-temperature stability, low-temperature flexibility, and intermediate-temperature fatigue resistance of paving binders and their mixture have been extensively studied [27,28]. In addition, some efforts have been made to investigate the curing kinetics of paving thermosets using thermal methods [15,29] and some studies have monitored the curing process using the Fourier Transform Infrared (FTIR) method [14,30]. However, the curing process investigation of paving thermosets is insufficient and rough. To our knowledge, there have been a few research studies that investigate curing kinetics via direct methods. In addition, studies on the mechanical and rheological performance evolution of thermosetting polymers for paving in pavement engineering can hardly be found in the literature. Consequently, the strength evolution of thermosetting pavements, such as epoxy asphalt concrete, is ambiguous and the opening time for traffic is typically determined using an empirical approach [31,32]. Therefore, it is essential to characterize the dynamic curing behaviors and understand the micro-mechanism of paving thermosets.

This study is organized as follows: Section 2 introduces the objectives; Section 3 presents the materials, sample fabrications, and methods; Section 4 reveals the results, discussion, and implications of the tests; and Section 5 concludes the main findings of this study.

2. Objectives

Considering the lack of research on paving thermosets' dynamic curing behaviors, this study aims to depict the chemical, rheological, and mechanical performance evolution of a novel paving binder, STR, which demonstrates better fatigue resistance and low-temperature toughness and shows promise as an alternative to EA for steel bridge deck paving. This study attempts to establish the curing kinetics via the FTIR spectroscopy test and characterize the curing-dependent rheological and weight loss properties of the thermoset. Specifically, the STR was cured at different temperatures with varying intervals and the evolution of the STR characteristic peaks were detected via FTIR test. Then, a suitable kinetic model was selected and established for STR. Afterward, the curing-dependent rheological properties, including the gel point and linear viscoelastic (LVE) region of STR, were obtained using rheology tests. Finally, the curing-induced weight loss of STR was monitored and an appropriate model was established to predict the weight loss of STR.

The results are beneficial for understanding the dynamic curing process of STR. Practically, the curing kinetic model and rheological property evolution can help to determine the construction parameters, such as the allowable duration and compaction time. In addition, the curing schemes in field conditions can be determined. By combining the weight loss characteristics, the potential curing shrinkage of super-tough resin concrete (STRC) can be evaluated. Finally, the curing characteristics and cured performance can be combined to better estimate the applicability of potential paving thermosets.

3. Materials and Methods

In this study, STR was purchased from a local supplier. STR is manufactured by mixing the main agent with the curing agent with a weight ratio of 1:1. The main properties of the two components are listed in Table 1.

3.1. FTIR Test

The FTIR test is capable of being used to monitor the variation in characteristic bands [33] in thermosets and therefore understand the curing process. Studies have also confirmed the comparability of the kinetics established via FTIR to other methods [34–36]. In this study, a Nicolet iS10 infrared spectrometer in attenuated total absorbance mode from the Thermo Scientific Corporation (Shanghai, China) was utilized to monitor the curing process. The scanning range was from 400 to 4000 cm^{-1} with a spectral resolution of 4 cm^{-1}.

Before calculating the characteristic peak areas, baseline corrections were conducted, and the spectra after baseline correction are demonstrated in Figure 2.

Table 1. Typical properties of the main agent and curing agent.

Properties	Main Agent	Curing Agent
	Results	Results
Appearance	Clear liquid	Light yellow liquid
Brookfield Viscosity (23 °C, mPa·s)	656	223
Specific Gravity (23 °C, g/cm^3)	1.13	1.02

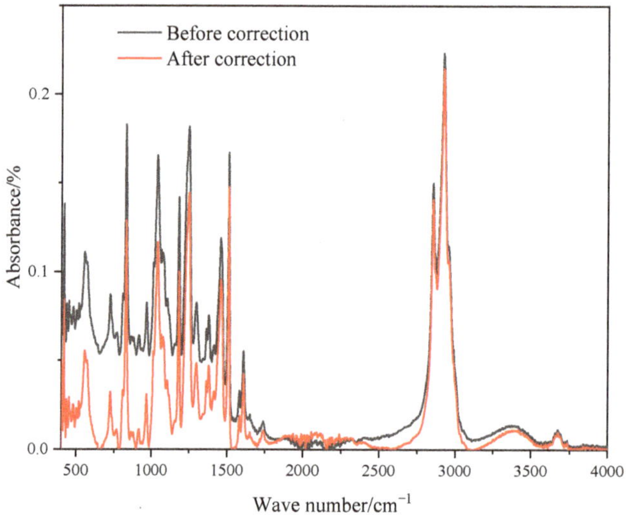

Figure 2. Example of baseline correction.

3.2. *Rheological Tests*

In this study, rheological tests, including the time sweep test and LVE region test, were conducted using an Anton Paar MCR102 dynamic shear rheometer (DSR, Anton Paar (Shanghai) Trading Co., LTD, Shanghai, China). Separated by the gel point, the STR binder may exhibit varying statuses and stress-strain responses: below the gel point, the STR binder can be regarded as a viscous liquid; above the gel point, the STR binder turns into a viscoelastic solid. Consequently, the rheological test protocols should be adjusted according to the solid–liquid states of the STR binder.

3.2.1. Sample Preparation

During the curing process, the STR sample may transform from a viscous liquid to a viscoelastic solid. For the viscous liquid, a traditional parallel plate can be utilized for the rheological tests, as shown in Figure 3a. However, after the gel point, a thermosetting three-dimensional network gradually formed, which is irreversible. Therefore, the viscoelastic solid STR cannot be remelted and reshaped like the asphaltic binder [7]. To characterize the rheological properties of the solid STR, a solid torsion bar fixture was adopted. The solid fixture requires the samples to be tall cylinders, with three fixed diameters: 8, 11.5, and 15 mm. STR demonstrates a relatively high shear modulus at low temperatures. Considering the torque limitation of the DSR equipment, the STR sample was fabricated with a diameter of 8 mm.

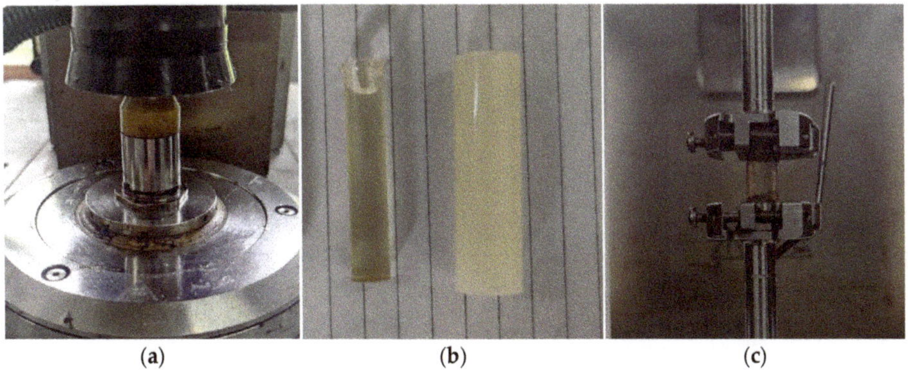

Figure 3. Rheological test equipment and sample preparation: (**a**) parallel plate; (**b**) solid super-tough resin (STR) sample; (**c**) solid torsion bar fixture.

To prepare such a sample, the homogenous STR mixture was poured into a cylindrical silicone tube with an inner diameter of 8 mm, as shown in Figure 3b. After the designed curing process, the STR sample was extracted and truncated with a height of 20 mm. Then, steel rings were glued to both ends of the STR cylinder. Finally, the STR sample was clamped on the fixture for the following tests, as shown in Figure 3c.

3.2.2. Time Sweep Test

A time sweep test was conducted to investigate the shear modulus evolution of STR during the curing process and, consequently, determine the curing duration corresponding to the gel point at varying curing temperatures. In this study, the time sweep test mainly focused on the modulus evolution near the gel point rather than the whole curing process. At the gel point, the three-dimensional network of the STR initiates but can hardly be considered solid yet. Hence, a parallel plate with a diameter of 25 mm was utilized to conduct the time sweep test of STR. The gap of the parallel plate was 1 mm. The time sweep test was conducted at 35, 50, 65, and 80 °C. During the test, the shear strain was kept constant at 1% with a loading frequency of 1 Hz.

3.2.3. Linear Viscoelastic (LVE) Region Test

Based on existing studies, the stress-strain response of a viscoelastic material may vary depending on the loading strain or stress amplitude [37,38]. When the loading strain is within the LVE region, the shear modulus is independent of the loading strain and remains constant. In this study, the LVE regions of STR samples with varying curing schemes were determined using the strain sweep test. The test was conducted at 25 °C with a loading frequency of 1.59 Hz. The parallel plate and solid torsion bar fixture were used for the testing before and after the gel point, respectively.

3.3. Weight Loss Monitoring

During the curing reaction process of STR, weight loss of the reactant was observed. Therefore, an electronic balance was used to continuously monitor the variation in the mixture weight. Specifically, the curing temperatures included 35, 50, 65, and 80 °C, which cover the conventional service or curing temperatures used in the field.

3.4. Experimental Schemes

In this study, chemical, rheological, and weight loss tests were conducted to comprehensively characterize the performance evolution of STR during the curing process. Details of the curing schemes and testing parameters are summarized in Table 2.

Table 2. Details of experimental schemes.

Tests	Curing Schemes		Test Schemes		
	Curing Temperatures/°C	Curing Durations/h	Test Temperatures/°C	Strain/%	Frequency/Hz
FTIR Test	35, 50, 65, 80	0–480	25	-	-
Weight Loss Test			35, 50, 65, 80	-	-
Time Sweep Test		0–12	35, 50, 65, 80	1	1
LVE Test		0–720	25	0.001–20/ 1×10^{-5}–0.02	1.59

4. Results and Discussion

4.1. Results of FTIR Test

4.1.1. Curing Degree under Defined Curing Schemes

The FTIR test was used to monitor the chemical evolution of STR and, therefore, follow the curing reaction. As shown in Figure 4, in the curing process, the intensity of the characteristic peak (-NCO) ranging from 2181 to 2359 cm^{-1} gradually decreases and finally disappears, which indicates the completion of the curing reaction. In contrast, the intensity of the characteristic peak (C=O) ranging from 1667 to 1795 cm^{-1} remains nearly unaffected throughout the polymerization process. Based on the Lambert-Beer principle, the area of a specified absorbance peak is proportional to the concentration of the corresponding chemical composition. However, according to the existing study [39], the intensity of an absorbance peak is highly dependent on the sample thickness. To eliminate such interference, the absorbance peak of C=O is selected as the reference peak, and the normalized absorbance index (*AI*) of the -NCO peak is defined in Equation (1). Hence, the conversion of the -NCO peak or curing degree α can be calculated using Equation (2).

$$AI = \frac{A_{-NCO}}{A_{C=O}} \quad (1)$$

$$\alpha_t = \frac{AI_0 - AI_t}{AI_0} \quad (2)$$

where A_{-NCO} and $A_{C=O}$ are the absorbance peak areas of -NCO and C=O, respectively; AI_0 is the normalized absorbance index at the beginning; and AI_t is the normalized absorbance index after a certain curing duration.

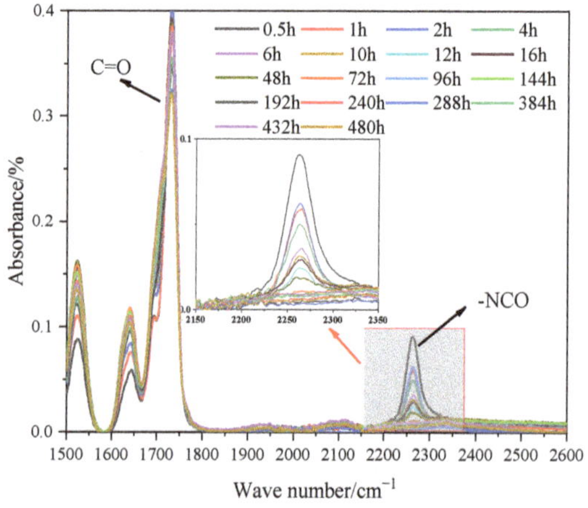

Figure 4. Evolution of STR characteristic peak during curing.

According to Equation (2), the absorbance index and curing degree of STR at varying curing temperatures can be calculated and are shown in Figure 5a,b.

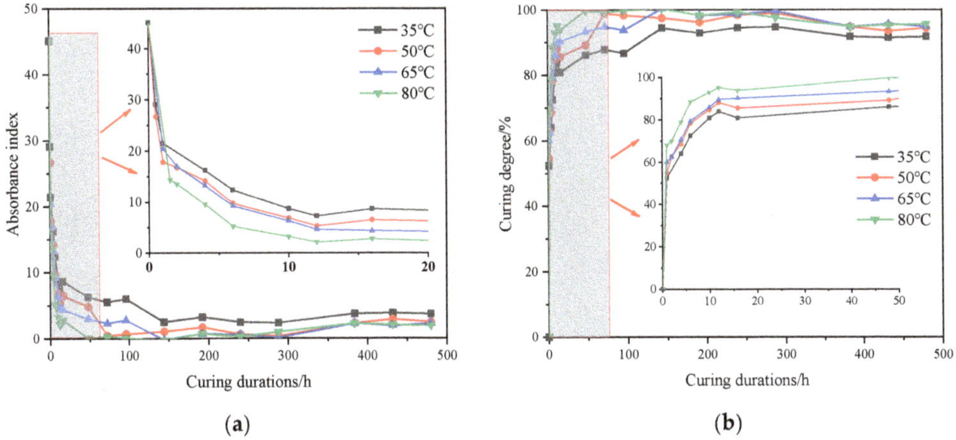

Figure 5. Evolution of STR curing parameters: (**a**) variation in absorbance index; (**b**) variation in curing degree.

As demonstrated in Figure 5, with the prolonging of curing duration, the absorbance index of the -NCO characteristic peak presents steadily decreasing trends at four curing temperatures. Consequently, the curing degrees or chemical band conversions increase significantly with the curing durations. Particularly, at the initial stage, both the absorbance index and curing degree change dramatically, which indicates the relatively high curing rates of STR in the early phase. For instance, when cured for 10 h, the *AI*s at four temperatures roughly decrease from 45 to 10; in addition, a remarkable curing rate can be observed, and the corresponding curing degrees at four curing temperatures all exceed 80%. Then, with the proceeding of the curing reaction, the curing degrees continue to increase, but with a considerably lower gradient, and level off gradually. Specifically, when STR is cured for 144 h, the curing degree reaches a plateau regardless of the curing temperature, which may imply the approximate completion of the curing reaction. These results imply that with the proceeding of the curing process, the three-dimensional network gradually forms and the movement of the branched or linear chains is restricted. Consequently, the reaction rate decreases gradually, and the reaction between reactants may transform from a chemical-controlled mode to a diffusion-controlled mode.

According to Figure 5, significant effects of curing temperature on the curing reaction can be observed. Generally, increased curing temperatures are beneficial to enhancing the curing rate and consequently increasing the curing degree. As revealed in Figure 5b, with the same curing durations, the curing degrees of STR at 80 °C are always the highest, followed by that of STR at 65 and 50 °C. In contrast, the curing degree of STR at 35 °C is the lowest. For instance, it takes 10 h for STR cured at 35 °C to reach an 80% curing degree. However, when cured at 80 °C, it takes 6 h to reach a curing degree of 80%. Furthermore, even when cured for 480 h, the curing degree of STR can hardly reach 100%, especially at low curing temperatures. When cured at 35 °C, the ultimate curing degree is lower than 95%. In contrast, when cured at 65 and 80 °C, the final curing degree approaches 100%. Hence, increased curing temperatures could promote chemical conversion and increase the ultimate curing degree.

4.1.2. Curing Kinetics

Currently, numerous phenomenological models [23] have been proposed to manifest the curing kinetics of thermosets, including the *n*th-order model, autocatalytic model, and Kamal model, as shown in Equations (3)–(5), respectively.

$$\frac{d\alpha}{dt} = k(1-\alpha)^n \tag{3}$$

$$\frac{d\alpha}{dt} = k\alpha^m(1-\alpha)^n \tag{4}$$

$$\frac{d\alpha}{dt} = (k_1 + k_2\alpha^m)(1-\alpha)^n \tag{5}$$

$$k = Ae^{-E_a/RT} \tag{6}$$

where α is the curing degree; t is the reaction duration; m and n are the curing reaction orders; k, k_1, and k_2 are the reaction rate constants; A is the pre-exponential factor; E_a is the curing reaction activation energy; R is the gas constant 8.314 J/(mol·K); and T is the absolute temperature.

According to the above equations, for the *n*th-order model, the maximum curing reaction rate occurs at the beginning; afterward, the reaction rate decreases gradually. In contrast, the curing reaction rate of the autocatalytic model begins at 0, and then, increases to the maximum. The Kamal model is a generalized *n*th-order model and autocatalytic model. Based on Figure 5b, the differential values of the curing degree were calculated as shown in Figure 6.

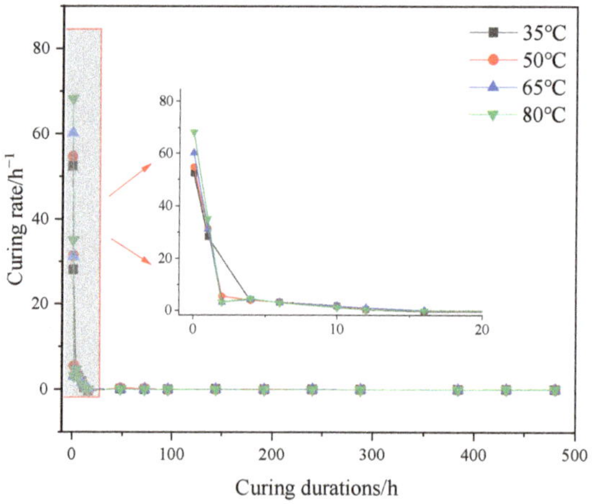

Figure 6. Evolution of curing reaction rate.

According to the curing reaction rate shown in Figure 6, the curing reaction reaches the maximum rate at the beginning. Then, due to the formation of a polymer network, the movements of the branched and linear chains are restricted, and consequently, the reaction rate gradually levels off. The curing behaviors of STR are in accordance with the *n*th-order reaction mechanism. Therefore, the curing reaction of STR is speculated to be the *n*th-order reaction. Based on the curing degree data and the corresponding curing rate, Equation (3) was used to fit the data, and the fitting parameters are listed in Table 3.

Table 3. Fitting parameters of the *n*th-order model.

Parameters	35 °C	50 °C	65 °C	80 °C
k	53.997	56.200	60.953	68.932
n	1.663	1.543	1.550	1.448

According to Table 3, both the reaction rate constant k and reaction order n vary at different temperatures, especially for k. The results indicate that increased curing temperatures can enhance the reaction rate constant k. The data further explain the aforementioned results whereby at high curing temperatures, the curing degree is higher than that of STR at low curing temperatures. As shown in Equation (6), the reaction rate constant k correlates with the curing temperature and follows the Arrhenius law. Equation (6) can be rewritten into Equation (7). As shown in Equation (7), ln$k$ and $1/T$ demonstrate a linear relationship. Hence, linear fitting was conducted based on Table 3. The results are shown in Figure 7a.

$$\ln k = \ln A - \frac{E_a}{RT} \tag{7}$$

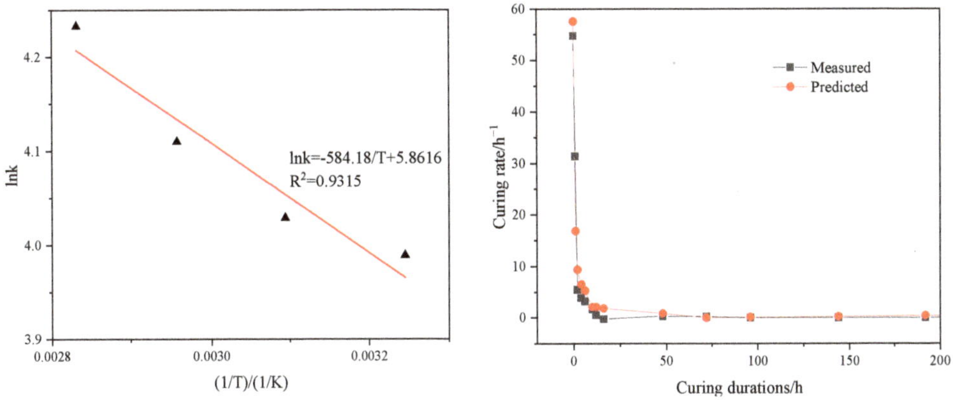

(a) (b)

Figure 7. Results of linear fitting and validation: (**a**) linear fitting; (**b**) result validation.

According to the linear fitting results, the activation energy E_a and pre-exponential factor A can be calculated from the slope and intercept, respectively. Then, E_a and A are determined as 4.86 kJ/mol and 351.29/h. In addition, the *n*th-order is taken as the average of n at four temperatures, as listed in Table 3. Hence, the curing reaction model of STR is described in Equation (8). Consequently, the predicted curing reaction rate at 50 °C can be calculated and compared with the test results, as shown in Figure 7b. The established *n*th-order kinetic model can help to determine the curing degree of STR at varying curing temperatures and durations and, consequently, understand the dynamic curing behaviors of STR.

$$\frac{d\alpha}{dt} = 351.29(1-\alpha)^{1.551}e^{-\frac{584.18}{T}} \tag{8}$$

4.2. Results of Time Sweep Test

In this study, the time sweep test was used to understand the rheological property evolution of STR and consequently determine the gel point at varying curing temperatures. Many methods have been proposed to determine the gel time [40,41]. In this research, the gel point was defined as the intersection point of the storage modulus and the loss modulus in a time sweep test using DSR equipment [42]. The results of the time sweep test at different curing temperatures are demonstrated in Figure 8.

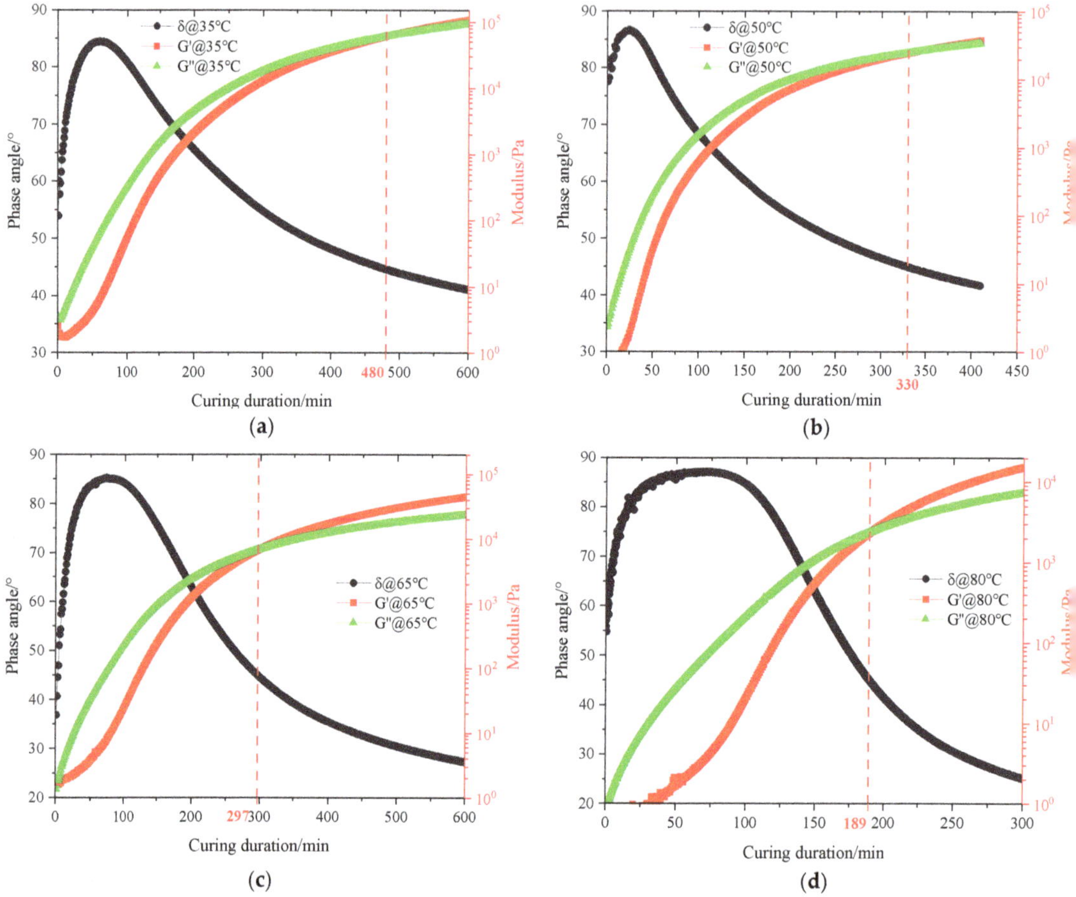

Figure 8. Results of time sweep test at different temperatures: (**a**) 35 °C; (**b**) 50 °C; (**c**) 65 °C; (**d**) 80 °C.

As revealed in Figure 8, with the prolonging of curing duration, the storage modulus and loss modulus both demonstrate significantly increased trends regardless of the curing temperature, which is attributed to the proceeding of the curing reaction. At the beginning of the curing process, the loss modulus is always higher than the storage modulus. Therefore, the stress–strain response is mainly dominated by the viscous properties initially. After the gel point time, the storage modulus is higher than the loss modulus, which indicates the transformation of STR from a viscous liquid to a viscoelastic solid. In addition, during the curing process, an induction period can be observed in storage modulus evolution. The increment rate of the storage modulus starts roughly at 0, and then, reaches the maximum according to the slope of the storage modulus evolution curve.

Apart from the evolution of the modulus, the phase angle of STR also demonstrates an interesting phenomenon. Generally, at the initial stage, the phase angle increases considerably and reaches a peak at about 85°. Thereafter, the phase angle decreases significantly due to the enhancement of the curing degree. According to the time sweep test results, the curing time to the cross point of the storage modulus and loss modulus, namely, gel time, can be determined, as summarized in Table 4. In addition, based on the established curing reaction model in the previous section, the curing degree of the gel point can be calculated, as shown in Table 4.

Table 4. Gel point time and curing degree of STR.

Temperatures	35 °C	50 °C	65 °C	80 °C	Average
Gel Point Time/min	480	330	297	189	-
Gel Point Curing Degree/%	76.69	76.01	75.06	75.164	75.73

As shown in Table 4, with the increase in temperature from 35 to 80 °C, the gel point time decreases significantly from 480 to 189 min. These results indicate that the elevated curing temperature is beneficial to promoting the curing reaction rate and shortening the gel point time of thermosets. Moreover, although a decrease in gel point time is observed, the gel point curing degree does not present an obvious difference. Regardless of the curing temperature, the gel point curing degree concentrates around 75.73%. A previous study [23] has also confirmed the finding that the curing temperatures show insignificant effects on the gel point curing degree. In addition, the gel point time can be correlated with the curing temperature using Equation (9).

$$\ln t_{gel} = c + \frac{E_a}{RT} \tag{9}$$

where t_{gel} is the gel point time; c is the equation constant; E_a is the activation energy for the gel point equation; R is the gas constant; and T is the absolute temperature.

By linear fitting $\ln t_{gel}$ with $1/T$, the equation for predicting the gel point time can be established. Therefore, linear fitting was conducted based on Table 4 with a correlation coefficient of 0.948. According to the fitting results, the slope of the equation is −0.6339 and the intercept is 2097.8. Consequently, the gel point time can be predicted using Equation (10). Based on Equation (10), the gel point time at a specified curing temperature can be calculated, which is beneficial for understanding the pre- and post-gelation characteristics of STR.

$$t_{gel} = \exp(\frac{2097.8}{T} - 0.6339) \tag{10}$$

4.3. Results of LVE Region Test

The specimens for the LVE region test were cured at 50 °C for different durations from 1 to 720 h. The LVE region test was conducted at 25 °C, and the loading strain varied from 0.001% to 2% or 0.1% to 20% depending on the curing durations of the STR samples. The LVE region is determined as the strain at which the complex shear modulus reduces to 95% of the initial value [38]. The results of the LVE region test are summarized in Figure 9.

According to Figure 9, with the increase in curing duration, the complex shear modulus reveals an apparent upward trend even after curing for 720 h. However, based on the FTIR test results, the curing reaction is approximately accomplished after 100 h at 50 °C. Therefore, the evolutions of STR's chemical composition and mechanical properties are inconsistent. In addition, the LVE region demonstrates a significant decreasing trend, especially when the curing duration increases from 2.5 to 4 h. When the STR samples are cured for fewer than 4 h, the LVE region exceeds 20%, which is remarkably higher than that of traditional asphalt. However, with the prolonging of curing durations, the LVE region decreases sharply, which indicates that the STR gradually converts from a viscous liquid to a viscoelastic solid. These results may be attributed to the branching of thermosetting polymers, which limits the free movement of polymer chains.

It is clear that, when cured for 10 h at 50 °C, the LVE region demonstrates a stable tendency and concentrates around 1%. After this, the increases in curing duration can hardly alter the LVE region. This phenomenon means that after 10 h of curing process, a thermosetting three-dimensional network is basically formed, and the movement of branched chains is confined. According to the previous section, the gel point time at 50 °C is about 5.5 h. Based on Figure 9c, it should be noted that the inflection point of LVE region evolution seems to be near the gel point. This interesting finding may further confirm the definition of the gel point and the apparent rheological property evolution at this point.

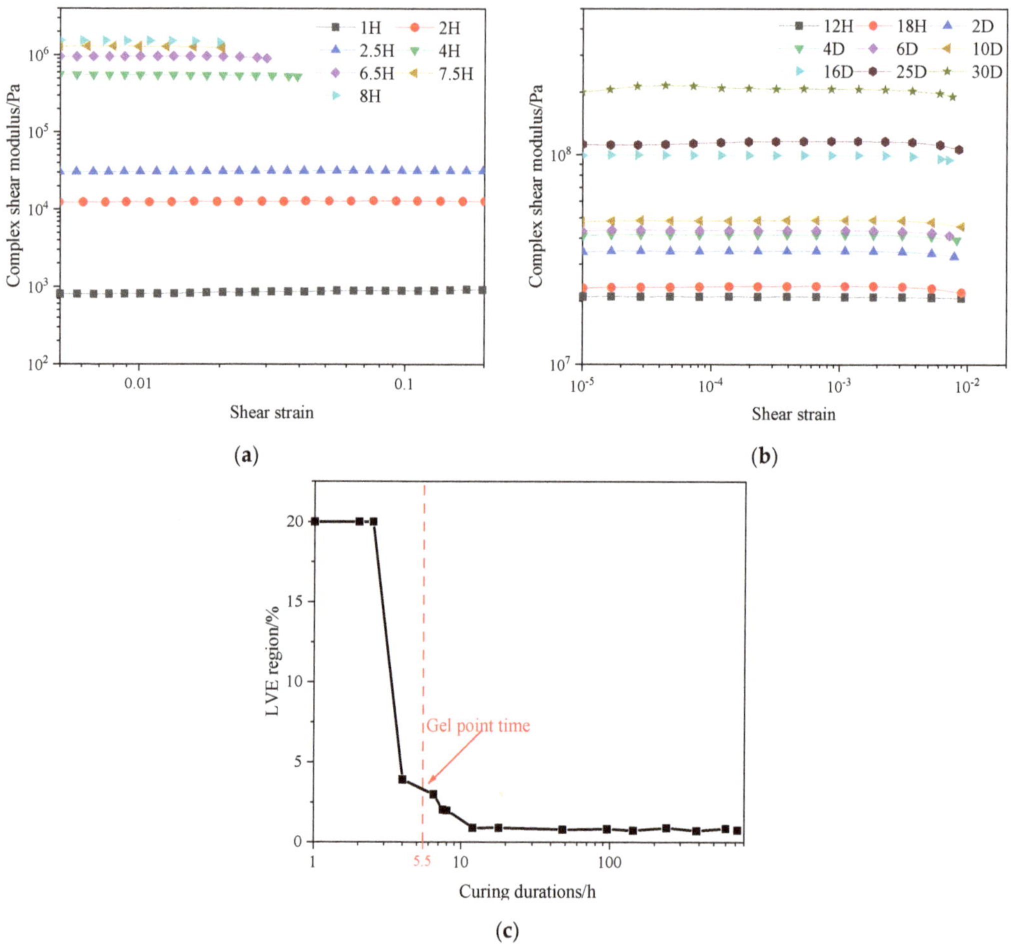

Figure 9. Results of the linear viscoelastic (LVE) region test: (**a**) short-term curing; (**b**) long-term curing; (**c**) relationship between curing durations and LVE region.

4.4. Weight Loss Monitoring

During the curing reaction process of STR, weight loss of the reactant is observed. Therefore, an electronic balance was used to continuously monitor the variation in the mixture weight. Specifically, the curing temperatures included 35, 50, 65, and 80 °C, which cover the conventional service or curing temperatures used in the field. The normalized results of the curing-induced weight loss are presented in Figure 10.

As revealed in Figure 10a, with the prolonging of curing durations, the weight of the reactant demonstrates a significantly decreasing trend, especially for the specimens cured at higher temperatures. The observed weight loss of STR may be attributed to the reaction-induced volatilization of the curing agent. Specifically, after being cured for 400 h, the weight loss of the specimen cured at 35 °C is about 8%; in contrast, when cured at 80 °C, the weight loss approaches 20%. These results indicate that the elevated curing temperatures may boost the curing reaction rate and increase the evaporation of the reactant.

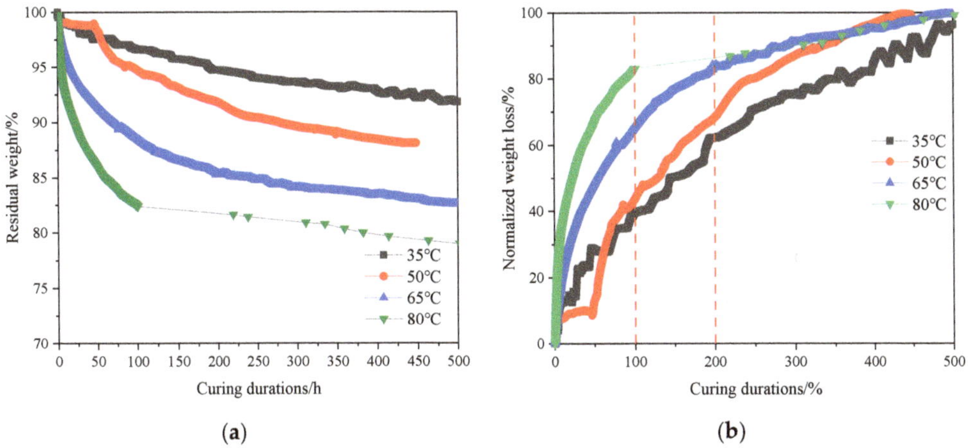

Figure 10. Results of weight loss monitoring: (**a**) evolution of residual weight; (**b**) evolution of weight loss.

In addition, the rates of STR weight loss at different curing temperatures all show decreasing trends. The weight loss of STR at the beginning of the curing reaction is higher than that of STR at the end of the curing reaction, especially for the specimens at high curing temperatures. As shown in Figure 10b, when cured for 100 h, the weight loss of STR cured at 35 and 50 °C is approximately 40% of the total weight loss. In contrast, the weight losses of STR cured at 65 and 80 °C exceed 60% and 80% by total weight loss, respectively. Therefore, the initial weight loss occupies most of the total weight loss during the whole curing process.

To predict the weight loss of STR at varying curing temperatures for different curing durations, the Maxwell equation was utilized to correlate the weight loss data, as shown in Equation (11). The fitting results are summarized in Figure 11 and Table 5.

$$W = A + B\exp(-\frac{t}{C}) \tag{11}$$

Table 5. Parameters of fitting results.

Temperatures/°C	*A*	*B*	*C*	R^2
35	90.83	8.5	271.18	0.99
50	86.49	13.82	204.1	0.99
65	82.97	14.02	110.96	0.99
80	80.39	16.07	45.81	0.97
Relationship between *T* and *A*, *B*, *C*	$A = -0.23 \times T + 98.53$ $R^2 = 0.99$	$B = 0.15 \times T + 4.32$ $R^2 = 0.84$	$C = 452.89 - 5.13 \times T$ $R^2 = 0.99$	-

In the above equation, *W* is the percentage weight loss of STR, %; *t* is the curing duration, h; and *A*, *B*, and *C* are the fitting constants.

Based on the fitting results listed in Figure 11 and Table 5, the proposed Maxwell equation could correlate well with the weight loss data. Consequently, the weight loss of

STR in the curing process can be predicted in varying curing temperature and duration combinations, as shown in Equation (12).

$$W = -0.23T + 98.53 + (0.15T + 4.32)\exp\left(-\frac{t}{452.89 - 5.13T}\right) \quad (12)$$

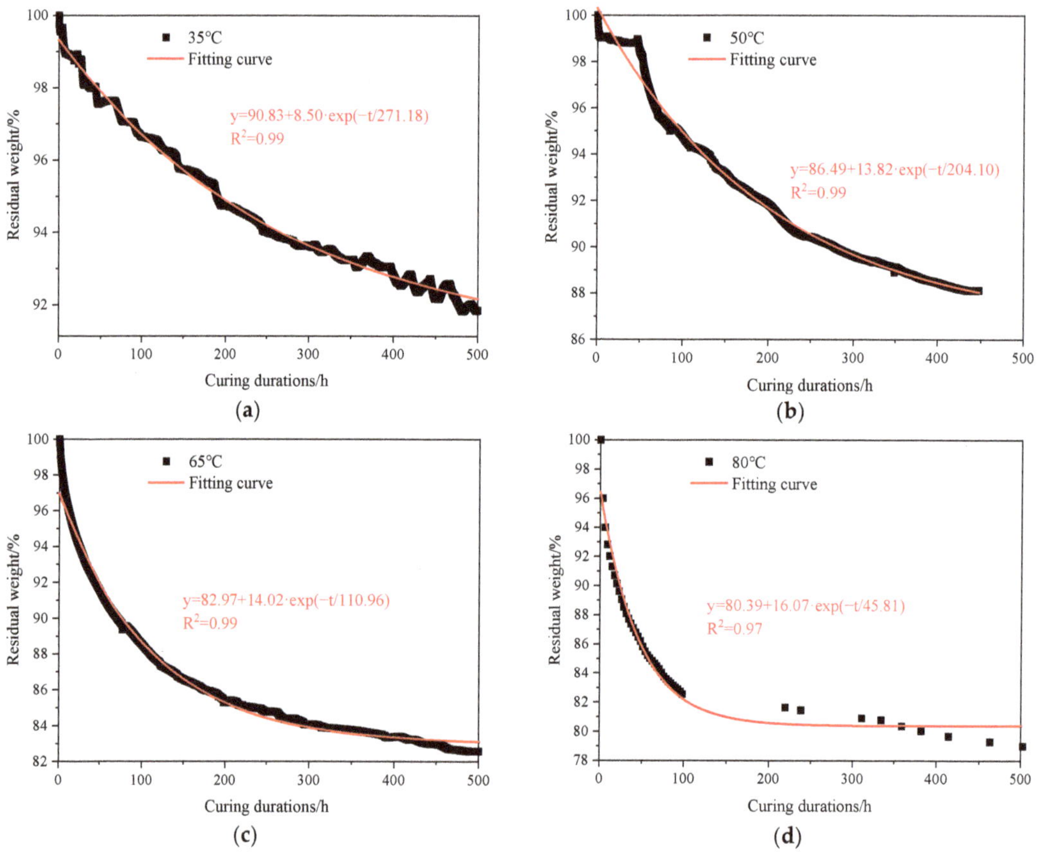

Figure 11. Results of nonlinear fitting: (**a**) 35 °C; (**b**) 50 °C; (**c**) 65 °C; (**d**) 80 °C.

5. Conclusions

In this study, chemical, rheological, and weight loss tests were conducted to comprehensively characterize the performance evolution and understand the dynamic curing behaviors of STR during the curing process. The main conclusions drawn are as follows.

(1) The curing degrees increase significantly with the curing durations, particularly at the initial stage. Specifically, when cured for 10 h, the curing degrees increase remarkably at four curing temperatures, all exceeding 80%. When cured for 144 h, the curing degree reaches a plateau regardless of the curing temperature. Moreover, increased curing temperature can also increase the curing rate and ultimate curing degree.

(2) The curing reaction of STR reaches the maximum rate at the beginning. Then, the reaction rate gradually levels off. Therefore, the curing reaction of STR is speculated to be the *n*th-order reaction. By fitting the test results, the *n*th-order curing kinetic model with an *n*th-order of 1.551 was established.

(3) With the increase in temperature from 35 to 80 °C, the gel point time decreases significantly from 480 to 189 min. However, the gel point curing degree does not present an obvious difference and concentrates around 75.73%.
(4) The LVE region demonstrates an obvious decreasing trend, especially when the curing duration increases from 2.5 to 4 h. When cured at 50 °C for more than 10 h, a three-dimensional network is basically formed and the LVE region demonstrates a stable tendency concentrating around 1%.
(5) Increased curing duration and temperature can boost the weight loss of reactants. After 400 h, the weight losses of specimens cured at 35 and 80 °C are about 8% and 20%, respectively. In addition, the initial weight loss occupies most of the total weight loss during the whole curing process. Finally, the Maxwell equation was proposed to predict the weight loss of STR with a specified curing scheme.

According to the results of this study, the curing kinetic model and rheological property evolution could help to determine construction parameters, such as the allowable duration and compaction time. Finally, the curing characteristics and cured performance could be combined to better estimate the applicability of STR and similar potential thermosets for steel bridge deck paving.

Author Contributions: Conceptualization, Y.H., F.N. and J.J.; methodology, Y.H., Z.Z. and J.T.; formal analysis, Y.H., J.T. and J.J.; investigation, Y.H., Z.Z. and J.T.; resources, F.N. and S.Z.; writing—original draft preparation, Y.H.; writing—review and editing, Y.H. and J.J.; supervision, F.N. and J.J.; funding acquisition, F.N. and S.Z. All authors have read and agreed to the published version of the manuscript.

Funding: This work was supported by the National Natural Science Foundation of China (52078131) and the Postgraduate Research and Practice Innovation Program of Jiangsu Province, China (KYCX21_0134). This research was also financially supported by a collaborative project with the China Railway 19th Bureau Group No. 3 Engineering Co., Ltd.

Institutional Review Board Statement: Not applicable.

Informed Consent Statement: Not applicable.

Data Availability Statement: All data are included in the article.

Conflicts of Interest: The authors declare no conflict of interest.

eferences

Zhu, Z.; Xiang, Z.; Li, J.; Huang, Y.; Ruan, S. Fatigue behavior of orthotropic bridge decks with two types of cutout geometry based on field monitoring and FEM analysis. *Eng. Struct.* **2020**, *209*, 109926. [CrossRef]
Han, Y.; Zhang, Z.; Tian, J.; Ni, F.; Gu, X. Long-Term In Situ Performance Evaluation of Epoxy Asphalt Concrete for Long-Span Steel Bridge Deck Pavement. *Coatings* **2023**, *13*, 545. [CrossRef]
Jia, X.; Huang, B.; Chen, S.; Shi, D. Comparative investigation into field performance of steel bridge deck asphalt overlay systems. *KSCE J. Civ. Eng.* **2016**, *20*, 2755–2764. [CrossRef]
Medani, T.O. Design Principles of Surfacings on Orthotropic Steel Bridge Decks. Ph.D. Thesis, Technische Universiteit Delft, Delft, The Netherlands, 2006.
Farreras-Alcover, I.; Chryssanthopoulos, M.K.; Andersen, J.E. Data-based models for fatigue reliability of orthotropic steel bridge decks based on temperature, traffic and strain monitoring. *Int. J. Fatigue* **2017**, *95*, 104–119. [CrossRef]
Xiang, Q.; Xiao, F. Applications of epoxy materials in pavement engineering. *Constr. Build. Mater.* **2020**, *235*, 117529. [CrossRef]
Han, Y.; Ni, F.; Jiang, J.; Tian, J.; Zhang, Z.; Zhou, Z.; Dong, Q. Rheological investigation of a super-tough resin-based binder for steel bridge deck paving. *Constr. Build. Mater.* **2023**, *384*, 131441. [CrossRef]
Han, Y.; Tian, J.; Ding, J.; Shu, L.; Ni, F. Evaluating the storage stability of SBR-modified asphalt binder containing polyphosphoric acid (PPA). *Case Stud. Constr. Mater.* **2022**, *17*, e01214. [CrossRef]
Raisi, A.; Asefnejad, A.; Shahali, M.; Doozandeh, Z.; Kamyab Moghadas, B.; Saber-Samandari, S.; Khandan, A. A soft tissue fabricated using a freeze-drying technique with carboxymethyl chitosan and nanoparticles for promoting effects on wound healing. *J. Nanoanalysis* **2020**, *7*, 262–274. [CrossRef]
Lesueur, D. Polymer modified bitumen emulsions (PMBEs). In *Polymer Modified Bitumen*; Elsevier: Amsterdam, The Netherlands, 2011; pp. 25–42.
Li, X.; Li, J.; Wang, J.; Yuan, J.; Jiang, F.; Yu, X.; Xiao, F. Recent applications and developments of Polyurethane materials in pavement engineering. *Constr. Build. Mater.* **2021**, *304*, 124639. [CrossRef]

12. Shundo, A.; Yamamoto, S.; Tanaka, K. Network formation and physical properties of epoxy resins for future practical applicatio *JACS Au* **2022**, *2*, 1522–1542. [CrossRef]
13. Cividanes, L.S.; Simonetti, E.A.; Moraes, M.B.; Fernandes, F.W.; Thim, G.P. Influence of carbon nanotubes on epoxy resin cu reaction using different techniques: A comprehensive review. *Polym. Eng. Sci.* **2014**, *54*, 2461–2469. [CrossRef]
14. Xie, H.; Li, C.; Wang, Q. A critical review on performance and phase separation of thermosetting epoxy asphalt binders and bor coats. *Constr. Build. Mater.* **2022**, *326*, 126792. [CrossRef]
15. Apostolidis, P.; Liu, X.; Kasbergen, C.; van de Ven, M.; Pipintakos, G.; Scarpas, A. Chemo-rheological study of hardening of epo: modified bituminous binders with the finite element method. *Transp. Res. Rec.* **2018**, *2672*, 190–199. [CrossRef]
16. Yu, L.; Chen, H.; Wang, H. Tracking and Analysis of Anti-slip Performance Evolutionary of Epoxy Asphalt Pavement on Ste Bridge Decks. *J. Munic. Technol.* **2022**, *40*, 29–34.
17. Jamshidi, A.; White, G.; Kurumisawa, K. Functional and field performance of epoxy asphalt technology–state-of-the-art. *Ro Mater. Pavement Des.* **2023**, *24*, 881–918. [CrossRef]
18. Tian, J.; Luo, S.; Lu, Q.; Liu, S. Effects of epoxy resin content on properties of hot mixing epoxy asphalt binders. *J. Mater. Civ. E* **2022**, *34*, 04022145. [CrossRef]
19. Apostolidis, P.; Liu, X.; van de Ven, M.; Erkens, S.; Scarpas, T. Kinetic viscoelasticity of crosslinking epoxy asphalt. *Transp. R Rec.* **2019**, *2673*, 551–560. [CrossRef]
20. Zhang, G. Study on Curing Deformation and Residual Stress of Variable Stiffness Composites. Ph.D. Thesis, Wuhan University Technology, Wuhan, China, 2019.
21. Gupta, A.M.; Macosko, C.W. Modeling strategy for systems with both stepwise and chainwise chemistry: Amine-epoxy networ with etherification. *J. Polym. Sci. Part B Polym. Phys.* **1990**, *28*, 2585–2606. [CrossRef]
22. Wise, C.W.; Cook, W.D.; Goodwin, A.A. Chemico-diffusion kinetics of model epoxy-amine resins. *Polymer* **1997**, *38*, 3251–32 [CrossRef]
23. Lucio, B.; de la Fuente, J.L. Kinetic and thermodynamic analysis of the polymerization of polyurethanes by a rheological meth *Thermochim. Acta* **2016**, *625*, 28–35. [CrossRef]
24. Boulkadid, M.K.; Touidjine, S.; Trache, D.; Belkhiri, S. Analytical methods for the assessment of curing kinetics of polyuretha binders for high-energy composites. *Crit. Rev. Anal. Chem.* **2022**, *52*, 1112–1121. [CrossRef] [PubMed]
25. Apostolidis, P.; Liu, X.; Erkens, S.; Scarpas, T. Oxidative aging of epoxy asphalt. *Int. J. Pavement Eng.* **2022**, *23*, 1471–14 [CrossRef]
26. Huang, W.; Guo, W.; Wei, Y. Thermal effect on rheological properties of epoxy asphalt mixture and stress prediction for brid deck paving. *J. Mater. Civ. Eng.* **2019**, *31*, 04019222. [CrossRef]
27. Liu, M.; Hu, J.; Sun, J.; Li, Y.; Luo, S. Characterization of Roadway Epoxy Asphalt Binder with Different Epoxy Contents. *J. Mat Civ. Eng.* **2023**, *35*, 04023144. [CrossRef]
28. Sun, J.; Huang, W.; Lu, G.; Luo, S.; Li, Y. Investigation of the performance and micro-evolution mechanism of low-conte thermosetting epoxy asphalt binder towards sustainable highway and bridge decks paving. *J. Clean. Prod.* **2023**, *384*, 1355 [CrossRef]
29. Zhang, Y.; Pan, X.; Sun, Y.; Xu, W.; Pan, Y.; Xie, H.; Cheng, R. Flame retardancy, thermal, and mechanical properties of mixe flame retardant modified epoxy asphalt binders. *Constr. Build. Mater.* **2014**, *68*, 62–67. [CrossRef]
30. Si, J.; Li, Y.; Yu, X. Curing behavior and mechanical properties of an eco-friendly cold-mixed epoxy asphalt. *Mater. Struct.* **20** *52*, 81. [CrossRef]
31. Huang, W.; Guo, W.; Wei, Y. Prediction of Paving Performance for Epoxy Asphalt Mixture by Its Time- and Temperatur Dependent Properties. *J. Mater. Civ. Eng.* **2020**, *32*, 04020017. [CrossRef]
32. Xu, J.; Ma, B.; Mao, W.; Wang, X. Strength characteristics and prediction of epoxy resin pavement mixture. *Constr. Build. Mat* **2021**, *283*, 122682. [CrossRef]
33. Zhang, W.; Li, Q.; Wang, J.; Zhao, Z.; Jin, D. Physical and chemical evaluation of adhesion recovery property of aged high-viscosi asphalt based on specialized composite rejuvenators. *J. Clean. Prod.* **2023**, *414*, 137758. [CrossRef]
34. Li, Y.; Li, J.; Ma, S.; Luo, Y. Different catalytic systems on hydroxyl-terminated GAP and PET with poly-isocyanate: Curin kinetics study using dynamic in situ IR spectroscopy. *Int. J. Polym. Anal. Charact.* **2016**, *21*, 495–503. [CrossRef]
35. Xu, L.; Li, C.; Ng, K.S. In-situ monitoring of urethane formation by FTIR and Raman spectroscopy. *J. Phys. Chem. A* **2000**, *10* 3952–3957. [CrossRef]
36. Cong, P.; Luo, W.; Xu, P.; Zhang, Y. Chemical and physical properties of hot mixing epoxy asphalt binders. *Constr. Build. Mat* **2019**, *198*, 1–9. [CrossRef]
37. Wekumbura, C.; Stastna, J.; Zanzotto, L. Destruction and recovery of internal structure in polymer-modified asphalts. *J. Mat Civ. Eng.* **2007**, *19*, 227–232. [CrossRef]
38. Jiang, J. Investigation of the Fatigue Property and Internal Structure of Asphalt Mixtures Based on Multiscale Methods. Ph. Thesis, Southeast University, Nanjing, China, 2019.
39. Kincal, D.; Özkar, S. Kinetic study of the reaction between hydroxyl-terminated polybutadiene and isophorone diisocyanate bulk by quantitative FTIR spectroscopy. *J. Appl. Polym. Sci.* **1997**, *66*, 1979–1983. [CrossRef]
40. Martínez-Miranda, M.; García-Martínez, V.; Gude, M. Gel point determination of a thermoset prepreg by means of rheolog *Polym. Test.* **2019**, *78*, 105950. [CrossRef]

Hu, H.; Li, S.; Wang, J.; Zu, L.; Cao, D.; Zhong, Y. Monitoring the gelation and effective chemical shrinkage of composite curing process with a novel FBG approach. *Compos. Struct.* **2017**, *176*, 187–194. [CrossRef]
Shah, D.U.; Schubel, P.J. Evaluation of cure shrinkage measurement techniques for thermosetting resins. *Polym. Test.* **2010**, *29*, 629–639. [CrossRef]

coatings

MDPI

Article

Investigation of Fatigue Performance for In-Service Asphalt Pavement Mixtures Using Push-Pull Fatigue Test

Zhanming Zhang [1,2], Duo Xu [2], Yunlong Fan [1], Hui Ma [1], Zihao Ding [1], Jiwang Jiang [2,*], Jie Li [1] and Tao Ma [2]

1 Jiangsu Highway Engineering Maintenance Technology Co., Ltd., Nanjing 211106, China; zhanming974@163.com (Z.Z.)

2 School of Transportation, Southeast University, Nanjing 210018, China; duoxu@seu.edu.cn (D.X.)

* Correspondence: jiang_jiwang@seu.edu.cn

Abstract: With the increasing service life of asphalt pavements, fatigue cracking has become a typical distress that significantly affects the structural lifespan of in-service asphalt pavements. To address the issues of high variability in laboratory fatigue tests and the poor correlation between test results and actual pavement cracking conditions, this study collected field core samples from in-service highways with varying degrees of cracking. The study employed push-pull fatigue tests on asphalt mixtures and linear amplitude sweep tests on extracted and recovered asphalt to evaluate fatigue performance. Quantitative relationships between transverse crack spacing and fatigue indices were established. With the inclusion of laboratory-compacted asphalt mixtures, the relationship between transverse crack spacing and the average pseudo-stiffness degradation rates was well described by a logarithmic function (R^2 = 0.928), while the apparent damage capacity achieved a stronger correlation (R^2 = 0.945). The findings demonstrate that the push-pull fatigue test, based on a viscoelastic continuum damage model, shows high stability and repeatability, with the coefficients of variation for the indices consistently below 11%. Compared to the average pseudo-stiffness degradation rates, the apparent damage capacity demonstrates better discriminative capability as an evaluation index for the intermediate-temperature fatigue performance of asphalt mixtures. The fatigue life from linear amplitude sweep tests on extracted and recovered asphalt from core samples shows a high degree of consistency with the patterns observed in various parameters of push-pull fatigue tests. The findings confirm the reliability of the push-pull fatigue test for evaluating intermediate-temperature fatigue performance of asphalt mixtures and provide the necessary theoretical support for establishing a fatigue performance evaluation system for in-service asphalt pavements.

Keywords: asphalt mixture; fatigue performance; push-pull fatigue test; linear amplitude sweep test; viscoelastic continuum damage model

1. Introduction

During long-term service, asphalt pavements develop various distresses, among which cracking severely compromises the structural integrity and functional performance of the pavement. This often necessitates frequent medium-to-large-scale maintenance activities, resulting in significant waste of non-renewable resources [1]. Asphalt pavement cracks are generally classified into bottom-up cracks and top-down cracks. It is widely recognized in the academic community that top-down cracks are primarily caused by fatigue failure under intermediate-temperature conditions induced by cyclic vehicular loading [2–5]. In China, the design service life of high-grade asphalt pavements is typically 15 years. Considering both service duration and the cumulative axle load applications, a large number of in-service highway asphalt pavements have reached or are nearing the end of their design life. Consequently, the risk of top-down fatigue failure in the asphalt top layer has increased.

To address the issue of fatigue cracking in asphalt pavement mixtures, various laboratory evaluation methods have been proposed, including the four-point bending fatigue test, semicircular bending fatigue test, indirect tensile fatigue test, and uniaxial push-pull fatigue test [6–10]. Among these, the four-point bending fatigue test is recommended in the Chinese specifications for asphalt and asphalt mixture testing as a method for estimating the fatigue life of asphalt mixtures. However, obtaining standard beam specimens from in-service pavements is extremely challenging and can potentially damage the pavement structure. Additionally, the results of this test are influenced by loading mode, temperature, and frequency, making it difficult to accurately reflect actual fatigue cracking conditions. As a result, the application of the four-point bending fatigue test in evaluating the fatigue performance of in situ asphalt mixtures remains controversial. In contrast, the semicircular bending and indirect tensile fatigue tests are widely used for evaluating the fatigue performance of in-service pavements due to the ease of specimen preparation and the simplicity of the required equipment [11–15]. However, these methods have limitations in mechanical analysis, as fatigue performance predictions primarily rely on phenomenological empirical models. Therefore, there is an urgent need for an evaluation method that is based on a clear mechanical model and can accurately reflect the actual service conditions.

To this end, researchers have developed a fatigue performance analysis method based on the Viscoelastic Continuum Damage Model (VECDM). Using uniaxial testing modes for asphalt mixtures, the VECD model quantifies fatigue damage during testing by employing the elastic-viscoelastic correspondence principle and Schapery's work potential theory, effectively eliminating the influence of the viscous components of the material [16,17]. Researchers have also proposed a simplified VECD model (S-VECDM) to evaluate the fatigue damage characteristics of asphalt mixtures under various loading conditions [18], and its predictive capabilities under different stress modes and strain levels have been validated [19]. The model analyzes the impact of damage accumulation at the microscopic scale on overall performance based on macroscopic observations, providing a more comprehensive approach to fatigue damage analysis. As a result, it has become one of the most advanced fatigue analysis methods.

As a crucial component of asphalt mixtures, asphalt binder provides adhesion and the necessary strength to resist traffic loads [20,21]. However, under prolonged loading, damage accumulation in the binder leads to a reduction in the modulus and strength of the mixture, ultimately shortening its fatigue life [22,23]. The Superpave specification evaluates the fatigue resistance of asphalt binders using the fatigue cracking factor $G^*\sin\delta$, a linear viscoelastic parameter. However, this approach struggles to capture the actual damage resistance in the nonlinear viscoelastic region. To address this, Bahia et al. [24] proposed the Time Sweep (TS) test, which uses a dynamic shear rheometer (DSR) to apply repeated cyclic loading of fixed amplitude to asphalt binders for fatigue damage evaluation. However, this method is time-consuming and yields results that are often difficult to reproduce. To improve efficiency, Bahia et al. [25] further developed the Linear Amplitude Sweep (LAS) test, which applies cyclic loading with linearly increasing strain amplitude at a fixed frequency to assess the fatigue performance of asphalt binders. The LAS test is more efficient than traditional methods and, when combined with the VECD model, provides a more accurate representation of fatigue damage. Studies based on LAS testing have shown that increased aging significantly reduces the fatigue life of asphalt binders [26], while the dosage of warm-mix additives has a notable impact on fatigue performance [27]. Zhang et al. [28] proposed three new indicators based on LAS testing—strain tolerance, strain energy tolerance, and the average decay rate of integrity at failure—which exhibit stronger correlations with actual pavement fatigue performance compared to traditional fatigue performance parameters. Since fatigue cracks primarily originate and propagate within the asphalt binder phase, the VECD approach to studying the relationship between binder and mixture performance can better clarify the mechanisms of fatigue failure [29,30]. Based on the research discussed above, the LAS test results for asphalt binders can be used to validate the effectiveness of push-pull fatigue tests for evaluating the fatigue performance of asphalt mixtures.

This study focuses on sections of in-service highway asphalt pavements with significant differences in fatigue cracking distress to characterize the fatigue performance under real service conditions. The push-pull fatigue test data were modeled using the viscoelastic continuum damage (VECD) model to investigate the quantitative relationships between the average pseudo-stiffness degradation rates (D^R) and the apparent damage capacity (S_{app}) indices with asphalt pavement crack spacing. Additionally, the LAS tests on asphalt extracted from core samples were used to validate the trends observed in the push-pull fatigue tests. From a material performance perspective, this research deepens the understanding of fatigue cracking mechanisms in existing asphalt pavements. It provides valuable references for the development and innovation of maintenance technologies aimed at extending the service life and ensuring the long-term durability of asphalt pavements.

2. Materials and Methods

2.1. Field Core Sampling

Core samples were taken from the asphalt pavement of highways in Jiangsu Province, where significant variations in crack spacing between different sections were observed. The core sampling locations were chosen from the wheel path areas of the intact driving lanes, with four cores taken from each section, spaced 25 m apart. The core sampling information is shown in Table 1. Additionally, laboratory-prepared fresh asphalt mixtures were used for experimental analysis as the initial state for material performance development, labeled as the "lab-compacted group" in Table 1.

Table 1. Information on field coring sections.

No.	Service Age (Year)	Type of Top Layer Mixture	Transverse Crack Spacing (m)	Transverse Crack Spacing Range (m)
Section 1	18	AC-13 [1]	7.32	5~10
Section 2	13	AC-13	17.83	10~20
Section 3	13	SMA-13 [2]	24.72	20~30
Section 4	21	AC-13	33.44	30~50
Section 5	5	SMA-13	62.92	50~70
Lab-compacted group 1	/	AC-13	/	/
Lab-compacted group 2	/	SMA-13	/	/

[1] AC stands for the dense-graded asphalt concrete in Chinese specifications for the design of highway asphalt pavement; 13 is the nominal maximum aggregate size of the mix. [2] SMA stands for stone mastic asphalt.

Figure 1 illustrates the core sampling process and typical core samples. All sections feature an asphalt mixture layer with a semi-rigid base course structure. The height of the asphalt mixture layer in the cylindrical core samples ranges from 170 to 180 mm (top layer: 40 mm, middle layer: 60 mm, bottom layer: 70–80 mm), with a diameter of 150 mm. The grading type of the top layer mixture is either AC-13 or SMA-13, both using PG70-22 SBS-modified asphalt.

Figure 1. Field coring procedure and typical cores.

2.2. Specimen Preparation

This study performed dynamic modulus and push-pull fatigue tests on the top layer asphalt mixture of the core samples. Given the limited thickness of the top layer in the pavement, scaled-down specimens were prepared. Previous studies have shown that small cylindrical scaled specimens with a diameter of 38 mm and a height of 110 mm exhibit similar damage characteristic curves to full-sized specimens, effectively representing the fatigue performance of the pavement core samples [31,32]. This study adopted this size and selected two core samples from each section, drilling a total of four cylindrical specimens for parallel testing. The specific preparation process for the specimens is shown in Figure 2.

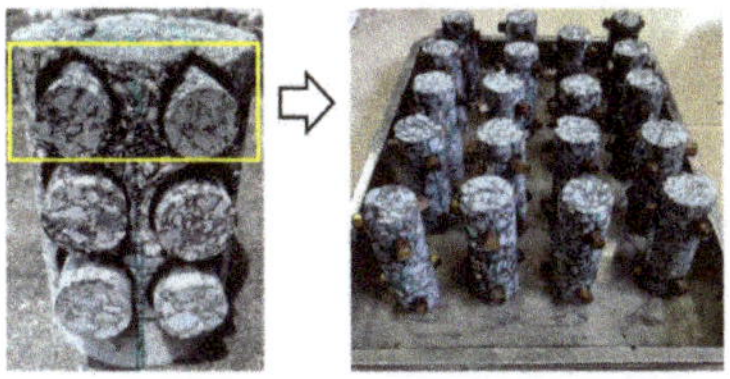

Figure 2. Preparation of scaled-down specimens of asphalt mixtures from field cores.

The fresh asphalt mixtures used as the lab-compacted group were prepared using a gyratory compactor, with mixture types selected to match those of the actual sections, namely AC-13 and SMA-13, with binder content of 5.1% and 5.8%, respectively. The asphalt used was freshly prepared SBS-modified asphalt that had undergone short-term aging, with specific technical parameters provided in the following section. To ensure that the gradation of the lab-compacted group mixtures was consistent with the field core samples, the grading curves of the two types of lab fresh mixtures were determined based on the pavement construction data. Figure 3 illustrates the gradation curves of the two laboratory-compacted mixtures. The gradation curves of AC-13 and SMA-13 reveal key differences, with SMA-13 containing higher amounts of coarse aggregate and mineral filler. In contrast, AC-13 features more fine aggregate and a smoother gradation. Finally, twenty-eight scaled cylindrical specimens were used for push-pull fatigue tests, consisting of four specimens per highway section and the laboratory-compacted samples.

As shown in Figure 4, the asphalt binder from the top layer of the core samples was extracted and recovered using a Swiss Buchi R215 rotary evaporator, with trichloroethylene as the solvent. After allowing the solution containing trichloroethylene and asphalt to settle for 24 h, the supernatant was filtered through specially designed filter paper to remove residual mineral powder before extraction. The extracted asphalt was poured into polytetrafluoroethylene molds to produce asphalt specimens suitable for the DSR parallel plate test, with dimensions of 8 mm in diameter and 2 mm in thickness.

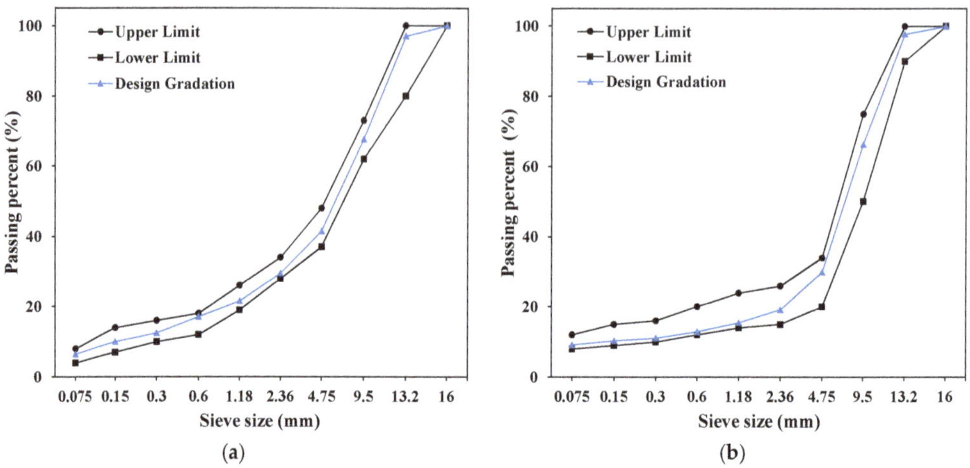

Figure 3. Design gradation of asphalt mixtures: (**a**) AC-13; (**b**) SMA-13.

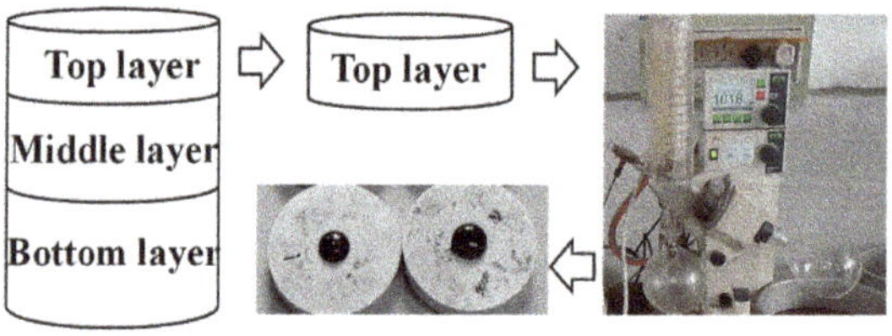

Figure 4. Extraction and recovery of field-aged asphalt binders.

The Superpave grade of the lab-compacted group's fresh SBS-modified asphalt was PG70-22, and its technical parameters are listed in Table 2. To simulate the aging effects during mixing and paving of asphalt mixtures, the fresh SBS-modified asphalt underwent short-term aging treatment prior to testing. Fifteen binder specimens were taken from field cores (three per section), along with three short-term aged SBS-modified asphalt specimens, totaling eighteen specimens.

Table 2. The technical specifications of SBS-modified asphalt in the lab-compacted group.

Basic Technical Indicators	SBS Modified Asphalt
Penetration/(25 °C, 0.1 mm)	54
Ductility/(5 °C, cm)	29
Softening Point/°C	77
Dynamic Viscosity/(60 °C, Pa·s)	3982
Elastic Recovery/(25 °C, %)	91

2.3. Testing Method

Before conducting the push-pull fatigue tests, dynamic modulus tests were performed to determine the mechanical properties of the asphalt mixture within the linear viscoelastic range under intermediate temperature conditions. As shown in Figure 5a, the dynamic modulus test was carried out using the UTM-25 (Universal Testing Machine-25) system. The testing temperatures were set at 4 °C, 20 °C, and 38 °C, with sinusoidal loading frequencies of 25, 10, 5, 1, 0.5, and 0.1 Hz. During the test, the dynamic modulus and phase angle of the specimens were recorded. Two parallel specimens were tested, and the average of the results was calculated.

(a) (b)

Figure 5. (**a**) Dynamic modulus and (**b**) push-pull fatigue tests of asphalt mixtures.

As shown in Figure 5b, the push-pull fatigue test was conducted in strain-controlled mode, with a loading strain of 300 με and a frequency of 10 Hz. To minimize the effects of viscoplasticity, the test temperature was set at 18 °C [7,33]. During the first phase of the test, a strain of 75 με was applied for 5 s at 10 Hz within the linear viscoelastic range of the mixture to determine the initial material parameters and identify any variability among specimens within the same group. After a 20-min interval, the second phase of the test was conducted at the pre-set loading strain.

The point at which the phase angle decreased significantly during the test was defined as the failure point, and the fatigue life of the specimen was determined accordingly. The test results were analyzed using the simplified viscoelastic continuum damage (S-VECD) model [31]. A power function was used to fit the damage characteristic curve, which describes the relationship between pseudo-stiffness (C) and damage (S), as shown in Equation (1):

$$C = 1 - C_{11}S^{C_{12}} \quad (1)$$

where C_{11} and C_{12} are the fitting parameters of the model.

Based on the damage characteristic curve, the average pseudo-stiffness degradation rate (D^R) is adopted as an evaluation index, representing the average decrease in pseudo-stiffness of the asphalt mixture during loading until failure. This index is independent of temperature, loading mode, and strain amplitude, and is less affected by test variability [18], making it suitable for characterizing the fatigue performance of asphalt pavement core samples. A larger D^R value indicates stronger fatigue damage resistance of the material. The calculation method is shown in Equation (2):

$$D^R = \frac{\int_0^{N_f} (1 - C)dN}{N_f} \quad (2)$$

where C represents the pseudo-stiffness of the asphalt mixture, with an initial value of 1; n represents the number of loading cycles; N_f represents the fatigue life of the specimen, defined as the number of load cycles at failure.

Based on the D^R criterion of the S-VECD model, Wang et al. [34] proposed a new fatigue parameter, the apparent damage capacity (S_{app}). The primary objective is to incorporate the accumulated damage (S) within the material, simultaneously considering the effects of modulus and toughness in the evaluation of fatigue performance. When the pseudo-stiffness C equals $(1 - D^R)$, the corresponding S value on the damage characteristic curve is defined as the apparent fatigue damage resistance, i.e., S_{app}. A larger S_{app} value indicates stronger fatigue resistance of the material. After a semi-empirical correction related to the dynamic modulus, its calculation method is shown in Equation (3):

$$S_{app} = \frac{1}{10000} \frac{\left(\frac{\alpha_T^{\frac{C_{12}}{\alpha+1}}}{C_{11}} D^R\right)^{\frac{1}{C_{12}}}}{|E^*|^{\frac{\alpha}{4}}} \tag{3}$$

where S_{app} is apparent damage capacity in GPa^{-1}; α_T represents the time-temperature equivalence shift factor; α is a material constant determined from the maximum slope of the relaxation modulus in a logarithmic scale; $|E^*|$ is the dynamic modulus of the material at 10 Hz and the target temperature, expressed in GPa. Additionally, a coefficient of 1/10,000 is applied to control the S_{app} value within the range of 0 to 40.

The Linear Amplitude Sweep (LAS) test is used to evaluate the fatigue performance of asphalt binders under intermediate temperature conditions, with the test temperature set at 18 °C, consistent with the push-pull fatigue test for mixtures. The test employs an 8 mm spindle, with the parallel plate gap set to 2 mm. The DSR equipment and LAS testing procedure are shown in Figure 6.

(**a**) (**b**)

Figure 6. (**a**) Anton Paar MCR 102 DSR; (**b**) LAS test.

The test consists of two stages. In the first stage, a frequency sweep test is conducted at 18 °C with a strain of 0.1%, and the loading frequency ranges from 0.2 Hz to 30 Hz to obtain the material properties of the asphalt binder within the linear viscoelastic region. In the second stage, a linear amplitude sweep is performed with a constant loading frequency of 10 Hz and strain control increasing linearly from 0.1% to 30%, to measure the fatigue damage resistance characteristics of the asphalt binder.

The VECD model is used to analyze the LAS test data and predict the fatigue life of the binder through a fatigue life prediction equation. The fatigue life prediction equations for asphalt binder are presented in Equations (4)–(7):

$$N_f = A(\gamma_{max})^B \tag{4}$$

$$A = \frac{f \times \left(D_f\right)^k}{k \times (\pi C_1 C_2)^{\alpha_b}} \tag{5}$$

$$B = 2\alpha_b \tag{6}$$

$$k = 1 + \alpha_b \times (1 - C_2) \tag{7}$$

where N_f represents the fatigue life of the asphalt binder; γ_{max} is the predicted maximum strain value of the pavement structure, which is set to 5% in this study; A and B are fatigue performance parameters; f is the loading frequency; D_f is the reduction in the initial complex shear modulus under the peak shear strain; C_1 and C_2 are the fitting parameters of the model; α_b is the characteristic parameter of asphalt in its undamaged state.

In the VECD model, the parameters A and B are typically used to evaluate the fatigue performance of asphalt binders, both being positively correlated with the fatigue resistance and modulus of the asphalt. However, parameter A is highly influenced by test variability, and parameter B struggles to capture the nonlinear behavior of asphalt. In comparison, the fatigue life N_f, derived from A and B, provides a more comprehensive and accurate assessment of the fatigue characteristics of asphalt materials [28]. Given that aged asphalt is more sensitive to stress and strain, higher strain levels are better suited to reflect its fatigue performance. Therefore, this study selects the fatigue life of asphalt sample under a 5% strain level ($N_{f\_}5\%$) as the parameter for evaluating fatigue performance.

3. Results and Discussion

3.1. Push-Pull Fatigue Test

The average pseudo-stiffness degradation rates (D^R) for top layer core samples from each section and the two types of lab-compacted asphalt mixtures are shown in Table 3. As observed from Table 3, the D^R values of top layer core samples from different sections vary significantly. The D^R value of Section 5, which has the largest crack spacing, is 1.82 times higher than that of Section 1, which has the smallest crack spacing. A higher D^R value indicates stronger fatigue cracking resistance of the mixture. It is evident that in sections with more severe field cracking and smaller crack spacing, the D^R values are lower. This demonstrates a strong correlation between the observed cracking conditions of asphalt pavements and the fatigue resistance of the mixtures, as represented by the D^R index.

Table 3. Results of D^R.

No.	Transverse Crack Spacing (m)	D^R		
		Mean	Standard Deviation	Coefficient of Variation
Section 1	7.32	0.33	0.027	8.3%
Section 2	17.83	0.38	0.029	7.5%
Section 3	24.72	0.41	0.027	6.7%
Section 4	33.44	0.44	0.021	4.9%
Section 5	62.92	0.59	0.038	6.4%
AC-13 (Lab-compacted)	/	0.69	0.026	3.7%
SMA-13 (Lab-compacted)	/	0.72	0.029	4.0%

The two groups of fresh asphalt mixtures prepared in the laboratory used SBS-modified asphalt subjected to short-term aging, which can be considered representative of the performance state of asphalt mixtures after paving and compaction but before traffic loading and long-term aging. Further comparisons reveal that the D^R values of the fresh asphalt mixtures are significantly higher than those of the top layer core samples. This indicates that the repeated traffic loads and aging effects experienced during long-term service significantly reduce the fatigue resistance of the field mixture, leading to fatigue cracking on the pavement surface.

To further investigate the quantitative relationship between pavement cracking conditions and the fatigue performance of asphalt mixtures, the relationship between transverse crack spacing and the D^R index for each highway section was plotted in Figure 7a. The blue lines indicate the fitted curves, while the triangles represent the data points. The results show that the relationship can be well described by a logarithmic function, with an R^2 value of 0.869. This quantitative relationship indicates that in sections with lower crack spacing (i.e., denser cracks), the D^R index exhibits more significant variations. In other words, the D^R index provides better differentiation for asphalt mixtures with poorer fatigue performance.

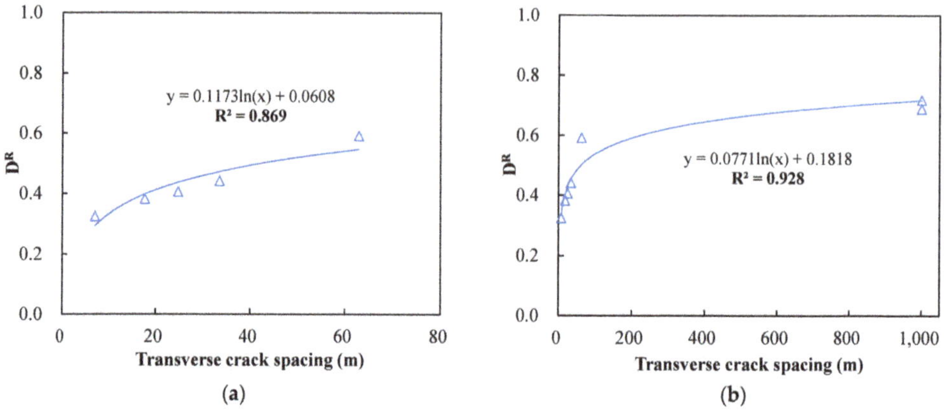

Figure 7. Relationship between crack spacing and D^R (**a**) excluding fresh mixes; (**b**) including fresh mixes.

To account for the complete service life of asphalt pavement mixtures, it is necessary to define the cracking condition of laboratory-compacted asphalt mixtures. Based on the range of the core sampling sections and the standard length of routine pavement maintenance units (1 km), this study defines the crack spacing of laboratory-compacted asphalt mixtures as 1000 m. Considering that this material represents newly paved asphalt pavements with no visible cracks (or extremely large crack spacing), this assumption is deemed reasonable.

After including the laboratory-compacted asphalt mixtures, the relationship between transverse crack spacing and the D^R index was plotted in Figure 7b. The logarithmic function fit shows an improved R^2 of 0.928. This indicates that the average pseudo-stiffness degradation rate (D^R) obtained from the push-pull fatigue test correlates strongly with actual pavement cracking conditions. Moreover, the D^R index shows potential for characterizing the fatigue performance of asphalt pavement mixtures across all service stages, starting from initial traffic opening.

Table 4 presents the apparent damage capacity (S_{app}) values for the top layer cores from each highway section and the two types of lab-compacted asphalt mixtures. A higher S_{app} value indicates greater fatigue resistance of the mixture. As shown in Table 4, the correlation between the S_{app} values and the crack spacing across the sections generally mirrors that observed for the D^R index. Except for the anomalously low S_{app} value in Section 4, the S_{app} values increase consistently with rising crack spacing, demonstrating a strong correlation. The lab-compacted asphalt mixtures, which have not been subjected to prolonged loading or aging, exhibit significantly higher S_{app} values compared to the top layer cores from all sections, aligning with the trend observed for the D^R index.

Similarly, the relationship between crack spacing and the S_{app} index for each highway section is plotted in Figure 8a. The blue lines indicate the fitted curves, while the squares represent the data points. The two variables exhibit a strong correlation that can be well-fitted by a logarithmic function, with R^2 = 0.929. This indicates that the S_{app} index is more sensitive to materials with poorer fatigue performance in terms of actual pavement cracking conditions.

Using the previously defined crack spacing for the lab-compacted asphalt mixtures (1000 m), the relationship between crack spacing and the S_{app} index, including the lab-compacted mixtures, is plotted in Figure 8b. The logarithmic fit improves compared to the case excluding the lab-compacted mixtures, with R^2 = 0.945. This demonstrates a clear trend in the S_{app} index across different crack spacing levels for the top layer asphalt mixtures. Combined with the evaluation results of the D^R index, it can be preliminarily inferred that the push-pull fatigue test effectively reflects the fatigue cracking performance of asphalt pavements under actual service conditions.

Table 4. Results of S_{app}.

No.	Transverse Crack Spacing (m)	S_{app}		
		Mean	Standard Deviation	Coefficient of Variation
Section 1	7.32	7.86	0.843	10.7%
Section 2	17.83	11.07	0.834	7.5%
Section 3	24.72	12.33	0.967	7.8%
Section 4	33.44	11.72	0.989	8.4%
Section 5	62.92	13.87	1.068	7.7%
AC-13 (Lab-compacted)	/	17.39	0.904	5.2%
SMA-13 (Lab-compacted)	/	19.53	1.060	5.4%

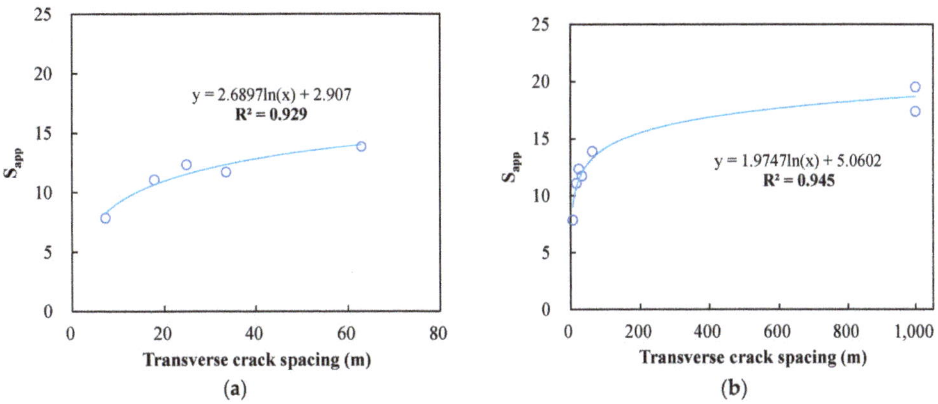

Figure 8. Relationship between crack spacing and S_{app} (**a**) excluding fresh mixes; (**b**) including fresh mixes.

Wang et al. [34] demonstrated that the D^R index primarily reflects the material's toughness, which is closely related to its energy absorption capacity. High-toughness materials can absorb and dissipate energy more effectively, thereby delaying the formation of fatigue cracks. On the other hand, the S_{app} index integrates the effects of both stiffness and toughness, simultaneously accounting for the material's modulus and energy absorption capacity. A higher modulus alone does not necessarily guarantee longer fatigue life, nor can the D^R value alone fully determine fatigue performance.

In their study, two mixtures with similar D^R values but significantly different modulus values were compared: C3 (control group) and R3 (high content reclaimed asphalt pavement group, with a higher modulus). Results showed that the R3 mixture exhibited better fatigue resistance under identical loading conditions and had a higher S_{app} index.

Field tests also revealed that the C3 pavement had more cracks [35]. Therefore, the S_{app} index, developed based on the D^R index, theoretically accounts for the stiffness of the mixture while considering its toughness. This allows for further differentiation of materials' fatigue performance in cases where D^R values are similar, providing better consistency with real-world pavement performance. This conclusion was confirmed in the fitting results of the previous section, where S_{app} showed a higher R^2 value than D^R, regardless of whether lab-compacted asphalt mixtures were included.

To further compare the sensitivity of different indices to changes in crack spacing, the variation of D^R and S_{app} indices with crack spacing is illustrated in Figure 9. Both indices are positively correlated with crack spacing, with the lab-compacted asphalt mixtures showing the highest values, indicating that sections with poorer fatigue resistance exhibit more severe cracking. Comparatively, the D^R index changes more gradually when crack spacing is less than 30 m, but shows a steeper increase in the 30–70 m range. This suggests that in larger crack spacing ranges (greater than 30 m), the improvement in crack conditions is increasingly influenced by the toughness of the surface mixture, as high-toughness mixtures effectively increase crack spacing. Conversely, the S_{app} index changes more rapidly in the 5–30 m range, reflecting that as fatigue damage progresses (with crack spacing below 30 m), S_{app} is more sensitive than D^R in capturing the decline in pavement performance.

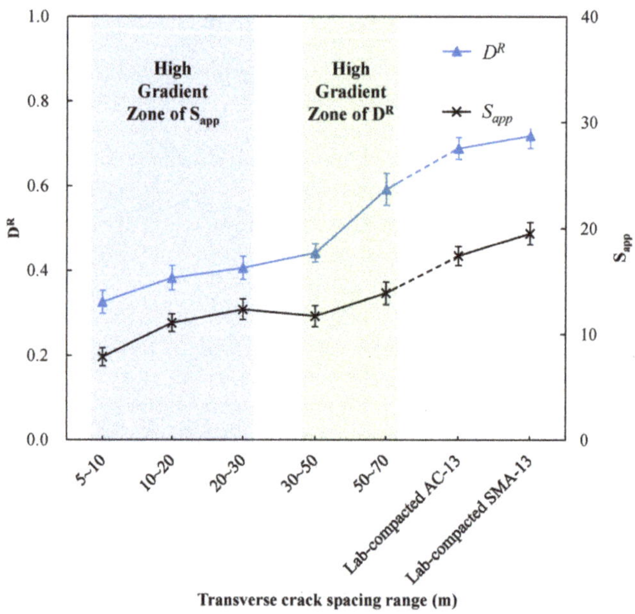

Figure 9. Comparison of D^R and S_{app}.

Comparing the lab-compacted AC-13 and SMA-13 mixtures in Figure 9, it was observed that the fatigue indices of the SMA-13 mixture were higher than those of the AC-13 mixture. The superior performance of SMA-13 can be attributed to its higher content of coarse aggregates and binder, which result in a more stable skeleton structure, greater durability, and improved fatigue resistance. To further evaluate the ability of the D^R and S_{app} indices to distinguish the fatigue performance of the two mixtures, a comparison was made not only between the lab-compacted mixtures but also between the top layer mixtures from different locations within sections 2 and 3 of the same highways. These two sections share the same service life and traffic direction, with similar stake numbers and traffic volume. The primary distinction lies in the type of top layer mixtures used.

As shown in Table 5, both the D^R and S_{app} indices of the SMA-13 mixture are higher than those of the AC-13 mixture, indicating that both indices effectively reflect the differences in fatigue performance between the two mixture types, with consistent evaluation results. To further quantify the discrimination ability of these two indices, the performance index ratio of AC-13 to SMA-13 for the same set of samples was calculated (see the AC-13/SMA-13 row in Table 5). Given that SMA-13 exhibits superior durability and fatigue resistance compared to AC-13, a smaller ratio indicates that the corresponding evaluation index has a better ability to differentiate between the two types of mixtures.

Table 5. Comparison of different types of mixtures.

Mixture Type	D^R		S_{app}	
	Top Layer of Cores	Lab-Compacted Mixtures	Top Layer of Cores	Lab-Compacted Mixtures
AC-13 (Section 2)	0.38	0.69	11.07	17.39
SMA-13 (Section 3)	0.41	0.72	12.33	19.53
AC-13/SMA-13	94.3%	95.9%	89.8%	89.0%

The comparison reveals that the S_{app} index demonstrates a stronger capacity to distinguish the fatigue performance of these mixtures. Therefore, when assessing the fatigue performance of mixtures with different gradation types, it is recommended to prioritize the S_{app} index.

To compare the stability and repeatability of the D^R and S_{app} indices, their coefficients of variation (COV) were analyzed (as shown in Figure 10). The COV of the D^R index for different road sections were all below 9%, indicating good stability of the D^R index. The lab-compacted asphalt mixtures had the lowest COV, not exceeding 4%, while the field core samples exhibited higher COV. This is primarily because the top layer asphalt mixtures in the field are subject to complex factors such as environmental and traffic loads, leading to lower stability in the test results compared to lab-compacted specimens. The COV of the S_{app} index for different sections were all below 11%. Similar to the D^R index, the lab-compacted asphalt mixtures exhibited the lowest COV, not exceeding 6%.

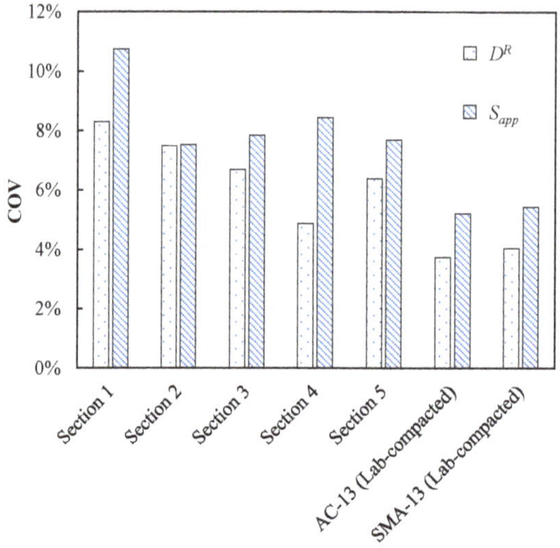

Figure 10. Comparison of Coefficient of Variation (COV) between D^R and S_{app}.

A comparison of the COV of the two indices revealed that the COV of the D^R index was consistently lower than that of the S_{app} index for the same materials. This may be related to the definition and calculation of the S_{app} index, which is influenced by multiple factors, including the D^R value of the specimen, the damage characteristic curve, and the representative dynamic modulus. These combined influences contribute to the slightly higher COV of the S_{app} index compared to the D^R index. Overall, both the D^R and S_{app} indices benefit from the uniaxial tension-compression testing mode and the S-VECD model, which are unaffected by loading mode, temperature, and stress/strain amplitude. The COV of all test groups were below 11%, effectively addressing the high variability and poor repeatability of traditional fatigue evaluation methods.

3.2. Validation Using Asphalt Binder LAS Test

The LAS test was conducted on the field-aged asphalt extracted from the top layer cores of each highway section, as well as on the short-term aged SBS-modified asphalt, to evaluate the fatigue cracking resistance of the asphalt binder at intermediate temperatures. The test results are presented in Table 6.

Table 6. Results of LAS tests.

No.	Transverse Crack Spacing (m)	$N_{f\_}5\%$		
		Mean	Standard Deviation	Coefficient of Variation
Section 1	7.32	226	13	5.8%
Section 2	17.83	603	28	4.7%
Section 3	24.72	737	32	4.3%
Section 4	33.44	425	21	4.9%
Section 5	62.92	1659	76	4.6%
Short-term aged SBS-modified asphalt	/	2352	75	3.2%

In the LAS test, a larger $N_{f\_}5\%$ indicates better fatigue performance of the asphalt binder at intermediate temperatures. The test results show that Section 4, with a service life exceeding 20 years, exhibits significant aging of the surface asphalt, resulting in a markedly low $N_{f}\_5\%$. Excluding Section 4, asphalt binders from sections with greater crack spacing demonstrate higher $N_{f}\_5\%$, indicating better resistance to fatigue cracking. The $N_{f}\_5\%$ of the short-term aged SBS-modified asphalt is significantly higher than that of the asphalt binder extracted from the cores. This highlights the effects of long-term field aging, which leads to changes in the physical and chemical properties of the asphalt binder, including hardening, increased viscosity, and reduced chemical components. These changes significantly degrade its fatigue performance. Aged asphalt exhibits increased brittleness and reduced strain capacity, making it more prone to crack initiation and propagation under repeated loading.

The above analysis demonstrates that the LAS test effectively reflects the fatigue performance of asphalt pavements at intermediate temperatures, with results showing a consistent trend with the fatigue parameters from uniaxial push-pull fatigue tests. The binder test further confirms the correlation between fatigue parameters from push-pull fatigue tests and pavement cracking severity, validating its reliability in evaluating the fatigue performance of in-service asphalt mixtures.

4. Conclusions

In this study, core samples were extracted from the typical sections of in-service asphalt pavements with varying cracking conditions on multiple highways. Push-pull fatigue tests were conducted on the mixtures to verify the relationship between test indicators and actual cracking performance and to establish a correlation between mixture fatigue parameters and pavement cracking conditions. Various evaluation indicators obtained from the push-pull fatigue tests were compared from multiple perspectives, and their applicability was analyzed. Furthermore, the LAS test results of asphalt binders were used to validate the consistency of the fatigue test indicators for asphalt mixtures. The main conclusions of this study are as follows:

1. The push-pull fatigue test method based on the S-VECD model provides reliable fatigue performance evaluation indicators, effectively assessing the intermediate-temperature fatigue damage characteristics of in-service asphalt mixtures. The fatigue indicators demonstrate a strong correlation with the actual cracking conditions of the pavement.
2. The apparent damage capacity indicator, S_{app}, accounts for both the modulus and toughness of the material, showing higher sensitivity to the fatigue performance of asphalt mixtures in heavily cracked pavement sections compared to the average pseudo-stiffness degradation rate (D^R). Additionally, S_{app} demonstrates superior differentiation in performance among mixtures with different gradation types. Based on extensive test results, its coefficient of variation remains below 11%, indicating good stability and repeatability.
3. The top layer gradation type significantly impacts the fatigue performance of asphalt mixtures. SMA-13 mixtures, with their more stable skeleton structure and higher binder content, exhibit superior fatigue performance than AC-13, as reflected by larger pavement crack spacing. In sections with smaller crack spacing, the fatigue performance of the mixtures deteriorates significantly, indicating that high stress concentrations and frequent loading intensify pavement damage.
4. The LAS test results further revealed the significant contribution of asphalt binder fatigue performance deterioration to the overall degradation of mixture performance. Across highway sections with varying cracking conditions, the fatigue life of asphalt binder demonstrated consistent trends with the fatigue parameters from mixture push-pull fatigue tests. This indicates that the decline in binder fatigue performance

is the primary cause of the reduction in asphalt mixture fatigue performance, effectively validating the correlation between push-pull fatigue test results and pavement cracking severity.

This study provides scientific evidence for assessing the fatigue performance of in-service asphalt pavements and designing maintenance strategies for crack distress, thereby extending pavement life. Given the variability of field material properties and environmental conditions, future research will investigate the fatigue behavior of field asphalt mixtures and binders under different service times, environmental factors, and loading conditions, further validating the correlation between laboratory fatigue tests and field performance.

Author Contributions: Conceptualization, Z.Z. and J.J.; Formal analysis, Z.Z., Y.F., H.M. and Z.D.; Funding acquisition, J.J.; Investigation, D.X., Y.F., H.M. and Z.D.; Methodology, Z.Z. and J.J.; Resources, J.J. and J.L.; Supervision, J.J. and T.M.; Writing—original draft, D.X. and J.J.; Writing—review & editing, Z.Z. and J.J. All authors have read and agreed to the published version of the manuscript.

Funding: This work was supported by the National Natural Science Foundation of China (Grant No. 52108421), Jiangsu Science and Technology Association Youth Science and Technology Talent Lifting Project (Grant No. JSTJ-2023-017), and Scientific Research Project of Jiangsu Communications Holding Co., Ltd. (Grant No. 2023JKY07).

Institutional Review Board Statement: Not applicable.

Informed Consent Statement: Not applicable.

Data Availability Statement: All data are included in the article.

Conflicts of Interest: Authors Zhanming Zhang, Yunlong Fan, Hui Ma, Zihao Ding, and Jie Li were employed by the company Jiangsu Highway Engineering Maintenance Technology Co., Ltd. The remaining authors declare that the research was conducted in the absence of any commercial or financial relationships that could be construed as a potential conflict of interest. The authors declare that this study received funding from Jiangsu Highway Engineering Maintenance Technology Co., Ltd. The funder had the following involvement with the study: study design, collection, analysis, the writing of this article, and the decision to submit it for publication..

References

1. Lyu, S.; Liu, C.; Qu, F.; Zheng, J. Test Methods and Characterization of Fatigue Performance of Asphalt Mixtures: A Review. *Chir J. Highw. Transp.* **2020**, *33*, 67–75.
2. Wu, S.; Wen, H.; Zhang, W.; Shen, S.; Mohammad, L.N.; Faheem, A.; Muhunthan, B. Field performance of top-down fatigu cracking for warm mix asphalt pavements. *Int. J. Pavement Eng.* **2016**, *20*, 33–43. [CrossRef]
3. Luo, X.; Luo, R.; Lytton, R.L. Energy-based mechanistic approach for damage characterization of pre-flawed visco-elasto-plast materials. *Mech. Mater.* **2014**, *70*, 18–32. [CrossRef]
4. Ziari, H.; Aliha, M.R.M.; Mojaradi, B.; Sarbijan, M.J. Investigating the effects of loading, mechanical properties and laye geometry on fatigue life of asphalt pavements. *Fatigue Fract. Eng. Mater. Struct.* **2019**, *42*, 1563–1577. [CrossRef]
5. Fang, C.; Guo, N.; Li, H.; Chu, Z.; Liu, T. Investigating the characterization of fatigue damage process for asphalt based o kinetics theory. *China J. Highw. Transp.* **2024**, 1–13.
6. Sagnol, L.; Quezada, J.C.; Chazallon, C.; Stoeckner, M. Effect of glass fibre grids on the bonding strength between two aspha layers and its Contact Dynamics method modelling. *Road Mater. Pavement Des.* **2019**, *20*, 1164–1181. [CrossRef]
7. Sabouri, M.; Kim, Y.R. Development of a Failure Criterion for Asphalt Mixtures Under Different Modes of Fatigue Loadin *Transp. Res. Rec.* **2014**, *2447*, 117–125. [CrossRef]
8. Ozer, H.; Al-Qadi, I.L.; Lambros, J.; El-Khatib, A.; Singhvi, P.; Doll, B. Development of the fracture-based flexibility index fc asphalt concrete cracking potential using modified semi-circle bending test parameters. *Constr. Build. Mater.* **2016**, *115*, 390–40 [CrossRef]
9. Manh Tuan, N.; Lee, H.J.; Baek, J. Fatigue Analysis of Asphalt Concrete under Indirect Tensile Mode of Loading Using Crac Images. *J. Test. Eval.* **2013**, *41*, 148–158. [CrossRef]
10. Cheng, H.L.; Liu, J.N.; Sun, L.J.; Liu, L.P. Critical position of fatigue damage within asphalt pavement considering temperatur and strain distribution. *Int. J. Pavement Eng.* **2021**, *22*, 1773–1784. [CrossRef]

Yuan, F.; Cheng, L.; Shao, X.; Dong, Z.; Zhang, L.; Wu, G.; He, X. Full-field measurement and fracture and fatigue characterizations of asphalt concrete based on the SCB test and stereo-DIC. *Eng. Fract. Mech.* **2020**, *235*, 107127. [CrossRef]
Du, H.; Ni, F.; Ma, X. Crack Resistance Evaluation for In-Service Asphalt Pavements by Using SCB Tests of Layer-Core Samples. *J. Mater. Civ. Eng.* **2021**, *33*, 04020418. [CrossRef]
Xu, D.; Ding, J.; Shu, L.; Ni, F.; Jiang, J.; Zhang, Z. In-situ tracking of multiscale fatigue degradation in the wearing course of asphalt pavement. *Constr. Build. Mater.* **2024**, *441*, 137588. [CrossRef]
Feng, D.; Cui, S.; Yi, J.; Chen, Z.; Qin, W. Evaluation Index of Low-temperature Asphalt Mixture Performance Based on Semi-circular Bending Test. *China J. Highw. Transp.* **2020**, *33*, 50–57.
Zhou, D.; Ma, Z.; Liu, L.; Sun, L. Study on Fatigue Performance of In-service Asphalt Pavement Based on Full-scale Accelerated Loading Test. *J. Highw. Transp. Res. Dev.* **2020**, *37*, 17–24.
Schapery, R.A. Correspondence principles and a generalized J integral for large deformation and fracture analysis of viscoelastic media. *Int. J. Fract.* **1984**, *25*, 195–223. [CrossRef]
Underwood, B.S.; Baek, C.; Kim, Y.R. Simplified Viscoelastic Continuum Damage Model as Platform for Asphalt Concrete Fatigue Analysis. *Transp. Res. Rec. J. Transp. Res. Board* **2012**, *2296*, 36–45. [CrossRef]
Wang, Y.; Kim, Y.R. Development of a pseudo strain energy-based fatigue failure criterion for asphalt mixtures. *Int. J. Pavement Eng.* **2019**, *20*, 1182–1192. [CrossRef]
Wang, Y.D.; Keshavarzi, B.; Kim, Y.R. Fatigue Performance Analysis of Pavements with RAP Using Viscoelastic Continuum Damage Theory. *KSCE J. Civ. Eng.* **2018**, *22*, 2118–2125. [CrossRef]
Shen, J. Climate zoning for road asphalt and asphalt mixture and the key technical index. *China J. Highw. Transp.* **1997**, *10*, 1–9.
Zhang, Q.; Xiao, X. Research review on constitutive model and micro structure of asphalt and asphalt mixture. *China J. Highw. Transp.* **2016**, *29*, 26–33.
Zeiada, W.A.; Underwood, B.S.; Pourshams, T.; Stempihar, J.; Kaloush, K.E. Comparison of conventional, polymer, and rubber asphalt mixtures using viscoelastic continuum damage model. *Road Mater. Pavement Des.* **2014**, *15*, 588–605. [CrossRef]
Li, H.; Luo, X.; Zhang, Y. Pseudo Energy-based Kinetic Characterization of Fatigue in Asphalt Binders. *China J. Highw. Transp.* **2020**, *33*, 115–124.
Bahia, H.U.; Zhai, H.; Zeng, M.; Hu, Y.; Turner, P. Development of binder specification parameters based on characterization of damage behavior (with discussion). *Assoc. Asph. Paving Technol.-Proc. Tech.* **2001**, *70*, 442–470.
Hintz, C.; Velasquez, R.; Johnson, C.; Bahia, H. Modification and Validation of Linear Amplitude Sweep Test for Binder Fatigue Specification. *Transp. Res. Rec.* **2011**, *2207*, 99–106. [CrossRef]
Zhang, H.; Xu, G.; Chen, X.; Wang, R.; Zhou, W. Fatigue Property of Aged Asphalt Binders Using Different Experimental Methods. *J. Build. Mater.* **2020**, *23*, 168–175.
Li, J. Effect of wax-based warm mix on rutting and fatigue properties of waste rubber powder modified asphalt. *J. China Foreign Highw.* **2023**, *43*, 230–235.
Zhang, R.; Sias, J.E.; Dave, E.V. Development of new performance indices to evaluate the fatigue properties of asphalt binders with ageing. *Road Mater. Pavement Des.* **2022**, *23*, 377–396. [CrossRef]
Cao, W.; Mohammad, L.N.; Barghabany, P. Use of Viscoelastic Continuum Damage Theory to Correlate Fatigue Resistance of Asphalt Binders and Mixtures. *Int. J. Geomech.* **2018**, *18*, 04018151. [CrossRef]
Safaei, F.; Castorena, C.; Kim, Y.R. Linking asphalt binder fatigue to asphalt mixture fatigue performance using viscoelastic continuum damage modeling. *Mech. Time-Depend. Mater.* **2016**, *20*, 299–323. [CrossRef]
Lee, K.C.; Pape, S.; Castorena, C.; Underwood, B.S.; Kim, Y.R. Strain-Level Determination Procedure for Small-Specimen Cyclic Fatigue Testing in the Asphalt Mixture Performance Tester. *Transp. Res. Rec.* **2019**, *2673*, 824–835. [CrossRef]
Kutay, M.E.; Gibson, N.; Youtcheff, J.; Dongre, R. Use of Small Samples to Predict Fatigue Lives of Field Cores Newly Developed Formulation Based on Viscoelastic Continuum Damage Theory. *Transp. Res. Rec.* **2009**, *2127*, 90–97. [CrossRef]
Ding, J.; Jiang, J.; Ni, F.; Dong, Q.; Zhao, Z. Correlation investigation of fatigue indices of fine aggregate matrix (FAM) and asphalt mixture containing reclaimed asphalt pavement materials. *Constr. Build. Mater.* **2020**, *262*, 120646. [CrossRef]
Wang, Y.D.; Underwood, B.S.; Kim, Y.R. Development of a fatigue index parameter, Sapp, for asphalt mixes using viscoelastic continuum damage theory. *Int. J. Pavement Eng.* **2022**, *23*, 438–452. [CrossRef]
Guo, X. Local Calibration of the MEPDG Using Test Track Data. Master's Thesis, Auburn University, Auburn, AL, USA, 2013.

coatings

MDPI

Article

Investigation of Factors Affecting the Intermediate-Temperatur Cracking Resistance of In-Situ Asphalt Mixtures Based on Semi-Circular Bending Test

Duo Xu [1], Fujian Ni [1,*], Hui Du [2], Zili Zhao [1], Jingling Wang [1] and Sheng Chen [3]

[1] School of Transportation, Southeast University, Nanjing 211189, China
[2] School of Civil Engineering, Fujian University of Technology, Fuzhou 350118, China
[3] Huai'an Boyan Civil Engineering Research Institute Co., Ltd., Huai'an 223003, China
* Correspondence: 101004746@seu.edu.cn

Abstract: Cracking is one of the main distresses in asphalt pavement. At present, few studies have been conducted on the cracking performance of asphalt mixtures from the field due to the difficulty of sample collection. Therefore, this study aims to assess the cracking resistance of in-service asphalt pavement at intermediate temperature using a large number of field cores in Jiangsu province, China. A semi-circular bending (SCB) test at 25 °C was conducted on field-cored samples covering three asphalt layers from 16 in-service road sections that represent a combination of influencing factors, including air void, mixture type, service age, cumulative number of equivalent single-axle loads (ESALs), and overload rate. The flexibility index (FI) and tensile strength were calculated from the experimental data as cracking performance evaluation indices. According to the analysis of variance results, at the top layer, ESALs and service age had a strong influence on cracking resistance. The decline rate of FI became slower with increasing ESALs. The most rapid decline in crack resistance with service age occurred on medium-traffic-level sections that served for over 14 years. At the middle layer, the overload rate replaced service age as a significant factor for FI. At the bottom layer, the air void was the only significant factor affecting the cracking resistance. In general, as the depth of layer increased, the effect of traffic load and service age decreased, whereas the effect of material properties increased. In addition, the FI and tensile strength were more sensitive to traffic load and air void, respectively.

Keywords: cracking resistance; intermediate temperature; semi-circular bending test; influencing factors; different asphalt layers

1. Introduction

Cracking is one of the major challenges experienced by asphalt pavements. The major cracking distresses in asphalt pavements can be categorized into reflection cracking, thermal cracking, and fatigue cracking [1]. Field fatigue cracking is caused by tensile stress concentrations produced by repeated traffic loads at intermediate temperatures. The presence of cracks increases the surface roughness, resulting in reduced ride quality and an increased risk of traffic accidents [2]. Without timely and appropriate maintenance or rehabilitation of the asphalt layer, cracking will result in severe performance degradation and the premature failure of pavements [3]. Therefore, to make reasonable rehabilitation decisions, it is important to assess the resistance to intermediate-temperature cracking of existing asphalt mixtures from in-service pavements.

Various testing methods have been proposed to characterize the cracking resistance of asphalt mixtures. These include the indirect tension (IDT) test [4,5], single-edge notch beam (SENB) test [6,7], disk-shaped compact tension (DCT) test [8], overlay test (OT) [9,10], Fenix test [11], dog-bone direct tension (DBDT) test [12], indirect tensile asphalt cracking test (IDEAL-CT) [13], and semicircular bending (SCB) test [3,14,15]. The pros and cons of each method and performance index were determined in many aspects [16–18]. As it is simple, reliable, and reproducible, the SCB test has drawn attention from researchers and was selected for this study to assess the cracking resistance of asphalt mixtures from field cores at intermediate temperatures. The SCB test uses monotonic loading at a constant rate and records a load–displacement curve for further evaluation [19]. Among the several different SCB testing modes, the semi-circular bending Illinois flexibility index (SCB-FI) test is widely recognized for its great capacity to distinguish the cracking performance of asphalt mixtures [15,20].

During the testing procedure of the SCB-FI and most other SCB tests, a notch was inserted along the symmetry axis of each specimen to produce a stress concentration [21,22]. However, notch cutting is costly and time consuming as well as producing unexpected testing variability. In terms of the fracture toughness of porous asphalt concrete at different temperatures combined with various displacement rates, notched SCB specimens have a larger variance than un-notched specimens [23]. The flexibility index (FI) results of notched specimens show a higher coefficient of variation (COV) than those of un-notched specimens (COV of 13.7%–41.2% versus 8.6%–32.0%) [24]. The higher variability of notched specimens may be owing to the difficulty in maintaining an identical position, width, and depth of the notch for different specimens. In addition, the material properties vary at the notch tip owing to the inhomogeneous characteristics of the asphalt mixtures, which may influence the crack propagation phase. In addition to lower variability, the un-notched SCB test can be employed to describe the complete damage process, including the crack-initiation phase [25]. Additionally, comparing the results of the un-notched and notched SCB tests, a strong linear correlation ($R^2 > 0.95$) exists in their FI values [24]. Therefore, the un-notched SCB test was selected in this study as a reasonable alternative to the conventional notched SCB test.

Several factors can contribute to the cracking performance of asphalt mixtures, including air void, binder performance grade (PG) and content, reclaimed asphalt pavement materials, aging, and environmental conditions [26–30]. Therefore, it is critical to identify and evaluate the factors that significantly affect the cracking performance of asphalt mixtures. Several studies have considered the effects of various factors on the cracking performance of asphalt mixture samples produced in the laboratory [26,31–33]. However, significant differences exist between laboratory-produced and field-cored samples, such as air void, compaction degree, production temperature, and homogeneity [34]. Therefore, it is necessary to determine the effects of various factors on in-service pavements by evaluating the cracking resistance of field-cored samples. Biligiri [35] compared the crack resistance and predicted service life of mixes in common pavement structures used in Sweden with those of mixes in modified pavement systems by utilizing SCB tests. Fang [36] investigated the fracture and fatigue characteristics of field core samples taken from a section of a highway in Jiangsu Province, China by using the SCB test and stereo-DIC. In terms of previous research on the influencing factors on field-cored sample properties, both mixture-design parameters and external conditions were considered for the statistical evaluation of the factors affecting high-temperature performance using the multi-sequenced repeated load test [37]. Significant factors that influence the tensile strength and fracture energy of field cores from the three asphalt layers were determined using low-temperature SCB tests [38,39].

The SCB test is now increasingly used for field core samples, due to its reasonable size, simple operation, and distinguishable index. However, as mentioned above, most of the studies mainly focus on a particular field case, or some laboratory-produced asphalt mixes. There is a lack of systematic studies on asphalt field materials with different service ages, traffic volumes, and material properties. This study uses the SCB test method and its related indexes to characterize the deterioration of the cracking resistance of asphalt field materials under the influence of different internal and external factors and reveal the significant factors affecting different layers of asphalt pavements.

In this study, the fracture potentials of 429 semicircular specimens from three asphalt layers of 16 in-service freeway sections with different structures, traffic loads, and service ages were assessed using the SCB test at 25 °C with a loading rate of 50 mm/min. The FI and tensile strength were chosen as cracking performance indices. Statistical analyses were conducted to verify the influence of material and external factors on the fracture resistance of in-service asphalt pavements.

2. Materials and Methods

2.1. Information about Field-Cored Samples

As shown in Table 1, field cores were drilled from 16 in-service road sections with diverse material properties and external conditions in Jiangsu Province, China. Table 1 lists the key information about the service age, pavement structure, mix type, and cumulative number of equivalent single-axle loads (ESALs) for each road section. The structure of all the sections is an asphalt mixture surface with a semi-rigid base built by cement-stabilized macadam or lime–fly ash macadam, which is the most common freeway structure in Jiangsu Province. The surface layer consists of top, middle, and bottom layers with thicknesses of approximately 4, 6, and 8 cm, respectively. The top and middle layers used the styrene-butadiene-styrene polymer-modified asphalt binder PG70-22 and the bottom layer used neat asphalt.

Table 1. Information about the road sections.

Road Sections	Service Age (Year)	Pavement Structure			ESALs (Million Times/ Lane)	Overload Rate (%)
		Top Layer	Middle Layer	Bottom Layer		
1	4.2	SMA-13 [1]	SUP-20 [2]	SUP-25	2.19	14.10
2	9.2	SMA-13	SUP-20	SUP-25	4.18	13.34
3	9.5	SMA-13	SUP-20	SUP-25	1.69	12.44
4	9.7	SMA-13	SUP-20	SUP-25	2.87	7.88
5	10.4	SMA-13	SUP-20	SUP-25	6.12	16.04
6	10.4	SMA-13	SUP-20	SUP-25	8.88	17.60
7	11.2	SMA-13	SUP-20	SUP-25	2.15	10.05
8	14.1	SMA-13	SUP-20	SUP-25	15.36	9.27
9	14.1	SMA-13	AC-20 [3]	AC-25	10.13	11.10
10	14.1	SMA-13	AC-20	AC-25	25.63	19.15
11	14.8	SMA-13	SUP-20	SUP-25	9.35	14.43
12	17.4	SMA-13	AC-20	AC-25	13.15	15.59
13	17.6	SMA-13	SUP-20	SUP-25	14.65	19.08
14	17.6	SMA-13	SUP-20	SUP-25	18.47	19.20
15	18.1	AK-13 [4]	AC-20	AC-25	25.90	21.68
16	18.6	AK-13	AC-20	AC-25	22.18	18.65

[1] SMA: stone mastic asphalt, [2] SUP: superpave volumetric design criteria, [3] AC: Chinese specifications for the design of highway asphalt pavements [40], [4] AK: a type of coarse-type dense graded asphalt mixture; 13/20/25 is the nominal maximum aggregate size of the mix.

Full-depth core samples of the asphalt surface course in the shape of a cylinder with a height of 18 cm and diameter of 15 cm were drilled at two-wheel tracks as shown in Figure 1. Figure 2a shows the coring procedure of asphalt field samples. The coring of all the sections was completed in the fourth quarter of 2021, with a total of 96 core samples used for this study. A typical core sample with three asphalt layers and a semi-rigid base, the field cores, and semi-circular specimens are shown in Figure 2b–d.

2.2. Testing Procedure

The untreated cylindrical core samples of the asphalt surface course were 150 mm in diameter and 170–180 mm in height. As shown in Figure 3, the top layer was cut according to its actual thickness, generally about 40 mm. The thicknesses of the middle and bottom layers were approximately 60 and 80 mm, respectively, and both were cut into cylinders of 50 mm thickness. The cylinders of all three layers were cut into two semicircles along the diameter to replicate each other. Nine semicircular specimens were discarded owing to variability in the cutting process, and 429 standard specimens were finally fabricated for subsequent SCB tests.

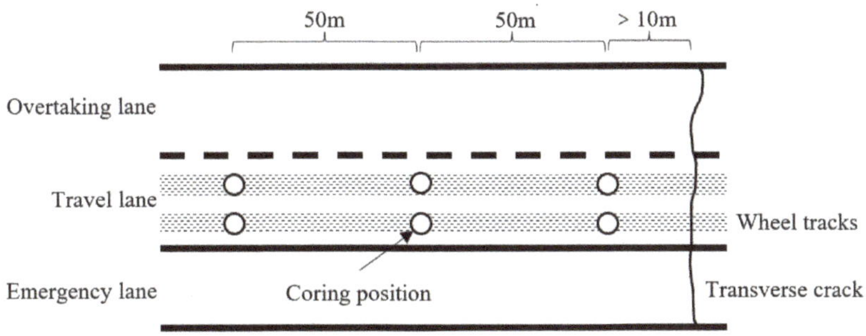

Figure 1. Schematic of coring position.

Figure 2. (**a**) Coring procedure; (**b**) A typical field core with three asphalt layers and a semi-rigid base; (**c**) Field cores; (**d**) Semi-circular specimens.

As shown in Figure 4, a UTM-25 (IPC, Melbourne, Australia) universal testing system with a bend test fixture was utilized in this study, which could deliver loads in compression with a resolution of 10 N and maximum capacity of 25 kN. The SCB test was conducted according to AASHTO TP 124 [21]. Figure 4c shows the schematic of the specimen under the SCB test. The distance between the two roller supports at the bottom of the sample is 0.8 times the diameter of the semi-circular specimen, which is 120 mm in this study. The top of the specimen is the load actuator. Before the test, the specimens were kept in an environmental chamber at 25 °C for more than 4 h to ensure that the internal temperature of the specimen also reaches 25 °C, which is the test temperature. The test was conducted using displacement control at a rate of 50 mm/min and the applied load was measured.

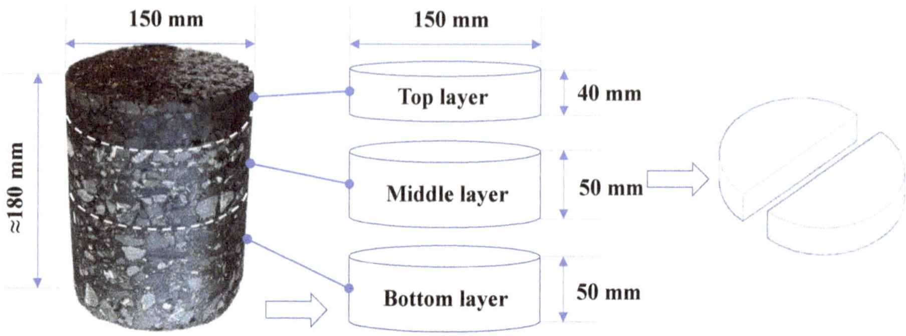

Figure 3. Schematic of SCB specimen preparation for the three asphalt layers.

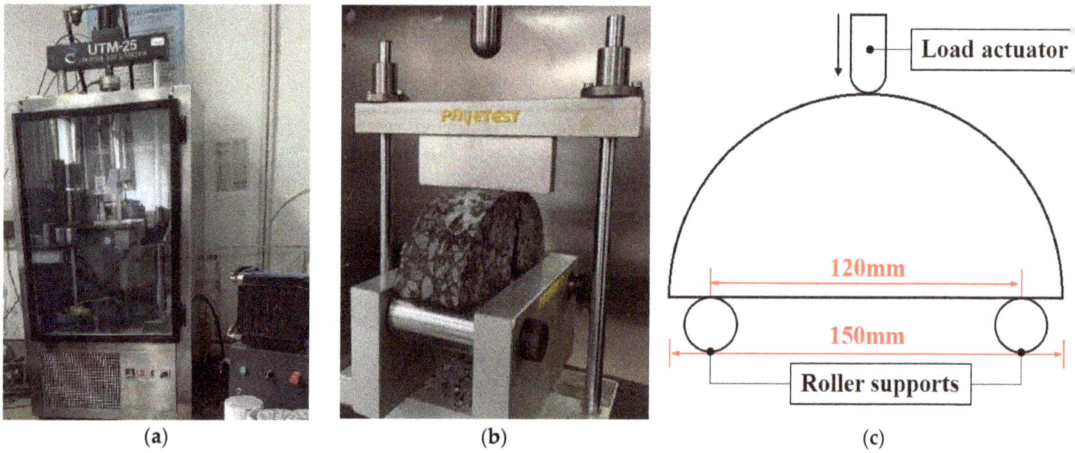

Figure 4. (**a**) UTM-25 universal testing system; (**b**) Bend test fixture; (**c**) Specimen under SCB test.

The FI is defined by Equation (1) using the parameters obtained from the load–displacement curve of the SCB test, as shown in Figure 5. A simple correction factor was established to eliminate the effect of thickness variations on the FI, as expressed in Equation (1) [41]. Considered as the evaluation index, the tensile strength (σ_t) of the un-notched SCB specimen geometry was determined using Equation (4) as follows [14]:

$$FI = \frac{G_f}{|m|} \times A \times \frac{B}{50}, \tag{1}$$

$$G_f = \frac{W_f}{A_{lig}} \times 10^6, \tag{2}$$

$$A_{lig} = 0.5 \times B \times D, \tag{3}$$

$$\sigma_t = \frac{4.8 \times F}{B \times D}, \tag{4}$$

where FI denotes the flexibility index, G_f (J/m^2) denotes the fracture energy, and m denotes the slope at the inflection point on the load–displacement curve after the peak point, as shown in Figure 5. A = 0.01 for unit conversion and scaling, B (mm) denotes the specimen thickness, and W_f (J) denotes the work of fracture, calculated as the area under the load–displacement curve using analytical integration. The pre-peak and post-peak load portions of the curve were fitted using a polynomial function and an exponential-based function before integration, where A_{lig} (mm^2) denotes the ligament area, σ_t (MPa) denotes the tensile strength, F (kN) denotes the peak load, and D (mm) denotes the specimen diameter.

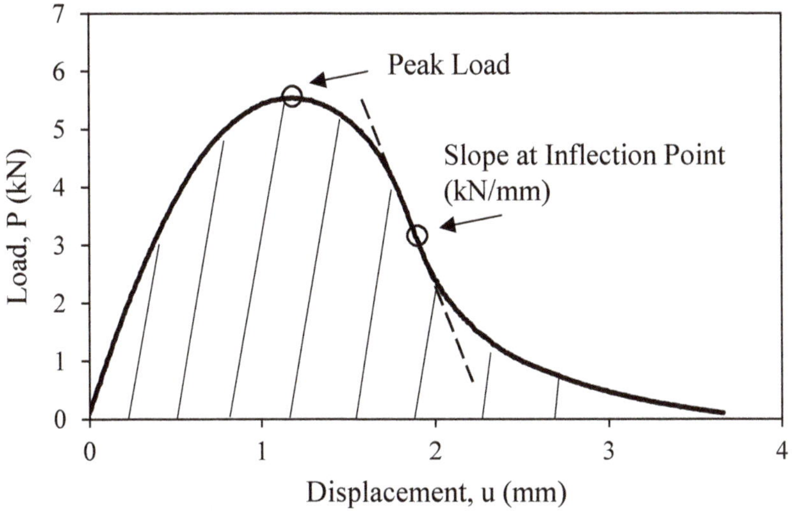

Figure 5. Typical load–displacement curve from the SCB test.

3. Results and Discussion

The distribution of the FI and tensile strength results are shown in Figure 6. In contrast, the tensile strength was much closer to a normal distribution, while the FI was mostly distributed below 10. The specimens with low fracture energy usually have a fast drop in the post-peak load–displacement curve, leading to a low FI value, which is the fracture energy divided by the slope.

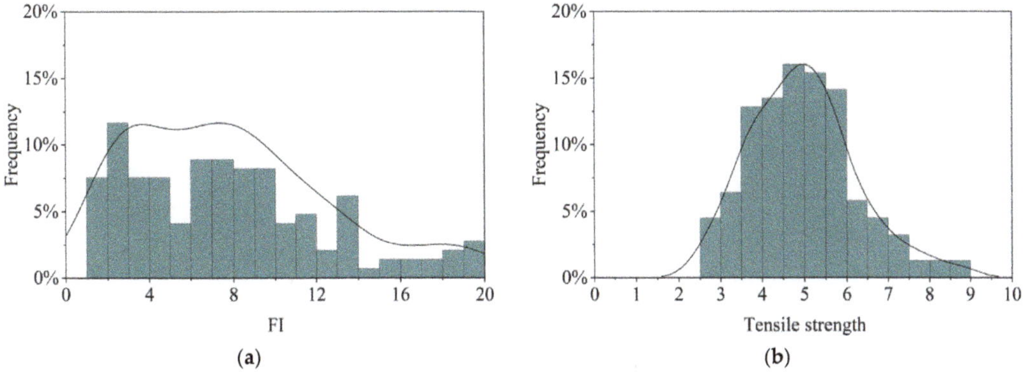

Figure 6. Distribution of (**a**) FI and (**b**) tensile strength.

Analysis of variance (ANOVA) is a valid method used to investigate whether different levels of control variables have significant effects on observation variables. In this study, ANOVA was performed to evaluate the effect of these factors on the cracking resistance of the three asphalt layer samples. A probability value (*p*-value) less than the significance level (0.05) indicated that the factor was statistically significant. The following five factors were considered: air void, mixture type, service age, ESALs, and overload rate. The overload rate, calculated as the percentage of axle load over 100 kN, ranged from 7.9% to 21.7% for the selected sections, indicating a serious overload problem.

3.1. Effect of Factors on the Top Layer

The ANOVA results for the FI and tensile strength of the top layers are presented in Table 2. In terms of the FI of the top layer, service age and ESALs were significant factors, with *p*-values less than 0.05. Another traffic load indicator, the overload rate, has a *p*-value close to 0.05, indicating that there is a certain effect of different levels of overload on the top layer's FI; however, it is less significant than the ESALs. With regard to tensile strength, only ESALs were a significant factor with a *p*-value less than 0.05. Owing to excessive compaction during construction and axial load, the air void of the top layer samples in this study was low and concentrated (3.4%–4.2%), which had an insignificant effect on the indices. The effects of ESALs and service age on FI, as well as the effect of ESALs on tensile strength, are analyzed in detail below.

Table 2. ANOVA results of indices for the top layer.

Index	Factors	Mean Square	F-Value [1]	*p*-Value [2]	Significance [3]
FI	Air void	1.2114	1.4624	0.2330	N
	Mixture type	0.9256	1.1174	0.2962	N
	Service age	3.5536	4.2899	0.0442 [4]	Y
	ESALs	13.6821	16.5171	0.0002 [4]	Y
	Overload rate	3.1338	3.7832	0.0582	N
Tensile strength	Air void	0.0362	0.0176	0.8953	N
	Mixture type	2.0054	0.9715	0.3304	N
	Service age	6.8863	3.3360	0.0754	N
	ESALs	15.0725	7.3016	0.0101 [4]	Y
	Overload rate	0.2629	0.1274	0.7231	N

[1] F-value, the ratio of mean squares, is used to determine the *p*-value. [2] Factor is significant if the *p*-value < 0.05. [3] Y indicates the factor is significant, while N indicates it is not significant. [4] *p*-value less than 0.05.

Figure 7 presents the detailed relationship between FI and the significant factors for the top layer. As shown in Figure 7a, although some of the data were scattered, the median values of FI declined continuously with increasing ESALs, indicating that there is significant fatigue performance degradation of the pavement material in the field under repeated load.

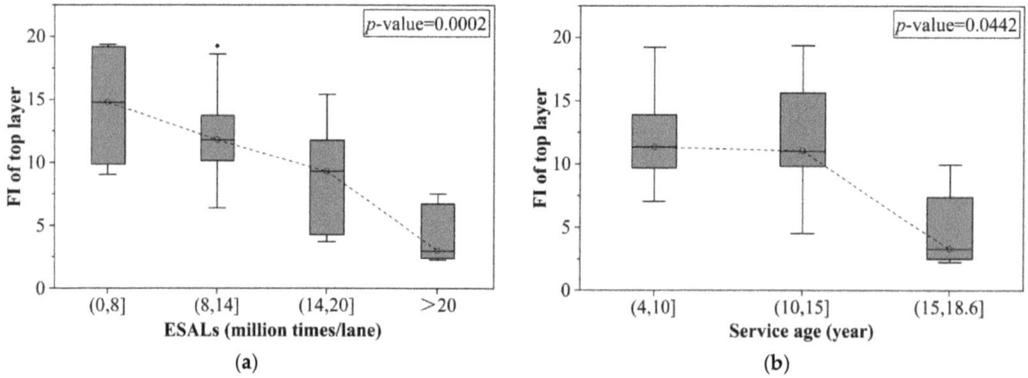

Figure 7. Relationship between FI and significant factors for the top layer: (**a**) FI versus ESALs and (**b**) FI versus service age.

Figure 7b reveals that the FI values of service age less than 15 years were mostly distributed between 9 and 16, while the FI of service age more than 15 years decreased significantly, with a reduction of about 70% in the median value. This indicates that environmental effects and load accumulation caused the cracking performance of the top layer material to decline sharply after more than 15 years of service.

The accumulation of ESALs is related to both the service age and the traffic level. The decline patterns were compared for samples of similar service age or ESALs to further validate the sensitivity of FI to the effects of service age and traffic level.

Two sample groups aged 14.1–14.8 and 17.4–18.6 years were selected to accurately investigate the relationship between FI and ESALs without the effect of service age. As shown in Figure 8, the FI curve for the 17.4–18.6 years sample consistently lies below the 14.1–14.8 years sample because of the longer service time. The FI of both groups keeps declining with increasing ESALs, while the rate of decline gradually decreases.

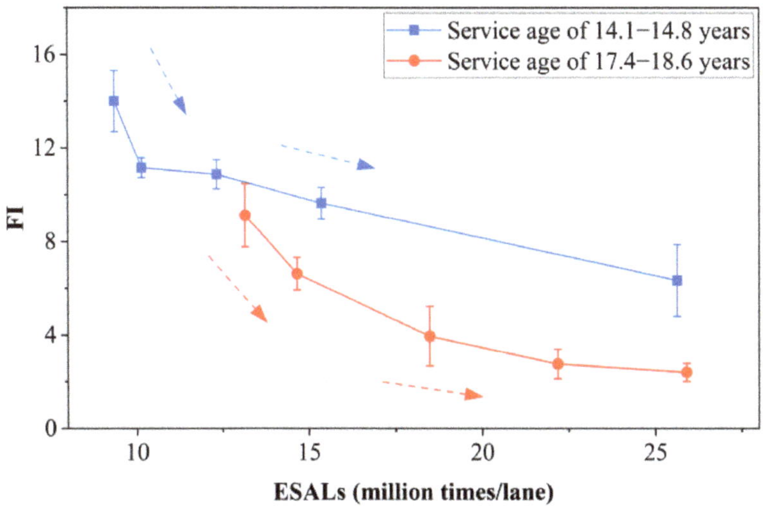

Figure 8. FI versus ESALs for samples of top layer with similar service ages.

In order to investigate the relationship between FI and service age, four groups of samples corresponding to different levels of ESALs were selected. The difference in ESALs within each group was less than one million times/lane to eliminate the effect of ESALs. Information about the four sample groups is shown in Table 3.

Table 3. Information about the four sample groups.

Group	Service Age (Years)	ESALs (Million Times/Lane)	Traffic Level
A	4.2–11.2	2.15–2.19	Light
B	10.4–14.8	8.88–9.35	Medium
C	14.1–17.6	14.65–15.36	Medium
D	14.1–18.1	25.63–25.9	Heavy

As seen in Figure 9, the FI curve of the group with more ESALs always lies below that of the group with fewer ESALs, showing a stepwise decline with age. The small graph in Figure 9 shows the absolute values of the slopes of the FI curves of the four groups, and the slope of Group C is much larger than for the other three groups. Group A and B had a service age of less than 10 years with intact or slightly damaged material in the top layer, so the FI declines the slowest with age. In the first 14 years of service for group D, with a heavy traffic level, the FI value declined to a relatively low level with axle load accumulation. Therefore, the FI of Group D declined more slowly with age than that of Group C, while still faster than that of group A and B because of the higher traffic level and service age.

In contrast, the top layer material of the sections with a medium traffic level represented by Group C showed the most significant decline in crack resistance with age after serving for around 14 years. The most probable reason for this result is that the designed life of expressways in Jiangsu Province is 15 years. For the top layer of the medium-traffic section, its cracking resistance declines slower than that of the heavy-traffic section during the designed life. Therefore, it can be seen in Figure 9 that the cracking resistance of the top layer of Group C is better than that of Group D around the designed life. After exceeding its designed life, the cracking resistance of the top layer of the medium-traffic road section decreases rapidly due to the long-term aging effect in the field.

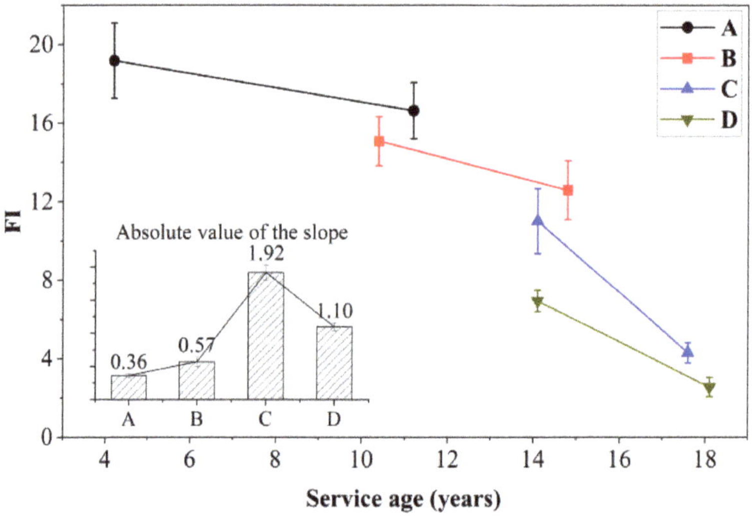

Figure 9. FI versus service age for samples of top layer with similar ESALs.

Figure 10 presents the relationship between tensile strength and ESALs. Compared with FI, the tensile strength declined with increasing ESALs at a much slower rate. From the smallest to the largest ESAL interval, the FI decreased by 79.57%; however, the tensile strength decreased by only 30.19%. Therefore, for the top layer samples, the FI was more sensitive to the ESALs than the tensile strength.

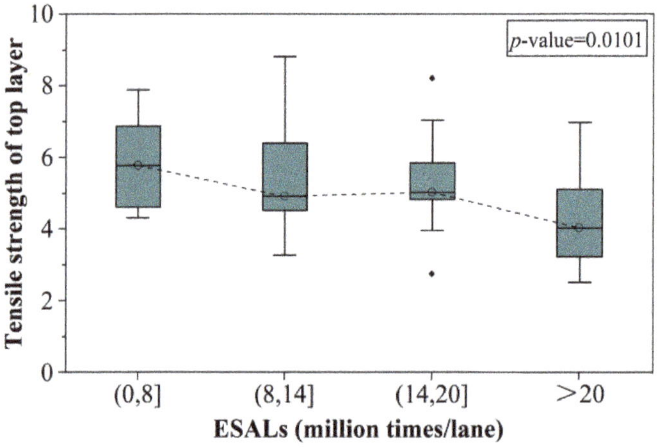

Figure 10. Relationship between tensile strength and ESALs for the top layer.

3.2. *Effect of Factors on the Middle Layer*

According to the ANOVA results of the FI for the middle layer, the ESALs and overload rate, with p-values of 0.0013 and 0.0156, respectively, were significant factors (p-values less than 0.05). In contrast to the top layer, service age, with a p-value of 0.0696, was not a significant factor affecting crack resistance. In addition, the p-values of the air void and mixture type were far beyond the significance level. In terms of tensile strength, only the air void had a p-value less than 0.05 (0.0029). The p-values of the other factors of tensile strength were significantly higher than 0.05. A particular analysis of the effect of ESALs and overload rate on FI, as well as the effect of air void on tensile strength, is presented below.

The FI of the middle layer showed a decreasing trend with ESALs as seen in Figure 11a. When the ESALs were between 8 and 20 million, the FI decreased slowly with ESALs in a stable period. When the ESALs were greater than 20 million, the median of FI decreased rapidly to 3.69, because the stability of the middle layer was greatly influenced by the development of top-down fatigue cracks under repeated axle load.

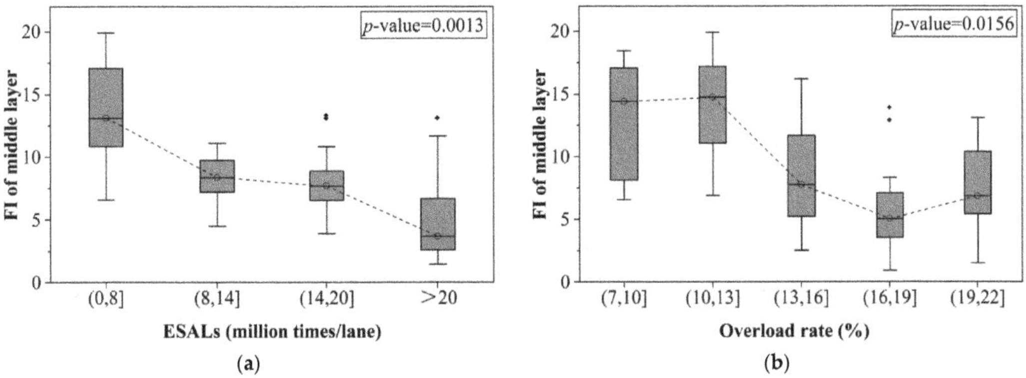

Figure 11. Relationship between FI and significant factors for the middle layer: (**a**) FI versus ESALs and (**b**) FI versus overload rate.

Compared with the top layer, the overload rate replaced service age as a significant factor in the FI of the middle layer. Under the protection of the top layer, the influence of external environmental conditions on the middle layer declined remarkably, weakening the effect of the service age. However, even with the protection of the top layer, a severely overweight axle load may still cause performance degradation. Therefore, for the middle layer, the difference in the overload rate had a more significant impact on FI than service age. According to the relationship between FI and overload rate presented in Figure 11b, although there are some fluctuations, the FI generally decreases as the overload rate increases. When the overload rate was approximately 10%–19%, the FI was highly sensitive to changes in the overload rate. Therefore, the cracking resistance of the middle layer of a heavy-traffic highway section with an overloaded rate over 10% declines faster than that of a light-traffic section with a low overload rate. For this kind of highway section, asphalt overlay is recommended as a maintenance method. The asphalt overlay shifts the original middle layer down in the overall structure of the pavement and slows down its performance degradation.

The air void of the SCB specimens was determined using AASHTO T 166 [42]. This is the only significant factor affecting the tensile strength of the middle layer. According to the correlation relationship displayed in Figure 12, the tensile strength gradually decreased with an increase in air void. The application of the linear fit is based on the relationship between tensile strength and air void found by previous studies [43,44]. This result is consistent with the fact that mixtures with low air void are stiffer than those with high air void, exhibiting a higher peak load and tensile strength in the SCB test [44].

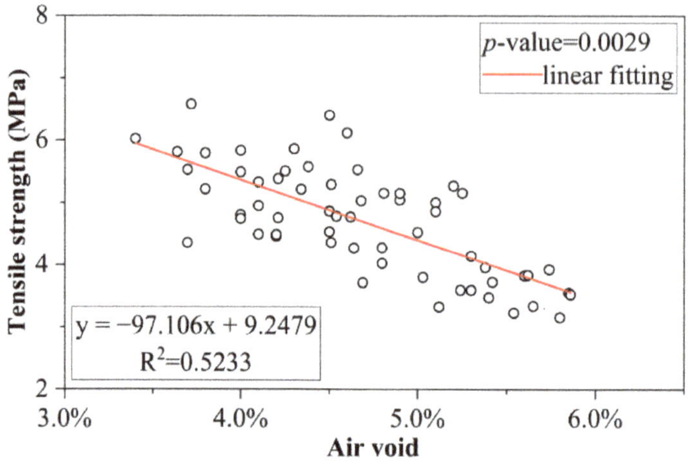

Figure 12. Relationship between tensile strength and air void for the middle layer.

3.3. *Effect of Factors on the Bottom Layer*

The ANOVA results revealed that the air void was the only significant factor for both the indices of the bottom layer. In addition, the *p*-value of the air void for the tensile strength was significantly smaller than that for the FI (<0.0001 vs. 0.0007). The specific effect of air void on both indices is analyzed below.

Compared with the top and middle layers, the bottom layer had the lowest position on the asphalt pavement surface course, and the environmental and load conditions were relatively tolerant. Therefore, ESALs, service age, and overload rate had slight effects on the performance of the bottom layers, which is consistent with the results of the ANOVA.

Figure 13 illustrates the correlation between the indices and the air void of the bottom layer. Kaseer [26] found a strong linear relationship between the FI and air void, which also existed in other studies [45–47]. Therefore, a linear fit was also performed for the values of FI and air void in this study. The tensile strength decreased with an increase in the air void, which was consistent with the law of the middle layer. However, FI showed an opposite trend. It is well known that more energy is consumed in the crack initiation and propagation phases of denser asphalt mixtures generated with more compaction energy. Therefore, the higher the air void, the lower the fracture energy [47]. Furthermore, the decrease in the peak load owing to the increase in the air void yielded a lower absolute value of m (post-peak slope). When calculating the FI, a decrease in fracture energy will decrease the FI; however, at the same time, a decrease in the absolute value of m will increase the FI with a more dominant effect [26]. Therefore, a tentative air void correction factor was used to adjust FI values based on the air void contents of the samples, as shown in Equation (5) [26]:

$$FI_{4\%} = FI \times \frac{4\%}{AV}, \tag{5}$$

where $FI_{4\%}$ is the corrected FI using 4% as a reference air void and AV is the air void content.

3.4. *Summary of Factors Influencing the Three Asphalt Layers*

According to the above analysis, the significant factors affecting the cracking resistance of the three asphalt layers are summarized in Table 4. It was found that the cracking performance of the top and middle layers was significantly affected by several factors, including ESALs, service age, overload rate, and air void. However, the bottom layer indices were affected only by air void. Regarding the FIs of the three layers, ESALs had a significant effect on the upper two layers but had no effect on the bottom layer. In addition, regarding the tensile strength, ESALs had a significant effect only on the top layer.

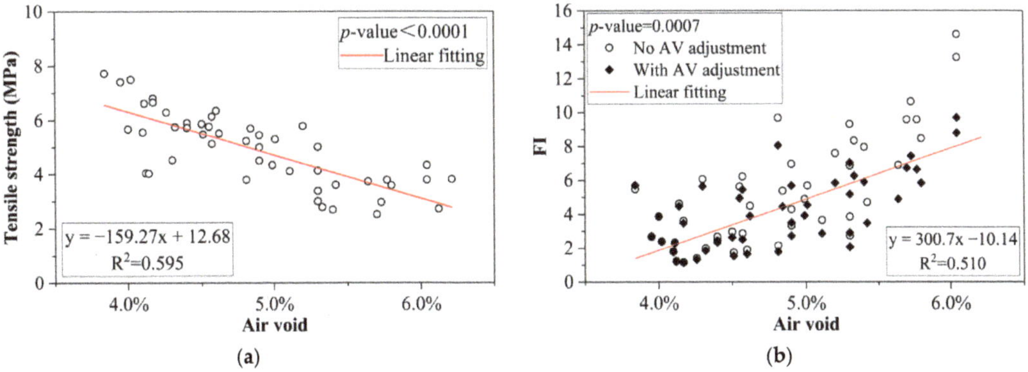

Figure 13. Relationship between significant factors and air void for the bottom layer: (**a**) tensile strength versus air void and (**b**) FI versus air void.

Table 4. Summary of factors affecting the cracking resistance for three asphalt layers.

Layers	Cracking Indices	Significant Factors	*p*-Value
Top	FI	Service age	0.0442
		ESALs	0.0002
	Tensile strength	ESALs	0.0101
Middle	FI	ESALs	0.0013
		Overload rate	0.0156
	Tensile strength	Air void	0.0029
Bottom	FI	Air void	0.0007
	Tensile strength	Air void	<0.0001

A multivariate technique was used to summarize the degree of linear relationship between ESALs and the FIs of the three layers. When an exact linear relationship exists between two variables, the correlation coefficient is 1 for a positive correlation and −1 for a negative correlation. When no linear relationship exists, the correlation coefficient tends to 0. Table 5 lists the correlation coefficient matrix. The degree of linear correlation between ESAL and FI was ranked as follows: top layer > middle layer > bottom layer. This indicates that there is no correlation between ESALs and the FI of the bottom layer because the coefficient is close to 0. The correlation results confirmed that traffic load only affects the crack resistance of the two upper asphalt layers.

Table 5. Multivariate correlations between ESALs and FIs of the three asphalt layers.

Correlation Item	ESALs	FI of Top Layer	FI of Middle Layer	FI of Bottom Layer
ESALs	1	−0.67382	−0.30739	0.03645
FI of top layer	−0.67382	1	0.179072	−0.11837
FI of middle layer	−0.30739	0.179072	1	−0.18856
FI of bottom layer	0.03645	−0.11837	−0.18856	1

The following Table 6 compares the effects of these five factors on cracking resistance at an intermediate temperature revealed by this study and previous studies in recent years. The current results on the effect of air void and mix type are consistent with previous studies. In the past, there was a lack of studies on the effects of external factors on different asphalt layers in the field.

Table 6. Comparison of effects of factors on cracking resistance with previous studies.

Factors	Cracking Resistance at Intermediate Temperature	
	Current Results	**Previous Studies**
Air void (↑) [1]	FI (↑) Tensile strength (↓) [2]	FI (↑) Tensile strength (↓) [26,38,44,45]
Mixture type	Non-sensitive	Non-sensitive [26,45]
Service age (↑)	↓ (top layer only)	Not found
ESALs (↑)	↓ (not for bottom layer)	Not found
Overload rate (↑)	↓ (middle layer only)	Not found

[1] ↑ indicates an increase in value, [2] ↓ indicates a decrease in value.

4. Conclusions

In this study, the intermediate-temperature cracking resistance of asphalt field mixtures from 16 freeway sections in Jiangsu Province, China was evaluated based on the SCB test. Five material properties and external conditions were considered in the statistical evaluation to reveal significant factors affecting the FI and tensile strength of the three asphalt layers. The key findings of this study are summarized below.

1. Regarding the top asphalt layer, ESALs had a significant effect on both indices, while service age only had a significant effect on FI. The FI declined with increasing ESALs, while the rate of decline gradually decreased. The most rapid decline in crack resistance with age occurred on medium-traffic-level sections that served for over 14 years.
2. For the middle asphalt layer, the FI was significantly affected by ESALs, followed by the overload rate. The FI was highly sensitive to an overload rate of approximately 10%–19%.
3. For the bottom asphalt layer, FI and tensile strength were only affected significantly by the air void.
4. Comparing the five factors, the effect of the traffic load decreased with the depth of the pavement structure. The air void gradually becomes a crucial factor with increasing depth. The current results on the effect of air void and mix type agree with the ones in previous studies. The effect of external factors on the cracking performance of different asphalt layers in the field was revealed, which has rarely been studied in the past.
5. Indicating the overall ductility of the asphalt mixture, the FI was more sensitive to the traffic load than the tensile strength. The tensile strength was more sensitive to the air void as a stiffness index. When the traffic load was a significant factor, both FI and tensile strength decreased as the traffic load increased. When the effect of the air void was dominant, the tensile strength decreased as the air void increased, whereas FI, counterintuitively, exhibited the opposite trend.

This study can be expanded by field samples from a broader array of freeway sections. In addition, it is of interest to establish a correlation between SCB cracking indices and real-world pavement cracking performance data. All these analyses will contribute to further investigations into the effects of various factors on the cracking of long-term-service asphalt pavements.

Author Contributions: Conceptualization, F.N. and D.X.; methodology, D.X., H.D. and Z.Z.; formal analysis, D.X. and H.D.; investigation, D.X., J.W. and S.C.; resources, F.N.; writing—original draft preparation, D.X.; writing—review and editing, H.D. and Z.Z.; supervision, F.N.; funding acquisition, F.N. All authors have read and agreed to the published version of the manuscript.

Funding: This research was funded by Jiangsu Communications Holding Co., Ltd., under several projects (No. 8521009107, No. 8521008990) and a Jiangsu Province Construction System Technology Project (No. 2022ZD099).

Institutional Review Board Statement: Not applicable.

Informed Consent Statement: Not applicable.

Data Availability Statement: All data are included in the article.

Conflicts of Interest: The authors declare no conflict of interest. The funders had no role in the design of the study; in the collection, analyses, or interpretation of data; in the writing of the manuscript; or in the decision to publish the results.

References

Lei, G.; Ni, F.; Braham, A.; Luo, H. Mixed-Mode cracking behavior of cold recycled mixes with emulsion using Arcan configuration. *Construct. Build. Mater.* **2014**, *55*, 415–422.

Al-Qadi, I.L. Pavement Damage Due to Conventional and New Generation of Wide-Base Super Single Tires. *Tire Sci. Technol.* **2005**, *33*, 210–226. [CrossRef]

Saha, G.; Biligiri, K.P. Fracture properties of asphalt mixtures using semi-circular bending test: A state-of-the-art review and future research. *Constr. Build. Mater.* **2016**, *105*, 103–112. [CrossRef]

Roque, R.; Buttlar, W.G. Development of a measurement and analysis system to accurately determine asphalt concrete properties using the indirect tensile mode. *Phys. Rev. D Part. Fields* **1992**, *83*, 2235–2239.

Kim, Y.R.; Wen, H. Fracture energy from indirect tension testing. *Asph. Paving Technol.* **2002**, *71*, 779–793.

Wagoner, M.P.; Buttlar, W.G.; Paulino, G.H. Development of a Single-Edge Notched Beam Test for Asphalt Concrete Mixtures. *J. Test. Eval.* **2005**, *33*, 1–13.

Kim, H.; Wagoner, P.M.; Buttlar, W.G. Micromechanical fracture modeling of asphalt concrete using a single-edge notched beam test. *Mater. Struct.* **2008**, *42*, 677–689. [CrossRef]

Wagoner, M.; Buttlar, W.; Paulino, G.; Blankenship, P. Investigation of the Fracture Resistance of Hot-Mix Asphalt Concrete Using a Disk-Shaped Compact Tension Test. *Transp. Res. Rec. J. Transp. Res. Board* **2005**, *1929*, 183–192. [CrossRef]

Zhou, J.; Scullion, T. *Upgraded Overlay Tester and Its Application to Characterize Reflection Cracking Resistance of Asphalt Mixtures*; FHWA/TX-04/0-4467-1; Texas Transportation Institute: Austin, TX, USA, 2003.

Zhou, F.; Sheng, H.; Chen, D.H.; Scullion, T. Overlay Tester: Simple Performance Test for Fatigue Cracking. *Transp. Res. Rec. J. Transp. Res. Board* **2007**, *2001*, 1–8. [CrossRef]

Jiménez, F.P.; Vidal, G.V.; Recasens, J.M.; Nieto, R.B. Fenix test: Development of a new test procedure for evaluating cracking resistance in bituminous mixtures. *Transp. Res. Rec. J. Transp. Res. Board* **2010**, *46*, 36–43. [CrossRef]

Koh, C.; Lopp, G.; Roque, R. Development of a Dog-Bone Direct Tension Test (DBDT) for Asphalt Concrete. In Proceedings of the 7th International RILEM Symposium on Advanced Testing and Characterisation of Bituminous Materials, Rhodes, Greece, 27 May 2009; pp. 585–596.

Zhou, F.; Im, S.; Sun, L.; Scullion, T. Development of an IDEAL Cracking Test for Asphalt Mix Design and QC/QA. *Road Mater. Pavement Des.* **2017**, *18*, 405–427. [CrossRef]

Molenaar, A.; Scarpas, A.; Liu, X.; Erkens, S. Semi-Circular Bending Test; Simple but Useful? *Asph. Paving Technol.* **2002**, *71*, 794–815.

Ozer, H.; Al-Qadi, I.L.; Lambros, J.; El-Khatib, A.; Singhvi, P.; Doll, B. Development of the fracture-based flexibility index for asphalt concrete cracking potential using modified semi-circle bending test parameters. *Constr. Build. Mater.* **2016**, *115*, 390–401. [CrossRef]

Braham, A.; Underwood, S. *State of the Art and Practice in Fatigue Cracking Evaluation of Asphalt Concrete Pavements*; Version 1.0; Association of Asphalt Paving Technologists: Lino Lakes, MN, USA, 2016.

Alkuime, H.; Tousif, F.; Kassem, E.; Bayomy, F. Review and evaluation of intermediate temperature monotonic cracking performance assessment testing standards and indicators for asphalt mixes. *Constr. Build. Mater.* **2020**, *263*, 120121. [CrossRef]

Sreedhar, S.; Coleri, E.; Haddadi, S.S. Selection of a Performance Test to Assess the Cracking Resistance of Asphalt Concrete Materials. *Constr. Build. Mater.* **2018**, *179*, 285–293. [CrossRef]

Elseifi, M.A. Modeling and evaluation of the cracking resistance of asphalt mixtures using the semi-circular bending test at intermediate temperatures. *Road Mater. Pavement Des.* **2012**, *13*, 124–139. [CrossRef]

Ozer, H.; Al-Qadi, I.L.; Singhvi, P.; Khan, T.; Rivera-Perez, J.; El-Khatib, A. Fracture Characterization of Asphalt Mixtures with High Recycled Content Using Illinois Semicircular Bending Test Method and Flexibility Index. *Transp. Res. Rec. J. Transp. Res. Board* **2016**, *2575*, 130–137. [CrossRef]

AASHTO TP 124-2018; Standard Method of Test for Determining the Fracture Potential of Asphalt Mixtures Using the Flexibility Index Test (FIT). American Association of State Highway and Transportation Officials: Washington, DC, USA, 2018.

Zhou, Z.; Gu, X.; Ni, F.; Li, Q.; Ma, X. Cracking Resistance Characterization of Asphalt Concrete Containing Reclaimed Asphalt Pavement at Intermediate Temperatures. *Transp. Res. Rec. J. Transp. Res. Board* **2017**, *2633*, 46–57. [CrossRef]

Molenaar, J.M.M. Performance Related Characterisation of the Mechanical Behavior of Asphalt Mixtures. Ph.D. Thesis, Delft University of Technology, Delft, The Netherlands, 2003.

Yan, C.; Zhang, Y.; Bahia, H.U. Comparison between SCB-IFIT, un-notched SCB-IFIT and IDEAL-CT for measuring cracking resistance of asphalt mixtures. *Constr. Build. Mater.* **2020**, *252*, 119060. [CrossRef]

Jiang, J.; Dong, Q.; Ni, F.; Zhao, Y.J. Effects of Loading Rate and Temperature on Cracking Resistance Characteristics of Asphalt Mixtures Using Nonnotched Semicircular Bending Tests. *J. Test. Eval.* **2019**, *47*, 4. [CrossRef]

26. Kaseer, F.; Yin, F. Development of an index to evaluate the cracking potential of asphalt mixtures using the semi-circular bendir test. *Constr. Build. Mater.* **2018**, *167*, 286–298. [CrossRef]
27. Kim, S.S.; Yang, J.J.; Etheridge, R.A. Effects of mix design variables on flexibility index of asphalt concrete mixtures. *Int. Pavement Eng.* **2020**, *21*, 1275–1280. [CrossRef]
28. Hanz, A.; Dukatz, E.; Retake, G. Use of Performance Based Testing for High RAP Mix Design and Production Monitorin *Asph. Paving Technol.* **2016**, *85*, 449–483. [CrossRef]
29. Zhu, Z.; Singhvi, P.; Espinoza-Luque, A.F.; Ozer, H.; Al-Qadi, I.L. Influence of mix design parameters on asphalt concrete agir rate using I-FIT specimens. *Constr. Build. Mater.* **2019**, *200*, 181–187. [CrossRef]
30. Brown, E.R. *Hot Mix Asphalt Materials Mixture Design, and Construction;* National Asphalt Pavement Association Research ar Education Foundation: Lanham, MD, USA, 2009.
31. Sreedhar, S.; Coleri, E. The effect of long-term aging on fatigue cracking resistance of asphalt mixtures. *Int. J. Pavement Eng.* **202** *23*, 308–320. [CrossRef]
32. Jahangiri, B.; Majidifard, H.; Meister, J.; Buttlar, W.G. Performance Evaluation of Asphalt Mixtures with Reclaimed Aspha Pavement and Recycled Asphalt Shingles in Missouri. *Transp. Res. Rec. J. Transp. Res. Board* **2019**, *2673*, 0361198119825e [CrossRef]
33. Batioja-Alvarez, D.; Lee, J.; Haddock, J.E. Understanding the Illinois Flexibility Index Test (I-FIT) using Indiana Asphalt Mixture *Transp. Res. Rec. J. Transp. Res. Board* **2019**, *2673*, 036119811984128. [CrossRef]
34. Koek, B.V.; Yilmaz, M.; Alatas, T.J. Evaluation of the Mechanical Properties of Field- and Laboratory-Compacted Hot-Mix Aspha *J. Mater. Civ. Eng.* **2014**, *26*, 9.
35. Biligiri, K.P.; Said, S.; Hakim, H.J. Asphalt mixtures' crack propagation assessment using semi-circular bending tests. *Int. Pavement Res. Technol.* **2012**, *5*, 209–217.
36. Yuan, F.; Cheng, L.; Shao, X.; Dong, Z.; He, X.J. Full-field Measurement and Fracture and Fatigue Characterizations of Aspha Concrete based on the SCB Test and Stereo-DIC. *Eng. Fract. Mech.* **2020**, *235*, 107127. [CrossRef]
37. Zhao, Z.; Jiang, J.; Ni, F. Factors Affecting the Rutting Resistance of Asphalt Pavement Based on the Field Cores Using Mul Sequenced Repeated Loading Test. *Constr. Build. Mater.* **2020**, *253*, 118902. [CrossRef]
38. Du, H.; Ni, F.; Zhao, Y. Characterization of the Influence Factors of Low-Temperature Fracture Properties for Serving Aspha Pavements Using Layer Core Samples. *J. Transp. Eng. Part B-Pavements* **2021**, *147*, 04021016. [CrossRef]
39. Du, H.; Ni, F.; Ma, X. Crack Resistance Evaluation for In-Service Asphalt Pavements by Using SCB Tests of Layer-Core Sample *J. Mater. Civ. Eng.* **2021**, *33*, 04020418. [CrossRef]
40. *JTG E20-2011;* Standard Test Methods of Bitumen and Bituminous Mixtures for Highway Engineering. Research Institute o Highway Ministry of Transport: Beijing, China, 2011.
41. Al-Qadi, I.; Ozer, H.; Lambros, J.; El Khatib, A.; Singhvi, P.; Khan, T.; Rivera, J.; Doll, B. *Testing Protocols to Ensure Performance High Asphalt Binder Replacement Mixes Using RAP and RAS;* Report No. FHWA-ICT-15-017; Illinois Center for Transportatio Urbana, IL, USA, 2015.
42. *AASHTO T 166-2013;* Standard Method of Test for Bulk Specific Gravity (Gmb) of Compacted Hot Mix Asphalt (HMA) Usin Saturated Surface-Dry Specimens. American Association of State Highway and Transportation Officials: Washington, DC USA, 2013.
43. Li, X.J.; Marasteanu, M.O. Using Semi Circular Bending Test to Evaluate Low Temperature Fracture Resistance for Aspha Concrete. *Exp. Mech.* **2010**, *50*, 867–876. [CrossRef]
44. Chen, X.; Solaimanian, M.J. Simple Indexes to Identify Fatigue Performance of Asphalt Concrete. *J. Test. Eval.* **2020**, *48*, 3999–401 [CrossRef]
45. Majidifard, H.; Jahangiri, B.; Rath, P.; Buttlar, W.G. Development of a Balanced Cracking Index for Asphalt Mixtures Tested i Semi-Circular Bending with Load-LLD. *Measurement* **2020**, *173*. [CrossRef]
46. Riveraperez, J.J. *Effect of Specimen Geometry and Test Configuration on the Fracture Process Zone for Asphalt Materials;* University o Illinois at Urbana-Champaign: Urbana, IL, USA, 2017.
47. Barry, M.K. *An Analysis of Impact Factors on the Illinois Flexibility Index Test;* University of Illinois at Urbana-Champaign: Urban IL, USA, 2016.

coatings

MDPI

ticle

tudy on Key Mechanical Parameters of High-Strength ;routing Material

eran Zhang [1,2,*], Baoyong Cao [2], Heng Zhang [2], Xiong Zhang [1] and Yan He [3]

[1] Key Laboratory of Advanced Civil Engineering Materials of Ministry of Education, School of Materials Science and Engineering, Tongji University, Shanghai 201804, China
[2] China Communications Construction Company Third Harbor Engineering Co., Ltd., Shanghai 200032, China
[3] School of Civil Engineering, Suzhou University of Science and Technology, Suzhou 215011, China
* Correspondence: 1610978@tongji.edu.cn

Abstract: In order to better design and calculate in infrastructures, it is necessary to clarify the key mechanical parameters of structural materials, such as axial compressive strength, elastic modulus and Poisson's ratio. High-strength grouting material (HSGM) have begun to be used as structural materials with the development of large and complex structures. A large number of test dates were used to analyze the relationship between the axial compressive strength and the cubic compressive strength of HSGM in the paper. ABAQUS software was used to model the specimens of axial compressive strength, and the strain cloud maps of concrete and HSGM were compared and analyzed. By considering HSGM as two-phase (sand and paste) composites, the relationship between elastic modulus of HSGM and mechanical parameters of component materials was derived, and the test results of the mechanical properties of HSGM with different ratios of sand to cement were used for verification. The test results show that the axial compressive strength of the HSGM is closer to the cubic strength than that of the concrete material, which accords with the finite element analysis results. The elastic modulus of high-strength grouting material conforms to the theoretical derivation of two-phase material. The material composition is one of the main factors affecting the elastic modulus. Poisson's ratio range of high-strength grouting material is 0.25 ± 0.01 by statistical analysis.

Keywords: high-strength grouting material; finite element analysis; axial compressive strength; modulus of elasticity; Poisson's ratio

1. Introduction

Cement-based material such as concrete, mortar and grout are widely used in structures because of its good compression performance and low-cost characteristics [1]. With the rapid development of cement-based material technology, high-strength grouting material can have the properties of large fluidity, super early strength and super high strength [2]. It has been a structural material in important engineering foundations such as offshore wind power connecting section not only a grouting material in equipment leveling [3]. The foremost problem is the fact that, due to the lack of data, high-strength grouting material is often considered as concrete in the design of structural foundations [4]. However, the upper limit of the strength range of conventional concrete is 80 MPa, while the final strength of HSGM can reach 120 MPa or higher [5]. In the foundation structure of major engineering such as offshore wind power, the mechanical parameters under pressure are important to the structural design calculation of the grout joint section [6]. The standard values of concrete axial compressive strength and elastic modulus used in design calculation have a certain model formula conversion relationship with the cubic strength, and Poisson's ratio also has a specified range [7]. There are obvious differences in composition and properties between the high-strength grouting material and concrete. To our knowledge, few studies have yielded mechanical parameters characteristic of HSGM. Therefore, it is urgent to study the key mechanical parameters of high-strength HSGM.

The standard solution to the problem is based on the verification of a model theoretical calculation and analysis of measured data. It is of interest to know whether HSGM still follows the rules of concrete. For our goal, we focus on three problems: what are the characteristics of HSGM axial compressive strength, what are the characteristics of HSGM elastic modulus and what is the range of PIFs genes he HSGM Poisson's ratio. In this paper, ABAQUS software is used to simulate the grout material and the concrete material test block, and the resulting stress cloud map is analyzed to clarify the difference between HSGM and concrete under axial compressive strength. The difference was verified by testing the axial compressive strength and cubic compressive strength of high-strength grouting material at different ages, and the formula coefficients between them were amended. The results of the elastic modulus and compressive strength are used to verify the elastic model of grout material which was derived by the Mori–Tanaka method and make clear the characteristics of the elastic modulus of grout material. Poisson's ratio range was calculated by a large number of test results. All these will provide a certain basis for the subsequent engineering design.

2. Experiments

2.1. Materials

The raw materials employed in this work were P.II 52.5 Portland cement from Conch; the main properties of the cement are given in Table 1: fine aggregate from Qingdao China with a fineness modulus is 2.6 and elastic modulus of parent rock is 75 GPa. The main properties of the fine aggregate are given in Table 2: high performance powder water-reducing agent from Yingshan PC100T. The main properties of the water reducing agent are given in Table 3. A physical drawing of cement, fine aggregate and agent are seen in Figure 1.

Table 1. The main properties of cement.

-	Density	Specific Surface Area	1-Day Compressive Strength	3-Day Compressive Strength	7-Day Compressive Strength	28-Day Compressive Strength
P.II52.5 Portland cement	3.02 g/cm^3	394 m^2/kg	23.1 MPa	39.3 MPa	50.8 MPa	57.9 MPa

Table 2. The main properties of fine aggregate.

-	Apparent Density	Bulk Density	Tight Bulk Density	Stone Powder Content	Water Absorption
Fine aggregate	2700 kg/m^3	1490 kg/m^3	1590 kg/m^3	1.5%	1.2%

Table 3. The main properties of water reducing agent.

-	Appearance	Moisture Content	pH	Bulk Density	Water Reduction Rate
Water reducing agent	Pale yellow to white	0.5%	7.2	450 kg/m^3	28%

Figure 1. Physical drawing of cement (**left**), fine aggregate (**middle**) and agent (**right**).

2.2. Experiment Design

Due to the high compressive strength, the water binder ratio of high-strength grouting materials is generally low; 0.2 was adopted to be the water-binder ratio in this experiment. The water reducing agent content was 0.4% of the cementing material. The apparent density of high-strength grouting material was set at 2400 kg/m^3. The dosage of cement, fine aggregate, water and admixture was calculated according to the apparent density, sand binder ratio and water-binder ratio parameters.

The axial compressive strength and cubic compressive strength of HSGM at different ages was tested. Since the structural grout materials have a higher strength at the later stage, in order to improve the data of the low-strength areas, the early strength data were included in the test range, and the strength of the data at each age of 1 day, 2 days, 3 days, 7 days, 14 days and 28 days were collected. There are many factors affecting the strength data of cement-based materials. In order to increase the accuracy of the data and eliminate the influence of strength fluctuations as far as possible, the test was repeated three times to test the axial compressive strength and cubic strength.

Specimens formed by structural grout materials with different sand binder ratios were used to verify the elastic modulus model which was calculated by two-phase theory. The cubic compressive strength and elastic modulus of HSGM with sand binder ratios of 1.75, 1.55, 1.35, 1.15, 0.95, 0.75, 0.55, 0.35 and 0 (pure pulp) at the age of 7 days were tested, respectively. The design of the experiment is given in Table 4.

Table 4. The design of the experiment.

No.	Water-Binder Ratio	Sand-Binder Ratio	Water Reducing Agent Content	Test Content
1	0.2	1.15	0.4%	Axial compressive strength, cubic compressive strength and Possion's ratio at 1 day, 2 days, 3 days, 7 days, 14 days and 28 days, three times
2	0.2	1.75	0.4%	Elastic modulus and cubic compressive strength at 7 days
3	0.2	1.55	0.4%	Elastic modulus and cubic compressive strength at 7 days
4	0.2	1.35	0.4%	Elastic modulus, cubic compressive strength and Possion's ratio at 7 days
5	0.2	1.15	0.4%	Elastic modulus and cubic compressive strength at 7 days
6	0.2	0.95	0.4%	Elastic modulus, cubic compressive strength and Possion's ratio at 7 days
7	0.2	0.75	0.4%	Elastic modulus and cubic compressive strength at 7 days
8	0.2	0.55	0.4%	Elastic modulus and cubic compressive strength at 7 days
9	0.2	0.35	0.4%	Elastic modulus and cubic compressive strength at 7 days
10	0.2	0	0.4%	Elastic modulus and cubic compressive strength at 7 days

2.3. Methods

The compressive strength test, elastic modulus test and Poisson's ratio test procedure was based on "Standard of Test Methods for Physical and Mechanical Properties of Concrete" (GB/T50081-2019) [8]. The loading device was the WHY-2000 electro-hydraulic servo pressure testing machine (Figure 2) manufactured by the Shanghai Hualong Test Instruments Corporation. During the elastic modulus and Poisson's ratio tests, the strain on the transverse and longitudinal deformation of the specimen was collected by a TDS-530 strain collector.

Figure 2. WHY-2000 electro-hydraulic servo pressure testing machine.

3. Results and Discussion

3.1. Relationship between Axial Compressive Strength and Cubic Compressive Strength of HSGM

3.1.1. Finite Element Modeling

Based on the ABAQUS software (version STANDARD 2020), the POLARIS-MesoConcrete plug-in was used to accomplish random aggregate production, aggregate group generation and aggregate group random delivery [9]. The maximum particle size of HSGM was 5 mm, and the particle size of concrete aggregate was a continuous gradation from 20 mm. The volume of aggregate in the whole specimen is shown in Table 5. Two-dimensional aggregate models of the axial strength samples of grouting material and concrete were generated respectively. The size of the sample is 300 mm × 100 mm.

Table 5. The volume of aggregate in HSGM and concrete.

-	Less Than 5 mm	5 mm–10 mm	10 mm–15 mm	15 mm–20 mm	Totol
Grouting material	42%	/	/	/	42%
Concrete	20%	9%	7%	6%	38%

As is shown in Figure 3, the green color of the base represents the cement paste, different colored spheres represent aggregate and the size of colored spheres is determined by Table 1.

In keeping with the reality, two steel plates were simulated, respectively, on the upper and lower parts of the specimen. A fixed friction force on the contact surface between steel plate and test block was set as 0.3, as shown in Figure 4. A load of 1100 kN was applied to the upper surface. The strain cloud diagram of HSGM and concrete was obtained through finite element analysis. The local image was cut in the middle of the specimen, as shown in Figure 5.

Concrete strength has strong size effect [10]. According to the Weibull distribution theory, when the size of the specimen increases, the volume of concrete increases, and thus the possibility of defects or large defects increases [11]. The damage and failure of concrete are developed from local stress concentration which is easy to occur in the inherent defects of concrete. The larger the structure size, the higher the possibility of large internal defects, the lower the ultimate compressive strength, and the more significant the size effect. For grout materials, its aggregate particle size is small, and at the same time, its flow performance is good, its gas content is low, so it has good filling property with the bubble and is easy to discharge, and the section structure is dense. The probability of large and serious defects is much lower than that of concrete. According to the strain cloud map (Figure 5), the deformation concentration phenomenon occurs more in the concrete test blocks, but less in the grout material test blocks, which is similar to the theoretical analysis results. The size effect of HSGM is lower than concrete, and the theoretical compressive strength of HSGM of different sizes would be much closer.

Figure 3. Schematic diagram of aggregate feeding of structural grout material (**left**) and concrete (**right**).

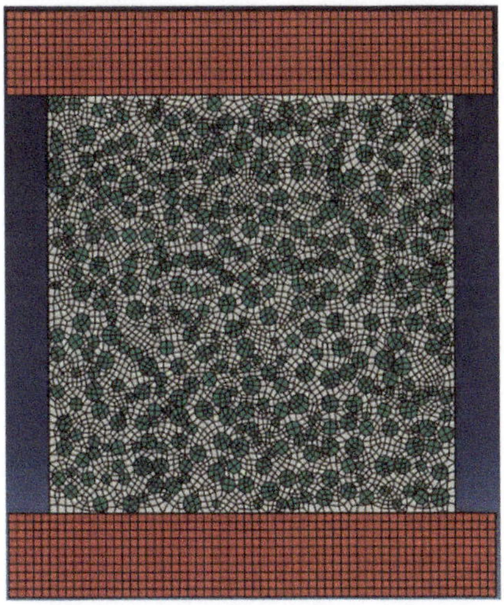

Figure 4. Test simulation diagram.

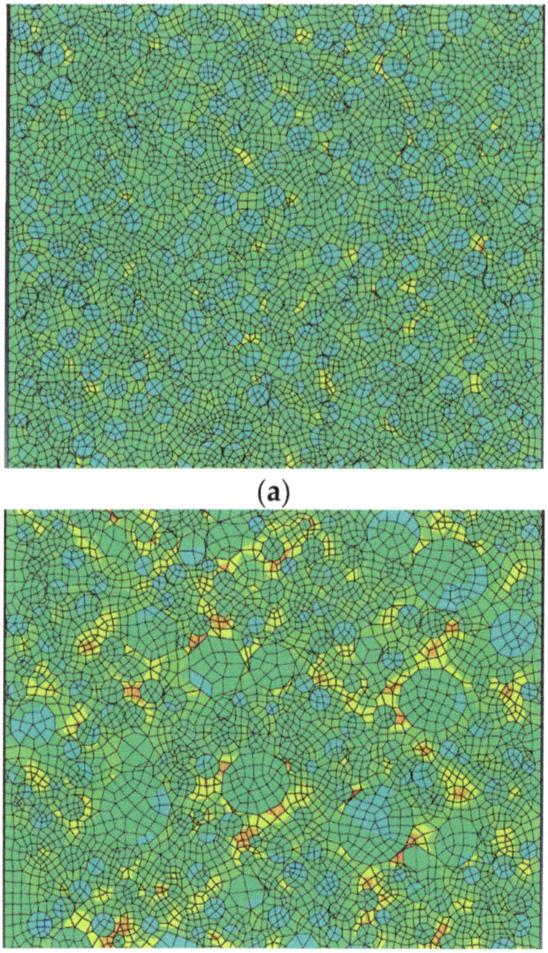

(a)

(b)

Figure 5. HSGM and concrete strain cloud diagrams. (**a**) HSGM strain cloud diagram, (**b**) concrete strain cloud diagram.

3.1.2. Formula for the Relationship between Axial Compressive Strength and Cubic Compressive Strength of HSGM

Figure 6 shows the relationship between the axial compressive strength and cubic compressive strength of concrete, ultra-high performance concrete in other papers [12–21], and also HSGM from Table 1 in this paper. The axial compressive strength of structural grouting increases with the increase in cubic strength, showing a linear growth trend similar to that of concrete. The axial compressive strength of HSGM is generally higher than that of concrete when the cubic compressive strength is the same. As shown in Figure 7, the coefficients of variation (CV) in different ages are close, which reflected the reliability of the data in this paper.

According to the test results, standard deviation processing was carried out on the strength data to determine the fluctuation range of each data. Since most of the concrete material's axial compressive strength and cubic compressive strength relationship formula adopted a linear relationship formula, the linear formula was used to fit the axial compressive strength and cubic compressive strength data of high-strength grouting material. We used 18 experimental units from Table 4 No.1 to build a linear regression, each experimental unit have three evaluation according to standard test procedure. The fitting results are shown in Figure 8. The fitting has a very high correlation coefficient which is 0.98, indicating that the fitting formula can well characterize the relationship between the axial compressive strength and the cubic strength of HSGM.

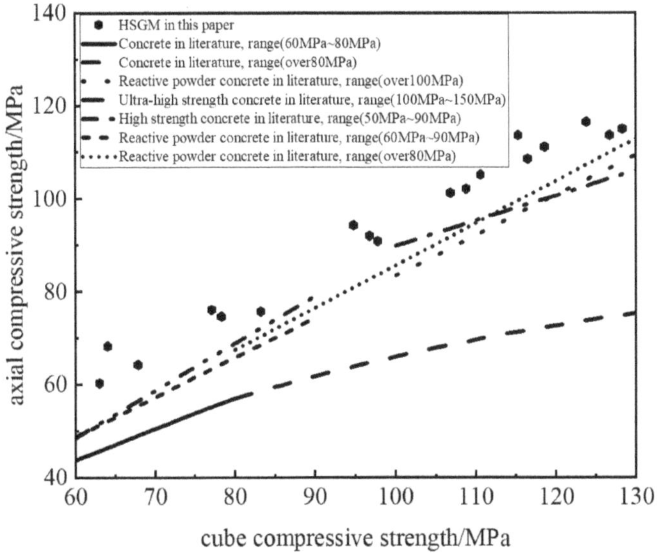

Figure 6. The relationship between axial strength and cubic strength of grout materials and concrete.

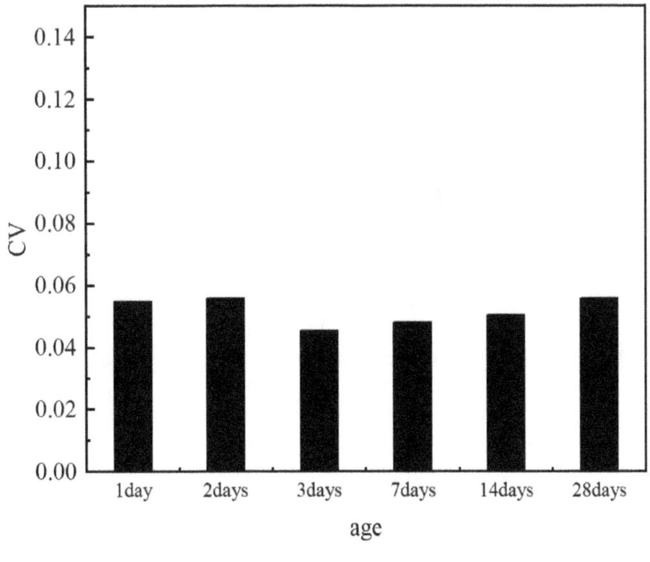

Figure 7. The coefficient of variation in different ages.

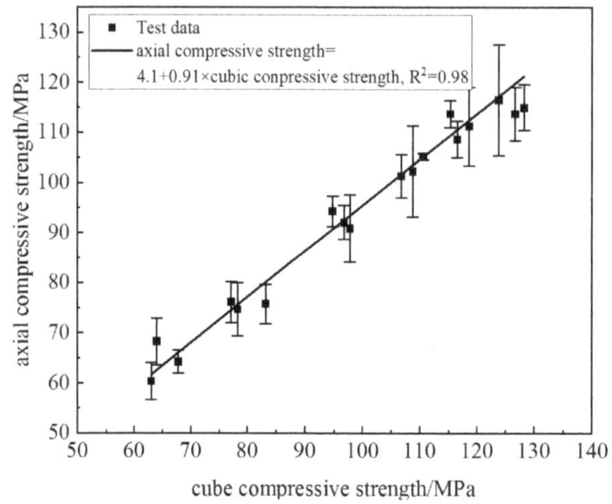

Figure 8. Fitting treatment of axial strength and cubic strength of HSGM.

Based on the above data analysis and fitting results, within the range of 60~120 MPa, the standard value formula of axial compressive strength of HSGM can be defined as

$$f_{ck} = 0.91 f_{cu,k} + 4.1 \quad (1)$$

where f_{ck} is the axial compressive strength of HSGM; $f_{cu,k}$ are cubic strength of structural grout material; 0.91 is the reduction coefficient; 4.1 is the correction coefficient.

In general, axial compressive strength is lower than cubic compressive strength in cement-based materials [22]. The gap between axial compressive strength and cubic compressive strength in HSGM is smaller than in concrete, the reduction coefficient of HSGM is higher than concrete which conforms to the finite element simulation results. The model formula of HSGM can better represent the mechanical characteristics of HSGM.

3.2. Relationship between Elastic Modulus and Cubic Compressive Strength of HSGM

3.2.1. Derivation of Elastic Modulus Model

The elastic modulus of HSGM can be deduced by using meso-hierarchy theory based on the distribution and performance parameters of component materials [23]. The grout material can be regarded as the combination of fine aggregate and cement paste. From the meso-level, it is assumed that the fine aggregate and cement paste in the grout material are elastic materials with the same properties in each phase. The bubbles are not considered because there are relatively few bubbles in the grout material. In addition, an interfacial transition zone is not considered because the interfacial transition zone of HSGM is relatively thin and has little effect [24]. The aggregate can be simplified into a sphere because of the aggregate for HSGM is relatively fine and its grain type is round. The aggregate can be seen as a hardened inclusion. The elastic modulus model of structural grout material can be predicted by combining with the theory of multi-phase inclusion.

Mori–Tanaka theory is a widely used multi-phase inclusion theory, which comprehensively considers the influence of distant strain on the interaction between medium [25]. The grout material model uses cement paste as the matrix phase and fine aggregate particles as the inclusion phase. As elastic material, its elastic modulus E, bulk modulus K, shear modulus G and Poisson's ratio v are related to each other [26]. By inputting the material parameters of aggregate into the model, the model relationship between the elastic modulus of HSGM, the elastic modulus of cement paste and the volume content of aggregate can be obtained:

$$E_c = \frac{9K_cG_c}{3K_c + G_c} \tag{2}$$

where E_c is the elastic modulus of HSGM, GPa; K_c is the volume modulus of HSGM, GPa; G_c is shear modulus of HSGM, GPa.

$$K_c = \frac{(18.02 + 15.35V_r + 0.221E_m - 0.221V_rE_m)E_m}{(25.96 - 25.96V_r + 0.319E_m + 0.375V_rE_m)} \tag{3}$$

where K_c is the volume modulus of HSGM, GPa; V_r is aggregate volume content, %; E_m is the elastic modulus of cement paste, GPa.

$$G_c = \frac{(5.764 + 6.236V_r + 0.082E_m - 0.082V_rE_m)E_m}{(14.52 - 14.52V_r + 0.206E_m + 0.191V_rE_m)} \tag{4}$$

where G_c is shear modulus of HSGM, GPa; V_r is aggregate volume content, %; E_m is the elastic modulus of cement paste, GPa.

3.2.2. Test Verification of Elastic Modulus

The volume content of aggregate in HSGM with different sand binder ratios can be calculated by density of material. The volume content and the cubic strength and compressive elastic modulus measured in the same age are shown in Table 6. It can be seen from the data in the table that the cubic strength of the specimen does not change significantly under different aggregate content, and the strength is basically around 115 MPa. This indicates that the volume content of aggregate has little influence on the strength at a low water binder ratio. Although the transition zone increased by the volume content of aggregate, the strength did not decrease, which verifies that the interfacial transition zone is no longer a weak area for HSGM.

Table 6. Strength and elastic modulus of grout materials with different sand to paste ratios.

No.	Sand to Paste Ratios	Volume Contents of Aggregate	Cubic Strength/MPa	Elastic Modulus/GPa
1	1.75	52%	114.7	42.1
2	1.55	49%	115.2	43.5
3	1.35	46%	109.1	42.2
4	1.15	42%	115.3	40.0
5	0.95	37%	115.4	39.3
6	0.75	32%	117.6	37.3
7	0.55	25%	116.1	33.4
8	0.35	18%	114.5	28.5
9	0	0%	116.5	27.0

The elastic modulus and Poisson's ratio of pure cement paste obtained in the test were substituted into Formula (5); the model relationship between the elastic modulus of structural grout material and the volume content of aggregate can be written as

$$E_c = 26.25 \times V_r^2 + 19.57 \times V_r + 27.36 \tag{5}$$

where E_c is the elastic modulus of structural grout material, GPa; V_r is aggregate volume content, %.

The derived model Formula (5) was placed in the elastic modulus and strength data graphs obtained by testing with different sand binder ratios, as shown in Figure 9.

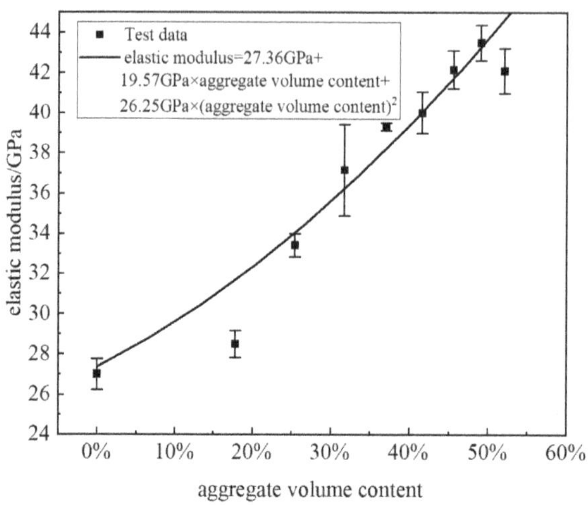

Figure 9. Relationship between aggregate content and elastic modulus of structural grout material.

Figure 9 is the result of 10 experimental units from Table 4, where each experimental unit contains three evaluated units. The average error rate is the average value of error rate contained in all experimental units. The error rate is calculated by the difference between the predicted value and the measured value divides the predicted value. As can be seen from Figure 9, when using the same material and aggregate with different volume contents, the model derived from the material property mechanism is close to the experimental data with the average error rate of just 3%, which proves the accuracy of the elastic modulus model of structural grout materials. At the same time, it also indicates that the elastic modulus is closely related to the aggregate content. The main reason is that the elastic modulus essentially reflects the load-deformation condition of the material at the elastic stage and the content change of its component materials will affect the overall deformation condition. It has been verified in the mechanical test of HSGM, indicating that the component materials of grouting material have a greater influence on the elastic modulus. Therefore, the strength should not be only considered when designing the elastic modulus in structural design, but also the composition and distribution of materials.

3.3. Selection of Poisson's Ratio

Poisson's ratio is the ratio of transverse deformation to longitudinal deformation when material is under stress. Most scholars believe that the compressive strength and tensile strength Poisson's ratio of concrete materials are the same, and Poisson's ratio is generally closely related to the composition of materials [27,28]. Poisson's ratio will fluctuate within a range around a certain point. Poisson's ratio of concrete in the specification is 0.2 [29]. The regression statistical method was adopted to make statistics on the compressive and tensile Poisson's ratio data obtained in the experiment in this paper, as shown in Table 7. By standard deviation analysis and weighted average calculate, a representative value range was obtained. The calculation formula of sample standard deviation was as follows:

$$S = \sqrt{\frac{\Sigma(X_i - \overline{X})^2}{n-1}} \tag{6}$$

where S is the sample standard deviation, X_i is the ith data, $\overline{X}$ is the average value of the whole data, and n is the number of the whole samples.

Table 7. Poisson's ratio statistics of structural grout materials.

No.	Compressive Poisson's Ratio	Tensile Poisson's Ratio
1	0.244	0.271
2	0.242	0.259
3	0.241	0.257
4	0.243	0.256
5	0.242	0.255
6	0.240	0.257
7	0.239	0.243
8	0.245	0.259
9	0.242	0.259
10	0.244	0.271
11	0.239	0.267
12	0.251	0.237
13	0.247	0.259
14	0.259	0.242
15	0.248	0.234
16	0.257	0.246
17	0.248	0.236
18	0.254	0.244
19	0.273	0.240
20	0.266	0.261
average value	0.248	0.253
standard deviation	0.009	0.011

The standard deviation of the compressive Poisson's ratio is 0.009 and the mean value of the compressive Poisson's ratio is 0.248. The standard deviation of the tensile Poisson's ratio is 0.011 and the mean value of the tensile Poisson's ratio is 0.253. After the combining of compressive and tensile Poisson's ratios, an average range of 0.25 ± 0.01 can be obtained. Comparing this range with actual data, as show in Figure 10. It can be seen from the figure that more than 95% points fall in this range, which is highly representative.

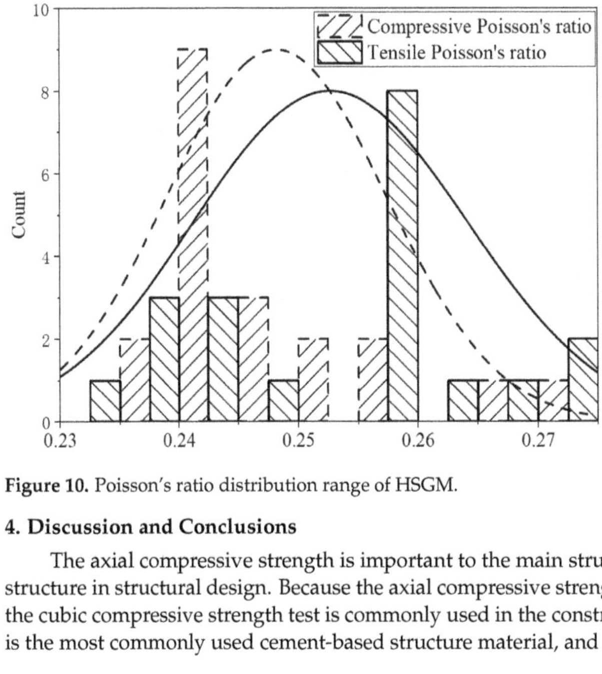

Figure 10. Poisson's ratio distribution range of HSGM.

4. Discussion and Conclusions

The axial compressive strength is important to the main structural force of the whole structure in structural design. Because the axial compressive strength test is not convenient, the cubic compressive strength test is commonly used in the construction process. Concrete is the most commonly used cement-based structure material, and the relationship between the axial compressive strength and cube compressive strength of concrete is relatively clear. HSGM has obvious differences in strength and maximum aggregate particle size to concrete.

The stress concentration phenomenon under pressure in HSGM is significantly less than that in concrete by the finite element analysis of HSGM and concrete axial compressive strength specimens under stress. Test dates show that the axial compressive strength of HSGM is generally higher than that of concrete when the cubic compressive strength is the same; the size effect of HSGM is not as obvious as concrete. The standard value formula between the axial compressive strength and cubic strength of high-strength grouting material can be defined.

The elastic modulus and Poisson's ratio reflect the deformation characteristics before specimen failure, and they can affect the deformation of the whole structure in the structure design. Under the same pressure, the deformation characteristics of aggregate and paste are different. The Mori–Tanaka model is very successful in predicting the effective properties of two-phase composites.

The relationship model between the elastic modulus of HSGM and the volume content of aggregate is given, by using cement paste as the matrix phase and fine aggregate particles as the inclusion phase with the theory of multi-phase inclusion. The results verify the accuracy of the model with the average error rate of just 3%. In elastic modulus design and calculation, the influence of material composition should be considered in high-strength grouting material.

Based on a large number of Poisson's ratio data of HSGM, the Poisson's ratio suitable for structural grout materials is given by statistical analysis, which is 0.25 ± 0.01. Poisson's ratio covers more than 95% of the measured data, and it is suggested to be used as Poisson's ratio for HSGM in structure design.

Author Contributions: Conceptualization, Y.Z.; methodology, Y.Z.; software, B.C.; validation, H.Z. and X.Z.; formal analysis, Y.H.; investigation, H.Z.; resources, B.C.; data curation, B.C.; writing—original draft preparation, Y.Z.; writing—review and editing, X.Z.; visualization, H.Z.; supervision, X.Z.; project administration, Y.H. All authors have read and agreed to the published version of the manuscript.

Funding: This research received no external funding.

Institutional Review Board Statement: Not applicable.

Informed Consent Statement: Not applicable.

Data Availability Statement: Not applicable.

Conflicts of Interest: The authors declare no conflict of interest.

References

1. Su, J.; Yang, Z.; Fang, Z. Size Effect of Concrete Compressive Strength with Different Aggregate Composition. *Adv. Concr. Struc* **2009**, *400–402*, 831–835. [CrossRef]
2. Lu, J.; Chen, J.; Gan, G.; Xu, F. Study on the Preparation of New Type High Capability and None Shrinkage Cement-base Grouting Material. *Mater. Rev.* **2016**, *30*, 123–129.
3. Wang, D.; Zhang, Y.; Sun, X.; Gao, J. Development of ultra-high performance cement-based grouting material and its research application in offshore wind power foundation construction. In Proceedings of the 2020 Fib Symposium: Concrete Structures fo Resilient Society, Shanghai, China, 22–24 November 2020; pp. 36–45.
4. Li, Z.X.; Chen, T.; Wang, X. Behavior of flat grouted connections subjected to lateral pressure and vertical load. *Constr. Buil Mater.* **2019**, *212*, 329–341. [CrossRef]
5. Fu, T.; Ren, X.Q.; Li, Y.; Wang, K.; Zhu, Z.X. Experimental study on the mechanical behavior of HTRB600E steel sleeve groutin based on UHPC. *Case Stud. Constr. Mater.* **2022**, *17*, e01656. [CrossRef]
6. Wang, X.; Chen, T.; Zhao, Q.; Yuan, G.K.; Liu, J.C. Fatigue Evaluation of Grouted Connections under Bending Moment in Offsho Wind Turbines Based on ABAQUS Scripting Interface. *Int. J. Steel Struct.* **2016**, *16*, 1149–1159. [CrossRef]
7. Chen, T.; Cao, C.C.; Zhang, C.H.; Wang, X.; Chen, K.; Yuan, G.K. Numerical modeling and parametric analysis of groute connections under axial loading. *Thin-Walled Struct.* **2020**, *154*, 106880. [CrossRef]

GB/T50081-2019; Standard of Test Methods for Physical and Mechanical Properties of Concrete. Ministry of Housing and Urban-Rural Development of the PRC, State Administration for Market Regulation: Beijing, China, 2019.
Cheng, S.-H.; Ren, Z.-G.; Li, P.-P.; Shangguan, J.-Y. Numerical Simulation of Concrete Aggregates with Arbitrary Shapes Based on ANSYS/LS-DYNA. *J. Wuhan Univ. Technol.* **2014**, *36*, 89–94. [CrossRef]
Yang, Q.W.; Du, S.G. Prediction of concrete cubic compressive strength using ANN based size effect model. *Comput. Mater. Contin.* **2015**, *47*, 217–236.
Kun, Z.; Bai Ru, L.; Xun An, Z.; Yi Hong, W.; Zhan, Q. Research on weibull distribution theory for cubic compressive strength test method of raw earth materials with different curing methods and time. *Key Eng. Mater.* **2021**, *896*, 129–140. [CrossRef]
Su, J.; Zhang, J.; Zhou, C.B.; Fang, Z. Size Effect on the Cubic Compressive Strength of Reactive Powder Concrete. *Rilem Proc.* **2016**, *105*, 585–589.
Zhou, J.J.; Pan, J.L.; Leung, C.K.Y.; Li, Z.J. Experimental study on mechanical behavior of high performance concrete under multi-axial compressive stress. *Sci. China Technol. Sic.* **2014**, *57*, 2514–2522. [CrossRef]
He, X.X.; Lin, S.Y.; Zhen, X.C. Experimental Study on Size Effect of Compressive Strength and Deformation Properties of Large Size Fly Ash Concrete. *Key Eng. Mater.* **2011**, *477*, 319–324. [CrossRef]
Bandeira, M.V.V.; Zydeck, R.C.; Kosteski, L.E.; Marangon, E. Concrete axial compressive strength with different loading directions and boundary conditions. *Materia* **2020**, *25*. [CrossRef]
FaghihKhorasani, F.; Kabir, M.Z.; AhmadiNajafabad, M.; Ghavami, K. Predicting compressive stress-strain curves of structural adobe cubes based on Acoustic Emission (AE) hits and Weibull distribution. *Int. J. Struct. Integr.* **2019**, *10*, 766–791. [CrossRef]
Xiong, M.X.; Liew, J.Y.R.; Wang, Y.B.; Xiong, D.X.; Lai, B.L. Effects of coarse aggregates on physical and mechanical properties of C170/185 ultra-high strength concrete and compressive behaviour of CFST columns. *Constr. Build. Mater.* **2020**, *240*, 117967. [CrossRef]
Abid, M.; Hou, X.M.; Zheng, W.Z.; Hussain, R.R. Effect of Fibers on High-Temperature Mechanical Behavior and Microstructure of Reactive Powder Concrete. *Materials* **2019**, *12*, 329. [CrossRef]
Ji, J.; Kang, W.; Jiang, L.Q.; Li, Y.H.; Ren, H.G.; Hao, S.X.; He, L.J.; Lin, Y.B.; Yu, C.Y. Mechanical Behavior of Reactive Powder Concrete Made From Local Material Subjected to Axial Pressure. *Front. Mater.* **2021**, *8*, 737646. [CrossRef]
Bandeira, M.V.V.; Torre, K.R.L.; Kosteski, L.E.; Marangon, E.; Riera, J.D. Influence of contact friction in compression tests of concrete samples. *Constr. Build. Mater.* **2022**, *317*, 125811. [CrossRef]
Wu, H.M.; Liu, K.; Yang, F.; Shen, B.; Ma, K.J.; Zhang, J.Y.; Liu, B. Experimental Mechanical Properties and Numerical Simulation of C80 Concrete with Different Contents of Stone Powder. *Materials* **2022**, *15*, 3282. [CrossRef]
Li, M.; Hao, H.; Cui, J.; Hao, Y.F. Numerical investigation of the failure mechanism of cubic concrete specimens in SHPB tests. *Def. Technol* **2022**, *18*, 1–11. [CrossRef]
Qiu, Y.R.; Zhang, G.X.; Yin, X.N. Parameters inversion of mortar in concrete considering the hardening process based on meso-mechanics. In Proceedings of the Third Technical Conference on Hydraulic Engineering, Hongkong, China, 13–14 December 2014; pp. 85–92.
He, Z.H.; Du, S.G.; Chen, D. Microstructure of ultra high performance concrete containing lithium slag. *J. Hazard. Mater.* **2018**, *353*, 35–43. [CrossRef] [PubMed]
Yang, C.C.; Huang, R. Double inclusion model for approximate elastic moduli of concrete material. *Cem. Concr. Res.* **1996**, *26*, 83–91. [CrossRef]
Krishnya, S.; Elakneswaran, Y.; Yoda, Y. Proposing a three-phase model for predicting the mechanical properties of mortar and concrete. *Mater. Today Commun.* **2021**, *29*, 102858. [CrossRef]
Mina, A.L.; Trezos, K.G.; Petrou, M.F. Optimizing the Mechanical Properties of Ultra-High-Performance Fibre-Reinforced Concrete to Increase Its Resistance to Projectile Impact. *Materials* **2021**, *14*, 5098. [CrossRef]
Yang, H.Q.; Rao, M.J.; Dong, Y. Influence study of extra-large stone limited size and content on full-graded concrete properties. *Constr. Build. Mater.* **2016**, *127*, 774–783. [CrossRef]
Aili, A.; Vandamme, M.; Torrenti, J.M.; Masson, B.; Sanahuja, J. Time evolutions of non-aging viscoelastic poisson's ratio of concrete and implications for creep of C-S-H. *Cem. Concr. Res.* **2016**, *90*, 144–161. [CrossRef]

coatings

MDPI

Article

Effect of Strain Levels on the Corrosion Resistance of an Enamel-Coated Steel Rebar

Fujian Tang [1,*], Hao Cui [1], Gang Li [1] and Shangtong Yang [2]

[1] School of Civil Engineering, Dalian University of Technology, Dalian 116024, China
[2] Department of Civil and Environmental Engineering, University of Strathclyde, Glasgow G1 1XQ, UK
* Correspondence: ftang@dlut.edu.cn

Abstract: The effect of strain levels on the corrosion resistance of an enamel-coated steel rebar is experimentally investigated in this study. Enamel coating was applied on the surface of a steel rebar by using the wet process. A strain gauge was attached on the surface of the coated steel rebar to record the strain levels and a plastic container was mounted for electrochemical corrosion tests. A stress-corrosion test set-up was designed to conduct corrosion and tensile tests simultaneously. The strain levels considered include 0 με, 300 με, 600 με, 900 με and 1200 με, and the electrochemical techniques employed include open circuit potential, linear polarization resistance and electrochemical impedance spectroscopy. The microstructure of the enamel coating was also examined with scanning electron microscopy. Results show that the enamel coating has a thickness of ~150 μm, and there are some air bubbles in the coating. The average corrosion current density of the uncoated steel rebar decreases from 18.64 $\mu A/cm^2$ to 14.39 $\mu A/cm^2$ in NaCl solution due to the generation of corrosion products. The corrosion current density of the enamel-coated steel rebar gradually increases from 0.49 $\mu A/cm^2$ when the strain is zero to 0.65 $\mu A/cm^2$ as strain reaches 1200 με, which is almost 40 times lower than that of the uncoated steel rebar. Impedance spectrum results show that the corrosion resistance of enamel coating decreases with an increase in the tensile strain level; however, it still protects steel rebar from corrosion to some degree.

Keywords: steel rebar corrosion; enamel coating; stress level; electrochemical impedance spectroscopy; linear polarization resistance

1. Introduction

The surface of steel rebars in reinforced concrete (RC) structures is initially covered with a protective passive film, which is formed due to the alkaline environment of the fresh concrete. However, with an increase of service life, this protective film gradually loses its function and corrosion occurs in the steel rebar due to carbonization of the concrete cover or diffusion of chloride from the environment [1–3]. Corrosion reduces the cross-sectional area and the mechanical properties of steel rebar [4], causes cracking, delamination and spalling of the concrete cover [5,6], leads to bond strength loss between the steel rebar and concrete [7], and finally results in a reduction in the carrying capacity and service life of RC structures [8,9].

Over the past decades, various methods have been developed to protect steel rebar in the RC structure from corrosion attack. One of the most effective and efficient methods is the use of a protective coating on the surface of steel rebar, as it can establish a physical barrier between the aggressive environment and the steel rebar under protection. Fusion bonded epoxy coating is one of the most widely used coatings for steel rebar. However, it reduces the bond strength with concrete and is prone to under-film corrosion if damaged during transportation or handling [10,11]. Hot-dip galvanized coating is also used in RC structures. However, chemical reactions might occur between the zinc coating and the concrete, especially in the curing stages of the concrete, and these chemical reactions produce hydrogen gas and increase the porosity of the interface between the galvanized steel rebar and the concrete, and consequently reduce the bond strength [12].

Vitreous enamel, an inorganic coating, is made by firing powdered glass onto the surface of a metal substrate at temperature between 750 °C and 850 °C. The oxides in the enamel powder would react with the element in the steel, such as carbon, to form a chemical bond between the enamel coating and the steel substrate, which prevents potential under-film corrosion [13]. Due to its glassy nature, the enamel coating has a shining and smooth surface, and is, thus, widely used for household and industrial applications. Moreover, enamel coating also demonstrates many advantages, such as a high resistance to abrasion, resistance to chemicals (i.e., both acidic and alkaline solutions), resistance to fire and high temperature and resistance to detergents [14]. Furthermore, its properties are flexible and can be adjusted by modifying the chemical composition of the frits [15]. For example, the corrosion resistance to alkaline solution shows a significant increase by adding zirconia, lithium oxides and milling additives based on quartz [16]; the resistance to acid can be enhanced by adding quartz and zirconium oxide [17]; the adhesion with substrate metal could be improved by adding NiO or CoO [18]; and crystallization treatment can improve the hardness of the enamel coating [19].

By adding cement powder or calcium silicate in the enamel frit, chemically reactive enamel coatings have been proposed for enhanced corrosion resistance and improved bond strength of steel reinforcement in concrete structures in the past decade. The microstructure, phase composition and corrosion resistance of enamel coating applied on smooth steel bars have been experimentally investigated, and results show that although the interconnected regions of calcium silicate particles provide a pathway for the penetration of aggressive chemicals, the chemically reactive enamel coating increases the corrosion resistance of steel rebars to some extent [20]. The long-term corrosion mechanism and rate of chemically reactive enamel coating applied on smooth steel bars embedded in mortar was also studied [21]. Different from smooth steel bar, deformed steel bars are widely used in reinforced concrete structures. However, the thickness of enamel coating is not uniformly distributed due to the presence of rebar deformation. Moreover, there is also some damage during transportation and handling at the construction site. Therefore, the short-term and long-term corrosion performance of chemically reactive enamel coating applied to a deformed steel rebar have also been studied; the effect of impact damage was considered and the performance was compared with fusion bonded epoxy coating [13,22]. In addition, the chloride threshold and adhesion strength with substate steel have also been investigated by pull-off tests, and results show that the addition of cement increased the adhesion strength with steel compared with pure enamel, and the chloride threshold of enamel-coated steel is greater than that of uncoated steel [23]. Regarding the bond strength with concrete, the presence of chemically reactive enamel coating reduces the porosity of the interfacial transition zone between steel and concrete, and results in an increase in the bond strength with concrete [24,25]. The microstructure and properties are dependent on the parameter of the enameling process, and, therefore, the effect of sintering temperature and coating cycles on the microstructure, phase composition and corrosion resistance have also been experimentally investigated [26–28].

In practical engineering applications, the enamel-coated steel rebar is in a state of tension or compression to resist the loadings imposed on the structures. For compression members, such as reinforced concrete columns, enamel coating is in compression and the corrosion resistance may not be an issue. However, for reinforced concrete members in bending or tension, the enamel-coated steel rebar is in a state of tension. As mentioned previously, enamel coating has a glassy nature and is brittle; therefore, some micro-cracks generated in the coating are possible if it is subjected to a larger deformation. However, the effect of tensile stress levels on the corrosion resistance of enamel-coated steel rebar has not been investigated and fully understood.

Therefore, this study aims to experimentally investigate the effect of tensile strain levels on the corrosion resistance of an enamel-coated steel rebar. A stress-corrosion test set-up was designed to conduct both electrochemical corrosion tests and tensile tests simultaneously. The enamel-coated steel rebar was first tensioned to different strain levels and then electrochemical corrosion tests, including open circuit potential, linear polarization resistance and electrochemical impedance spectroscopy, were conducted. For comparison, an uncoated steel rebar was also prepared and tested.

2. Materials and Methods

2.1. Preparation of the Enamel-Coated Steel Rebar

A HRB400 steel rebar with a dimeter of 12 mm was used in this study, and its chemical composition is shown in Table 1. The steel rebar was cut into ~300 mm long pieces, and the two ends were screw threaded for applying tensile strain during the corrosion tests. After the screw thread processing, the steel rebars were cleaned with a sand blaster and rinsed with acetone before enamel coating.

Table 1. Chemical composition of the steel rebar used in this study.

Elements	C	Si	Mn	P	S	Cr	Ni	Cu	Mo	V	Fe
wt.%	0.240	0.390	1.260	0.023	0.027	0.180	0.040	0.090	0.002	0.002	97.746

Commercially available enamel powder was used, and its chemical composition is shown in Table 2. The enamel powder was mixed with tap water with a weight ratio of 1.0:2.35 to form the enamel slurry. The cleaned steel rebars were dipped in the enamel slurry until the entire surface was fully covered with wet enamel. The wet enamel-coated steel rebars were moved into a furnace with a temperature of 60 °C for 10 min to remove moisture. During the preheating process, the enamel-coated steel rebars were rotated constantly to ensure a uniform enamel coating layer on the steel rebars. After the preheating process, the temperature of the furnace was gradually increased up to 840 °C and fired for 10 min. After 10 min, the temperature of the furnace decreased gradually to room temperature and, finally, the enamel-coated steel rebars were produced.

Table 2. Chemical composition of the enamel powder used in this study.

Oxides	SiO_2	Al_2O_3	B_2O_3	$NaNO_3$	CaF	KCO_3	Li_2CO_3	TiO_2	CoO	NiO	MnO_2	$NaCO_3$
wt.%	37.8	15.8	14.6	4.8	3.6	6.7	5.4	2.7	0.5	1.2	5.2	1.7

To control the strain levels during the stress-corrosion tests, a strain gauge was attached on the surface of the enamel-coated steel rebars, as shown in Figure 1a. In addition, a plastic container made of polyethylene terephthalate (PET, molecular formula $(C_{10}H_8O_4)_n$) was mounted with the steel rebar for electrochemical corrosion tests. A round hole was drilled in the bottom of the plastic container and the steel rebar passed through the hole, then epoxy resin was applied to seal the gap between the steel rebar and the plastic container. Although a small amount (~6 ng/L) of antimony (Sb) would migrate from the PET to the solution [29], its effect on the corrosion reaction of steel is insignificant. Moreover, it is easier to drill a hole in the bottom of a plastic container and fix it with the steel rebar compared to the glass container. The steel rebars mounted with a plastic container and ready for tests are shown in Figure 1b. To ensure the repeatability of the test results, three enamel (PE)-coated steel rebars and three uncoated (UN) steel rebars were prepared and tested.

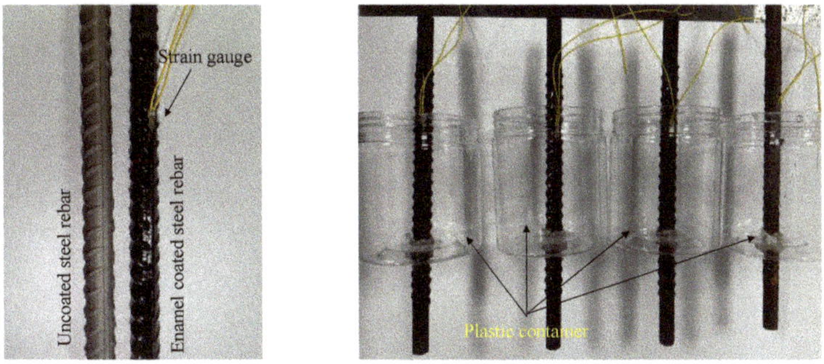

(**a**) (**b**)

Figure 1. (**a**) Uncoated and enamel-coated steel rebar; (**b**) three enamel steel rebars and one uncoated steel rebar mounted with a plastic container.

2.2. *Stress-Corrosion Test Set-Up*

Figure 2 shows the stress-corrosion test set-up. The two screwed ends of the steel rebar were fixed with two 20 mm thick steel plates by using screws. A hydraulic jack was placed between the two steel plates to apply force against the top plate. Regarding the corrosion test, the plastic container was filled with 3.5 wt.% NaCl solution which was made by mixing purified sodium chloride powder in de-ionized water. The loading was first applied to a designated strain level which was controlled by the strain gauge reading, and then electrochemical corrosion tests were conducted. The strain levels investigated in this study included 0 $\mu\varepsilon$, 300 $\mu\varepsilon$, 600 $\mu\varepsilon$, 900 $\mu\varepsilon$ and 1200 $\mu\varepsilon$. The electrochemical corrosion techniques employed include open circuit potential, linear polarization resistance and electrochemical impedance spectroscopy (EIS). A three-electrode system was applied including the steel rebar as the working electrode, a platinum sheet with 20 mm $\times$ 20 mm $\times$ 0.2 mm as the counter electrode and a saturated calomel electrode (SCE) as the reference electrode, as shown in Figure 2. Linear polarization resistance was conducted by applying a potential from -20 mV to 20 mV around the open circuit potential with a scan rate of 0.167 mV/s. EIS tests were conducted by applying a sinusoidal potential of 10 mV around the open circuit potential with frequency ranging from 100 kHz to 5 mHz with 5 points per decade of frequency.

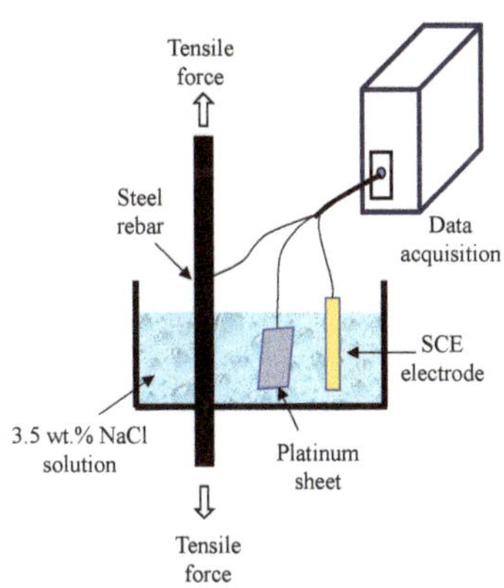

Figure 2. Schematic illustration of the stress-corrosion test set-up.

2.3. Microstructural Observation

To investigate the microstructure of the enamel coating, an additional enamel-coated steel rebar was fabricated in the laboratory. The enamel-coated steel rebar was first placed in a PVC pipe with an inside diameter of 20 mm and then filled with epoxy resin. After curing for 24 h, a ~50 mm long piece was cut with a band saw and sectioned in the middle into two halves with a machine. The cross-sections were polished successively with silicate carbide papers of grade 60, 180, 320, 480, 600, 800 and 1200, then rinsed with de-ionized water and acetone, and dried in an oven prior to microstructural observation with scanning electron microscopy (Hitachi SU5000, Tokyo, Japan).

3. Results and Discussion

3.1. Microstructure of the Enamel Coating

Figure 3 shows the cross-sectional SEM microstructure of the enamel coating investigated in this study. It has a thickness of approximately 150 μm and contains some air bubbles near the interface with the steel rebar. The presence of air bubbles is attributed to the chemical reactions between oxides and water in the enamel and carbon in the steel rebar at firing temperatures that that release carbon dioxide and hydrogen gases [23].

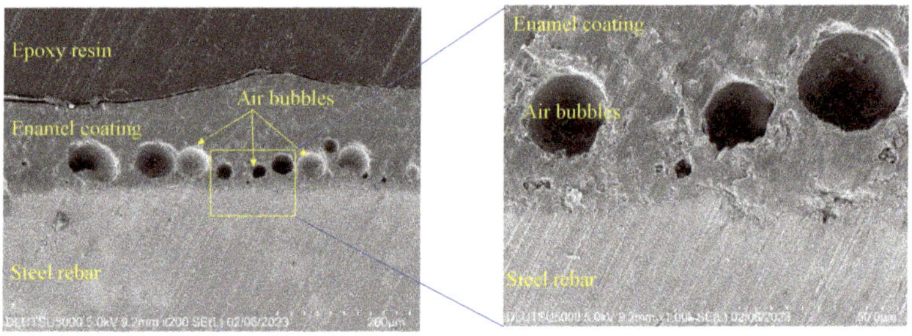

Figure 3. Microstructure of enamel coating.

3.2. OCP and Linear Polarization Resistance

The open circuit potential (OCP) of uncoated and three enamel-coated steel rebars immersed in 3.5 wt.% NaCl solution is shown in Figure 4. Each point represents the average of three specimens with an error bar representing one standard deviation. The average OCP of the uncoated steel rebar is around −625 mV/SCE when the tensile strain is zero, then decreases to be around −690 mV/SCE when the tensile strain increases to 300 με, and then remains stable as the tensile strain increases up to 1200 με. Regarding the enamel-coated steel rebars, the average OCP is around −660 mV/SCE, and the tensile strain level does not affect the OCP significantly.

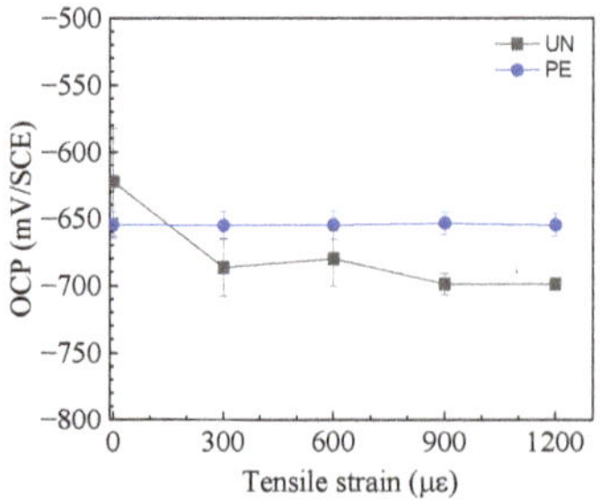

Figure 4. Open circuit potential of uncoated (UN) and enamel (PE)-coated steel rebars at different strain levels in 3.5 wt.% NaCl solution.

Figure 5a compares the polarization resistance of uncoated steel rebar with that of the enamel-coated steel rebar at different strain levels, which was obtained from the linear polarization resistance curve. The average polarization resistance of three uncoated steel rebars is 1.51 kΩ cm^2 when the tensile strain is zero, and it increases to 1.76 kΩ cm^2 when the tensile strain in the streel rebar reaches 300 με. With an increase in the tensile strain level, the polarization resistance gradually increases and finally reaches 1.85 kΩ cm^2 when the tensile strain in the streel rebar is 1200 με. The increase in the polarization resistance with an increase in strain levels is mainly attributed to the generation of corrosion products that accumulated on the steel rebar surface and slowed down the diffusion of oxygen. The average polarization resistance of the enamel-coated steel rebar is 54.43 kΩ cm^2 when the tensile strain is 0 με, and it decreases to 45.07 kΩ cm^2 when the tensile strain reaches 300 με, as shown in Figure 5a. With a further increase in the tensile strain, the polarization resistance continues to decrease and reaches 42.07 kΩ cm^2 when the tensile strain is 1200 με. The decrease in the polarization resistance with the increase in tensile strain level is mainly attributed to the micro-cracks generated in the enamel coating due to tensile stress in the steel rebar.

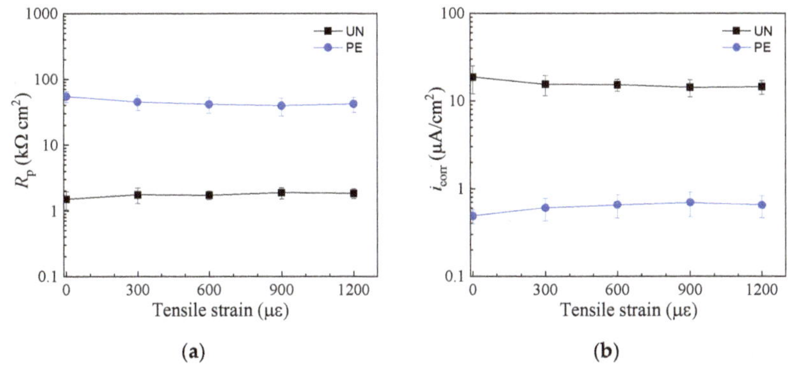

Figure 5. (**a**) Polarization resistance and (**b**) corrosion current density of the uncoated (UN) and enamel (PE)-coated steel rebar in 3.5 wt.% NaCl solution at different strain levels.

The corrosion current density of both uncoated and enamel-coated steel rebars at different tensile strain levels was calculated by using the following equation [21]:

$$i_{corr} = \frac{B}{R_p} \tag{1}$$

where R_p is the polarization resistance obtained from the linear polarization curve and B is the constant associated with the Tafel slopes; B = 26 mV was used in this study [23]. Figure 5b compares the average corrosion current density of uncoated and enamel-coated steel rebars. The corrosion current density of the uncoated steel rebar in 3.5 wt.% NaCl solution is 18.64 μA/cm^2 when the tensile strain is 0 με, and decreases to 15.45 μA/cm^2 and 14.39 μA/cm^2 when the tensile strain is 300 με and 1200 με, respectively. For enamel-coated steel rebars, the average corrosion current density is around 0.49 μA/cm^2 when the tensile strain is 0 με. The corrosion current density increases with an increase in the tensile strain in the enamel-coated steel rebar, and reaches 0.65 μA/cm^2 when the tensile strain is 1200 με. Although the corrosion resistance of the enamel-coated steel rebar decreases with an increase in the tensile strain, the enamel coating still provides the steel rebar with significant protection from corrosion.

3.3. EIS

Figure 6 shows the impedance spectrum of a representative uncoated steel rebar in 3.5.wt.% NaCl solution at different tensile strain levels in the format of both Nyquist and Bode plots. Depressed semi-circles are observed in the Nyquist plots, as shown in Figure 6a, and the radius of the semi-circle is approximately 750 Ωcm^2, as there is no tensile strain in the uncoated steel rebar. The radius of the semi-circle gradually increases with an increase in the tensile strain levels. One time constant is present in the Bode plots, as shown in Figure 6b, corresponding to the interfacial properties between the solution and the rebar steel, i.e., the double layer. The scatter symbols represent measured data, while the continuous line represents the results fitted by using the equivalent electrical circuit, as shown in Figure 7a. In Figure 7a, R_s is the solution resistance, R_{ct} is the charge transfer resistance and CPE_{dl} represents the non-ideal capacitive behavior of the double layer between the solution and the rebar steel where corrosion occurred. CPE is the abbreviation of the constant phase element, which is used to consider the non-homogeneity of the electrochemical system that mainly arises from irregularities on the steel surface, the surface roughness, fractal surface and irregular distribution of the applied potential [13]. The CPE can be mathematically expressed as [30]

$$Y_{CPE} = Y(j\omega)^n \tag{2}$$

where Y is the parameter with units of $\Omega^{-1}cm^{-2}sn$, ω is the angular frequency in rad/s and n reflects the deviated degree of the capacitance of the electrode from the ideal capacitor. The CPE is a capacitor when n = 1, a resistor when n = 0, an inductor when n = −1 and Warburg diffusion impedance when n = 0.5.

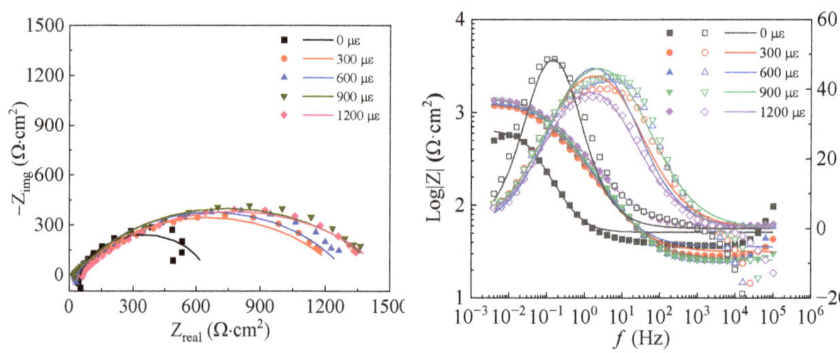

(a) **(b)**

Figure 6. Impedance spectrum of a representative uncoated steel rebar immersed in 3.5 wt.% NaCl solution at different strain levels: (**a**) Nyqusit plots and (**b**) Bode plots.

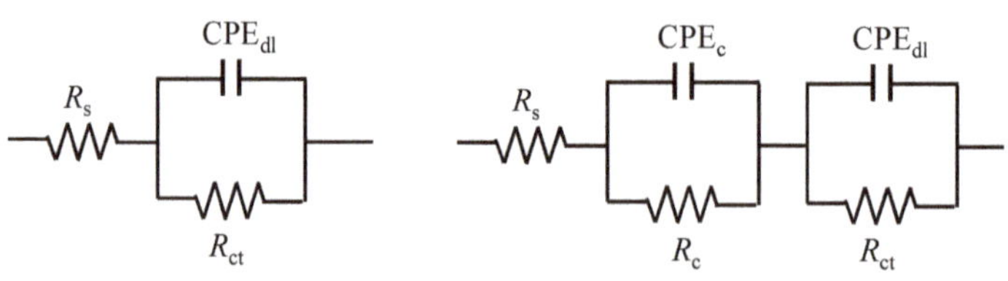

(a) **(b)**

Figure 7. Equivalent electrical circuit: (**a**) uncoated steel rebar and (**b**) enamel-coated steel rebar.

The impedance spectra of a representative enamel-coated steel rebar in 3.5 wt.% NaCl solution at different tensile strain levels are shown in Figure 8. As shown in Figure 8a, the magnitude of the impedance of the enamel-coated steel rebar is almost 60 times greater than that of the uncoated steel rebar. Different from that of the uncoated steel rebar (see Figure 6), two depressed semi-circles are observed for the enamel-coated steel rebar in the Nyquist plots, as displayed in Figure 8a, and, accordingly, two time-constants are present in the Bode plots, as shown in Figure 8b. The time-constant in the high frequency range is attributed to the properties of the enamel coating, while the time-constant in the low frequency range is associated with the double layer at the interface between the solution and the rebar steel. The equivalent electrical circuit used to fit the impedance spectrum of the enamel-coated steel rebar is shown in Figure 7b. In addition to the two parameters related to the double layer, two additional parameters associated with the enamel coating are introduced, in which CPEc represents the non-ideal capacitive behavior of the enamel coating and R_c is the enamel coating resistance. The fitted results are the continuous lines in Figure 8, indicating satisfactory fitting.

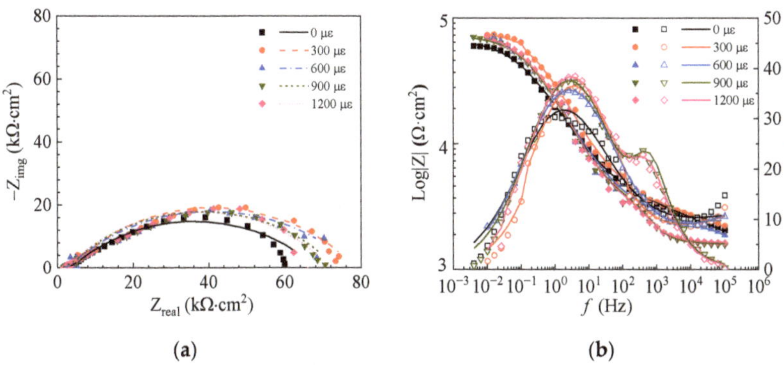

Figure 8. Impedance spectra of a representative enamel-coated steel rebar at different strain levels in the format of (**a**) Nyquist plots and (**b**) Bode plots.

Figure 9 shows the effect of tensile strain on the average enamel coating resistance, which was extracted from the fitted results of the equivalent electrical circuit. It can be observed that the average enamel coating resistance decreases from 10.44 kΩ cm^2 to 1.50 kΩ cm^2 as the tensile strain increases from 0 με to 1200 με. The dramatic decrease in the enamel coating resistance is attributed to the generation of micro-cracks in the enamel coating under tensile strain. The micro-crack provides a pathway for the solution to penetrate through and reach the steel rebar, and, consequently, the coating resistance decreases.

The charge transfer resistance of both the uncoated and the enamel-coated steel rebar specimens in 3.5 wt.% NaCl solution at different tensile strain levels is compared and shown in Figure 10. Once immersed in the solution, the average charge transfer resistance of the uncoated steel rebar is 1.21 kΩ cm^2 and it increases gradually up 1.72 kΩ cm^2 as the tensile strain of the steel rebar increases to 1200 με. For the enamel-coated steel rebar, the average charge transfer resistance increases from 55.05 kΩ cm^2 to 66.04 kΩ cm^2 as the tensile strain increases from 0 με to 600 με, and then gradually decreases to 57.42 kΩ cm^2 as the tensile strain reaches 1200 με. Despite of the variation of the charge transfer resistance of the enamel-coated steel rebar specimens, the R_{ct} of the enamel-coated steel rebar is still around 40 times greater than that of the uncoated steel rebar. Therefore, enamel coating still provides the steel rebar with strong protection against corrosion even when subjected to tensile strain, which is consistent with the results of polarization resistance, as shown in Figure 5a.

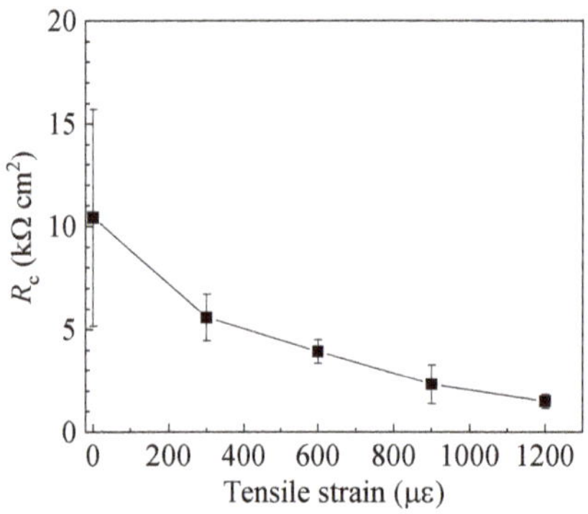

Figure 9. Effect of the tensile strain level on the enamel coating resistance.

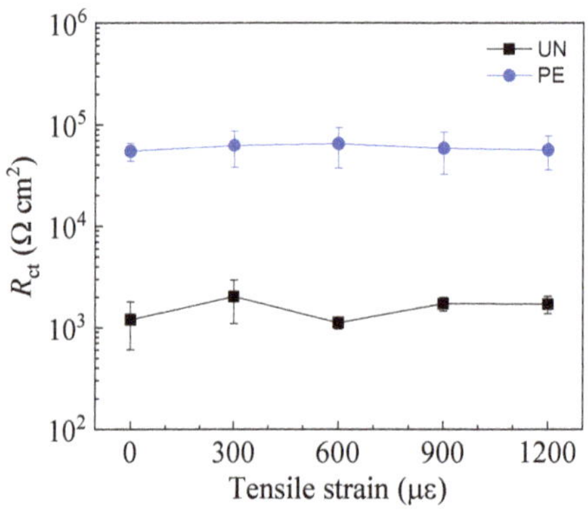

Figure 10. Comparison of the charge transfer resistance of the uncoated (UN) and enamel (PE)-coated steel rebar.

It is noted that the open circuit potential (OCP) of the enamel-coated steel rebar is close to that of the enamel-coated steel rebar, as shown in Figure 4, while a significant difference is observed regarding the polarization resistance (see Figure 5) and the impedance parameters (see Figure 10). This is because the OCP indicates the corrosion state of the steel rebars, while both the polarization resistance and the electrochemical impedance reflect the rate of corrosion, i.e., the rate of electron transfer between the steel rebar and the NaCl solution [17]. The close OCP between uncoated and enamel-coated steel rebars indicates that both are in a state of active corrosion; however, the corrosion rate of the enamel-coated steel rebar is significantly lower than that of the uncoated steel rebar, based on both polarization resistance and impedance results. In addition, the linear polarization resistance is comparable with electrochemical impedance spectroscopy, and both demonstrate the same degree of improvement in the corrosion resistance for the steel rebar after enamel coating.

In previous studies, the corrosion resistance of enamel coating was also investigated by coating on both smooth [20] and deformed steel rebars [13]. The results showed that the coating resistance of enamel, when applied on a smooth steel rebar, was 8.6 kΩ cm^2 [20], while it was 1.3 kΩ cm^2 when applied on a deformed steel rebar [20]. In this study, the coating resistance of enamel is 10.44 kΩ cm^2 when the tensile strain is 0 με, which is significantly higher than that in previous studies. This is mainly attributed to the different microstructure of the two enamel coatings. Many air bubbles were present and uniformly distributed in the enamel coating used in previous studies [13,20], while less air bubbles were observed and only concentrated near the interface with the rebar steel for the enamel coating used in this study, as shown in Figure 3. In addition, the coating resistance of low-temperature sintered enamel coatings was around 30.00 kΩ cm^2 [26], which is almost three times higher than that used in this study. This is because the thickness of the low temperature sintered enamel coating was around 300 μm, which is twice the thickness of the enamel coating investigated in this study.

4. Conclusions

The effect of tensile strain levels on the corrosion resistance of enamel-coated steel rebar in 3.5 wt.% NaCl solution was experimentally investigated in this study. The electrochemical techniques used include open circuit potential, linear polarization resistance and electrochemical impedance spectroscopy. The tensile strain levels investigated were 0 με, 300 με, 600 με, 900 με and 1200 με. The microstructure of the enamel coating was also examined. Based on the results and discussion, the following conclusions can be drawn:

(1) The enamel coating has a thickness of ~150 μm, and there are some air bubbles in the coating which are formed due to the release of carbon dioxide and hydrogen gases from chemical reactions between enamel oxides and carbon element in the steel rebar during the firing process.

(2) The corrosion current density of the uncoated steel rebar decreases from 18.64 μA/cm^2 to 14.39 μA/cm^2 in NaCl solution due to the generation of corrosion products. The average corrosion current density of the enamel-coated steel rebar is 0.49 μA/cm^2, which is around 40 times lower than that of the uncoated steel rebar. The corrosion current density of the enamel-coated steel rebar increases with an increase in the tensile strain level, and it reaches 0.65 μA/cm^2 when the tensile strain is 1200 με. The decrease in the corrosion resistance is attributed to the micro-cracks generated in the enamel coating under the tensile strain.

(3) The tensile strain reduces the enamel coating resistance, and it decreases from 10.44 kΩ cm^2 to 1.50 kΩ cm^2 as the tensile strain increases from 0 to 1200 με. The charge transfer resistance of the enamel-coated steel rebar remains around 40 times greater than that of the uncoated steel rebar when the tensile strain is 1200 με. Therefore, enamel coating still provides steel rebar with strong protection against corrosion subjected even to tensile strain.

Author Contributions: Conceptualization, F.T.; methodology, F.T. and H.C.; software, H.C.; validation, F.T., H.C. and S.Y.; formal analysis, H.C. and S.Y.; investigation, H.C.; data curation, H.C.; writing—original draft preparation, F.T.; writing—review and editing, G.L. and S.Y.; visualization, S.Y.; supervision, F.T. and G.L.; project administration, G.L.; funding acquisition, F.T. and G.L. All authors have read and agreed to the published version of the manuscript.

Funding: This study was supported by the National Natural Science Foundation of China (No. 52078099 and 52225804).

Institutional Review Board Statement: Not applicable.

Informed Consent Statement: Not applicable.

Data Availability Statement: Some or all data that support the findings of this study are available from the corresponding author upon reasonable request.

Conflicts of Interest: The authors declare no conflict of interest.

eferences

Angst, U.M. Challenges and opportunities in corrosion of steel in concrete. *Mater. Struct.* **2018**, *51*, 4. [CrossRef]
Andrade, C. Propagation of reinforcement corrosion: Principles, testing and modelling. *Mater. Struct.* **2019**, *52*, 2. [CrossRef]
Andrade, C.; Izquierdo, D. Statistical treatments of chloride threshold and corrosion propagation rate. *Corros. Mater. Degrad.* **2022**, *3*, 598–611. [CrossRef]
Tang, F.; Lin, Z.; Chen, G.; Yi, W. Three-dimensional corrosion pit measurement and statistical mechanical degradation analysis of deformed steel bars subjected to accelerated corrosion. *Constr. Build. Mater.* **2014**, *70*, 104–117. [CrossRef]
Li, C.; Yang, S. Prediction of concrete crack width under combined reinforcement corrosion and applied load. *ASCE J. Eng. Mech.* **2011**, *137*, 722–731. [CrossRef]

6. Andrade, C.; Cesetti, A.; Mancini, G.; Tondolo, F. Estimating corrosion attack in reinforced concrete by means of crack openin *Struct. Concr.* **2016**, *17*, 533–540. [CrossRef]
7. Tang, F.; Lin, Z.; Qu, H.; Chen, G. Investigation into corrosion-induced bond degradation between concrete and steel rebar wi acoustic emission and 3D laser scan techniques. *J. Infrastruct. Preserv. Resil.* **2022**, *3*, 5. [CrossRef]
8. Zhang, W.; Zhang, H.; Gu, X.; Liu, W. Structural behavior of corroded reinforced concrete beams under sustained loading. *Cons Build. Mater.* **2018**, *174*, 675–683. [CrossRef]
9. Li, C.; Yang, S.; Saafi, M. Numerical simulation of behavior of reinforced concrete structures considering corrosion effects c bonding. *ASCE J. Struct. Eng.* **2014**, *140*, 04014092. [CrossRef]
10. Manning, D.G. Corrosion performance of epoxy-coated reinforcing steel: North American experience. *Constr. Build. Mater.* **199** *10*, 349–365. [CrossRef]
11. Cairns, J.; Abdullah, R. Fundamental tests on the effect of an epoxy coating on bond strength. *ACI Mater. J.* **1994**, *91*, 331–338.
12. Tan, Z.; Hansson, C.M. Effect of surface condition on the initial corrosion of galvanized reinforcing steel embedded in concre *Corros. Sci.* **2008**, *50*, 2512–2522. [CrossRef]
13. Tang, F.; Chen, G.; Brow, R.K.; Volz, J.S.; Koenigstein, M.L. Corrosion resistance and mechanism of steel rebar coated with thre types of enamel. *Corros. Sci.* **2012**, *59*, 157–168. [CrossRef]
14. Rossi, S.; Russo, F.; Calovi, M. Durability of vitreous enamel coatings and their resistance to abrasion, chemical, and corrosion: review. *J. Coat. Technol. Res.* **2021**, *18*, 39–52. [CrossRef]
15. Rossi, S.; Zanella, C.; Sommerhuber, R. Influence of mill additives on vitreous enamel properties. *Mater. Des.* **2014**, *55*, 880–88 [CrossRef]
16. Izgagina, D.A.; Uglinskikh, M.Y.; Vlasova, S.G. Composition development and property study of alkali-resistance enamel for th protection of chemical apparatus. *Glass Ceram.* **2018**, *75*, 234–236. [CrossRef]
17. Jones, D.A. *Principles and Prevention of Corrosion*; Prentice Hall: Hoboken, NJ, USA, 1996.
18. Yang, X.; Jha, A.; Brydson, R.; Cochrane, R.C. An analysis of the microstructure and interfacial chemistry of steel-enamel interfac *Thin Solid Films* **2003**, *443*, 33–45. [CrossRef]
19. Wang, D. Effect of crystallization on the property of hard enamel coatings on steel substrate. *Appl. Surf. Sci.* **2009**, *255*, 4640–464 [CrossRef]
20. Tang, F.; Chen, G.; Volz, J.S.; Brow, R.K.; Koenigstein, M. Microstructure and corrosion resistance of enamel coatings applied t smooth reinforcing steel. *Constr. Build. Mater.* **2012**, *35*, 376–384. [CrossRef]
21. Tang, F.; Chen, G.; Volz, J.S.; Brow, R.K.; Koenigstein, M.L. Cement-modified enamel coating for enhanced corrosion resistance steel reinforcing bars. *Cem. Concr. Compos.* **2013**, *35*, 171–180. [CrossRef]
22. Tang, F.; Chen, G.; Brow, R.K. Chloride-induced corrosion mechanism and rate of enamel- and epoxy-coated deformed steel ba embedded in mortar. *Cem. Concr. Res.* **2016**, *82*, 58–73. [CrossRef]
23. Tang, F.; Cheng, X.; Chen, G.; Brow, R.K.; Volz, J.S.; Koenigstein, M.L. Electrochemical behavior of enamel-coated carbon steel i simulated concrete pore water solution with various chloride concentrations. *Electrochim. Acta* **2013**, *92*, 36–46. [CrossRef]
24. Yan, D.; Reis, S.; Tao, X.; Chen, G.; Brow, R.K.; Koenigstein, M. Effect of chemically reactive enamel coating on bonding streng at steel/mortar interface. *Constr. Build. Mater.* **2012**, *28*, 512–518. [CrossRef]
25. Moser, R.D.; Allison, P.G.; Williams, B.A.; Weiss Jr, C.A.; Diaz, A.J.; Gore, E.R.; Malone, P.G. Improvement in the geopolymer-t steel bond using a reactive vitreous enamel coating. *Constr. Build. Mater.* **2013**, *49*, 62–69. [CrossRef]
26. Yang, F.; Yan, D.; Tang, F.; Chen, G.; Liu, Y.; Chen, S. Effect of sintering temperature on the microstructure, corrosion resistanc and crack susceptibility of chemically reactive enamel (CRE) coating. *Constr. Build. Mater.* **2020**, *238*, 117720. [CrossRef]
27. Yan, D.; Xu, Z.; Qian, H.; Tang, F.; Liu, Y.; Huang, Z. Phase structure and corrosion resistance of multilayer low-temperatu sintered chemically reactive enamel coatings. *ASCE J. Mater. Civ. Eng.* **2020**, *32*, 04020249. [CrossRef]
28. Yan, D.; Qian, H.; Huang, Z.; Tang, F.; Chen, G.; Deng, J.; Liu, Y. Study of lower temperature-sintered enamel coating on steel bar Effect of coating cycles. *ASCE J. Mater. Civ. Eng.* **2020**, *32*, 04020262. [CrossRef]
29. Leechart, P.; Inthorn, D.; Thiravetyan, P. Effect of NaCl on antimony and phthalate compounds leached from PET bottles. *Wat Supply* **2015**, *15*, 766–772. [CrossRef]
30. Cordoba-Torres, P. Relationship between constant-phase element (CPE) parameters and physical properties of films with distributed resistivity. *Electrochim. Acta* **2017**, *225*, 592–604. [CrossRef]

coatings

MDPI

ticle

ong-Term In Situ Performance Evaluation of Epoxy Asphalt :oncrete for Long-Span Steel Bridge Deck Pavement

jin Han †, Zhu Zhang †, Jiahao Tian, Fujian Ni * and Xingyu Gu

School of Transportation, Southeast University, Nanjing 211189, China
* Correspondence: 101004746@seu.edu.cn
† These authors contributed equally to this work.

Abstract: Suitable evaluation of distress is beneficial to understanding the in situ performance of deck pavement. This study attempts to evaluate the long-term in situ performance of American ChemCo epoxy asphalt concrete on the Xihoumen Bridge (XHMB) after 12 years of service. The traditional performance indexes were adopted to reveal the performance of XHMB. Then, based on the typical distresses, a new pavement performance index (PPI) was proposed to characterize the authentic distress condition. Finally, the performance evaluation and evolution were conducted. According to the results, the rutting depth indexes and riding quality indexes of all lanes are higher than 97 and 94, respectively. The pavement condition indexes of the pass lanes and drive lanes in 2021 are greater than 94 and 86, respectively, which is contradictory to the distribution of numerous distresses on the pavement. According to the PPI results, the PPIs of the down direction pass lane are mostly 100. However, for the down direction drive lane, the PPIs of about 30% of segments are below 80 or 60. Finally, based on the limited data, the distress of American ChemCo epoxy asphalt concrete may initiate after serving for 4–5 years and then escalate after about 10 years.

Keywords: orthotropic steel bridge deck pavement; epoxy asphalt concrete; long-term evaluation; pavement performance index; distress evolution

1. Introduction

Bridges are designed to traverse an obstacle to connect the separated two sides [1]. The design and selection of the bridge type may be influenced by many factors. Particularly for an urban area, the bridge should be constructed as an attractive and user-friendly structure [1,2]. For a highway bridge, these functions may be limited and an orthotropic steel bridge deck (OSBD) is extensively utilized to communicate two separate terminals with a considerable distance. Compared with traditional concrete bridges, OSBD prevails because of its lightweight, superb structural strength, and long service life [3]. Additionally, the OSBD segments could be prefabricated, which is capable to shorten the construction period and equips the construction process with great flexibility. To isolate the steel deck surface from environmental corrosion and provide vehicles with a flat and skid-resistance layer, multilayered pavement structures are constructed on the steel deck.

The OSBD pavement system typically consists of the anticorrosion primer, the waterproof tack coat, and the surfacing layer from the bottom to the top [4]. The anticorrosion primer and the waterproof tack coat layer are designed to protect the steel deck from rusting and therefore prolong the service life of the infrastructure [5]. Additionally, the waterproof tack coat also functions to integrate the steel deck with the surfacing layer [6,7]. The surfacing layer is capable to carry, transfer, and distribute the vehicle loads from the pavement surface to the steel deck [8]. Differing from the traditional asphalt pavement, the OSBD pavement typically faces a more rigorous service environment, including complex stress conditions and wider service temperature ranges [3,9]. Considering the harsh field conditions, higher requirements are put forward for the mechanical performance of OSBD pavement.

Previously, various bituminous materials, including the guss asphalt (GA) concrete and (stone) mastic asphalt (S)MA, have been developed and utilized as the OSBD surfacing layer [3]. Those materials could be paved alone or together to form divergent pavement systems due to their excellent waterproofness, deformation compliance, and flexibility [10,11]. However, traditional bituminous materials and their concrete may melt and therefore deform permanently at high temperatures. The existence of this temperature instability distress, including rutting, bleeding, and shoving, undoubtedly restricts the application of GA and SMA in OSBD [12,13]. In addition, bituminous concrete is prone to ravel due to the coupling effects of traffic loads and harsh environmental conditions, including moisture, freeze–thaw conditions, and bitumen age [14–16].

Decades ago, epoxy asphalt concrete (EAC) was invented by the Shell Petroleum Company and then utilized as the pavement material on the San Mateo–Hayward Bridge orthotropic deck [17,18]. The epoxy asphalt is typically composed of thermosetting epoxy resin and thermoplastic base asphalt with a weight ratio of 1:1. According to previous studies, the base asphalt is evenly distributed in the continuous epoxy resin in the form of spherical particles [19]. The unique encapsulated structure may be beneficial to improving the epoxy resin network flexibility [20]. In 2000, EAC was first used in China on the Nanjing Second Yangtze River Bridge to release the severe distress of bituminous concrete [18,21]. In the following several years, EAC has prevailed as the OSBD surfacing material due to its outstanding high-temperature stability, mechanical strength, moisture damage resistance, and deformation compliance [20,22,23].

However, EAC is not problem-free. Premature failure may arise from inexperienced construction, improper temperature control, and inappropriate materials ratio [7]. Apart from this abnormal distress, fatigue cracks may occur on the pavement surface and then propagate into block cracks as a result of repeated vehicle loading [10]. The existence of distresses compromises the serviceability, including the flatness, skid resistance, and durability, of the OSBD surfacing pavement. To end this dilemma, scholars attempt to evaluate the performance of surfacing materials through laboratory tests and consequently forecast the in situ distress [24,25]. Nonetheless, the loading and constraint conditions in laboratory tests even for the full-scale test may differ from the actual situation in the field [26,27]. The laboratory test could characterize the mechanical properties of materials but could rarely capture the authentic response of those materials in the field.

The distress survey of OSBD pavement has also been conducted to clarify the evolution of typical distresses [12,28,29]. Unfortunately, these distress surveys usually cover a relatively short period and are consequently unable to convey the comprehensive evolution of pavement distresses. In addition, the distress survey may simply record the number, size, and severity of distresses and could hardly manifest the quantitative performance of OSBD pavement [21]. By contrast, in road asphalt pavement engineering, appropriate performance indexes were put forward to quantitatively characterize the service performance of pavement [30–32], which is beneficial to guiding the maintenance of pavement with severe distress [33–35]. However, the performance indexes adopted in road pavement engineering may not accommodate OSBD pavement due to the huge difference between steel bridges and roads.

To fill this gap, the in situ condition and performance evolution of Xihoumen Bridge (XHMB) deck pavement over twelve years are evaluated in this study. Specifically, based on the annual performance detection, the pavement performance of XHMB is evaluated using the traditional performance evaluation indexes, which were originally used in the road pavement field. Then, the applicability of the indexes to revealing the performance of OSBD pavement is analyzed. Afterward, a new pavement condition index (PPI) is proposed to characterize the authentic pavement condition. Finally, based on the PPI, the performance evaluation and partition of XHMB pavement are conducted. The organization of this study is illustrated in Figure 1.

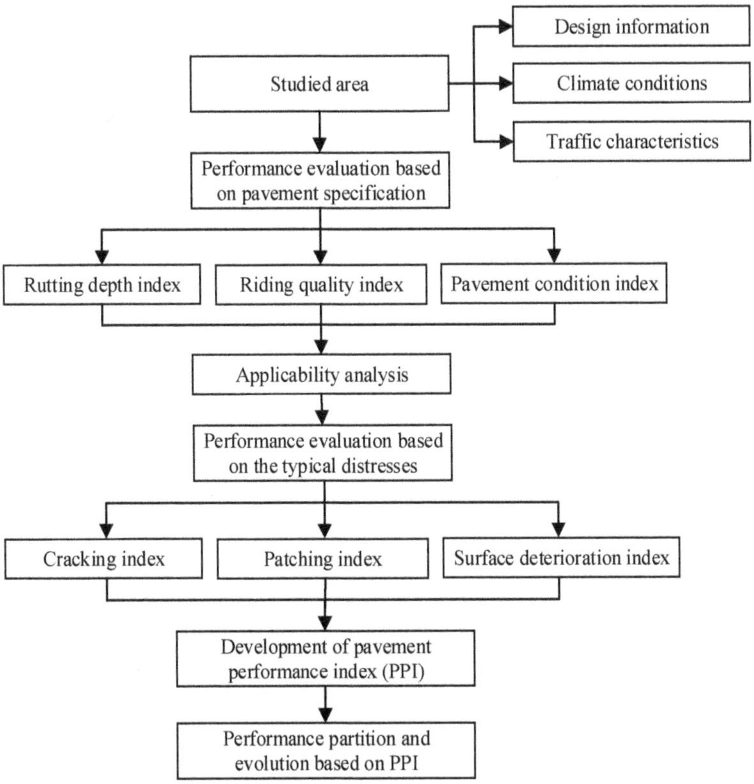

Figure 1. Organization and methodology.

2. Study Area

2.1. Location and Design Parameters

XHMB is a double-tower, double-cable unsymmetrical steel box girder suspension bridge, located in Zhejiang Province. The geographical location diagram and external view are shown in Figure 2a,b. XHMB is designed with a main span of 1650 m and was the first separate double steel box girder suspension bridge in the world. The detailed parameters of the bridge are listed in Table 1. Based on the field performance of several typical pavement systems constructed at home and abroad, the XHMB deck is paved with double-layered American ChemCo epoxy asphalt concrete with a thickness of 55 mm (30 mm upper layer + 25 mm bottom layer), as illustrated in Figure 2c.

Table 1. Main parameters of XHMB.

Bridge Name	Xihoumen Bridge
Structure type	Suspension bridge
Opening time	December 2009
Designed service life/year	15
Main span/m	1650
Total span/m	2230
Longitudinal gradient/%	2.5
Steel deck plate thickness/mm	14 (Strengthening position: 16)
Stiffener thickness/mm	8
Distance between adjacent stiffeners/mm	600

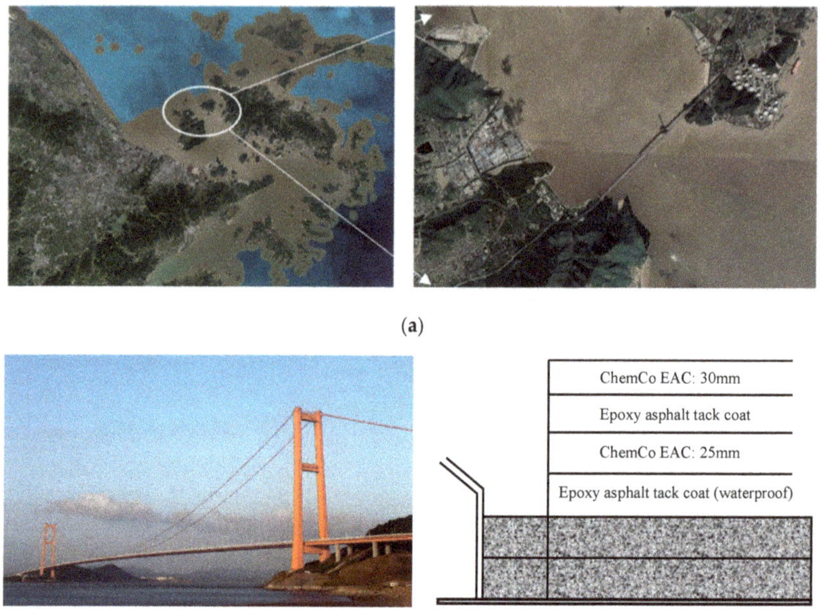

(a)

(b) (c)

Figure 2. Location and appearance of XHMB: (**a**) geographical location diagram; (**b**) external view; (**c**) pavement structure.

2.2. Climate

XHMB is located in a sea area frequently affected by typhoons. The hydrologic, geological, and climatic conditions of XHMB are complex. The region where XHMB is located belongs to the subtropical monsoon climate region, which is typically characterized by high temperatures and rainy in summer and low temperatures and dry in winter. According to local meteorological data, the annual average temperature is about 18 °C. The historical extreme maximum temperature is 34 °C, and the extreme minimum temperature is −5 °C. The average annual number of rainy days is as high as 200 days and the average annual precipitation is as high as 1800 mm. Detailed climatic data are provided in Figure 3. The severe environmental conditions put forward the strict requirements for the performance of OSBD surfacing materials.

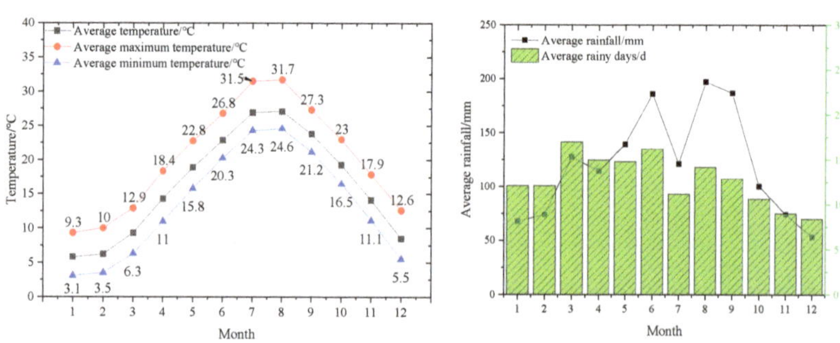

(a) (b)

Figure 3. Climatic information for XHMB: (**a**) temperature; (**b**) rainfall.

2.3. Traffic Volume

The XHMB was opened to traffic in 2009 and has undertaken heavy transit traffic volume since then. According to existing studies [6,36], fatigue cracking of the OSBD surfacing layer is mainly attributed to the repeated shear and tensile stresses induced by vehicles. Therefore, the traffic data of XHMB from 2010 to 2020 were collected and analyzed. The results are shown in Figure 4a. In this study, the direction from Ningbo to Zhoushan is denoted as the up direction and the direction from Zhoushan to Ningbo is denoted as the down direction. The direction and lane distribution of traffic was also studied and is presented in Figure 4b.

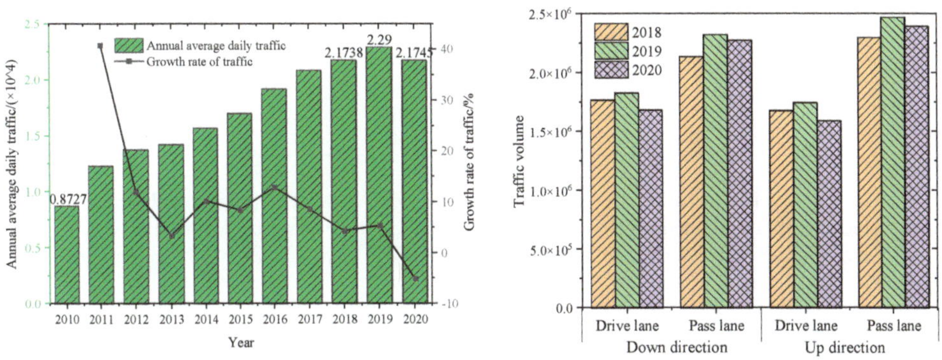

(a) (b)

Figure 4. Analysis of traffic volume: (**a**) variation in traffic volume; (**b**) direction distribution in traffic.

According to Figure 4a, since 2010, the annual average daily traffic (AADT) demonstrates a significant increasing trend, from 8727 in 2010 to 21745 in 2020. The annual average traffic growth is about 15%. In addition, the traffic growth rates year by year were also calculated, as shown in Figure 4. The traffic growth rates approximately range from 4% to 12% except for the growth rate in 2011 and 2020. Based on the traffic growth tendency, the traffic volume of XHMB may continuously rise in the coming years. As shown in Figure 4b, the traffic volume in the up direction is slightly higher than that in the down direction. By contrast, the traffic volume of the pass lane is significantly higher than that of the drive lane regardless of the direction. The lane distribution of traffic may ascribe to the driving habits of vehicles with different loads. Typically, vehicles with heavy loads prefer to drive in the drive lane at a relatively low speed, while vehicles with light loads usually drive on the pass lane at a high speed.

3. Performance Evaluation of XHMB Pavement

Currently, considering the lack of corresponding OSBD pavement evaluation specifications, the assessment methodology of road pavement performance was extensively utilized to reveal the OSBD pavement conditions by bridge maintenance companies. In this study, the evaluation methodology of road pavement performance was utilized to characterize the XHMB surfacing layer conditions. By comparing the evaluation indexes with the actual OSBD pavement performance, the applicability of such an evaluation scheme could be clarified.

After the XHMB was opened to traffic, an annual pavement performance inspection was conducted. The inspected performance includes rutting condition, flatness, and pavement surface condition, which are characterized by the rutting depth index (RDI), riding quality index (RQI), and pavement surface condition index (PCI), based on the Highway Performance Assessment Standards in China (JTG 5210-2018).

3.1. Rutting Depth Index (RDI)

The RDI is calculated according to Equation (1).

$$RDI = \begin{cases} 100 - RD & (RD \leq 10) \\ 90 - 3 \times (RD - 10) & (10 < RD \leq 40) \\ 0 & (RD > 40) \end{cases} \quad (1)$$

where RD is the rutting depth, mm.

According to the RDI value and the corresponding RD value, the rutting condition of pavement could be graded, as listed in Table 2.

Table 2. Grades of pavement rutting condition.

Grade	A	B	C	D	E
RDI	[100, 90]	(90, 80]	(80, 70]	(70, 60]	(60, 0]
RD/mm	[0, 10]	(10, 13.3]	(13.3, 16.7]	(16.7, 20]	>20

The XHMB was paved with double-layered American ChemCo epoxy asphalt concrete, which is well acknowledged as a thermal-setting material. The rutting condition of XHMB pavement was recently determined in November 2020. The results are shown in Figure 5.

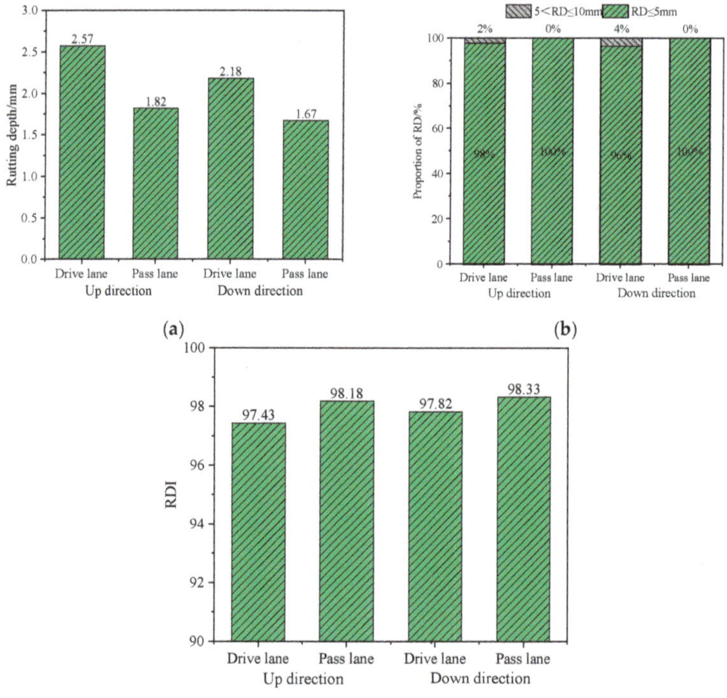

Figure 5. Rutting condition of XHMB pavement: (**a**) average RD; (**b**) RD distribution; (**c**) average RDI.

As illustrated in Figure 5a, the average RDs of the four lanes range from 1.67 to 2.57 mm. The RDs of drive lanes are slightly higher than that of pass lanes regardless of the driving directions. This may attribute to the loading difference between vehicles driving in the driving lane and pass lane. In addition, a mild difference in RDs between the up direction and down direction could also be observed.

To precisely characterize the RD distribution, the RDs of the four lanes were collected in 10 m units and the proportions of different RD intervals were calculated, as shown in Figure 5b. The RDs of pass lanes are all lower than 5 mm. By contrast, the RDs of a few sections in drive lanes range from 5 to 10 mm. The RD data imply that the RDIs of the XHMB all exceed 90 even for the section with relatively severe rutting distress. Finally, according to Figure 5c, the average RDIs of all four lanes are all higher than 95, which further indicates the outstanding rutting resistance of the XHMB pavement.

3.2. *Riding Quality Index (RQI)*

The RQI is calculated according to Equation (2).

$$RQI = \frac{100}{1 + 0.026 \times e^{0.65 \times IRI}} \tag{2}$$

where IRI is the international roughness index, m/km.

According to the RQI value and the corresponding IRI value, the flatness of pavement could be graded, as listed in Table 3.

Table 3. Grades of pavement flatness.

Grade	A	B	C	D	E
RQI	[100, 90]	(90, 80]	(80, 70]	(70, 60]	(60, 0]
IRI/mm	[0, 2.3]	(2.3, 3.5]	(3.5, 4.3]	(4.3, 5.0]	>5.0

The rutting condition of XHMB pavement was recently detected in November 2020. The results are shown in Figure 6.

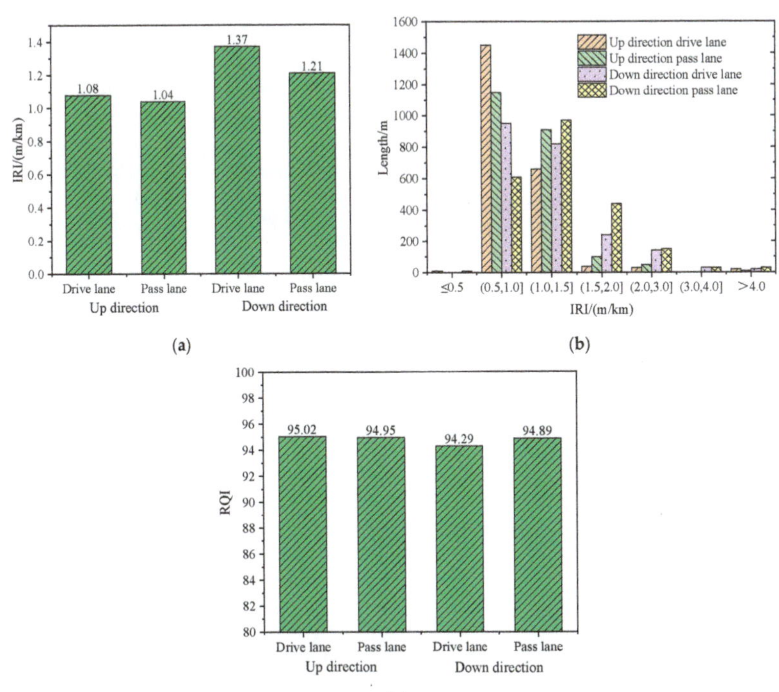

Figure 6. The roughness of XHMB pavement: (**a**) average IRI; (**b**) IRI distribution; (**c**) average RQI.

As shown in Figure 6a, the average IRIs of the four lanes ranges from 1.04 to 1.37 m/km, which indicates the good flatness of XHMB pavement. Note that the IRIs of drive lanes are a little larger than that of pass lanes for both directions. Additionally, there are some distinctions between IRIs of the up direction and down direction. Similar to the RD analysis, the IRI distribution was also studied based on the flatness data tested every 10 m. According to Figure 6b, for both directions, the IRI centered in the range (0.5, 1.5] m/km. Compared with the up direction, the IRI of down direction pavement is more to the right side of the figure, which further confirms the minor flatness of the up direction pavement. Figure 6c demonstrates the RQI of the four lanes. Consistent with the results of IRI, the average RQIs exceed 90 and concentrate around 95.

Combining the RDI and RQI results, the rutting resistance and flatness of the XHMB surfacing layer are in fairly good condition. Severe rutting distress and poor flatness were not observed. This phenomenon could be ascribed to the outstanding high-temperature stability of epoxy asphalt concrete even under strict in-service environmental conditions.

3.3. Pavement Surface Condition Index (PCI)

The PCI mainly concerns with the integrity of pavement through the visual survey. The occurrence of distresses, including map crack, longitudinal crack, transverse crack, bleeding, rutting, shoving, pothole, raveling, and patching, may decrease the PCI with diverse weighting coefficients according to the Highway Performance Assessment Standards in China (JTG 5210-2018). The PCI is determined according to Equations (3) and (4). In this study, the weighting coefficient and calculation of 21 types of distresses are not described in detail.

$$\mathrm{PCI} = 100 - 15 \times \mathrm{DR}^{0.412} \tag{3}$$

$$\mathrm{DR} = 100 \times \frac{\sum_{i=1}^{21} w_i \times A_i}{A} \tag{4}$$

where DR is the pavement distress ratio, %; w_i is the weighting coefficient of type i distress; A_i is the equivalent distress area, m^2; and A is the detected pavement area, m^2.

According to Equations (3) and (4), the average PCIs of the XHMB four lanes over several years were calculated and are shown in Figure 7. Note that the distress condition of XHMB was conducted by the automatic detection vehicle before 2021. Considering the identification accuracy of such a method, a detailed visual survey was carried out in June 2021 to determine the authentic distress condition of XHMB before maintenance treatment. The number, type, and severity of distresses were recorded to calculate the PCI.

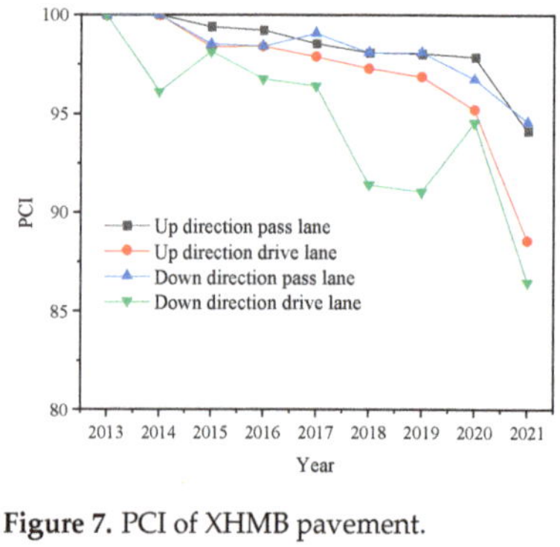

Figure 7. PCI of XHMB pavement.

According to Figure 7, with the increase in service durations, the PCI demonstrates a steadily decreased trend regardless of the driving direction and lanes except for some outliers. It should be noted that dramatic declines in PCIs are observed from 2020 to 2021. This may be attributed to the inconsistency of distress survey methods. The inspection accuracy of detection vehicles before 2021 may be lower than that of the visual survey in 2021, especially for the microcrack in XHMB pavement, as shown in Figure 8.

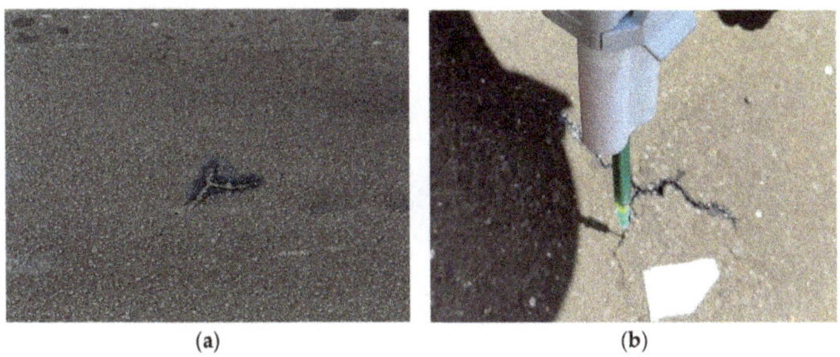

Figure 8. Irregular cracks on XHMB pavement. (**a**) chicken-feet crack, (**b**) irregular crack.

As shown in Figure 7, the PCIs of the two pass lanes in different directions are always higher than that of the drive lanes, which indicates the relatively severe deterioration of drive lanes and the negative impacts of heavy-duty loads on the durability of the OSBD surfacing layer. Although a continuous downward trend of PCI over years is observed, three out of the four lanes have PCIs greater than 95 for before 2021. Even for the down direction drive lane, the PCI before 2021 is still higher than 92. In addition, after the sharp decrease in PCI from 2020 to 2021, the PCIs of the pass lanes and drive lanes in 2021 are still greater than 94 and 86, respectively.

According to the PCI results, the XHMB pavement is in quite good condition and the distresses seem not apparent or severe. However, based on the latest visual survey, numerous distresses were observed on the XHMB pavement, especially on the drive lanes, as pictured in Figure 9. The spatial distribution of distresses in 2021 is drawn schematically in Figure 10.

Figure 9. Typical distresses of XHMB pavement: (**a**) cracks; (**b**) pothole and patching.

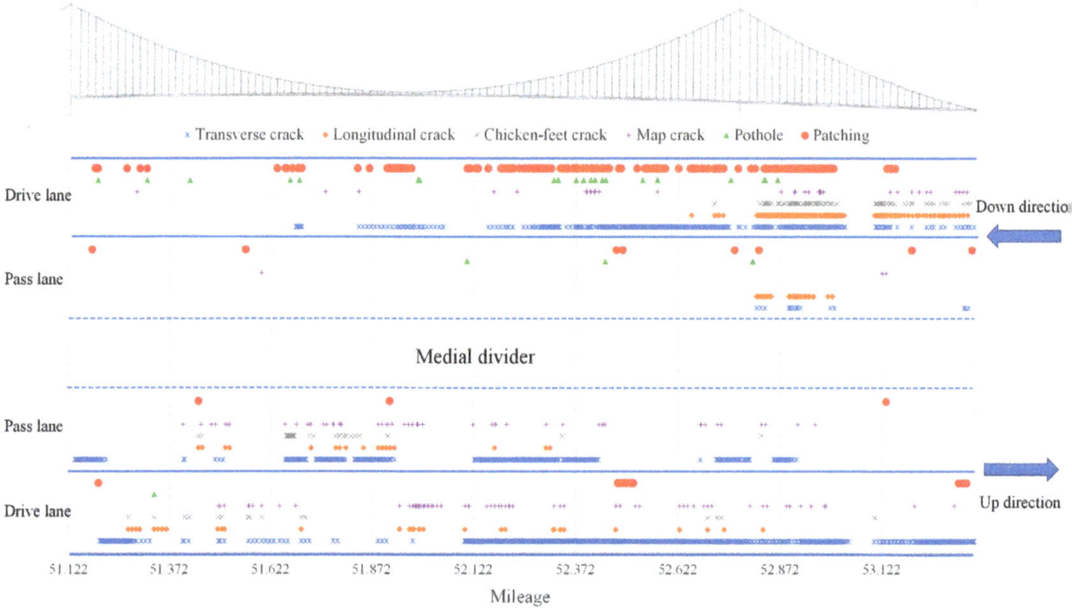

Figure 10. Spatial distribution of XHMB pavement distresses.

As demonstrated in Figure 9, short cracks emerge densely on the pavement surface. The separate short cracks may join together and evolve into map cracks without effective treatment. Then, water can permeate to the bottom of the pavement system and weaken the interface bonding strength between the pavement system and steel deck plate. Finally, the weakened area turns into a pothole, and patching is conducted to rehabilitate the service performance of the pavement. Based on Figure 10, the pass lane surface condition in both directions is significantly superior to that of the drive lane in both directions, especially for the down direction pass lane. In addition, numerous transverse cracks widely distribute in the drive lanes in both directions; patching is observed mainly in the drive lane in the down direction. By contrast, the map crack mainly concentrates on the up direction drive lane and pass lane. The occurrence of distresses with such high intensity is far from the good condition of XHMB pavement indicated by the PCI index.

The inconsistency between PCI and the actual condition of XHMB pavement may be attributed to the limited distresses types in XHMB pavement. The calculation of PCI in the current standard is based on the principle of point deduction. When the distress type is relatively single, although the pavement is in a poor condition, the PCI is still very high as long as other distresses are less. In addition, the PCI is typically calculated per kilometer, which is too broad to characterize the detailed performance of OSBD pavement with a 1000–2000 m span. Hence, the OSBD pavement condition should be evaluated based on the typical distresses and reveal the performance with a relatively high resolution.

4. XHMB Pavement Condition Evaluation Based on Typical Distresses

4.1. Typical Distresses of XHMB Pavement

The distresses type and number were recorded over several years and the data from 2017 to 2021 were used to determine the typical distress of XHMB pavement. The results are shown in Figure 11.

According to Figure 11, with the increase in service duration, the distresses number demonstrates a significant increase trend, especially from 2020 to 2021. Among the distresses, the cracks, including longitudinal, transverse, and chicken-feet cracks, occupy the primary part of the XHMB distresses during the 5 years that were studied, followed by patching and map crack. The raveling, rutting, and shoving distress could rarely be observed in XHMB EAC. Based on the survey results, it is concluded that the typical distress of XHMB pavement may consist of linear and areal damage. The linear deterioration includes longitudinal, transverse, (shown in Figure 9a), and chicken-feet cracks (shown in Figure 8). By contrast, the areal deterioration includes map crack, pothole, and patching (shown in Figure 9b).

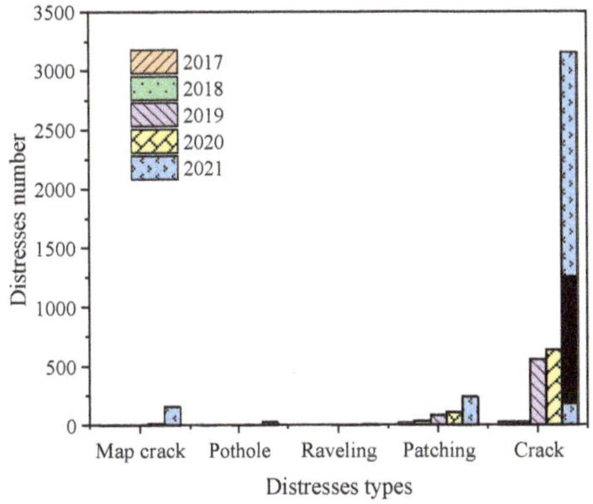

Figure 11. Distress conditions of XHMB pavement.

4.2. Pavement Performance Index (PPI)

The previous section describes the typical distresses, which are widely divergent from the road pavement. Therefore, a pavement performance index (PPI) is proposed to authentically evaluate the pavement condition of XHMB. Similar to PCI, the PPI is still determined using the deduction system. However, the evaluated pavement length decreases from 1 km in road pavement evaluation to 10 m in OSBD pavement. Hence, the OSBD pavement is separated into several segments with a length of 10 m and width of 3.75 m. This may be beneficial for capturing the detailed condition of OSBD pavement with a considerable resolution. Additionally, the PPI is mainly determined by three parts, including crack index (CI), patching index (PI), and surface deterioration index (SDI), rather than the 21 types of distresses regulated in Highway Performance Assessment Standards in China (JTG 5210-2018). The calculation of PPI is shown in Equation (5).

$$PPI = 0.3 \times CI + 0.3 \times PI + 0.4 \times SDI \quad (5)$$

where CI is the crack index; PI is the patching index; SDI is the surface deterioration index; and the numbers before CI, PI, and SDI are the weighting coefficients, which are determined by the expert scoring method.

4.2.1. Crack Index (CI)

As mentioned above, the cracks on XHMB pavement include longitudinal, transverse, and chicken-feet cracks. The CI is mainly adopted to characterize the effects of these linear cracks on the performance of XHMB pavement. As shown in Figure 9, the linear crack is short and narrow with considerable density. Therefore, both the length and the number of cracks are taken into consideration when calculating the CI. Considering the connected shape of chicken-feet cracks, the effects of chicken-feet cracks on pavement performance are multiplied by a weighting coefficient of 2.5. This means the equivalent number and length of the chicken-feet crack is 2.5 times the measured number and length. The CI is determined by interpolation based on Table 4. It should be noted that the threshold value listed in Table 4 to calculate the CI and the following PI and SDI is determined by the expert scoring method.

Table 4. Threshold of CI calculation.

Total Length of Cracks/m	Number of Cracks	CI
[0, 0.1]	[0, 1]	[100, 90]
(0.1, 0.3]	(1, 3]	(90, 80]
(0.3, 2.7)	(3, 27)	(80, 0]
≥2.7	≥27	0

The CI of an evaluated area is calculated based separately on the crack length and number, and the smaller CI among the two calculations is adopted to determine the PPI.

4.2.2. Patching Index (PI)

Patching mainly refers to the rehabilitation of severe distresses, such as potholes and map cracks. However, patching can hardly renew the performance of pavement suffering from distresses. The incompatibility between the original pavement materials and patching materials may generate a weak surface. In addition, the flatness of the patched area is always unsatisfactory, which may decrease the comfortability of driving vehicles. Hence, in this study, the patching is treated as relatively severe distress. The PI is determined by interpolation based on Table 5.

Table 5. Threshold of PI and SDI calculation.

Total Patching Ratio/%	PI and SDI
[0, 0.05]	[100, 90]
(0.05, 0.2]	(90, 80]
(0.2, 0.5]	(80, 70]
(0.5, 1]	(70, 60]
(1, 2]	(60, 40]
(2, 8]	(40, 0]
>8	0

Note that the patching ratio refers to the ratio of the patching area to the investigated area. As noted above, the evaluated area in this study refers to a segment with a length of 10 m and a width of 3.75 m.

4.2.3. Surface Deterioration Index (SDI)

Surface deterioration in this study mainly involves the map crack, raveling, and pothole distress. This distress is typically evolved from the longitudinal, transverse, and chicken-feet cracks. Owing to the existence of surface deteriorations, moisture may easily penetrate the interface between the concrete and the steel deck plate, which is detrimental to the durability of the steel bridge. Similar to patching distress, the surface deterioration can be categorized as areal distress and is significantly detrimental to pavement performance. However, compared with patching distress, the surface deterioration could be more harmful. Therefore, the threshold of SDI calculation is kept the same as the PI, as shown in Table 5.

4.3. *XHMB Pavement Performance Evaluation Based on PPI*

The PPIs of XHMB's four lanes were calculated according to the visual survey results from 2021. The results are shown in Figure 12.

As shown in Figure 12, the XHMB pavement conditions are revealed with a high resolution compared with the PCI. According to Figure 12, the PPIs for the pass lanes in both directions are obviously higher than that of the drive lanes. Among the four lanes, the performance of down direction pass lane is significantly superior to other lanes. To be specific, the PPIs of the down direction pass lane are mostly 100, which indicate no distress is observed in this lane. Although the PPIs of a few segments are lower than 90, the lowest PPI in this lane is still larger than 70. By contrast, the drive lanes demonstrate lower PPI and worse pavement conditions, especially for the down direction drive lane. The PPIs of a larger proportion of segments are below 80 or even below 60. Concerning the up direction drive and pass lanes, although the PPIs of some segments are lower than 80, the PPIs mainly range from 90 to 100. Figure 13 is provided to better understand the PPI distribution for the four lanes.

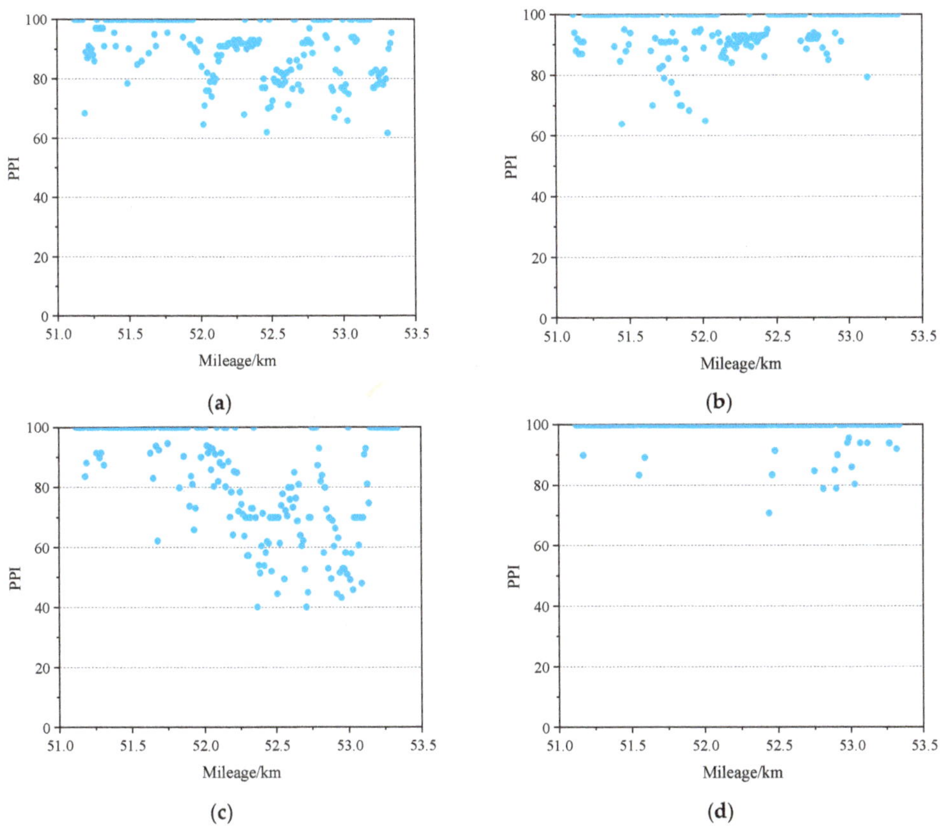

Figure 12. PPIs of XHMB pavement in 2021: (**a**) up direction drive lane; (**b**) up direction pass lane; (**c**) down direction drive lane; (**d**) down direction pass lane.

As revealed in Figure 13, the percentages of segments with PPI above 90 in four lanes are 65.47%, 84.75%, 52.47%, and 95.07%, respectively. To be specific, the percentages of segments with a PPI above 90 for the pass lanes in both directions center around 90%, which is dramatically higher than that of drive lanes. In addition, the percentages of segments with a PPI below 80 for the pass lanes in both directions are lower than 5%. By contrast, the percentages of up direction and down direction drive lanes are about 18% and 36%, respectively. For the down direction drive lane, 12.11% of the total span shares a PPI value below 60 and may face severe distresses.

According to Figures 12 and 13, most segments of the XHMB pavement are in good condition without the influence of distress, while there are also considerable segments suffering from serious distress, especially the segments in drive lanes. Therefore, XHMB pavement conditions are greatly divergent and varying maintenance treatments should be piecewise adopted to rehabilitate the XHMB pavement performance and prolong the service life.

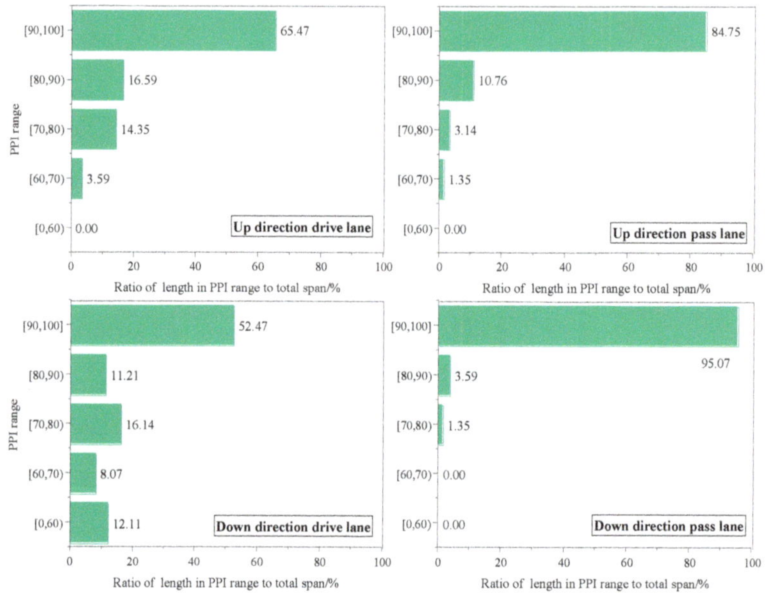

Figure 13. PPI distribution for the four lanes.

4.4. Performance Partition of XHMB Pavement

Considering the dispersion of the distress distribution, maintenance treatments should be determined and then conducted in sections. Based on the PPI, the schematic performance partition of XHMB pavement in 2021 was drawn and is shown in Figure 14.

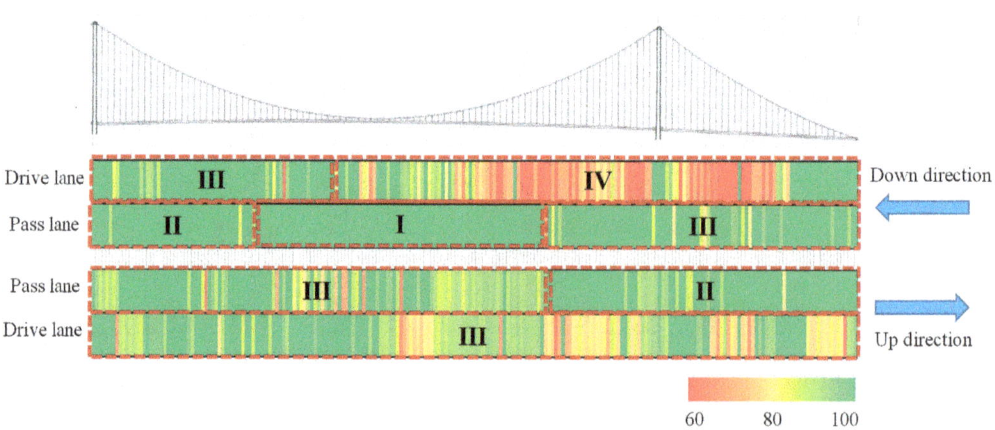

Figure 14. Performance partition of XHMB pavement based on PPI value.

As shown in Figure 14, the performance condition of XHMB pavement in 2021 is divided into four different grades: I, II, III, and IV, among which the I grade refers to the best performance. Generally speaking, the PPIs of considerable segments in the up and down direction drive lanes fall between 60 and 80. Hence, the drive lanes are categorized as type III and type IV. By contrast, the pass lanes are mainly divided into type II and type III. It should be noted that no distress was reported in the middle of the down direction pass lane. Therefore, the distress-free section is categorized as type I.

According to the performance partition of XHMB pavement, suitable maintenance measures could be adopted to enhance the pavement conditions. Pavements with different performance grades should be treated with varying maintenance programs. For type I pavement, a functional thin overlay could be utilized to improve the skid resistance and mitigate the degradation effects of environmental factors on the pavement materials. For type II pavement, sectional maintenance is necessary to prevent the deterioration of existing distress. For type III pavement and type II pavement with intensive distress, to make full use of the residual value of the bottom concrete, the upper layer should be milled and refilled with fresh concrete. Finally, concerning type IV pavement, to prevent the corrosion of steel deck plate, the pavement system should be milled and repaved.

4.5. Performance Evolution of XHMB Pavement

There are limited reports of tracking surveys for OSBD pavement conditions in China, especially for American ChemCo epoxy asphalt concrete deck pavement. To understand the in situ performance of American ChemCo epoxy asphalt concrete and long-spanned suspension bridge deck pavement, the performance variation for XHMB pavement was evaluated based on the PPI and distress parameters. The distress variation for XHMB pavement is shown in Figure 15a.

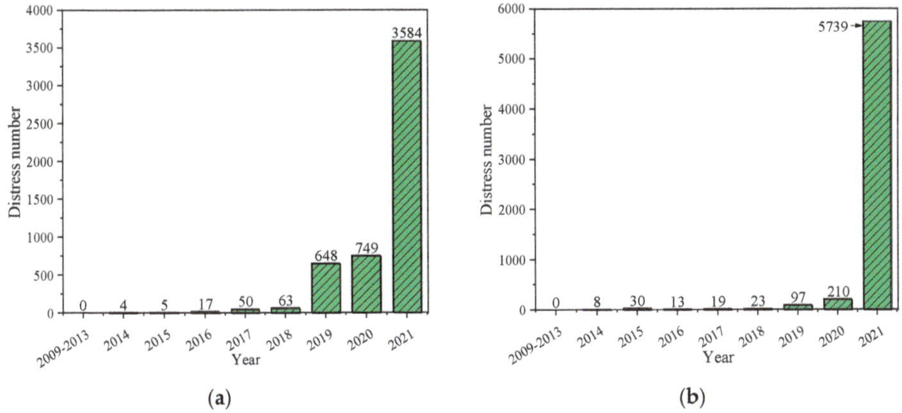

Figure 15. Variation of XHMB and JTB pavement distresses number: (**a**) Xihoumen Bridge (XHMB); (**b**) Jintang Bridge (JTB).

According to Figure 15a, the distress first appeared in 2014 after the XHMB was opened to traffic for 5 years. A similar phenomenon was also observed in other OSBD pavements in Zhoushan cross-sea bridge clusters, such as the Jintang Bridge (JTB) and Taoyaomen Bridge (TYMB). Variation in the JTB distress number is shown in Figure 15b. TYMB and JTB were opened to traffic in 2006 and 2009 and were originally paved with the same materials as XHMB. The TYMB and JTB are twin-tower, cable-stayed bridges with a total span of 888 and 1210 m, respectively. The TYMB and JTB share the same traffic loading and environmental conditions as XHMB. The distress occurred in the TYMB and JTB first after the two bridges were opened to traffic for 4 and 5 years. According to the limited data, the distress of American ChemCo epoxy asphalt concrete may initiate after the concrete served for 4–5 years, regardless of the bridge structures.

As shown in Figure 15a,b, from 2014 to 2018, the distress number demonstrates a steady upward trend for both XHMB and JTB. Then, an obvious increase in distress numbers could be observed both in XHMB and JTB from 2018 to 2019. Furthermore, after 2019, the distress number continued to rise exponentially. This phenomenon may indicate that after the American ChemCo epoxy asphalt concrete was opened to traffic for about 10 years, the distress shows an outbreak trend. At this point, necessary structural maintenance is needed to rehabilitate the pavement performance.

5. Conclusions

This study aims to evaluate the long-term in situ performance of XHMB pavement. The traditional performance evaluation indexes were adopted to reveal the pavement performance of XHMB. The applicability of those indexes to revealing the performance of OSBD pavement was analyzed. On this basis, a new pavement condition index, PPI, was proposed to characterize the authentic pavement condition. In addition, the performance evaluation, partition, and variation for XHMB were conducted. The main conclusions are shown as follows.

(1) After being opened to traffic for more than 10 years, the average RDs of XHMB pavement range from 1.67 to 2.57 mm, and the average RDIs of all four lanes are all higher than 95, which indicates the outstanding rutting resistance of the XHMB pavement.
(2) Similar to the rutting condition, the roughness of XHMB pavement is in good condition. In 2020, the average IRIs range from 1.04 to 1.37 m/km and the average RQIs of the four lanes concentrate around 95.
(3) The PCIs of the pass lanes and drive lanes in 2021 are greater than 94 and 86, respectively, which may imply favorable surface conditions and structural integrity of XHMB pavement. However, numerous distresses, including cracks, patching, and map cracking, widely distribute on the XHMB pavement with a close interval, which is far from the good surface condition reflected by the PCI. Hence, the PCI can hardly characterize the authentic surface conditions of OSBD pavement with a high resolution.
(4) The cracks occupy the primary part of the XHMB distresses, followed by patching and map cracking. Therefore, a pavement performance index (PPI) consisting of the crack index, patching index, and surface deterioration index, was developed.
(5) According to the PPI results, the XHMB pavement conditions are revealed with a high resolution compared with the PCI. The performance of the pass lane is superior to the drive lane. To be specific, the PPIs of the down direction pass lane are 100 mostly. By contrast, for the down direction drive lane, the PPIs for about 30% of the segments are below 80 or even below 60.
(6) Based on the divergent XHMB pavement conditions, the performance condition of XHMB pavement in 2021 was divided into four different grades, with varying maintenance treatments required to rehabilitate the pavement performance.
(7) According to the distress quantity of XHMB and the adjacent JTB, the distress of American ChemCo epoxy asphalt concrete may initiate after the concrete has served for 4–5 years. After the concrete was opened to traffic for about 10 years, the distress shows an outbreak trend. At this point, necessary structural maintenance is needed to rehabilitate pavement performance.

Author Contributions: Conceptualization, F.N., Y.H. and Z.Z.; methodology, Y.H., Z.Z. and J.T.; formal analysis, Y.H., Z.Z. and J.T.; investigation, Y.H., Z.Z. and J.T.; resources, F.N.; writing—original draft preparation, Y.H. and Z.Z.; writing—review and editing, Y.H. and Z.Z.; supervision, F.N. and X.G.; funding acquisition, F.N. and X.G. All authors have read and agreed to the published version of the manuscript.

Funding: This work was supported by the National Natural Science Foundation of China (52078131), the science and technology project of Anhui Communications Holding Group Co., Ltd. (JKKJ-2019-06), and the Postgraduate Research and Practice Innovation Program of Jiangsu Province (KYCX21_0134).

Institutional Review Board Statement: Not applicable.

Informed Consent Statement: Not applicable.

Data Availability Statement: All data are included in the article.

Conflicts of Interest: The authors declare no conflict of interest.

eferences

Salamak, M.; Fross, K. Bridges in urban planning and architectural culture. *Procedia Eng.* **2016**, *161*, 207–212. [CrossRef]

Corticelli, R.; Pazzini, M.; Mazzoli, C.; Lantieri, C.; Ferrante, A.; Vignali, V. Urban Regeneration and Soft Mobility: The Case Study of the Rimini Canal Port in Italy. *Sustainability* **2022**, *14*, 14529. [CrossRef]

Zhao, Y.; Ni, F.; Zhou, L.; Jiang, J. Performance evaluation of long-span suspension bridge pavement based on long-term maintenance data. *J. Mater. Civ. Eng.* **2020**, *32*, 04019363. [CrossRef]

Jia, X.; Huang, B.; Chen, S.; Shi, D. Comparative investigation into field performance of steel bridge deck asphalt overlay systems. *KSCE J. Civ. Eng.* **2016**, *20*, 2755–2764. [CrossRef]

Liu, X.; Medani, T.; Scarpas, A.; Huurman, M.; Molenaar, A. Experimental and numerical characterization of a membrane material for orthotropic steel deck bridges: Part 2. *Finite Elem. Anal. Des.* **2008**, *44*, 580–594. [CrossRef]

Zhang, X. Emergency and Long-term Measures of Steel Deck Pavement Distresses at BRT Station. *J. Fail. Anal. Prev.* **2017**, *17*, 1081–1089. [CrossRef]

Bocci, E.; Canestrari, F. Analysis of structural compatibility at interface between asphalt concrete pavements and orthotropic steel deck surfaces. *Transp. Res. Rec.* **2012**, *2293*, 1–7. [CrossRef]

Medani, T.O. Design Principles of Surfacings on Orthotropic Steel Bridge Decks. Ph.D. Thesis, Delft University of Technology, Delft, The Netherlands, 2006.

Farreras-Alcover, I.; Chryssanthopoulos, M.K.; Andersen, J.E. Data-based models for fatigue reliability of orthotropic steel bridge decks based on temperature, traffic and strain monitoring. *Int. J. Fatigue* **2017**, *95*, 104–119. [CrossRef]

Chen, X.; Qian, Z.; Liu, X.; Lei, Z. State of the art of asphalt surfacings on long-spanned orthotropic steel decks in China. *J. Test. Eval.* **2012**, *40*, 1252–1259. [CrossRef]

Han, Y.; Cui, B.; Tian, J.; Ding, J.; Ni, F.; Lu, D. Evaluating the effects of styrene-butadiene rubber (SBR) and polyphosphoric acid (PPA) on asphalt adhesion performance. *Constr. Build. Mater.* **2022**, *321*, 126028. [CrossRef]

Jia, X.; Huang, B.; Bowers, B.F.; Rutherford, T.E. Investigation of tack coat failure in orthotropic steel bridge deck overlay: Survey, analysis, and evaluation. *Transp. Res. Rec.* **2014**, *2444*, 28–37. [CrossRef]

Luo, S.; Lu, Q.; Qian, Z.; Wang, H.; Huang, Y. Laboratory investigation and numerical simulation of the rutting performance of double-layer surfacing structure for steel bridge decks. *Constr. Build. Mater.* **2017**, *144*, 178–187. [CrossRef]

Han, Y.; Jiang, J.; Ni, F.; Dong, Q.; Zhao, X. Effect of cohesive and adhesive parameters on the moisture resistance of thin friction course (TFC) with varying mix design parameters. *Constr. Build. Mater.* **2020**, *258*, 119420. [CrossRef]

Han, Y.; Jiang, J.; Ding, J.; Zhao, Z.; Ma, X.; Ni, F. Investigation of the raveling potential of thin friction course (TFC) under freeze–thaw conditions. *Mater. Struct.* **2021**, *54*, 39. [CrossRef]

Han, Y.; Ding, J.; Han, D.; Zhao, Z.; Ma, X.; Ni, F. Evaluating the thermal aging-induced raveling potential of thin friction course (TFC). *Constr. Build. Mater.* **2022**, *321*, 126160. [CrossRef]

Pauli, A.T.; Huang, S.-C. Relationship between asphalt compatibility, flow properties, and oxidative aging. *Int. J. Pavement Res. Technol.* **2013**, *6*, 1–7.

Huang, W.; Qian, Z.; Chen, G.; Yang, J. Epoxy asphalt concrete paving on the deck of long-span steel bridges. *Chin. Sci. Bull.* **2003**, *48*, 2391–2394. [CrossRef]

Tian, J.; Luo, S.; Lu, Q.; Liu, S. Effects of Epoxy Resin Content on Properties of Hot Mixing Epoxy Asphalt Binders. *J. Mater. Civ. Eng.* **2022**, *34*, 04022145. [CrossRef]

Chen, C.; Eisenhut, W.O.; Lau, K.; Buss, A.; Bors, J. Performance characteristics of epoxy asphalt paving material for thin orthotropic steel plate decks. *Int. J. Pavement Eng.* **2020**, *21*, 397–407. [CrossRef]

Luo, S.; Lu, Q. Condition survey and analysis of first epoxy asphalt concrete pavement on orthotropic bridges in China. A ten-year review. *J. Southeast Univ.* **2011**, *27*, 417–422.

Liu, Y.; Qian, Z.; Shi, X.; Zhang, Y.; Ren, H. Developing cold-mixed epoxy resin-based ultra-thin antiskid surface layer for steel bridge deck pavement. *Constr. Build. Mater.* **2021**, *291*, 123366. [CrossRef]

Zhang, Z.; Li, J.; Wang, Z.; Long, S.; Jiang, S.; Liu, G. Preparation and performance characterization of a novel high-performance epoxy resin modified reactive liquid asphalt. *Constr. Build. Mater.* **2020**, *263*, 120113. [CrossRef]

Zhao, Y.; Jiang, L.; Zhou, L. Improvement of bending fatigue test for asphalt surfacing on orthotropic steel bridge deck. *Adv. Civ. Eng. Mater.* **2016**, *5*, 325–336. [CrossRef]

Zhao, Y.; Jiang, J.; Ni, F.; Zhou, L. Fatigue cracking resistance of engineered cementitious composites (ecc) under working condition of orthotropic steel bridge decks pavement. *Appl. Sci.* **2019**, *9*, 3577. [CrossRef]

Park, H.M.; Choi, J.Y.; Lee, H.J.; Hwang, E.Y. Performance evaluation of a high durability asphalt binder and a high durability asphalt mixture for bridge deck pavements. *Constr. Build. Mater.* **2009**, *23*, 219–225. [CrossRef]

Kim, T.W.; Baek, J.; Lee, H.J.; Lee, S.Y. Effect of pavement design parameters on the behaviour of orthotropic steel bridge deck pavements under traffic loading. *Int. J. Pavement Eng.* **2014**, *15*, 471–482. [CrossRef]

28. Yin, C.; Zhang, H.; Pan, Y. Cracking mechanism and repair techniques of epoxy asphalt on steel bridge deck pavement. *Trans Res. Rec.* **2016**, *2550*, 123–130. [CrossRef]
29. Ding, J.; Luo, S. A sixteen-year review of condition survey and analysis of steel deck pavement on Jiangyin Yangtze River Brid *Mater. Sci. Forum* **2016**, *873*, 91–95. [CrossRef]
30. Dong, Q.; Chen, X.; Dong, S.; Ni, F. Data Analysis in Pavement Engineering: An Overview. *IEEE Trans. Intell. Transp. Syst.* **202** *23*, 22020–22039. [CrossRef]
31. Yao, L.; Leng, Z.; Jiang, J.; Ni, F. Modelling of pavement performance evolution considering uncertainty and interpretability: machine learning based framework. *Int. J. Pavement Eng.* **2022**, *23*, 5211–5226. [CrossRef]
32. Li, H.; Jiang, J.; Ni, F. Factors Affecting Maintenance Probability and Resurfacing Thickness Based on the Pavement Manageme System. *Adv. Civ. Eng.* **2020**, *2020*, 8817081. [CrossRef]
33. Yao, L.; Dong, Q.; Jiang, J.; Ni, F. Deep reinforcement learning for long-term pavement maintenance planning. *Comput.-Aided C Infrastruct. Eng.* **2020**, *35*, 1230–1245. [CrossRef]
34. Yao, L.; Leng, Z.; Jiang, J.; Ni, F. Large-Scale Maintenance and Rehabilitation Optimization for Multi-Lane Highway Aspha Pavement: A Reinforcement Learning Approach. *IEEE Trans. Intell. Transp. Syst.* **2022**, *23*, 22094–22105. [CrossRef]
35. Yao, L.; Dong, Q.; Ni, F.; Jiang, J.; Lu, X.; Du, Y. Effectiveness and cost-effectiveness evaluation of pavement treatments usir life-cycle cost analysis. *J. Transp. Eng. Part B Pavements* **2019**, *145*, 04019006. [CrossRef]
36. Hoang, V.H.; Nguyen, Q.T.; Tran, A.T.; Tran, T.C.H.; Do, T.A. Mechanical behavior of the asphalt wearing surface on an orthotrop steel bridge deck under cyclic loading. *Case Stud. Constr. Mater.* **2022**, *16*, e00836. [CrossRef]

coatings

MDPI

rticle

nvestigation of the Wet and Thermal Conditions Effect on the Micro-Scale Characteristics of Interfacial Transition Zone of Porous Asphalt Mixture

ong Zhao [1], Xiang Ma [2,*], Hao Wang [1] and Chaolin Zhang [2]

1 Zhong Jiao Jian Ji Jiao Expressway Investment Development Co., Ltd., Shijiazhuang 050000, China
2 College of Civil Engineering, Nanjing Forestry University, Nanjing 210037, China
* Correspondence: max@njfu.edu.cn

Abstract: Complex environmental factors can significantly influence the micro-properties of porous asphalt (PA) mixture. Therefore, the effects of short-term and long-term thermal aging and moisture immersion on the micromechanics properties, micro-morphology, and chemical element distributions of the interfacial transition zone (ITZ) of PA mixture were studied to reveal the mechanisms by which environmental degradation impact performance by means of nanoindentation (NI), backscattering scanning electron microscope (BSEM), and energy-dispersive X-ray spectroscopy (EDS) analysis techniques. The results show that the ITZ is not the softest part of the PA mixture, and the order of modulus is aggregate > ITZ > asphalt mastic. ITZ thickness is about 10–20 μm. Thermal aging has negligible effect on the width of ITZ. With increasing aging degree, the asphalt mastic and ITZ modulus increase, while water damage reduces the modulus of ITZ and slightly increases the width of ITZ. Thermal aging has little effect on the morphology of ITZ, while water damage will lead to microcracks and micropores in ITZ. Compared with thermal aging, water damage has a greater effect on the morphology of ITZ and leads to softening of the interfacial transition zone and asphalt mastic phase. The width of ITZ obtained by EDS line scanning is basically consistent with that of nanoindentation. Overall, external environmental factors have a more significant effect on the mechanical properties than the micro-morphology of ITZ. The outcomes of this research provide a better understanding of the impact of the service environment on the microscopic characteristics of PA mixture.

Keywords: porous asphalt mixture; interfacial transition zone; nanoindentation; backscatter scanning electron microscope; energy-dispersive X-ray spectroscopy analysis techniques; micro-scale characteristics

1. Introduction

Porous asphalt mixture (PA) has the advantages of drainage, noise reduction and water mist reduction due to its high void ratio (18%–22%) [1,2]. However, compared with dense-graded mixture, PA mixture with a high void ratio is more easily damaged by media such as sunlight, air and water in the environment [3–6], having a negative impact on the asphalt–aggregate microscopic interface, thus leading to a decline in the macro performance of the mixture [7]. It is necessary to analyze the effects of the external environment on the performance of PA mixture at the micro level, so as to gain insight with the aim of improving its structural durability.

In the process of the mixing and compaction of PA mixture, a series of physical, chemical and mechanical actions between asphalt and filler occur around the surface of the aggregate, thus forming a narrow area around the aggregate, which is called the interfacial transition zone (ITZ) [8]. The microstructure affects the macroscopic characteristics of PA [9–11], and ITZ is regarded as a weak area because of its relatively low strength [12]. The ITZ between the aggregate and the asphalt mastic has an important effect on the macroscopic mechanical behavior of the PA mixture, because the generation and development of damage in the PA mixture mainly occur in these low-strength zones [13–15]. In addition, studies have shown that during concrete freezing and thawing, ITZ exhibits higher deformation than asphalt mastic and aggregates due to its higher porosity and weaker strength [16]. Determining the properties of the interfacial transition zone (ITZ) between asphalt mastic and aggregate is helpful for understanding some of the mechanisms of asphalt pavement fracture behavior [17]. It is difficult to study ITZ using traditional experimental methods due to the scale of ITZ being at the micro level. In recent years, with the development of nanoindentation technology, it has become possible to use nanoindentation (NI) to directly measure the width and nano-mechanical properties of ITZ [18,19]. In addition, backscatter scanning electron microscopy (BSEM) is an effective tool for visualizing ITZ [20], and is generally used in conjunction with energy-dispersive X-ray spectroscopy (EDS) to qualitatively reveal the chemical characteristics of ITZ [21,22]. According to NI analysis, Huang [23] found that the thickness and elastic modulus of ITZ varied with different components in the recycled concrete aggregate, and ITZ adjacent to the porous components (cement mastic and brick) had greater thickness and lower elastic modulus, and verified the accuracy of the thickness measured by nanoindentation by BSEM and EDS. Zhu [13] also used NI to study the interfacial transition zone between different aggregates and asphalt mastic. It was found that the width of ITZ was 5–20 μm, and the accuracy of NI test results was verified by EDS.

To date, many studies have successfully been performed characterizing the interfacial transition zone of asphalt concrete mixture using different microscopic testing methods, but few studies have addressed the effect of wet and thermal aging on the interfacial transition zone of PA mixture. Therefore, determining the properties of the interfacial transition zone of PA mixture and the influence of external environmental factors is helpful for revealing the influencing mechanism of PA mixture durability [24].

The main objective of this study is to investigate the effect of the wet and thermal conditions on the micro-scale characteristics of ITZ of porous asphalt mixture. To achieve this objective, PA samples were prepared and subjected to different environmental factors, and NI, BSEM and EDS were used to study the micro-mechanics, micro-morphology and chemical properties of the interfacial transition zone between asphalt mastic and aggregate in PA mixture.

2. Materials and Test Methods

2.1. Material Properties and Gradation Design

The raw materials of the PA mixture include SBS-modified asphalt, basalt aggregate, and limestone mineral powder, and their technical properties are able to meet the requirements of Technical Specification for Construction of Highway Asphalt Pavement (JTG F40-2004). The detailed technical properties of asphalt are shown in Table 1. During the mixing process of the PA mixture, high-viscosity modifier is added, with a mixing amount of 8% of asphalt mass [25], in order to improve the durability of PAC. The gradation composition of the mixture is shown in Table 2, and the optimal asphalt–stone ratio is 4.6%.

Table 1. Basic technical properties of SBS-modified asphalt.

Technical Properties	Detection Result
Penetration (25 °C, 100 g, 5 s)/0.1 mm	55.5
Ductility (5 °C, 5 cm/min)/cm	37.5
Softening point/°C	78.5
Dynamic viscosity (60 °C, Pa·s)	4841

Table 2. Mixture gradation.

Mesh Size (mm)	16	13.2	9.5	4.75	2.36	1.18	0.6	0.3	0.15	0.075
Pass rate (%)	100	92.2	60.7	20.5	14.1	11.3	7.6	6	5.3	4.6

2.2. Nanoindentation Test

2.2.1. Sample Preparation

Firstly, the method of accelerated aging under high temperature with forced ventilation found in the Standard Test Methods of Bitumen and Bituminous Mixtures for Highway Engineering (JTG E20-2011) was used to prepare the original, short-term aging and long-term aging gyratory compaction samples of PAC-13. The preparation process of that nanoindentation sample is shown in Figure 1, and the specific preparation method is as follows:

(1) Cutting. Cut a 10-mm-thick slice in the vertical middle of the mixture, and then cut a 10 mm × 10 mm × 10 mm cube of the mixture from the middle of the slice as shown in Figure 1a. To reduce the effect of cutting on the internal structure, the sample was cured in low temperature condition at −20 °C for 24 h before cutting.

(2) Curing of epoxy resin. The nanoindentation specimen requires high flatness, so the surface of the specimen needs to be polished. In order to facilitate polishing and prevent the specimen from tilting during the test, the mixture cube block needs to be embedded in the epoxy resin. First, put the mixture test block into a silica gel mold with a diameter of 20 mm, then mix the epoxy resin and curing agent in a ratio of 3:1, and pour them into the silica gel mold after the bubbles basically disappear as shown in Figure 1b.

(3) Grind and polish. After the resin is cured, use a metallographic grinder to polish the surface of the test sample as shown in Figure 1c, remove the epoxy resin wrapped on the surface, and expose the test surface. The test sample should be polished step by step with 180-, 400-, 800-, 1200-, 1600- and 2000-mesh abrasive paper for 5 min for each grade of abrasive paper. After polishing, spray diamond spray polishing agent on the surface of the test sample, and then further polish on polishing cloth for 20 min. Finally, the sample was cleaned in an ultrasonic cleaner for 3 min to remove dust, diamond particles, and the like. Put it into a sample storage box and air dry it for later use as shown in Figure 1d.

(4) Cure preservation in water bath. In order to study the influence of water on the interfacial transition zone between asphalt and aggregate, the samples of unaged PA mixture prepared according to the above steps need to be cured in water bath, and the curing condition is to soak in water at 60 °C for 1 day. After air drying, it can be used for testing.

After the above preparation, four kinds of sample were obtained under the action of external environment, including: as-received, short-term-aged, long-term-aged and high-temperature water-bath-treated specimens.

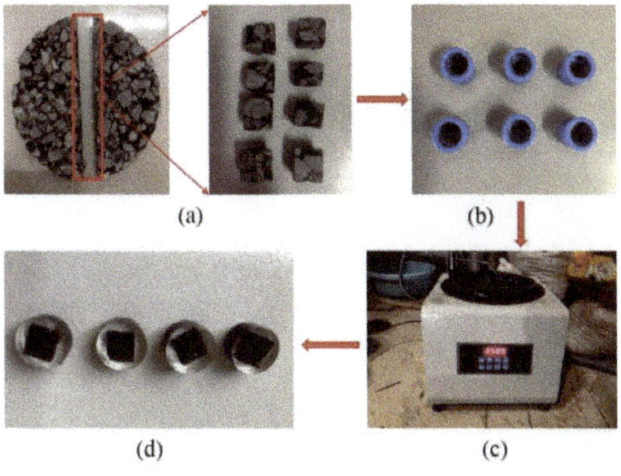

Figure 1. Preparation process of nanoindentation sample: (**a**) cutting, (**b**) curing, (**c**) grind and polish, (**d**) prepared sample.

2.2.2. Test Method

In this paper, a micro nanoindentation instrument produced by Nano Force Company of America (Nanomechanics, Inc., Oak Ridge, TN, USA.) is used to conduct the nanoindentation test. As shown in Figure 2, a Berkovich indenter, which is more suitable for asphalt materials, is used as the indenter. Firstly, the area of interest (AOI) of asphalt and aggregate is found by transmission electron microscope, as shown in Figure 3, and 10×4 grid points (column interval 25 μm, row interval 10 μm) are arranged on AOI as nanoindentation test points. The influences of residual stress and deformation on the next measuring point are eliminated. In accordance with the relevant research results [22], the nanoindentation parameters in this paper were a loading rate of 0.01 mN/s, a maximum load of 0.3 mN, a full load time of 200 s, and the same load control mode was selected for asphalt, asphalt–aggregate interfacial transition zone and aggregate.

(**a**) Appearance (**b**) Sample table

Figure 2. Nanoindentation equipment.

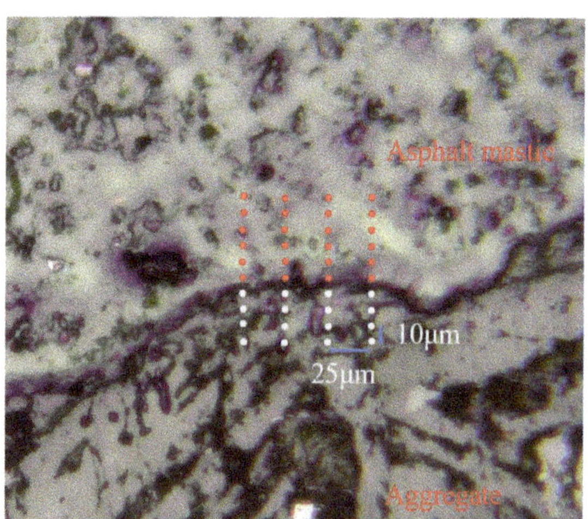

Figure 3. Nanoindentation grid setting. (Red points are on asphalt mastic and white points are on aggregate).

2.2.3. Calculation Principle

A typical load–displacement curve obtained from a nanoindentation experiment is shown in Figure 4, which consists of three stages: a loading phase, a full load phase and an unloading phase. The Young's modulus can be calculated from the unloading curve:

$$E_r = \frac{1}{\beta} \cdot \frac{S}{2} \cdot \frac{\sqrt{\pi}}{\sqrt{A_c}} \tag{1}$$

In the formula, S is the contact stiffness, obtained from the initial slope of the unloading curve, β is the geometric correction factor, where $\beta = 1.034$ for the Berkovich indenter used in this study, A_c is the contact area at peak maximum load, which is related to the contact depth h_c at the half angle of the Berkovich indenter as shown in Equation (2), where $\theta = 65.27°$. P_{max} is the maximum load, h_{max} is the maximum indentation depth, ε is a geometric constant, $\varepsilon = 0.75$.

$$A_c = 3\sqrt{3}h_c^2 tan^2\theta \tag{2}$$

$$h_c = h_{max} - \varepsilon\frac{P_{max}}{S} \tag{3}$$

E_r is the composite modulus of the indenter and the sample. For the independent analysis of the test material, the composite modulus is converted to the elastic modulus. The elastic modulus of the sample can be obtained from Equation (5).

$$\frac{1}{E_r} = \frac{1 - v_s^2}{E_s} + \frac{1 - v_i^2}{E_i} \tag{4}$$

$$E_s = \left(1 - v_s^2\right) \cdot \left[\frac{1}{E_r} - \frac{1 - v_i^2}{E_i}\right]^{-1} \tag{5}$$

In the formula, ν_s represents the Poisson's ratio of the sample, ν_i represents the Poisson's ratio of the indenter, the elastic modulus of the indenter used in this study is taken as $E_i = 1140$ Gpa, $\nu_i = 0.07$, the Poisson's ratio ν_s of the sample has little influence on the calculation result of the elastic modulus, and $\nu_s = 0.3$ [13] is adopted in this study.

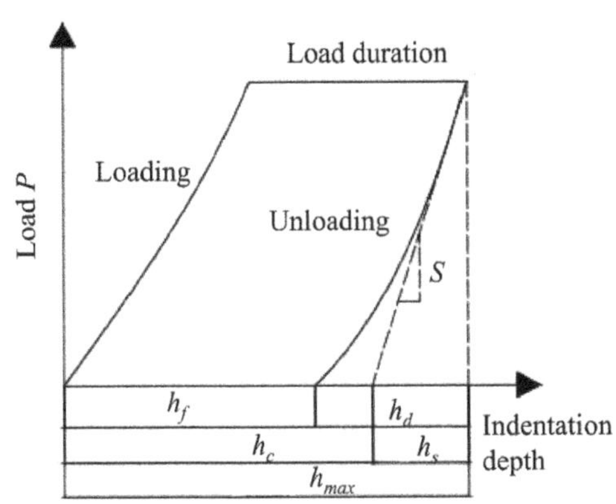

Figure 4. A typical load–displacement curve of nanoindentation. (*hf*: remaining indentation depth).

2.3. *BSEM and EDS Tests*

The morphology of the interfacial transition zone was obtained by scanning electron microscopy (SEM, Hitachi Limited, Tokyo, Japan) in backscattered electron (BSE) mode. The sample preparation method is similar to that of the NI sample. After the epoxy resin is cast, it is cut into cylinders with a height of 5 mm. Since the BSEM test does not require surface flatness, no subsequent grinding is required. The pixel brightness of the BSE image corresponds to the atomic number of the measured phase below the sample surface. Therefore, the voids are the darkest, followed by asphalt and aggregate, so that the BSE image has a clear distinction between asphalt and aggregate. It should be noted that, due to the non-conductivity or poor conductivity of the sample, in order to avoid the impact of the "charging effect", the sample needs to be sprayed with gold before testing.

The elemental composition and distribution of the sample can be characterized using energy-dispersive X-ray spectroscopy (EDS) detection (Oxford Instruments Nanotechnology Tools Limited, Oxford, UK). In this study, EDS test was carried out with BSE images to obtain the basic mapping of the interfacial transition zone. There are three kinds of EDS test methods: point scanning, line scanning and surface scanning. In this study, surface scanning was used to obtain the distribution of elements in the interfacial transition zone, and then line scanning was used to study the thickness of the interfacial transition zone between asphalt and aggregate with the change in element concentration.

3. Test Results and Analysis of ITZ

3.1. Micromechanical Properties

Taking the original sample as an example, the test section is near the ITZ, as shown in Figure 5. The elastic modulus of each effective indentation was calculated, and a two-dimensional modulus contour map of the test area was drawn, as shown in Figure 6. The asphalt mastic phase is a blue area with the lowest modulus; the red area is the aggregate phase with the highest modulus; the green area in the middle is the asphalt mastic aggregate interfacial transition zone (ITZ), whose modulus is between that of the asphalt mastic and the aggregate. Compared with asphalt mastic, the mechanical properties of ITZ in PA mixture are closer to those of aggregate [26,27]. This may be due to the combination of asphalt and filler (the size is within the scope of nanometers or no more than a few microns) to fill the micro surface pores of the aggregate and penetrate into the aggregate, forming a mechanical interlock. In addition, due to the chemical bonding between asphalt and aggregate and electrostatic force, a harder and stronger asphalt stone composite structure, namely the interfacial transition zone (ITZ), is formed around the aggregate surface than asphalt mastic. It can be seen from the two-dimensional equivalent modulus diagram that the width of the ITZ is roughly between 10–20 μm.

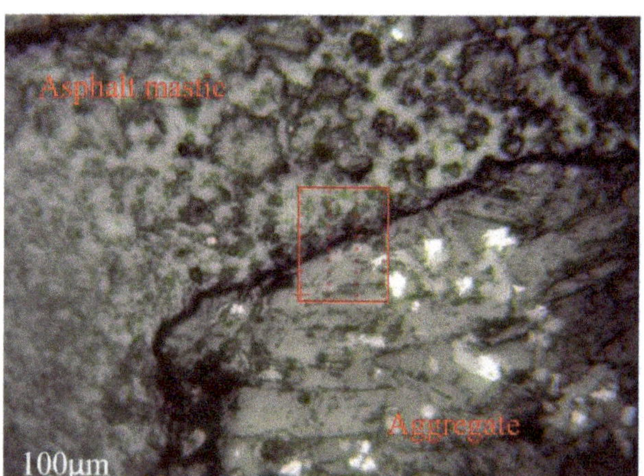

Figure 5. Test section.

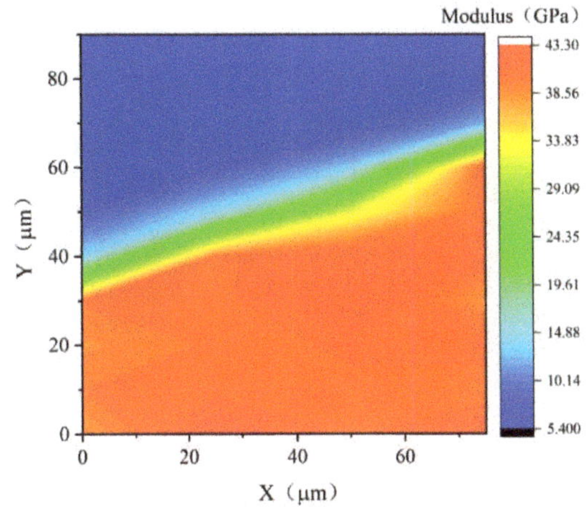

Figure 6. Equivalent modulus map of the test section.

The elastic modulus is used to determine the thickness and mechanical properties of the interfacial transition zone. In the change trend chart of the elastic modulus, the "sudden drop" zone in the middle is generally identified as the interfacial transition zone, from which the thickness of the interfacial transition zone can be determined. Analyzing 40 data points consisting of 10 rows and four columns, the average value of each row of data was counted, and the curve graph of modulus was drawn, as shown in Figure 7. The width of ITZ before and after the aging of the PA mixture is similar, at about 18 μm. In addition, the width of ITZ increased slightly to about 21 μm after water immersion, which may be caused by the loosening of ITZ caused by water damage. The average value of modulus of each phase was calculated, as shown in Table 3. With increasing aging degree, the modulus of asphalt mastic phase and ITZ increased, but this had little effect on the modulus of aggregate phase, which was due to the increase in asphalt and hardening of asphalt caused by aging. This shows that aging has little effect on ITZ width, mainly with respect to asphalt mastic phase and ITZ mechanical properties. Contrary to the aging effect, the modulus of asphalt mastic phase and ITZ decreased after the action of water. It can be seen that the damage of water will lead to the softening of the interfacial transition zone and asphalt mastic.

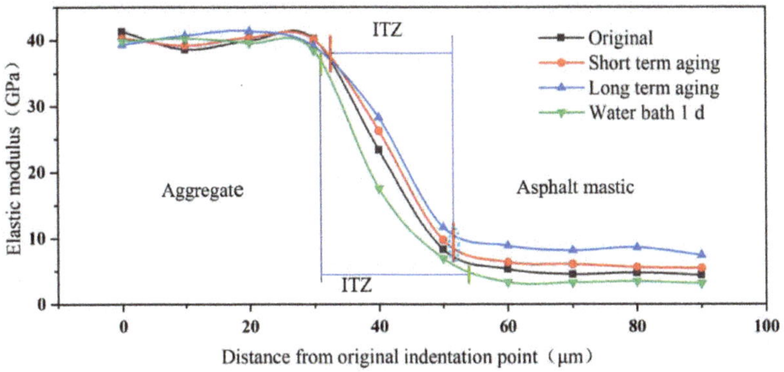

Figure 7. The trend of average elastic modulus.

Table 3. Average modulus of each phase under different working conditions.

Project	Original			Short-Term Aging			Long-Term Aging			60 °C Water Bath 1 d		
	AM	ITZ	BA	AM	ITZ	BA	AM	ITZ	BA	AM	ITZ	BA
Modulus (GPa)	4.72	15.81	40.00	5.82	17.46	39.81	8.21	20.36	40.12	3.28	12.21	39.54

Note: AM stands for asphalt mastic; BA stands for basalt.

3.2. Micro Morphology Characteristics

In this study, the BSEM test is used to calibrate and photograph the interfacial transition zone of samples under four working conditions (original sample, short-term aging, long-term aging and immersion). As shown in Figure 8, the interface between the interfacial transition zone and the aggregate is obvious, but there is no obvious interface between the interfacial transition zone and the asphalt mastic. There is no obvious difference in the interfacial transition zone between the original sample and the short-term aging sample. The interfacial transition zone and asphalt of asphalt mastic phase of long-term aging samples have wrinkle phenomenon and rough surface. This phenomenon is caused by the long-term aging of asphalt, the degradation of butadiene, and the destruction of long-chain polymer to form a shorter SBS polymer structure, which makes it impossible to fully surround filler particles, and the volatilization of light components reduces the lubricity of asphalt, which leads to a little microcracks can be seen in the ITZ. For the sample immersed in water bath at 60 °C for 1 h, the morphology of the sample is obviously different from that of the original sample, which is mainly reflected in the appearance of microspores in the asphalt mastic phase and the interfacial transition zone (ITZ), as shown by the yellow area, and the appearance of micro cracks in the ITZ. The asphalt mastic phase is separated from the aggregate phase, as shown by the blue area. In addition, it is observed that the asphalt in the ITZ decreases and becomes thinner, and more filler surfaces are exposed. The main reason for the above phenomenon is that the erosion of water forms a water film on the surface of the aggregate, resulting in an asphalt–water–aggregate system. The accumulation of water leads to the enhancement of the isolation effect of the water molecular film on the adhesion of the asphalt–aggregate [28], which reduces the adhesion properties of the asphalt to the aggregate, resulting in the stripping of the asphalt mastic phase and the aggregate phase. It can be seen that, compared with aging, water damage has a greater impact on the morphology of the interfacial transition zone.

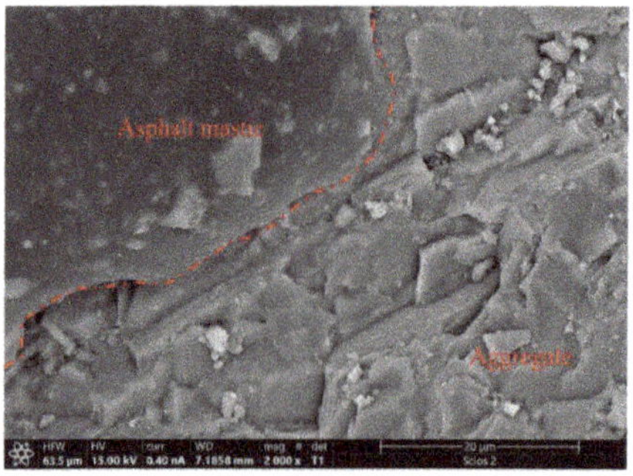

(**a**) Original

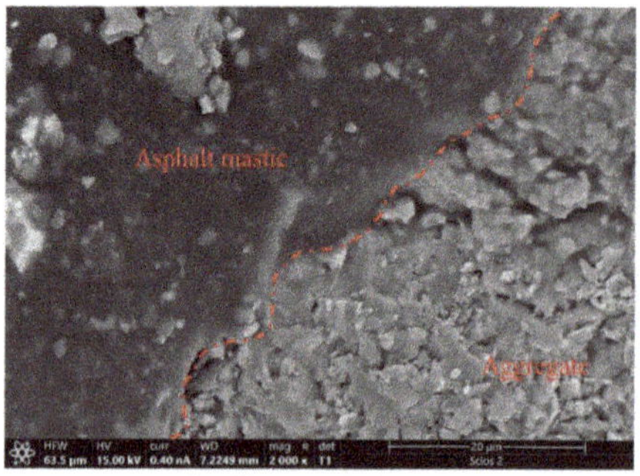

(**b**) Short term aging

Figure 8. *Cont.*

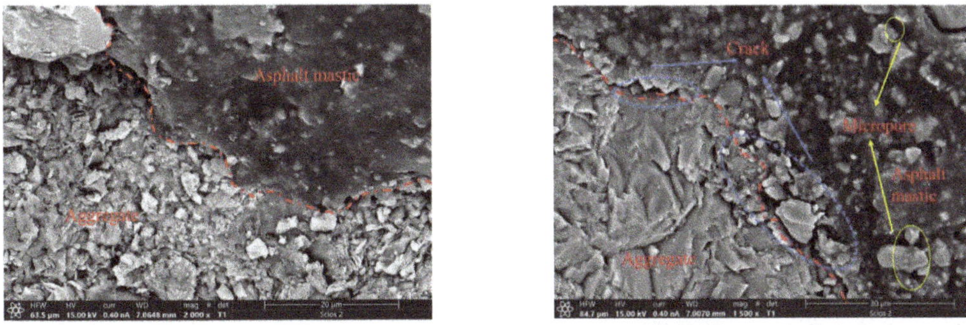

(**c**) Long-term aging (**d**) 60 °C water bath for 1 d

Figure 8. BSE images of the interfacial transition zone using SEM.

3.3. Chemical Element Distribution

Due to the possible mismatch between nanomechanical properties and inhomogeneities, the determination of ITZ thickness based on nanoindentation test results alone may not be sufficient. To obtain more accurate thicknesses, ITZ in PA mixtures was identified using elemental analysis. Figure 9 is an element distribution diagram of the interfacial transition region under different working conditions. The brightness of the color represents the content of the element in the element mapping image. The brighter the color, the higher the content. It can be seen from Figure 9 that the asphalt mastic phase has the phenomenon of high calcium and low silicon, which is mainly due to the fact that the filler used in this study is limestone powder (with high calcium carbonate content), so the phenomenon of high calcium and low silicon occurs. The aggregate is basalt (with high silica content), so the aggregate phase is characterized by high silica and low calcium. Aging has no obvious effect on this phenomenon, but the effect of high temperature water bath can increase the content of calcium element in asphalt mastic phase, which also indicates that the effect of water will damage asphalt, reduce the adhesion between asphalt and filler, and expose the surface of filler, thus increasing the content of calcium element in mapping image, which is also consistent with the phenomenon observed in BSE image. After long-term aging, the brightness of O element in asphalt mastic increases slightly and the brightness of C element decreases, which indicates that the content of O element increases and the content of C element decreases after long-term aging, which is caused by asphalt aging, while short-term aging has little effect.

(**a**) Original (**b**) Short Term Aging (**c**) Long Term Aging (**d**) 60 °C Water Bath 1 d

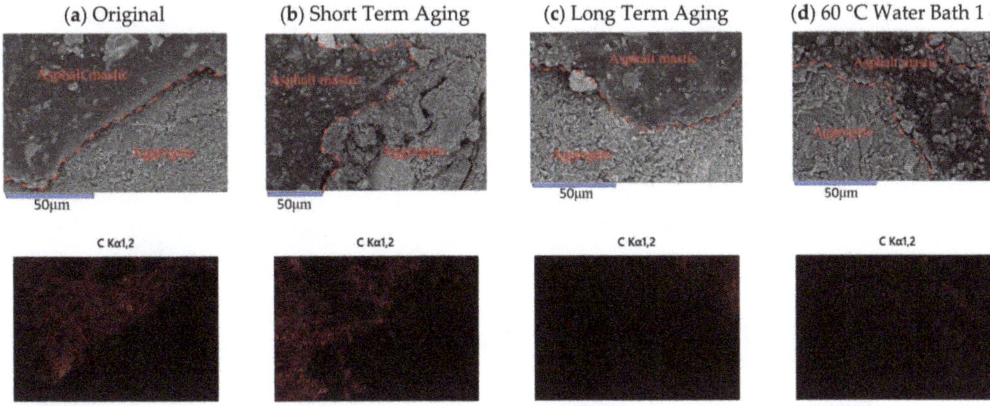

Figure 9. *Cont.*

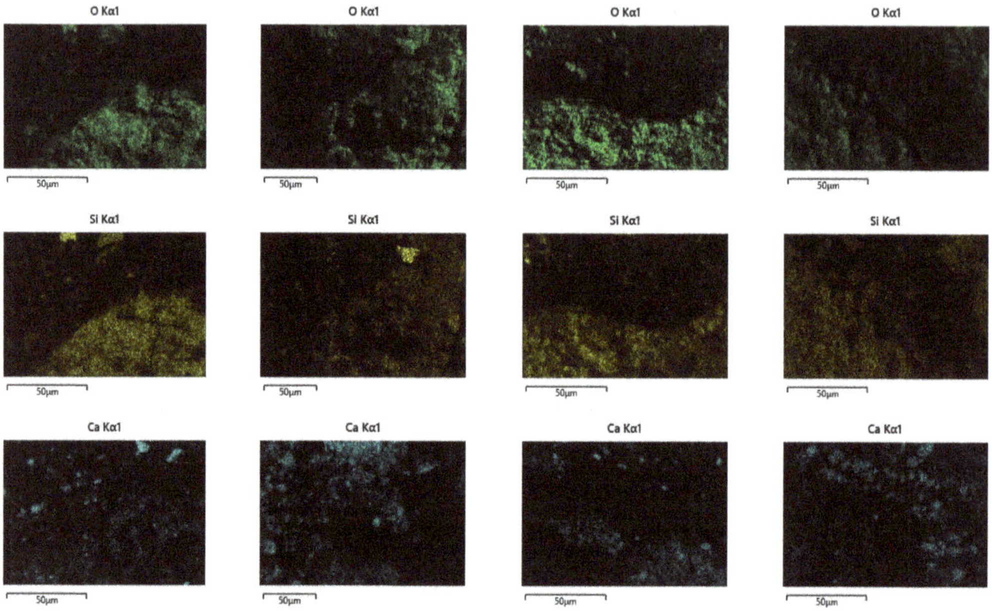

Figure 9. EDS elemental maps of interfacial transition zones.

However, the mapping image can only demonstrate the element distribution, and cannot effectively identify the thickness of the interfacial transition zone. In this study, the length of the interfacial transition zone is 70 μm, in order to determine the thickness of the interfacial transition zone according to the changes in the number of elements.

Figure 10 shows the change in the quantity of main elements (C, Ca, Si and O) along the line of samples under different working conditions. Since each sample is made of the same material and the aggregate is basalt (with high silicon dioxide content), the Si element in the aggregate phase fluctuates at a high level, gradually decreasing when it reaches the interfacial transition zone, and then fluctuates at a low level in the asphalt mastic phase, and the O element shows the same trend. The trend for Ca and C is the opposite. Because the filler used is limestone (with high calcium carbonate content), the amount of Ca element fluctuates at a low level in the aggregate phase gradually increases in the interfacial transition zone, and then fluctuates at a higher level in the asphalt mastic phase. It can be seen from Figure 10 that the number of elements in the ITZ is between that of aggregate and asphalt mastic. According to the change in element content, the width of the ITZ is about 20 μm for the original sample, short-term aging sample and long-term aging sample, and about 23 μm for the sample immersed in water at 60 °C for 1 d. The above results are basically consistent with those obtained by nanoindentation. Therefore, the external environmental factors have little influence on the ITZ width, mainly on its mechanical properties.

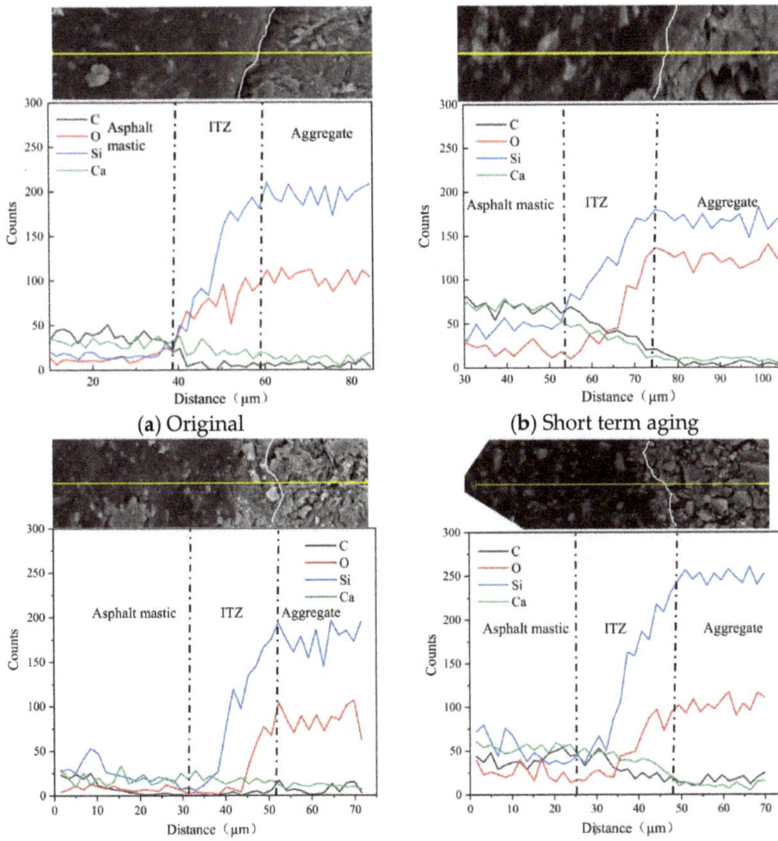

(a) Original

(b) Short term aging

(c) Long term aging

(d) 60 °C water bath for 1 d

Figure 10. EDS line scanning results across aggregates, ITZ and asphalt mastic.

4. Conclusions and Findings

The microscopic mechanical properties, microscopic morphology characteristics and chemical element distribution of the interfacial transition zone of PA mixture were studied on the basis of nanoindentation, BSEM and EDS tests, and the influence of wet and thermal environments on the microscopic properties of the interfacial transition zone was analyzed. The following conclusions and findings can be drawn:

(1) Nanoindentation grid arrangement is able to effectively reflect the three-phase performance distribution law of asphalt mastic-ITZ-aggregate. The interfacial transition zone is not the softest part of the PA mixture, and the modulus can be ordered as aggregate > ITZ > asphalt mastic. According to the equivalent modulus diagram, it is found that the thickness of ITZ is roughly between 10 and 20 μm, which is consistent with previous research conclusions.

(2) Aging has no effect on the width of ITZ, mainly affecting its mechanical properties. With increasing degree of aging, the modulus of asphalt mastic and ITZ increased. Water damage slightly increased the width of ITZ, which may be due to the loosening of ITZ caused by water, and the decreasing modulus of asphalt mastic and ITZ caused by water.

(3) There was no obvious difference between the BSE images of the interfacial transition zone of the original sample and the short-term aging pattern. The interfacial transition zone of the long-term aging sample and the asphalt of the asphalt mastic phase have wrinkles and rough surfaces; micropores and microcracks appear in the asphalt mastic phase and

the interfacial transition zone of the water-immersed sample, and the erosion of water reduces the adhesion performance between asphalt and aggregate, resulting in the peeling of the asphalt mastic phase and aggregate phase. In addition, the high-temperature water bath caused asphalt loss of ITZ. Compared with aging, water damage has a greater impact on the morphology of the interfacial transition zone.

(4) The amount of elements in the interfacial transition zone is between the aggregate and the asphalt mastic, and the width of the interfacial transition zone can be well characterized by line scanning. The width of the interface transition zone of samples for the original and aged samples was about 20 μm, while that for the water immersed samples was about 23 μm, whereby the results obtained from EDS analysis were almost the same as the results obtained by nanoindentation.

Notably, the thermal and moisture characteristics have obvious influence with modulus and morphology of ITZ, which have been assumed to be the reason for the stripping of PA. Therefore, for future research, an investigation of the relationship between the difference of ITZ and the micro properties, especially the durability of PA, is required.

Author Contributions: Methodology, X.M.; validation, D.Z. and X.M.; formal analysis, C.Z.; investigation, D.Z.; resources, H.W.; data curation, D.Z. and C.Z.; writing—original draft preparation, C.Z.; writing—review and editing, X.M.; visualization, H.W.; supervision, X.M.; funding acquisition, H.W. All authors have read and agreed to the published version of the manuscript.

Funding: This research was funded by the Hebei Provincial Transportation Scientific Research Project (NO. TH1-202021).

Institutional Review Board Statement: Not applicable.

Informed Consent Statement: Not applicable.

Data Availability Statement: The data related to this manuscript can be available on reasonable request from corresponding authors.

Conflicts of Interest: The authors have no conflict of interest regarding this manuscript.

References

1. Ma, X.; Jiang, J.; Zhao, Y.; Wang, H. Characterization of the interconnected pore and its relationship to the directional permeabili of porous asphalt mixture. *Constr. Build. Mater.* **2021**, *269*, 121233. [CrossRef]
2. Ma, X.; Zhou, P.; Wang, L.; Jiang, J.; Wang, J. Internal structure change of porous asphalt concrete under coupled conditions load, moisture and temperature. *Constr. Build. Mater.* **2022**, *314*, 125603. [CrossRef]
3. Yu, H.; Zhu, Z.; Zhang, Z.; Yu, J.; Oeser, M.; Wang, D. Recycling waste packaging tape into bituminous mixtures towar enhanced mechanical properties and environmental benefits. *J. Clean. Prod.* **2019**, *229*, 22–31. [CrossRef]
4. Xiao, F.; Herndon, D.A.; Amirkhanian, S.; He, L. Aggregate gradations on moisture and rutting resistances of open graded frictic course mixtures. *Constr. Build. Mater.* **2015**, *85*, 127–135. [CrossRef]
5. Wang, X.; Gu, X.; Ni, F.; Deng, H.; Dong, Q. Rutting resistance of porous asphalt mixture under coupled conditions of hig temperature and rainfall. *Constr. Build. Mater.* **2018**, *174*, 293–301. [CrossRef]
6. Cui, Y.; Ma, X.; Sun, X. Effects of hydro-thermal coupling on adhesion between asphalt and aggregate. *J. Railw. Sci. Eng.* **2018**, *15*, 614–6
7. Jiang, J.; Leng, Z.; Yang, B.; Lu, G.; Tan, Z.; Han, M.; Dong, Z. Penetration mechanism of the emulsion-based rejuvenat in damaged porous asphalt mixture: Microstructure characterization and 3D reconstruction. *Mater. Des.* **2022**, *221*, 1110 [CrossRef]
8. Hu, J.; Liu, P.; Huang, Q.; Qian, Z.; Luo, S. Research on interfacial zone failure of asphalt mixture mixed with recycled aggregat *Constr. Build. Mater.* **2022**, *319*, 126113. [CrossRef]
9. Król, J.B.; Khan, R.; Collop, A.C. The study of the effect of internal structure on permeability of porous asphalt. *Road Mat Pavement Des.* **2018**, *19*, 935–951. [CrossRef]
10. Asfour, S.; Bernardin, F.; Toussaint, E. Experimental validation of 2D hydrothermal modelling of porous pavement for heatir and solar energy retrieving applications. *Road Mater. Pavement Des.* **2020**, *21*, 666–682. [CrossRef]
11. Hu, J.; Qian, Z.; Liu, P.; Wang, D.; Oeser, M. Investigation on the permeability of porous asphalt concrete based on microstructu analysis. *Int. J. Pavement Eng.* **2020**, *21*, 1683–1693. [CrossRef]
12. Brand, A.S.; Roesler, J.R. Bonding in cementitious materials with asphalt-coated particles: Part I–The interfacial transition zor *Constr. Build. Mater.* **2017**, *130*, 171–181. [CrossRef]
13. Zhu, X.; Yuan, Y.; Li, L.; Du, Y.; Li, F. Identification of interfacial transition zone in asphalt concrete based on nano-scale metrolo techniques. *Mater. Des.* **2017**, *129*, 91–102. [CrossRef]

4. Yi, J. Study on Freeze-Thaw Damage Characteristics of Porous Asphalt Mixtures Based on Interfacial Behaviors. Ph.D. Thesis, Harbin Institute of Technology, Harbin, China, 2012.
5. Yuan, Y.; Zhu, X.; Li, L.; Wang, H. Effect of the interfacial zone on the tensile-damage behavior of an asphalt mixture containing MSWI bottom ash aggregates. *J. Mater. Civ. Eng.* **2017**, *29*, 04016269. [CrossRef]
6. Sicat, E.; Gong, F.; Ueda, T. Experimental investigation of the deformational behavior of the interfacial transition zone (ITZ) in concrete during freezing and thawing cycles. *Constr. Build. Mater.* **2014**, *65*, 122–131. [CrossRef]
7. Wang, L. The Research for Influencing Mechanism of Interface Bonding Property between Asphalt and Aggregate. Master's Thesis, Hefei University of Technology, Hefei, China, 2016.
8. Hosan, A.; Shaikh FU, A.; Sarker, P.; Aslani, F. Nano-and micro-scale characterisation of interfacial transition zone (ITZ) of high volume slag and slag-fly ash blended concretes containing nano SiO_2 and nano $CaCO_3$. *Constr. Build. Mater.* **2021**, *269*, 121311. [CrossRef]
9. Fu, J.; Xiong, D.; Li, Z.; Ding, Q.; Jia, D. Micron scratch test and parameter analysis of aggregate-asphalt mortar interfacial Transition zone. *J. Build. Mater.* **2022**, *1*. Available online: https://kns.cnki.net/kcms/detail/31.1764.TU.20220117.1304.015.html (accessed on 6 February 2023).
0. Hu, H. Study on Performance of Recycled Coarse Aggregate Asphalt Mixture and Microscopic Characteristics of Interface Transition Zone. Master's Thesis, Yangzhou University, Yangzhou, China, 2021.
1. Li, W.; Xiao, J.; Sun, Z.; Kawashima, S.; Shah, S.P. Interfacial transition zones in recycled aggregate concrete with different mixing approaches. *Constr. Build. Mater.* **2012**, *35*, 1045–1055. [CrossRef]
2. Gao, Y.; Zhu, X.; Corr, D.J. Characterization of the interfacial transition zone of CNF-Reinforced cementitious composites. *Cem. Concr. Compos.* **2019**, *99*, 130–139. [CrossRef]
3. Huang, Q.; Qian, Z.; Hu, J.; Zheng, D.; Chen, L.; Zhang, M.; Yu, J. Investigation on the properties of aggregate-mastic interfacial transition zones (ITZs) in asphalt mixture containing recycled concrete aggregate. *Constr. Build. Mater.* **2021**, *269*, 121257. [CrossRef]
4. Xing, S. Research on Durability of Drained Asphalt Pavement Based on Asphalt Aggregate Interface Theory under Dynamic Water Pressure. Master's Thesis, Southeast University, Nanjing, China, 2020.
5. Wang, X.; Gu, X.; Dong, Q.; Wu, J.; Jiang, J. Evaluation of permanent deformation of multilayer porous asphalt courses using an advanced multiply-repeated load test. *Constr. Build. Mater.* **2018**, *160*, 19–29. [CrossRef]
6. Barbhuiya, S.; Caracciolo, B. Characterisation of asphalt concrete using nanoindentation. *Materials* **2017**, *10*, 823. [CrossRef] [PubMed]
7. Karki, P.; Yuya, P.A.; Kim, Y.R.; Turner, J.A. Nanomechanical properties of constituent phases in bituminous mixtures. *J. Mater. Civ. Eng.* **2016**, *28*, 4016090. [CrossRef]
8. Liu, Z. Moisture Diffusion Behavior and Damage Characteristics of Asphalt Mixture within Multi-Scale Domain. Ph.D. Thesis, Harbin Institute of Technology, Harbin, China, 2020.

coatings

MDPI

Article

The Influence of CO_2-Cured Incinerated Waste Fly Ash on the Performance of Reactive Powder Concrete

Jianhu Xu [1], Hui Wang [2,*], Wanzhen Wang [2] and Feiting Shi [2]

1 Zhejiang Institute of Communications, Hangzhou 311112, China
2 School of Civil Engineering and Geographic Environment, Ningbo University, Ningbo 315215, China; shifeiting@ycit.cn (F.S.)
* Correspondence: wanghui4@nbu.edu.cn

Abstract: Incinerated waste fly ash is a toxic solid, which can cause serious harm to the environment. CO_2-cured incinerated waste fly ash may be useful in decreasing the toxicity of waste fly ash and improving the corresponding mechanical properties of cement-based material with incinerated waste fly ash. Meanwhile, this technology can play a certain role in reducing the content of CO_2 in the atmosphere. In this study, the influence of CO_2-cured incinerated waste fly ash on the rheological parameters (the slump flow and plastic viscosity) and the setting time of fresh reactive powder cement concrete (RPC) is investigated. The flexural and compressive strengths of hardened RPC standard cured for 1 day, 3 days, and 28 days are measured. The leached amounts of Cr and Zn immersed in water for 6 months are measured. The scanning electron microscope photos, thermogravimetric analysis curves, and mercury intrusion curves are obtained. Our results show that the slump flow, the setting time, and the flexural and compressive strengths increased, and the plastic viscosity decreased by adding the waste fly ash with the maximum varying rates of 12.1%, 41.7%, 41.3%, and 61.2%, respectively. CO_2 curing on the waste fly ash can increase the setting time and the flexural and compressive strengths with the maximum varying rates of 19.2%, 13.1%, and 14.2%. The effect of CO_2-cured waste fly ash on the mechanical strengths of RPC is quite limited.

Keywords: CO_2-cured; incinerated waste fly ash; rheological parameters; mechanical properties; scanning electron microscope

1. Introduction

Environmental protection, energy conservation, and the rationality of resources have become important topics in today's society [1,2]. However, with the progress of science and technology, the development of the economy, and the continuous improvement of human material demand, a large amount of domestic garbage has been produced. Traditional domestic waste is incinerated and landfilled [3,4]. The incinerated waste fly ash contains a certain amount of toxic metal elements, which may pollute soil and water [5,6]. A lot of efforts have to be made to solve this problem.

The application of incineration waste fly ash (WFA) in production practice provides ideas for treating this solid waste [7]. Mandpe has confirmed that the WFA can be used in the production of fertilizers [8]. As pointed out in some journals, the WFA possesses a large amount of cementious active substances, which may be advantageous to the mechanical performances of cement-based materials. Ferraro has reported that the lightweight aggregates made from municipal solid WFA through single and double-step pelletization processes show better mechanical strengths than the lightweight aggregates made by clay [9]. The volcanic ash effect of WFA can improve the mechanical strength and the durability of cement concrete [10]. Cui et al. [11] has reported that the reasonable addition (5%~25% by mass ratio of total binder materials) of WFA can improve the mechanical strengths of reactive powder cement concrete (RPC). Meanwhile, the corrosion resistance of reinforced RPC is increased by mixing the WFA [12]. These things considered, when the RPC with WFA is cured in higher curing temperatures, the mechanical strengths are significantly improved. Even though the incinerated WFA is proven to improve the mechanical strength and corrosion resistance of RPC or other cement-based materials, the problem of toxic release during use has always puzzled many researchers [13]. Based on this reason, methods to reduce the toxicity of WFA need to be developed.

CO_2 curing on raw materials and cement concrete is proven to improve the corresponding mechanical strength [14]. The long-term performances of alkali-activated cementitious materials, magnesium oxysulfate paste, ceramisite cement concrete, and RPC are improved by CO_2 curing [15–17]. The flexural and compressive strengths of RPC are increased to 113.4%~136.1% and 115.8%~141.3% of the standard cured RPC through the use of CO_2 curing for 24 h with the CO_2 concentration of 8% by volume of the total air [18]. Moreover, CO_2 curing can effectively improve the corrosion resistance of the inner steel bars of RPC [19,20]. When CO_2 curing is provided in the RPC, the mechanical strengths' loss rates are 13.1%~25.6%, after 300 NaCl freeze–thaw cycles, while the mechanical strengths' loss rates are 26.8%~37.1% after 30 NaCl dry–wet alternations. However, when the standard curing is provided, the corresponding strengths' loss rates are 33.6%~45.3% and 41.2%~43.8.1% after 300 NaCl freeze–thaw cycles and 30 NaCl dry-wet alternations [21,22]. CO_2 curing is proven to solidify toxic substances in some solid wastes [23]. If CO_2 curing is used in the waste fly ash, the toxicity is likely to be reduced. The CO_2-cured waste fly ash may improve the mechanical properties and reduce the toxicity of raw materials [24,25]. Meanwhile, the application of CO_2 in the WFA provides another way to consume this hazardous gas. However, few journals have reported on this topic.

In this research, the WFA is cured by CO_2 for 24 h. The slump flow, plastic viscosity, and setting time of fresh RPC paste is combined with the WFA, whose mass ratio by the total mass of mineral admixtures are 0%, 25%, 50%, 75%, and 100%, respectively. The corresponding flexural strength, compressive strength, and the drying shrinkage are measured. The scanning electron microscope photos, thermogravimetric analysis curves, and mercury intrusion curves are applied in revealing the mechanism of macro performance. This study will contribute to reducing the content of hazardous waste, reducing exhaust gas pollution, and improving the environment.

2. Experimental Section

2.1. Raw Materials

The granulated blast furnace slag powder (GGBS) and superfine silica fume (SF) were provided by Lingshou County Qiangdong Mineral Products Processing Factory, Shijiazhuan, China. Additionally, the SF showed a specific surface area of 14.8 m^2/g and a density of 2.2 g/cm^3. The density and the specific surface area of the GGBS were 2.91 g/cm^3 and 436.2 m^2/g, respectively. Incinerated waste fly ash manufactured by the Shanghai waste incineration plant was used. The incinerated waste fly ash was obtained after the incineration process. The incinerated waste fly ash (WFA) was cured in the TH-2 concrete carbonization test box with 8% CO_2, which was provided by Shengshi Huike Testing Equipment Co., Ltd., Shanghai, China. The WFA was raw ash, which was directly used after firing. Some WFA was cured in the TH-2 concrete carbonization test box for 1 day. The ordinary Portland (P·O) cement was offered to us by Wenshui County Sunshine Cement Co., Ltd., Luliang, China. P·O cement showed the strength grade of 42.5 MPa, normal cement. The ultrafine Grade I fly ash. provided by Nanjing Hongqian Environmental Protection Engineering Co., Ltd., Nanjing, China, was used in this study. Lingshou Tuolin (Shijiazhuang, China) mineral products processing plant's quartz sands with the particle sizes of 0.68~1.21 mm, 0.33~0.61 mm and 0.14~0.31 mm were used as aggregate in this study. A polycarboxylic water-reducing agent with a water-reducing rate of 40% was used for adjusting the fresh RPC's flowability. Tables 1 and 2 show the chemical composition and particle size distribution of powder materials, respectively.

Table 1. The particle size distribution of the cementitious materials (%).

Types	Particle Size/μm						
	0.3	0.6	1	4	8	64	360
WFA	0.14	0.51	2.3	17.4	31.2	97.3	100
P·O cement	0.1	0.31	2.6	15.1	28.7	93.5	100
GGBS	0.04	0.12	3.3	19.4	35.2	98.2	100
SF	31.3	58.5	82.4	100	100	100	100
Quartz sand	0	0	0	0	0.04	21	100

Table 2. The chemical composition of the cementitious materials (%).

Types	SiO_2	Al_2O_3	Fe_xO_y	MgO	CaO	SO_3	K_2O	Na_2O	Ti_2O	CdO	Cr_2O_3	PbO	CuO	ZnO	Loss on Ignition
WFA	22.4	4.4	0.9	-	20.2	9.56	6	4.3	10.2	0.07	0.09	0.09	0.09	0.5	21.2
P·O cement	20.8	5.6	3.8	1.8	62.1	2.8	-	-	-	-	-	-	-	-	3.1
GGBS	34	14.9	0.5	9.8	36.9	0.3	3.6	-	-	-	-	-	-	-	-
SF	90.8	0.21	0.62	0.23	0.44	0.2	7.5	-	-	-	-	-	-	-	-
Quartz sand	99.8	-	0.2	-	-	-	-	-	-	-	-	-	-	-	-

2.2. Sample Manufacturing

Table 3 shows the mixing proportions of the RPC specimens. All powder materials were put into UJZ-15 and stirred for 30 s. After that, the uniformly mixed solution with a water-reducing agent and water was added and mixed for another 210 s. Once the mixing was finished, the fresh RPC paste was applied in the measurement of the slump flow, plastic viscosity, and yield shear stress. After the measurement of the fresh RPC paste's working properties, the specimens with size of $40 \times 40 \times 160$ mm^3 were prepared for the determination of flexural strength, compressive strength, and drying shrinkage rate. Meanwhile, specimens with a size of $100 \times 100 \times 100$ mm^3 were applied in the measurement of leaching of the toxic metal ions. The manufacturing process of RPC specimens are shown in Figure 1.

Table 3. Mixture design of RPC per one cubic meter (kg).

Water	P·O Cement	WFA	SF	GGBS	Quartz Sand	Water-Reducer
244.4	740.7	0	370.3	111.1	977.9	16.3
244.4	740.7	92.6	277.7	111.1	977.9	16.3
244.4	740.7	185.2	185.2	111.1	977.9	16.3
244.4	740.7	277.7	92.6	111.1	977.9	16.3
244.4	740.7	370.3	0	111.1	977.9	16.3

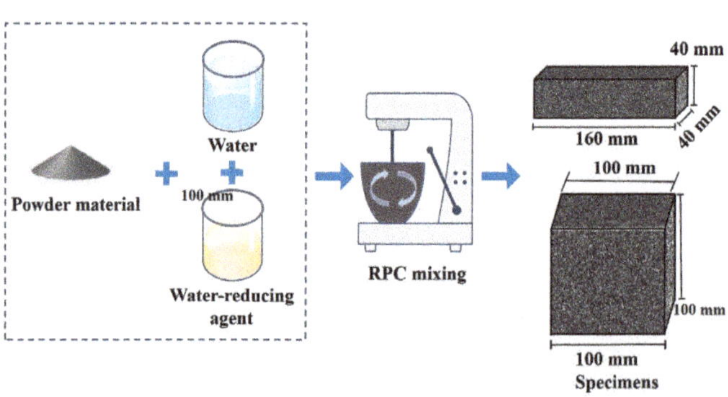

Figure 1. The manufacturing process of RPC specimens.

2.3. *Measurement of Basic Performances*

The jumping table method is used for measuring the slump flow of the fresh RPC paste. The Brookfield rotational rheometer (Beiying Electronic Technology Co., Ltd., Shanghai, China) was used for testing the plastic viscosity and yield shear stress. ZKS-100 mortar setting time tester (North China Road Construction Instrument, Cangzhou, China) was used to obtain the setting time of fresh RPC paste. The measuring process can refer to Chinese standard JGJ/T70-2009 and references [26–28]. The flexural and compressive strengths of specimens cured in the standard curing environment (temperature of 20 ± 2 °C and relative humility of 98.4%) were tested by carbon material bending and a compression machine provided by Jinan Liling Testing Machine Co., Ltd., Jinan, China. The flexural loading speed and the compressive loading speed were 0.05 kN/s and 2.4 kN/s, respectively. The sample manufacturing and measuring processes are described in GB/T 17671-1999 Chinese standard [29]. The shrinkage rod with an ATORN electronic dial indicator manufactured by Henderson Machinery Technology Co., Ltd., Beijing, China, supporting the middle of one end of the rectangular specimen, was applied in the measurement of dry shrinkage rate. When the specimen's length varied, the value of the length change was read out by the dial indicator. Then, the dry shrinkage rate was obtained. Before the leaching measurement of the toxic metal ions was obtained, specimens were immersed in deionized water for 6 months. The deionized water, immersing the specimens each month, was moved to the RIS Intrepid ER/S ICP atomic emission spectrometer for the measurement of toxic metal ions. In this investigation, the average of six specimens' testing parameters were used in each experiment. Figure 2 shows the measuring process of mechanical strength.

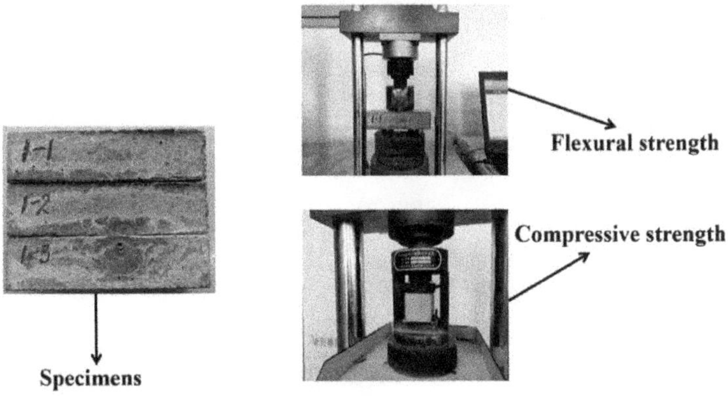

Figure 2. The flexural and compressive strengths of RPC.

2.4. *Thermal Analysis and Scanning Electron Microscope*

The WFA and the harden RPC specimens with incinerated WFA were prepared for the measurement of the thermogravimetric (TG) analysis and the scanning electron microscope (SEM). Before the TG analysis, all samples were ground into powder and moved to the WCR-123 microcomputer differential thermal analyzer (Beiguang Hongyuan Instrument Co., Ltd., Beijing, China) to obtain the thermogravimetric curves. After being sprayed gold, samples were moved to the FEI Quanta200 field emission scanning electron microscope (Shousi Zhixin Technology Co., Ltd., Beijing, China) in order to acquire SEM photos.

2.5. *Mercury Intrusion Analysis*

Mercury intrusion curves were obtained by Antonpa mercury intrusion meter provided by Antonpa (Shanghai) Trading Co., Ltd., Shanghai, China. The diameters of the samples used in the experiment ranged from 3.1 to 5.6 mm. The samples were removed from the cores of the specimens.

3. Results and Discussion

3.1. Rheological Parameters and Setting Time of Fresh RPC

Figure 3 shows the slump flow of the fresh RPC. The slump flow of the fresh RPC increased with the increasing dosages of WFA. This was attributed to the fact that the fineness of fly ash was higher than that of the WFA. The fly ash with a larger specific surface area could absorb more free water. Therefore, the slump flow of fresh RPC with WFA was higher. Moreover, the CO_2-cured WFA could decrease the slump flow of fresh RPC due to the fact that the CO_2 can react with alkaline substances in the WFA, forming carbonates, resulting in the reduction of material porosity and increasing the slump flow [30–32].

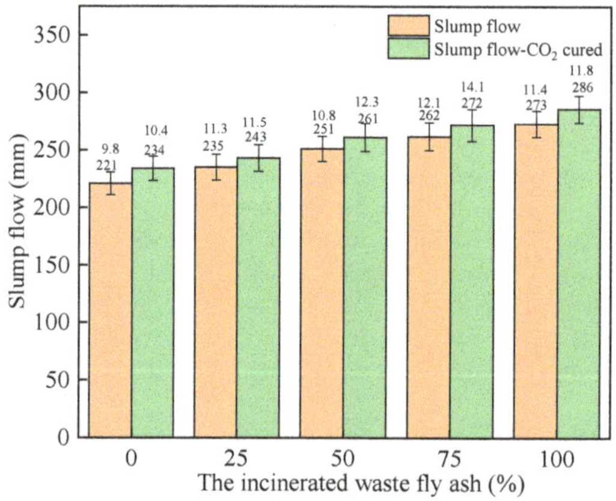

Figure 3. The slump flow of the fresh RPC.

The plastic viscosity of the fresh RPC is shown in Figure 4. The results of the plastic viscosity were the opposite to the fresh RPC's slump flow. The addition of WFA had a reducing effect on the plastic viscosity. Moreover, the CO_2-cured WFA could further decrease the plastic viscosity of fresh RPC. The slump flow of fresh RPC could be increased by WFA with the increasing rates of 10.3%~12.1%. Additionally, the decreasing rates of the plastic viscosity of fresh RPC were 21.7%~47.5%. The increasing rates of slump flow and the decreasing rates of plastic viscosity by CO_2 curing on WFA were 7.8%~11.2% and 13.4%~31.2%.

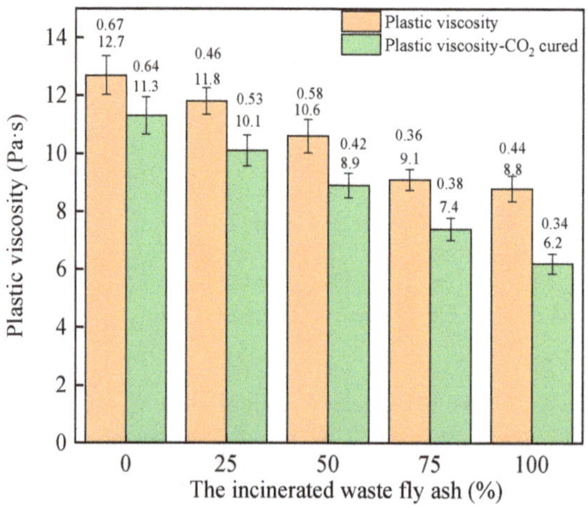

Figure 4. The plastic viscosity of the fresh RPC.

Figure 5 shows the setting time of the fresh RPC. The setting time increased with the increasing dosage of WFA. This was attributed to the fact that the fly ash could absorb a higher dosage of free water than that of WFA, due to the fact that a higher dosage of free water can increase the distance between cement particles and water [33–35]. Therefore, more time was taken to form a skeleton structure, which led to an increase in the setting time. Moreover, as observed in Figure 5, the CO_2-cured WFA can increase the setting time of fresh RPC. This is because CO_2 curing can reduce the pores of WFA, resulting in a higher amount of free water and eventually leading to an increase in the setting time of fresh RPC [36]. The increased rates of setting time, caused by the increased dosage of WFA and the CO_2 curing on WFA were 16.3%~41.7% and 11.3%~19.2%, respectively.

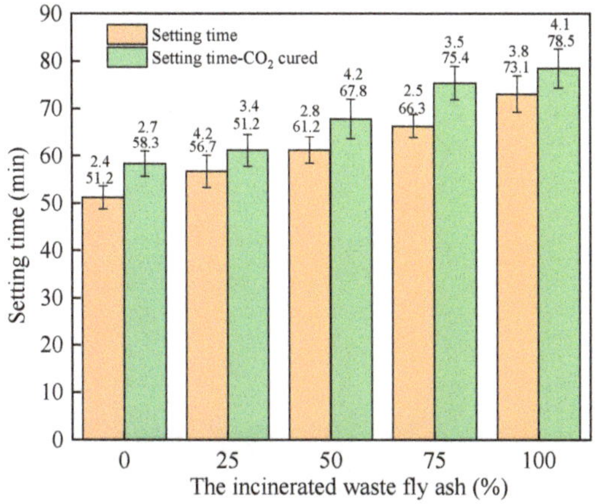

Figure 5. The setting time of the fresh RPC paste.

3.2. Mechanical Performance

The flexural and compressive strengths of RPC with WFA are shown in Figure 6a,b. All specimens were cured for 1 day, 3 days, and 28 days. The flexural and compressive strengths of RPC increased with the curing time, due to the increased hydration degree. The mechanical strengths of RPC were increased by the addition of WFA, which is because the active substances in WFA can promote the secondary hydration of cement, leading to an increase in the hydration degree and the amount and compactness of hydration products, resulting in higher mechanical strength [37,38]. The specimens cured for 1 day show 35.1%~41.3% flexural strengths of the specimens cured for 28 days. Meanwhile, the compressive strengths of specimens cured for 1 day are 48.6%~61.2% of the specimens cured for 28 days.

The mechanical strengths of RPC with CO_2-treated WFA are shown in Figure 6c,d. The flexural strength and compressive strength were increased by adding more CO_2-treated WFA. This was ascribed to the improved secondary hydration of cement by CO_2-treated incinerated WFA [39,40]. The specimens cured for 1 day show 28.3%~39.2% flexural strengths of the specimens cured for 28 days. Meanwhile, the compressive strengths of specimens cured for 1 day are 43.4%~57.1% of the specimens cured for 28 days. Comparing Figure 6c,d with Figure 6a,b, the flexural and compressive strengths of RPC with conventionally treated WFA are 93.1%~97.6% and 91.2%~96.5% of the RPC with CO_2-treated incinerated waste fly ash. This can be ascribed to the forming of carbonate by CO_2 curing, which can reduce the pores of inner incinerated waste fly ash, thus improving the mechanical strengths of RPC [41]. Generally, the increasing effect of CO_2 is inapparent. This may be ascribed to the fact that CO_2 curing can decrease the pores of raw materials, which can improve the mechanical strength [42]. However, the CO_2 curing can reduce the activity of fly ash and garbage fly ash, resulting in strength reduction.

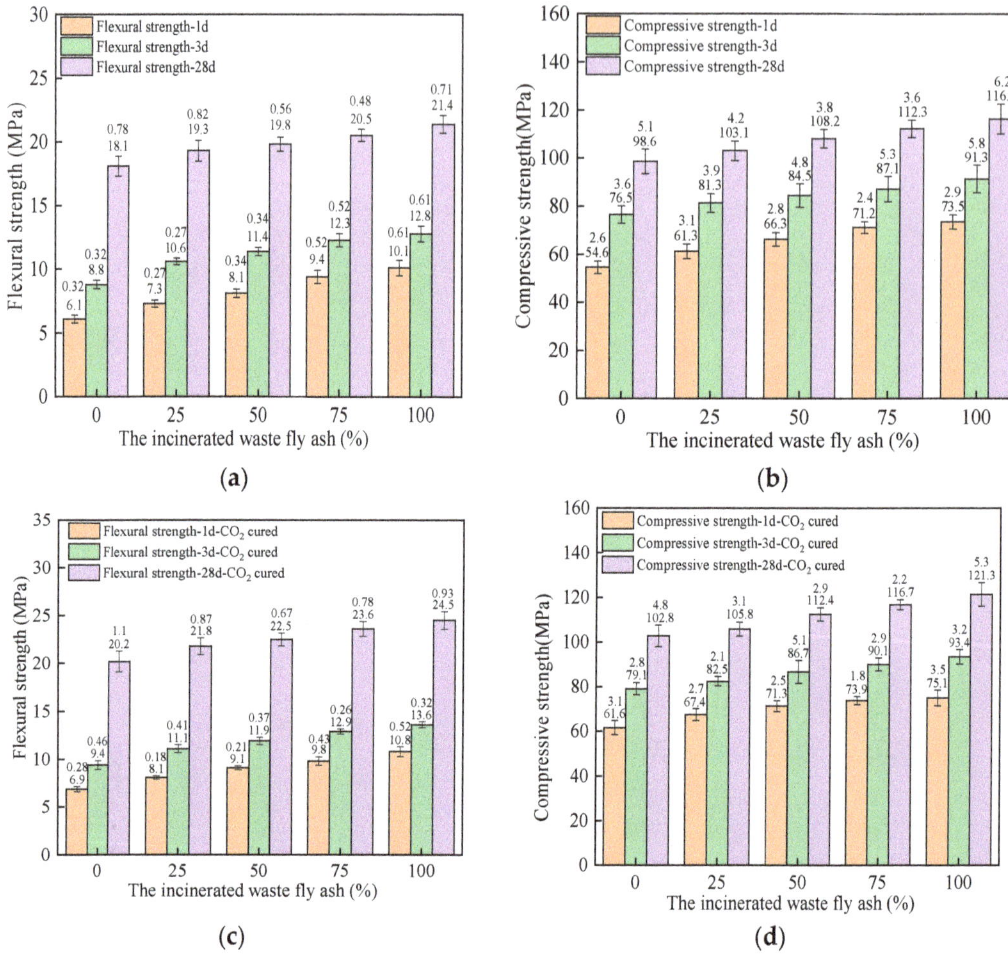

Figure 6. The mechanical strengths of RPC with conventionally treated incinerated waste fly ash. (**a**) Flexural strength. (**b**) Compressive strength. (**c**) Flexural strength. (**d**) Compressive strength.

The increasing rates of mechanical strengths are shown in Figures 7 and 8. As observed in Figures 7 and 8, the increasing rates of mechanical strengths increase with the increasing dosages of WFA and decrease with increased curing age. This is because the WFA can improve the mechanical strengths of RPC. However, the WFA can increase the hydration degree of cement in RPC and the improvement in earlier curing age is higher than that in later curing age [43,44]. Additionally, the increasing rates of mechanical strengths of RPC with CO_2-cured WFA are lower than those of RPC with conventionally WFA, due to the decreased activity of WFA by CO_2 curing [45]. The increasing rates of flexural and compressive strengths of RPC by WFA are 6.6%~65.6% and 4.6%~34.6%.

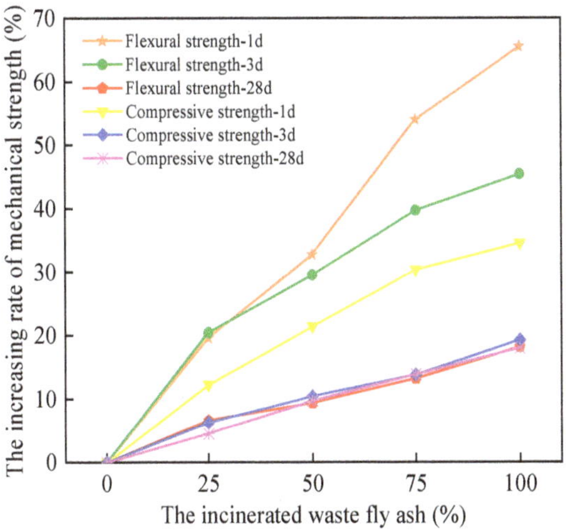

Figure 7. The increasing rates of mechanical strengths of RPC with conventionally treated incinerated waste fly ash.

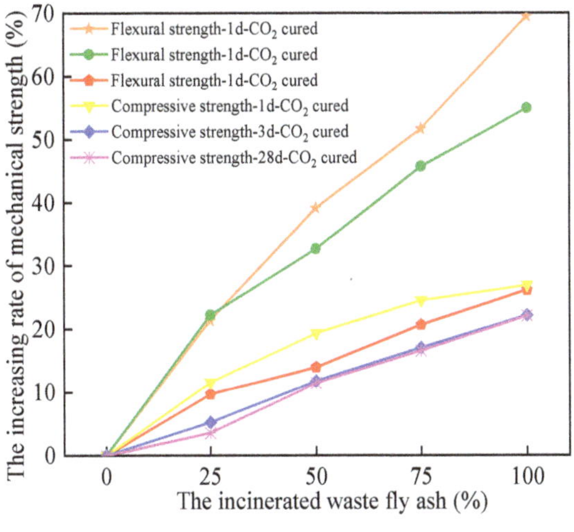

Figure 8. The increasing rates of mechanical strengths of RPC with CO_2 treated incinerated waste fly ash.

3.3. *The Leaching of Toxic Metal Elements*

Figure 9 shows the leaching amount of toxic metal elements (Cr and Zn). The leaching amount of toxic metal elements increased with the immersing time and the addition of WFA. The CO_2 curing can decrease the leaching amount of toxic metal elements by 76.1% and 71.3%. This is because the WFA contains some toxic metal elements (Cr and Zn), which, when immersed in the water for several months, the toxic metal elements seepage through pores of RPC, thus increasing the content of toxic metal elements [46,47], while the CO_2 curing can decrease the amount of leached toxic metal elements. Comparing Figure 9a with Figure 9b, the amount of leached Zn is higher than that of leached Cr.

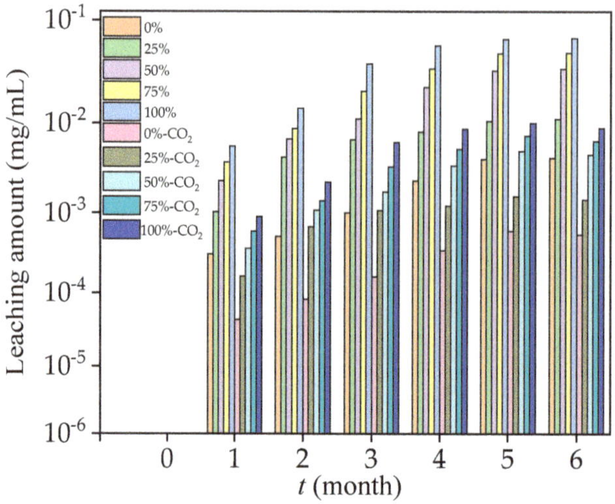

(**a**)

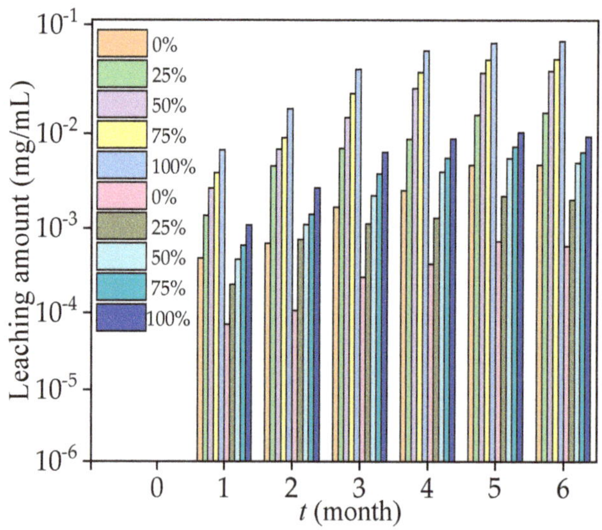

(**b**)

Figure 9. The leaching amount of toxic metal elements. (**a**) The leaching amount of Cr. (**b**) The leaching amount of Zn.

3.4. Microscopic Analysis

Figure 10 shows the thermogravimetric analysis curves of WFA. Moreover, some WFA is CO_2-cured. As observed in Figure 10, the TG decreases in three steps. In the first step (32.03~104.02 °C), the TG obviously decreases. The WFA can absorb some free water in the air. The first reduction of TG is attributed to the evaporation of free water. Meanwhile, as the temperature increases from 104.02 to 581.3 °C, the TG of WFA decreases. The CaO in the incinerated waste fly ash can react with the water in the air, forming $Ca(OH)_2$. The increasing temperature can accelerate the decomposition of $Ca(OH)_2$. Finally, when the temperature ranges from 581.3 to 712.4 °C, the TG continues to reduce due to the resolved $CaCO_3$, formed by the reaction of $Ca(OH)_2$ and CO_2.

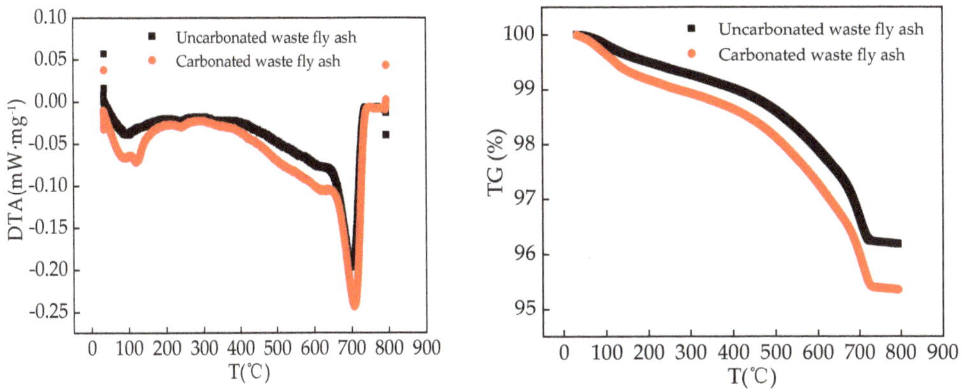

Figure 10. Thermogravimetric analysis curves of incinerated waste fly ash particles.

The SEM photos of WFA particles are shown in Figure 11. As observed from Figure 11, the uncarbonated WFA particle shows smooth particle. However, the carbonated WFA particle shows coarse particles. This is attributed to the formed $CaCO_3$ via the reaction of CaO and CO_2. The $CaCO_3$ absorbs on the surface of the WFA particle, which can prevent the leakage of harmful substances from the WFA.

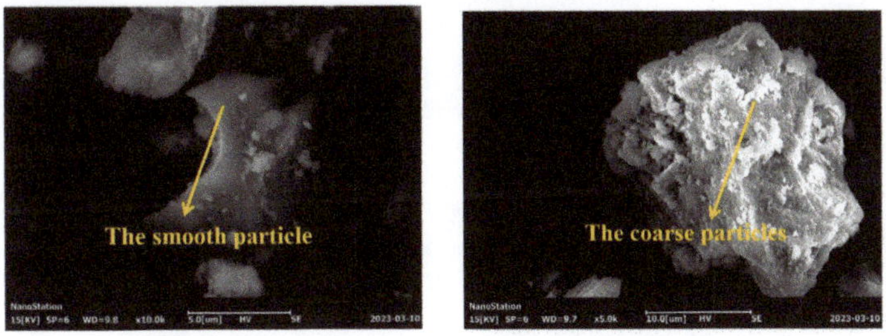

(a) **(b)**

Figure 11. The SEM photos of incinerated waste fly ash particles. (**a**) The incinerated waste fly ash particle (**b**) The carbonated incinerated waste fly ash particle.

The SEM photos of RPC with WFA are shown in Figure 12. As observed in Figure 12a, the flocculent hydration product and crack are found in Figure 12a. As shown in Figure 12b, the addition of carbonated WFA has improved the compactness of hydration products, which indicates that the CO_2-cured WFA can increase the mechanical strength by improving the hydration products' compactness. As obtained from Figures 11 and 12, more nanoscale calcium carbonate is generated on the surface of the fly ash after carbonization. This part of calcium carbonate can adsorb cement ions to promote the early hydration of cement and improve early strength. Moreover, the calcium carbonate on the surface of WFA can effectively strengthen the interface between WFA and cement, reducing the formation of cracks. The addition of carbonized WFA reduces the content of rod-shaped hydration products and promotes the transformation of rod-shaped products into dense products. Therefore, the mechanical strengths of RPC are increased by adding the CO_2-cured WFA.

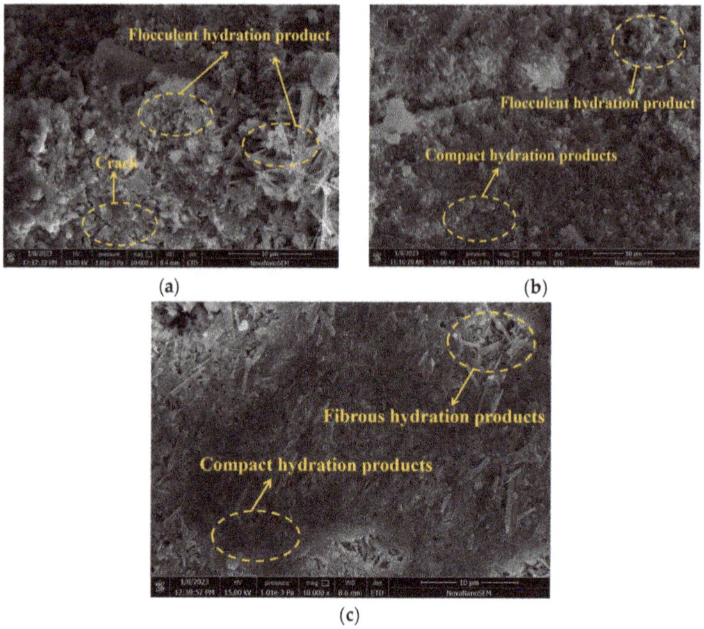

Figure 12. The SEM photos of RPC with incinerated waste fly ash. (**a**) RPC with incinerated waste fly ash (**b**) RPC with 50% carbonated incinerated waste fly ash. (**c**) RPC with 100% carbonated incinerated waste fly ash.

The mercury intrusion curve of RPC with WFA are shown in Figure 13. As illustrated in Figure 13, the main pore diameter of the RPC with WFA ranges from 1.37~4.11 $\times$ 10^5 nm. It can be found in Figure 13, that the increase in mass ratio of CO_2-cured WFA leads to a decrease in the total volume of pores (The curve areas can be used to characterize the total volume of pores). Additionally, the increasing dosages of CO_2-cured WFA decrease the volume ratio of pores with a large diameter. These results indicate that CO_2-cured WFA can improve the mechanical strength of RPC.

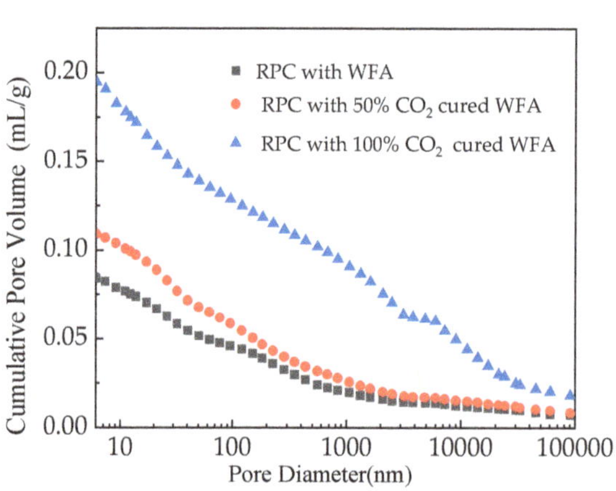

Figure 13. Relationship between pore size and mercury intake.

4. Conclusions

In this study, the influence of incinerated waste fly ash and the corresponding CO_2 curing on the rheological parameters and setting time of fresh RPC are systematically researched. The mechanical properties and intrinsic mechanism are studied. The conclusions can be summarized as follows:

The addition of incinerated waste fly ash can increase the slump flow of fresh RPC with the increasing rates of 10.3%~12.1%. The increasing content of incinerated waste fly ash can lead to decreasing the plastic viscosity with the decreasing rates of 21.7%~47.5%. The mechanical strengths of RPC are improved by adding the incinerated waste fly ash. The increasing rates of incinerated waste fly ash on the mechanical strengths of RPC is decreased by the curing age. The increasing rates of flexural and compressive strengths of RPC caused by incinerated waste fly ash are 6.6%~65.6% and 4.6%~34.6%.

The addition of CO_2-cured incinerated waste fly ash can increase the slump flow and decrease the plastic viscosity with the varying rates of 7.8%~11.2% and 13.4%~31.2%. The addition of incinerated waste fly ash and the effect of CO_2 curing on incinerated waste fly ash can increase the setting time of fresh RPC with the varying rates of 16.3%~41.7% and 11.3%~19.2%. The CO_2-cured incinerated waste fly ash can further improve the mechanical strengths of RPC by decreasing the pore volumes and diameters.

The leaching toxic metal elements (Cr and Zn) are increased with the immersing time, while the amount of Zn is higher than that of the leached Cr. CO_2 curing can decrease the leaching amount of toxic metal elements by 76.1% and 71.3%.

Author Contributions: Conceptualization, H.W. and J.X.; methodology, W.W.; software, H.W.; validation, H.W., J.X. and W.W.; formal analysis, F.S.; investigation, W.W.; resources, H.W.; data curation, J.X.; writing—original draft preparation, H.W.; writing—review and editing, H.W.; visualization, W.W.; supervision, F.S.; project administration, W.W.; funding acquisition, W.W. All authors have read and agreed to the published version of the manuscript.

Funding: The Project of Science and Technology for Public Welfare of Ningbo City (Grant No. 2022S179) and Zhejiang Provincial Natural Science Foundation [LY22E080005].

Institutional Review Board Statement: Not applicable.

Informed Consent Statement: Not applicable.

Data Availability Statement: Not applicable.

Conflicts of Interest: The authors declare no conflict of interest.

References

Avila-López, U.; Almanza-Robles, J.M.; Escalante-García, J. Investigation of novel waste glass and limestone binders using statistical methods. *Constr. Build. Mater.* **2015**, *82*, 296–303. [CrossRef]

Chen, L.; Wang, Y.; Wang, L.; Zhang, Y.; Li, J.; Tong, L.; Hu, Q.; Dai, J.-G.; Tsang, D.C.W. Stabilisation/solidifification of municipal solid waste incineration fly ash by phosphate-enhanced calcium aluminate cement. *J. Hazard. Mater.* **2021**, *408*, 124404. [CrossRef] [PubMed]

Ren, J.; Hu, L.; Dong, Z.; Tang, L.; Xing, F.; Liu, J. Effect of silica fume on the mechanical property and hydration characteristic of alkali-activated municipal solid waste incinerator (MSWI) fly ash. *J. Clean. Prod.* **2021**, *295*, 126317. [CrossRef]

Tian, X.; Rao, F.; Morales-Estrella, R.; Song, S. Effects of aluminum dosage on gel formation and heavy metal immobilization in alkali-activated municipal solid waste incineration fly ash. *Energy Fuels* **2020**, *34*, 4727–4733. [CrossRef]

Zhang, Y.; Wang, L.; Chen, L.; Ma, B.; Zhang, Y.; Ni, W.; Tsang, D. Treatment of municipal solid waste incineration flfly ash: State-of-the-art technologies and future perspectives. *J. Hazard. Mater.* **2021**, *411*, 125132. [CrossRef] [PubMed]

Alderete, N.; Joseph, A.; Van den Heede, P.; Matthys, S.; De Belie, N. Effective and sustainable use of municipal solid waste incineration bottom ash in concrete regarding strength and durability. *Resour. Conserv. Recycl.* **2021**, *167*, 105356. [CrossRef]

Xing, J.; Tang, Q.; Gan, M.; Ji, Z.; Fan, X.; Sun, Z.; Chen, X. Co-treatment of municipal solid waste incineration fly ash (MSWI FA) and municipal sludge: A innovative method to improve sludge dewatering with fly ash dichlorination. *J. Environ. Manag.* **2023**, *332*, 117403. [CrossRef]

Mandpe, A.; Yadav, N.; Paliya, S.; Tyagi, L.; Yadav, B.R.; Singh, L.; Kumar, S.; Kumar, R. Exploring the synergic effect of fly ash and garbage enzymes on biotransformation of organic wastes in in-vessel composting system. *Bioresour. Technol.* **2021**, *322*, 124557. [CrossRef]

9. Ferraro, A.; Ducman, V.; Colangelo, F.; Korat, L.; Spasiano, D.; Farina, I. Production and characterization of lightweight aggregate from municipal solid waste incineration fly-ash through single and double-step pelletization process. *J. Clean. Prod.* **202** *383*, 135275. [CrossRef]
10. Ramesh Kumar, G.; Bharani, S.; Anupriya, R.; Jeevitha, V.; Gnanajothi, G. Influence of electronic waste and fly ash in strength an durability properties of concrete. *Mater. Today Proc.* **2022**, *62*, 2230–2234. [CrossRef]
11. Cui, L.; Wang, H. Influence of waste fly ash on the rheological properties of fresh cement paste and the following electrica performances and mechanical strengths of hardened specimens. *Coatings* **2021**, *11*, 1558. [CrossRef]
12. Cui, L.; Wang, H. Research on the mechanical strengths and the following corrosion resistance of inner steel bars of RPC with ric husk ash and waste fly ash. *Coatings* **2021**, *11*, 1480. [CrossRef]
13. Du, Y.; Hao, W.; Shi, F.; Wang, H.; Xu, F.; Du, T. Investigations of the mechanical properties and durability of reactive powde concrete containing waste fly ash. *Buildings* **2022**, *12*, 560. [CrossRef]
14. Witoon, T.; Numpilai, T.; Nijpanich, S.; Chanlek, N.; Kidkhunthod, P.; Cheng, C.K.; Ng, K.; Vo, D.-V.N.; Ittisanronnachai, S Wattanakit, C.; et al. Enhanced CO_2 hydrogenation to higher alcohols over K-Co promoted In_2O_3 catalysts. *Chem. Eng. J.* **202** *431*, 133211. [CrossRef]
15. Sharma, R.; Kim, H.; Pei, J.; Jang, J.G. Dimensional stability of belite-rich cement subject to early carbonation curing. *J. Build. Eng* **2023**, *63*, 105545. [CrossRef]
16. Xu, F.; Chang, R.; Zhang, D.; Liang, Z.; Wang, K.; Wang, H. Improvement of CO_2-cured sludge ceramsite on the mechanica performances and corrosion resistance of cement concrete. *Materials* **2022**, *15*, 5758. [CrossRef] [PubMed]
17. Peng, L.; Yang, J.; Wang, H.; Jin, X. The influence of CO_2 Curing on the mechanical performance and the corresponding chlorid ion resistance of alkali-activated compound mineral admixtures. *Coatings* **2022**, *12*, 1920. [CrossRef]
18. Khan, R.I.; Ashraf, W.; Olek, J. Amino acids as performance-controlling additives in carbonation-activated cementitious material *Cement Concr. Res.* **2021**, *147*, 106501. [CrossRef]
19. Siddique, S.; Naqi, A.; Jang, J.G. Influence of water to cement ratio on CO_2 uptake capacity of belite-rich cement upon exposur to carbonation curing. *Cem. Concr. Compos.* **2020**, *111*, 103616. [CrossRef]
20. Liang, J.; Zhu, H.; Chen, L.; Qinglin, X.; Ying, G. Rebar corrosion investigation in rubber aggregate concrete via the chlorid electro-accelerated test. *Materials* **2019**, *12*, 862. [CrossRef]
21. Ogawa, Y.; Bui, P.T.; Kawai, K.; Sato, R. Effects of porous ceramic roof tile waste aggregate on strength development an carbonation resistance of steam-cured fly ash concrete. *Construct. Build. Mater.* **2020**, *236*, 117462. [CrossRef]
22. Khan, R.I.; Haque, M.I.; Ashraf, W.; Shah, S.; Saleh, N. Role of biopolymers in enhancing multiscale characteristics of carbonatio cured cementitious composites. *Cem. Concr. Compos.* **2022**, *134*, 104766. [CrossRef]
23. Sarmiento, L.E.M.; Clavier, K.A.; Townsend, T.G. Trace element release from combustion ash co-disposed with municipal soli waste. *Chemosphere* **2020**, *252*, 126436. [CrossRef] [PubMed]
24. Song, Z.; Zhang, X.; Tan, Y.; Zeng, Q.; Hua, Y.; Wu, X.; Li, M.; Liu, X.; Luo, M. An all-in-one strategy for municipal solid wast incineration fly ash full resource utilization by heat treatment with added kaolin. *J. Environ. Manag.* **2023**, *329*, 117074. [CrossRef
25. Lv, Y.; Yang, L.; Wang, J.; Zhan, B.; Xi, Z.; Qin, Y.; Liao, D. Performance of ultra-high-performance concrete incorporatin municipal solid waste incineration fly ash. *Case Stud. Constr. Mater.* **2022**, *17*, e01155. [CrossRef]
26. Wong, G.; Fan, X.; Gan, M.; Ji, Z.; Ye, H.; Zhou, Z.; Wang, Z. Resource utilization of municipal solid waste incineration fly ash i iron ore sintering process: A novel thermal treatment. *J. Clean. Prod.* **2020**, *263*, 121400. [CrossRef]
27. *JGJ/T 70-2009*; Standard for Test Method of Basic Properties of Construction Mortar. Ministry of Housing and Urban-Rura Development of the People's Republic of China: Beijing, China, 2009.
28. Zhu, J.; Qu, Z.; Liang, S.; Li, B.; Du, T.; Wang, H. The macro-scopic and microscopic properties of cement paste with carbo dioxide curing. *Materials* **2022**, *15*, 1578. [CrossRef]
29. *GB/T 17671-1999*; Method of Testing Cements-Determination of Strength. The State Bureau of Quality and Technical Supervisio Beijing, China, 1999.
30. Wang, S.; Guo, W.; Bai, Y.; Pan, H.; Qiu, Y.; Xue, C.; Zhao, Q. Preparation and characterization of mortar specimens based o municipal solid waste incineration fly ash-activated slag. *J. Build. Eng.* **2023**, *69*, 106254. [CrossRef]
31. Liu, J.; Wang, Z.; Xie, G.; Li, Z.; Fan, X.; Zhang, W.; Xing, F.; Tang, L.; Ren, J. Resource utilization of municipal solid wast incineration fly ash—Cement and alkali-activated cementitious materials: A review. *Sci. Total Environ.* **2022**, *852*, 15825 [CrossRef]
32. Dong, P.; Liu, J.; Wang, H.; Yuan, H.; Wang, Q. Sustainable municipal solid waste incineration fly ash (MSWIFA) alkali-activate materials in construction: Fabrication and performance. *Nanobuild* **2022**, *14*, 43–52. [CrossRef]
33. Tang, P.; Florea, M.V.A.; Spiesz, P.; Brouwers, H.J.H. Application of thermally activated municipal solid waste incineration (MSW bottom ash fines as binder substitute. *Cem. Concr. Compos.* **2016**, *70*, 194–205. [CrossRef]
34. AL-Ameeri, A.S.; ImranRafifiq, M.; Tsioulou, O.; Rybdylova, O. Impact of climate change on the carbonation in concrete due t carbon dioxide ingress: Experimental investigation and modelling. *J. Build. Eng.* **2021**, *44*, 102594. [CrossRef]
35. Smirnov, V.G.; Manakov, A.Y.; Dyrdin, V.V.; Ismagilov, Z.R.; Mikhailova, E.S.; Rodionova, T.V. The formation of carbon dioxid hydrate from water sorbed by coals. *Fuel* **2018**, *228*, 123–131. [CrossRef]
36. Qin, L.; Gao, X.; Chen, T. Influence of mineral admixtures on carbonation curing of cement paste. *Constr. Build. Mater.* **201** *212*, 653–662. [CrossRef]

. Li, Z.; Zhang, W.; Jin, H.; Fan, X.; Liu, J.; Xing, F.; Tang, L. Research on the durability and sustainability of an artificial lightweight aggregate concrete made from municipal solid waste incinerator bottom ash (MSWIBA). *Const. Build. Mater.* **2023**, *365*, 129993. [CrossRef]

. Liu, B.; Yang, L.; Shi, J.; Zhang, S.; Yalçınkaya, Ç.; Alshalif, A.F. Effect of curing regime on the immobilization of municipal solid waste incineration fly ash in sustainable cement mortar. *Environ. Pollut.* **2023**, *317*, 120839. [CrossRef]

. Xiang, J.; Qiu, J.; Wang, F.; Li, Z.; Gu, X. Utilization of bioactivated incineration bottom ash in cement binder for mortar harmless treatment and performance improvement. *J. Build. Eng.* **2022**, *57*, 104980. [CrossRef]

. Tripathy, A.; Acharya, P.K. Characterization of bagasse ash and its sustainable use in concrete as a supplementary binder-A review. *Const. Build. Mater.* **2022**, *322*, 126391. [CrossRef]

. Matalkah, F. Recycling of hazardous medical waste ash toward cleaner utilization in concrete mixtures. *J. Clean. Prod.* **2023**, *400*, 136736. [CrossRef]

. Zha, X.; Ning, J.; Saafi, M.; Dong, L.; Dassekpo, J.B.M.; Ye, J. Effect of supercritical carbonation on the strength and heavy metal retention of cement-solidified fly ash. *Cem. Concr. Res.* **2019**, *120*, 36–45. [CrossRef]

. Ding, Y.; Xi, Y.; Gao, H.; Wang, J.; Wei, W.; Zhang, R. Porosity of municipal solid waste incinerator bottom ash effects on asphalt mixture performance. *J. Clean. Prod.* **2022**, *369*, 133344. [CrossRef]

. Xie, W.; Li, H.; Yang, M.; He, L.; Li, H. CO_2 capture and utilization with solid waste. *J. Clean. Prod.* **2022**, *3*, 199–209. [CrossRef]

. Dong, P.; Ding, W.; Yuan, H.; Wang, Q. 3D-printed polymeric lattice-enhanced sustainable municipal solid waste incineration fly ash alkali-activated cementitious composites. *Dev. Built. Environ.* **2022**, *12*, 100101. [CrossRef]

. Wu, X.; Gu, F.; Su, C.; Wang, W.; Pu, K.; Shen, D.; Long, Y. Preparing high-strength ceramsite from ferronickel slag and municipal solid waste incineration fly ash. *Ceram. Int. Part A* **2022**, *48*, 34265–34272. [CrossRef]

. Assi, A.; Bilo, F.; Zanoletti, A.; Ponti, J.; Valsesia, A.; La Spina, R.; Zacco, A.; Bontempi, E. Zero-waste approach in municipal solid waste incineration: Reuse of bottom ash to stabilize fly ash. *J. Clean. Prod.* **2020**, *245*, 118779. [CrossRef]

coatings

MDPI

Article

Utilization of Fly Ash and Red Mud in Soil-Based Controlled Low Strength Materials

Xianghui Kong [1,*], Gaoqiang Wang [1], Shu Rong [2], Yunpeng Liang [1], Mengmeng Liu [1] and Yanhao Zhang [3,*]

[1] School of Transportation Engineering, Shandong Jianzhu University, Jinan 250101, China; wanggq97@163.com (G.W.); liangyunpeng99@163.com (Y.L.); liumengmeng992022@163.com (M.L.)

[2] Jinan Kingyue Highway Engineering Company Limited, Jinan 250014, China; rongs2019@163.com

[3] School of Municipal & Environmental Engineering, Shandong Jianzhu University, Jinan 250101, China

* Correspondence: kongxh@sdjzu.edu.cn (X.K.); zhyh@sdjzu.edu.cn (Y.Z.)

Abstract: To reduce the harm caused by industrial solid waste to the environment, and solve the problem of excavated soil disposal in buried pipeline management projects, this study proposes a method to produce soil-based controlled low strength materials (CLSM) by using industrial solid wastes (including fly ash and red mud) as partial replacements for cement. The properties of CLSM were characterized in terms of flowability, unconfined compressive strength, phase composition and microstructure. The test results showed that fly ash could significantly improve the flowability of CLSM, while red mud had more advantages for the strength development. When 20% fly ash and 30% red mud were combined to replace cement, the fluidity of CLSM was 248 mm, and the unconfined compressive strength (UCS) at 3, 7 and 28 days was 1.08, 1.49 and 3.77 MPa, respectively. The hydration products of CLSM were mainly calcium silicate hydrate gels, ettringite and calcite. Fly ash provided nucleation sites for cement hydration, while the alkali excitation of red mud promoted the dissolution of SiO_2 and Al_2O_3 in fly ash. The filling and gelation of hydration products make the microstructure dense, which improves the mechanical properties of the mixture.

Keywords: controlled low strength materials; fly ash; red mud; excavated soil; mix design; microscopic properties

1. Introduction

With the development of municipal pipeline management projects, a large amount of spoil generated from construction sites in various regions cannot be effectively resolved. Excavated soil for reuse is considered to have a positive impact on both the engineering economy and the environment [1–3]. The construction of traditional backfill materials has problems such as small operation space and a dead angle on the contact surface with the pipeline, which makes it difficult to guarantee the compaction quality [4]. Therefore, it is important to develop a backfill material with compactness and ease of construction by using excavated soil.

Controlled low strength materials (CLSM), also known as flow backfill materials with self-leveling and self-compacting properties, can replace traditional compacted backfill [5,6]. CLSM is often applied to bridge approaches, retaining wall backs, pipe trench backfilling and other projects [7]. In general, the flowability of CLSM needs to be between 200 and 300 mm, and the 28-day UCS is required to be less than 8.3 MPa [8]. The strength can be controlled effectively by adjusting the dosage of cement. Considering excavability, the UCS strength is no higher than 2.1 MPa [9]. CLSM does not require high strength, so a large number of non-standard materials are allowed in its production [10].

Currently, the mix proportion of CLSM is not constrained [11]. Wu and Tsai [12] used rubber particles as fine aggregate instead of traditional sand and found that non-sand CLSM was not flowable. Kuo and Gao [13] developed a CLSM using the bottom ash of a municipal solid waste incinerator as aggregate which had sufficient load-bearing capacity with 28-day UCS of 6.25 ~ 5.87 MPa. Yan et al. [14] proposed to produce CLSM with dredged sediment and bottom ash. Their study showed that the formation of calcium silicate hydrate (C-S-H) gels was the main reason for the strength improvement, which also had a positive effect on heavy metals immobilization. The CLSM they produced had a good general performance while achieving a waste utilization rate of 80%. Kim et al. [15] used steel slag as a new aggregate to develop CLSM. They found that the flowability and strength of CLSM decreased with the increase of steel slag content. They also pointed out that steel slag particle size also has a different effect on the properties of the mixture.

Cement hydration produces a large amount of cementitious material that allows CLSM to set and harden [16]. However, the production of cement is a process with high energy consumption and high emissions. Therefore, the application of industrial solid waste as supplementary cementitious material (SCM) in CLSM is also attracting attention. The inclusion of industrial solid waste in CLSM production is intended to control performance development while reducing production costs [17]. Fly ash has been proved to be effective in improving the fluidity of CLSM, which is greatly related to its spherical shape [18]. Moreover, fly ash can improve the strength of mixture by reducing porosity and improving the interlocking mechanism between aggregates, especially in the late curing period [19,20]. Alkaline activation is helpful to stimulate the vitality of fly ash and improve the strength of mixture [21]. However, other admixtures are sometimes added to CLSM to meet engineering applications, such as those with a rapid hardening property.

Red mud is a solid waste produced by the aluminum industry, with an annual output of up to 70 million tons in China [22]. For a long time, the strong alkalinity and high content of heavy metals in red mud have made it challenging to utilize. However, the main components of red mud contain many volcanic ash active substances such as SiO_2 and Al_2O_3, and the high alkalinity of red mud can also be used as a source of alkali to activate cementitious materials [23]. Cheng et al. [24] developed a new type of cementitious material using red mud and evaluated its physical and chemical properties. The new cementitious material took a shorter time to reach the peak heat of hydration and its microstructure was more compact. After adding red mud, the voids of the cementitious material were filled and the mechanical properties were improved. This bonding property of red mud means it was thought it could be added to the production of CLSM as SCM [25,26].

In this study, fly ash and red mud were used as replacement materials for cement, and excavated soil was used as fine aggregate to produce the CLSM mixture. Twenty different mix ratios were used to explore the effects of fly ash and red mud replacing cement on the macro properties of CLSM. The alternative types were divided into independent alternative and co-alternative. Taking high fluidity and early strength required by CLSM as control indexes, the optimal ratio of fly ash and red mud to replace cement was studied. More importantly, the interaction mechanism between fly ash and red mud is analyzed by microcosmic means. This provides a new idea for the utilization of fly ash and red mud resource engineering.

2. Materials and Methods

2.1. Raw Materials

In this study, excavated soil was used as fine aggregate, cement was used as the main cementing agent, and fly ash and red mud were used as supplementary cementitious materials. Excavated soil was taken from a municipal engineering pipeline construction in Shandong Province. After removing vegetation and debris, the soil was sealed in bags and placed in a cool room to reduce the effect of ambient temperature and humidity on the soil. Excavated soil was tested for particle analysis and basic physical properties according to JTG 3430-2020 [27]. Figure 1 shows the grain size distribution of natural soil. Other physical properties of the soil are summarized in Table 1.

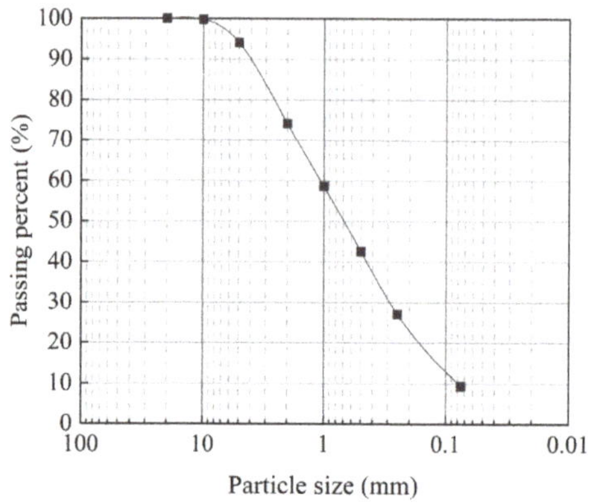

Figure 1. Grain size distribution of the natural soil.

Table 1. Physical properties of soil.

Physical Properties	Value
Natural water content (%)	10.8
Liquid limit (%)	48.9
Plastic limit (%)	22.9
Plasticity index	26.0
Specific gravity	2.74
Particles (< 0.075 mm) (%)	9.34

The cement was P.O 42.5 cement produced by a company in Shandong Province (Jiuqi Building Materials Co., LTD, Shandong, China), and its specific surface area was 0.36 m^2/g. As the main cementing material, the hydration of cement produces a large number of cementing substances which could improve the strength of the mixture. The fly ash used in this study was taken from a power plant in Henan Province. The specific surface area of the fly ash was 0.43 m^2/g. The red mud was produced by the industrial Bayer aluminum process from a red mud dump in Zhengzhou City, Henan Province (Yixiang New Materials Co., LTD, Zhengzhou, China). Its specific surface area was 0.68 m^2/g and the pH value was 12.57. The appearances of raw materials and the microscopic images of the particle surface are shown in Figure 2.

Figure 2. Appearance and microscopic images of raw materials.

The chemical compositions of the raw materials are listed in Table 2, which were measured by X-ray fluorescence (XRF, S8 TIGER, Bruker Corporation, Karlsruhe, Germany). According to the measurement results, it can be seen that the compositions of excavated soil were mainly SiO_2 and Al_2O_3, but the cement was dominated by CaO and SiO_2. The high content of SiO_2 and Al_2O_3 in fly ash and red mud gave them the ability to partially replace cement.

Table 2. Chemical compositions of excavated soil, cement, fly ash and red mud.

Compound	Excavated Soil (%)	Cement (%)	Fa (%)	Rm (%)
SiO_2	70.41	31.40	54.90	28.66
CaO	1.50	42.10	3.96	3.37
Al_2O_3	15.92	12.70	28.10	23.32
Fe_2O_3	4.59	4.19	5.38	26.19
K_2O	2.77	1.17	2.94	1.42
Na_2O	1.58	0.62	0.78	12.40
MgO	1.95	3.28	0.77	0.32
TiO_2	0.78	0.66	1.44	1.91
MnO	0.08	0.18	0.06	0.06
ZnO	0.01	0.01	/	0.01
SO_3	0.07	3.32	0.68	1.84

2.2. Sample Preparation and Mix Design

All raw materials needed to be dried and weighed by a digital balance with 0.1 g accuracy. The excavated soil used was passed through a 5 mm mesh screen. The soil, cement, fly ash and red mud were put into the mixer and dry mixed for 30 s. Subsequently, calculated water was added to them and mixing was continued for 2 min. After mixing was completed, the CLSM slurry was partially poured into the flowability test tool and the remaining slurry was molded in a 70.7 mm × 70.7 mm × 70.7 mm cube test mold. The specimen prepared for molding was pre-cured at room temperature for 24 h and then demolded. The samples were conditioned in a curing box at a temperature of 20 °C and 97% humidity.

A comprehensive test program was designed to evaluate the effect of fly ash and red mud on CLSM, as shown in Table 3. A total of 20 different mix designs are included. The presence of Fa0 (or Rm0) in the named species of the mixture means that the CLSM performance characteristics when red mud (or fly ash) is separately studied. The amount of water and soil in all mixtures was fixed. The replacement amount of fly ash was increased from 0% to 30%, and the replacement amount of red mud was increased from 0% to 40%. The replacement amount is the proportion of the mass of fly ash or red mud replacing the mass of cement.

2.3. Testing Methods

2.3.1. Flowability Test

The flowability of CLSM was tested according to ASTM D 6103 [28]. A cylinder with a height of 150 mm and an inner diameter of 75 mm was used. The cylinder was lifted as soon as the CLSM slurry filled it and the mixture was allowed to diffuse freely on the plane. When the diffusion range was no longer increasing, the watershed size was measured with a ruler in both vertical directions. The average of the two measurements was specified as the fluidity of the CLSM. The accuracy was 1 mm.

2.3.2. Unconfined Compressive Strength Test

The UCS test for CLSM was performed in accordance with JGJ/T 70-2009 [29]. The curing ages of test samples were 3, 7 and 28 days, respectively. The test instrument was a pavement material strength tester (LD127, Guanghui test instrument co., LTD, Cangzhou, China), with the accuracy of 0.01 MPa. The base plate was pressurized at a lifting rate of 1 mm per minute on the sample, and the maximum strength value of the sample at the time of destruction was recorded. Each test included three specimens, and the average value of the test results of the three specimens was taken as the final value.

Table 3. Mix proportions for CLSM mixtures.

Mix Name	Fa/Binder (%)	Rm/Binder (%)	Weight Fraction				
			Water	Soil	Binder		
					Cement	Fa	Rm
Fa0-Rm0	0	0	0.58	1	0.29	0	0
Fa0-Rm10		10			0.26	0	0.03
Fa0-Rm20		20			0.23	0	0.06
Fa0-Rm30		30			0.20	0	0.09
Fa0-Rm40		40			0.17	0	0.12
Fa10-Rm0	10	0	0.58	1	0.26	0.03	0
Fa10-Rm10		10			0.23	0.03	0.03
Fa10-Rm20		20			0.20	0.03	0.06
Fa10-Rm30		30			0.17	0.03	0.09
Fa10-Rm40		40			0.15	0.03	0.12
Fa20-Rm0	20	0	0.58	1	0.23	0.06	0
Fa20-Rm10		10			0.20	0.06	0.03
Fa20-Rm20		20			0.17	0.06	0.06
Fa20-Rm30		30			0.15	0.06	0.09
Fa20-Rm40		40			0.12	0.06	0.12
Fa30-Rm0	30	0	0.58	1	0.20	0.09	0
Fa30-Rm10		10			0.17	0.09	0.03
Fa30-Rm20		20			0.15	0.09	0.06
Fa30-Rm30		30			0.12	0.09	0.09
Fa30-Rm40		40			0.09	0.09	0.12

2.3.3. XRD Test

X-ray diffraction (XRD, PANalytical B.V., EA Almelo, The Netherlands) was used to identify the phase composition. The samples for XRD were taken from samples cured for 28 days. Samples were dried and passed through a 0.075 mm mesh screen. A Cu target was used for 5° to 90° scanning at 10° per minute.

2.3.4. SEM Test

The microstructure of the mixture determines the macroscopic properties of the sample. Therefore, a scanning electron microscope (SEM, SU5000, Hitachi High-tech Company, Tokyo, Japan) was used to analyze the CLSM microstructure. The samples were cured for 28 days and cut into 10 mm × 5 mm × 5 mm block shapes. The spray-gold process was necessary because of the weak conductivity of the samples.

3. Results and Discussion

3.1. Flowability Analysis

Self-leveling properties require the high flowability of CLSM mixtures. The effects of fly ash and red mud on the flowability of CLSM are shown in Figure 3. From Figure 3, it can be seen that the increase of fly ash content improved the flowability of the mixture, while the red mud had the opposite effect. The flowability of Fa20-Rm0 was the largest among all the mix proportion samples. In this replacement, the ball of fly ash played an effective role in promoting the flowability increase. However, the excessive admixture was not conducive to the development of CLSM workability. Fly ash replacement continued to increase, resulting in a lower flowability. As the amount of red mud replacement increased from 0% to 40%, the mixture flowability decreased from 230 to 175 mm. Fluidity below the minimum limit (200 mm) occurred at 30% and 40% red mud replacement. The explanation for this phenomenon is that the specific surface area of red mud was much larger than that of cement, and the increase in the amount of water adsorbed on the surface led to a decrease in the flowability of the mixture. The free water in CLSM was reduced by the replacement of cement with red mud, hence more water was required in the mix to meet the high flowability requirement.

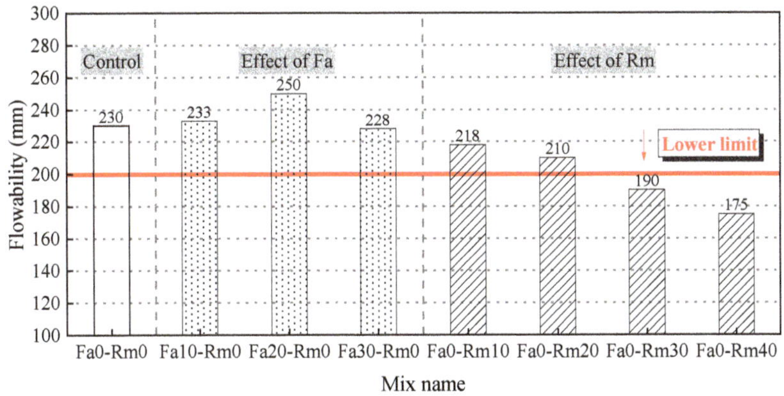

Figure 3. Flowability of CLSM with independent replacement of Fa and Rm.

The effects of combined fly ash and red mud replacement cement on CLSM flowability are shown in Figure 4. According to the trend of the curves in the figure, it can be seen that the effects of fly ash and red mud on flowability under combined replacement had the same pattern as that in independent replacement. The data comparison revealed that the difference in fluidity between Fa10-Rm10 and Fa10-Rm40 was 38 mm, between Fa20-Rm10 and Fa20-Rm40 was 25 mm, and between Fa30-Rm10 and Fa30-Rm40 was 35 mm. Therefore, in spite of the presence of red mud, the 20% fly ash replacement amount still had a significant improvement in flowability. Only Fa10-Rm40, Fa30-Rm30 and Fa30-Rm40 had flowability below the limit. This was due to the fact that fly ash replacement at 10% and 30% had less effect on the increase of fluidity than replacement at 20%. Combined with the high red mud replacement, the superimposed effect of the double negative effect reduced the flowability. It should be noted that the CLSM produced in this research used 100% soil as fine aggregate. Compared with sand, soil-based CLSM is more prone to segregation at high fluidity and is unable to harden the mixture quickly.

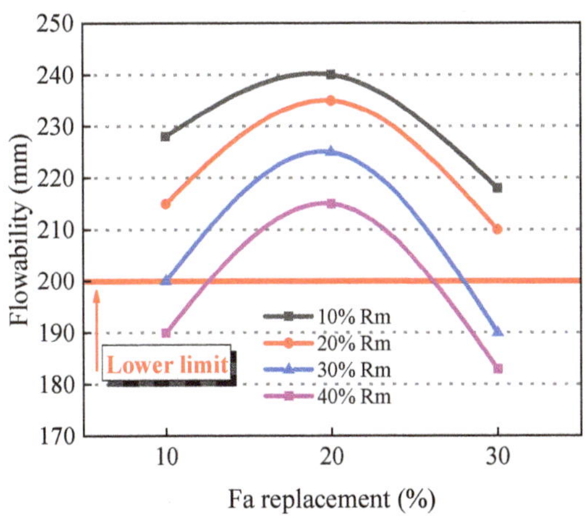

Figure 4. Flowability of CLSM with co-replacement of Fa and Rm.

3.2. UCS Analysis

The effects of independent replacement of fly ash and red mud on the strength of CLSM are shown in Figure 5. It was a common phenomenon that the strength of all mixtures increased with curing age. Analysis of the test results revealed that the CLSM strength kept decreasing as the amount of fly ash replacement increased. For instance, the 3, 7 and 28 days UCS of Fa20-Rm0 were 0.62, 1.09 and 3.73 MPa. Compared with the control group (Fa0-Rm0), the strength decreased by 45%, 39% and 19%, respectively. CLSM strength increased first and then decreased with the increase in red mud replacement amount. The UCS of Fa0-Rm30 always remained maximum at all ages. The 3, 7 and 28 days UCS of Fa0-Rm30 were 1.25, 2.24 and 5.11 MPa, respectively. The strength growth rates were 11%, 26% and 12% compared with Fa0-Rm0. Based on the strength change, there was a threshold value for the red mud content in CLSM, and 30% was the optimal amount of replacement when red mud was independently affected. The most significant factor affecting the strength change is the content of cement in the mixture, thus controlling the cement content is an important way to achieve controllability of CLSM strength.

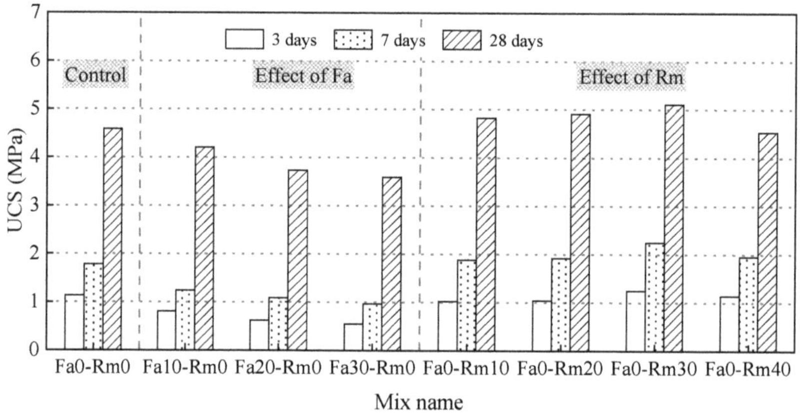

Figure 5. UCS of CLSM with independent replacement of Fa and Rm.

Figure 6 shows the strength change of CLSM for combined replacement with fly ash and red mud at 3, 7 and 28 days of curing age, respectively. Due to the presence of red mud, the CLSM strength tended to increase at first and then decrease with the growth of fly ash replacement. The strength of CLSM is very sensitive to the amount of fly ash replacement [30,31]. The test results showed that a 20% fly ash replacement amount was beneficial to the UCS development of CLSM. The deterioration effect with a high replacement amount of fly ash on strength was obvious in long-term curing. This showed that fly ash had a finite effect on the development of strength. Considering rapid construction, sufficient attention needs to be paid to the early strength (3 and 7 days) of CLSM [32]. With 20% fly ash replacement, the 30% red mud replacement was beneficial for the UCS development for 3 days and 7 days of the mixture. The small particle size and large specific surface area of red mud can reduce the hardening time and improve the early strength of the mixture by reducing the amount of free water in the slurry. At 28 days of curing, there was a significant strength difference at a high amount of fly ash replacement. Therefore, to meet the application requirements of high flowability and fast hardness of CLSM, Fa20-Rm30 was considered in this study as the optimal mix proportion for the combined replacement of fly ash and red mud.

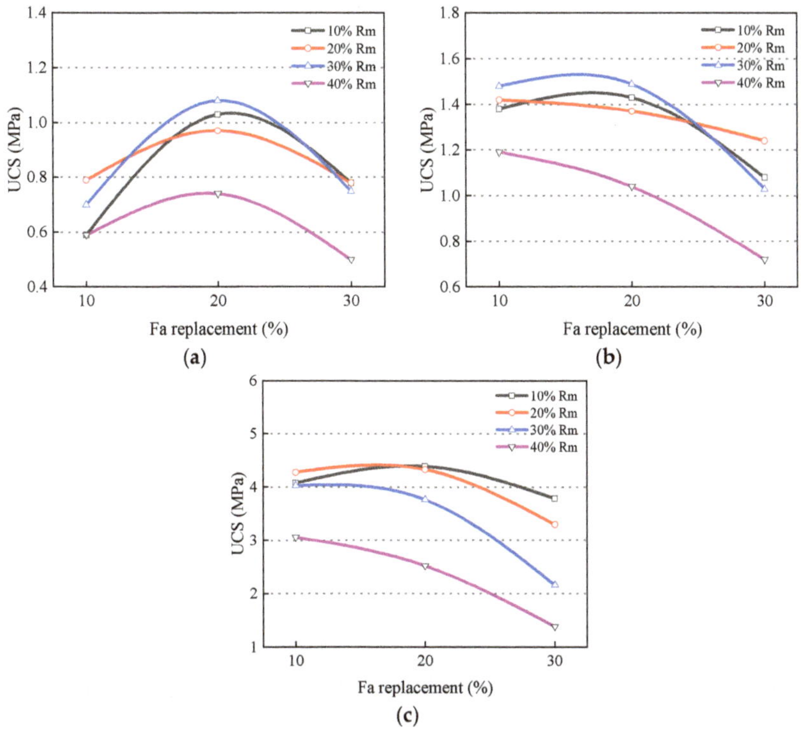

Figure 6. UCS of CLSM with co-replacement of Fa and Rm (**a**) 3 days; (**b**) 7 days; (**c**) 28 days.

3.3. XRD Analysis

The XRD patterns of Fa0-Rm0, Fa20-Rm0, Fa0-Rm30, Fa20-Rm30 are shown in Figure 7. Since the fine aggregate of CLSM was excavated soil, the quartz diffraction peak in the mixture was large. The hydration products were mainly C-S-H gels, and ettringite. There are two types of C-S-H gels: Jennite type and Tobermorite type. They are distinguished by the difference in Ca/Si ratios, which are larger in Jennite C-S-H gels [33]. The C-S-H diffraction peak in Figure 7 was the Jennite type, which had the lowest intensity in Fa20-Rm0 and the highest in Fa0-Rm30. The replacement of cement with fly ash reduced Ca/Si ratios in the mixture, which caused the C-S-H shift from Jennite type to Tobermorite type and consequently reduced the peak intensity.

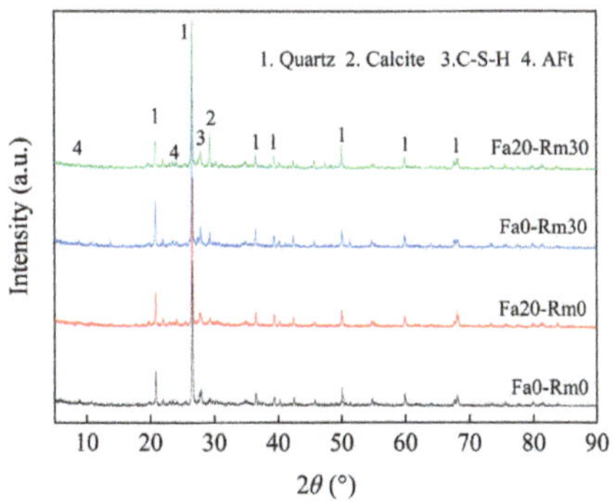

Figure 7. XRD patterns of CLSM at different mixes.

The addition of red mud to the mixture increased the alkalinity of the pore solution, while fly ash increased the concentration of SO_4^{2-}. This would promote the hydration reaction in the direction of AFt production and lead to an increase in its yield. Consequently, the intensity of the AFt diffraction peak was higher in the CLSM containing solid waste compared with Fa0-Rm0.

The source of calcite diffraction peak strength included two parts, one was soil, and the other was carbonization of hydration products. Compared with Fa0-Rm0 and Fa20-Rm0, Fa0-Rm30 and Fa20-Rm30 had a high calcite content. Ca^{2+} combines with CO_3^{2-} to produce a $CaCO_3$ crystal framework, which is slightly soluble in water. $CaCO_3$ precipitated out in the form of sediment and no longer participated in the reaction. Ban et al. [34] found that hydration products in polymers could coexist with $CaCO_3$. The development of strength depends on the strength of the hydration product itself and the compactness of the bonding particles. It should be noted that the amount of calcite in Fa20-Rm30 was significantly higher than in Fa0-Rm30. This indicated that more carbonate in the red mud was involved in the reaction after further reduction of the cement content [35].

3.4. SEM Analysis

SEM images can clearly show the microscopic morphology of CLSM. The effects of fly ash and red mud on the microstructure of CLSM were clarified. Fa0-Rm0, Fa20-Rm0, Fa0-Rm30 and Fa20-Rm30 were studied in SEM images, as shown in Figure 8. The structural differences in these microscopic images were in accordance with the changes in the strength of the components of the cementitious material.

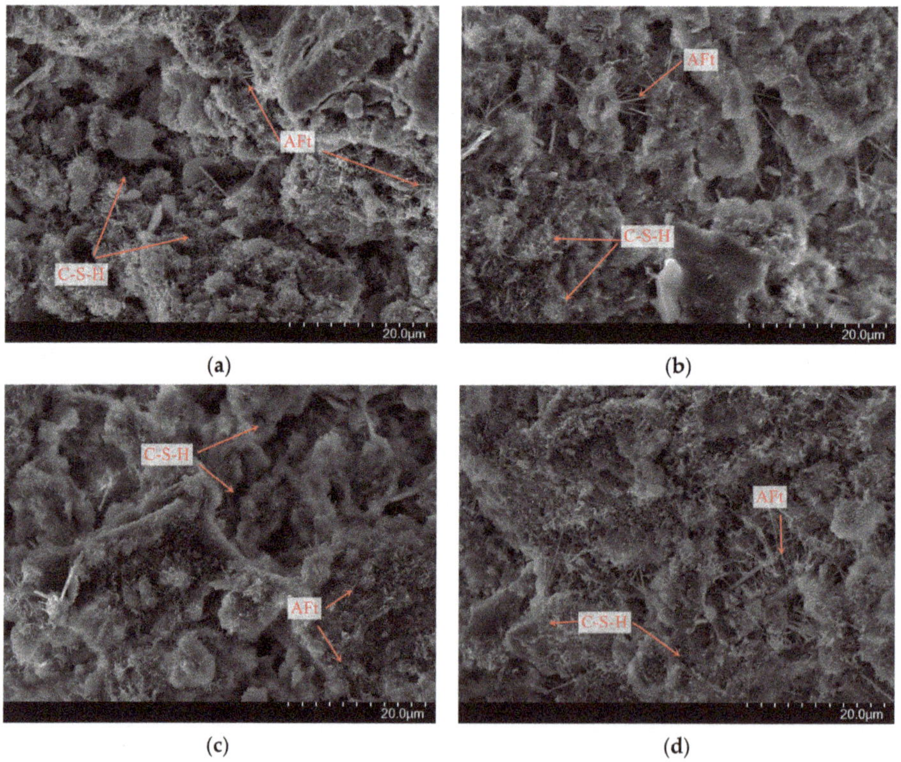

Figure 8. SEM images of CLSM at different mixes (**a**) Fa0-Rm0; (**b**) Fa20-Rm0; (**c**) Fa0-Rm30; (**d**) Fa20-Rm30.

The mixture was observed to be filled with a large amount of C-S-H gel and Aft, and formed a dense network structure in Fa0-Rm0. The strength of the mixture was improved by using C-S-H gels as the main cementing phase to connect other hydration products, which provided the bonding force for the soil particles [31]. Unlike the other groups, some hexagonal crystal structures were found in the mixture. This may have been caused by the partial conversion of AFt to AFm due to the decrease of SO2-4 concentration in the pore solution. A large number of C-S-H gels were found attached to the surface of the soil and fly ash particles in Fa20-Rm0. Fly ash provided good nucleation sites for C-S-H gels. The cement was no longer wrapped and the hydration process could proceed more deeply. The soil particles which can be seen in Fa0-Rm30 were covered by large areas of hydration products. This was due to the high alkalinity of the red mud promoting more C-S-H gels and AFt production. The hydration products can effectively reduce the harmful impact of macropores on the mixture structure. In addition, Ghalehnovi et al. [36] reported that the cementitious properties of red mud are not as good as cement, but its fine particles are more effective in filling the pores.

Cement in Fa20-Rm30 accounted for only 50% of all cementitious materials. Therefore, the total amount of hydration products was less than in the other groups. In addition to the gels, calcite also had a positive effect on structural compactness. The particle filling effect of fly ash and red mud can refine the pore structure. The high content of SiO_2 and Al_2O_3 in fly ash was excited by the high alkalinity of red mud. More hydration products were used to fill the internal voids of the CLSM to reduce the sample porosity, which was conducive to CLSM strength maintenance. In spite of the cement, the content was only half of that of Fa0-Rm0, and obvious pores did not appear (Figure 8d). Consequently, the synergistic effect of fly ash and red mud contributed to maintaining the good mechanical properties of CLSM mixtures even if the cement content was reduced.

4. Conclusions

In this study, fly ash and red mud as SCM to replace part of the cement, and excavated soil was used as fine aggregate to make CLSM with good self-leveling and self-compacting properties. The flowability, UCS, phase composition and microstructure of CLSM mixture were investigated by changing the replacement amount of fly ash and red mud. The main conclusions of this study are as follows:

(1) The effects of fly ash and red mud on CLSM performance were very different. Fly ash improved the flowability, while red mud was more beneficial for strength growth of the CLSM mixture. With the increase of fly ash replacement, the fluidity of CLSM first increased and then decreased, while the UCS of CLSM gradually decreased. When the fly ash replacement cement mass was 20%, the fluidity of CLSM reached an inflection point. With the increase of red mud replacement amount, the fluidity of CLSM gradually decreased until it was not easy to flow, and when the replacement amount was 30%, the UCS of CLSM showed an inflection point. When the cement was replaced by fly ash and red mud at the same time, the mixture mix proportion with optimal engineering performance was Fa20-Rm30. Its fluidity was 248 mm, and the UCS at 3, 7 and 28 days was 1.08, 1.49 and 3.77 MPa, respectively. The mixture met both high flowability and fast hardening requirements.

(2) Hydration products of CLSM cured for 28 days included C-S-H gels and ettringite. The C-S-H gels were the main cementing phase that connected other hydration products to form a dense mesh structure. The strong alkaline environment generated by red mud in contact with water not only promoted the hydration of cement but also had a positive effect on stimulating the activity of SiO_2 and Al_2O_3 in fly ash. The high content of carbonate in red mud promoted the carbonization reaction. The resulting calcium carbonate could coexist with other hydration products, increasing the strength of the mixture. Fly ash provided good nucleation sites for hydration products, and a large number of hydration products attached to the particle surface to refine the pore structure.

Author Contributions: Conceptualization, X.K. and Y.Z.; methodology, G.W. and S.R.; validation, G.W., Y.L. and M.L.; formal analysis, G.W.; investigation, S.R; resources, X.K and Y.Z.; data curation, S.R.; writing original draft, X.K. and G.W.; writing review, Y.Z.; supervision, X.K. All authors have read and agreed to the published version of the manuscript.

Funding: This research was funded by Key Research and Development Project of Shandong (Grant No. 2020CXGC011404).

Institutional Review Board Statement: Not applicable.

Informed Consent Statement: Not applicable.

Data Availability Statement: The data presented in this study are available on request from the corresponding author.

Conflicts of Interest: The authors declare no conflict of interest.

References

1. Chittoori, B.; Puppala, A.J.; Raavi, A. Strength and stiffness characterization of controlled low-strength material using nativ high-plasticity clay. *J. Mater. Civ. Eng.* **2014**, *26*, 04014007. [CrossRef]
2. Arulrajah, A.; Ali, M.M.Y.; Piratheepan, J.; Bo, M.W. Geotechnical properties of waste excavation rock in pavement subbas applications. *J. Mater. Civ. Eng.* **2012**, *24*, 924–932. [CrossRef]
3. Cabalar, A.F.; Alosman, S.O. Influence of rock powder on the behaviour of an organic soil. *Bull. Eng. Geol. Environ.* **2021**, *8*(8665–8676. [CrossRef]
4. Ling, T.C.; Kaliyavaradhan, S.K.; Poon, C.S. Global perspective on application of controlled low-strength material (CLSM) fo trench backfilling—An overview. *Constr. Build. Mater.* **2018**, *158*, 535–548. [CrossRef]
5. Siddique, R. Utilization of waste materials and by-products in producing controlled low-strength materials. *Resour. Conser Recycl.* **2009**, *54*, 1–8. [CrossRef]
6. Pujadas, P.; Blanco, A.; Cavalaro, S.; Aguado, A. Performance-based procedure for the definition of controlled low-strengt mixtures. *J. Mater. Civ. Eng.* **2015**, *27*, 06015003. [CrossRef]
7. Do, T.M.; Kim, Y.; Dang, M.Q. Influence of curing conditions on engineering properties of controlled low strength material mad with cementless binder. *KSCE J. Civ. Eng.* **2017**, *21*, 1774–1782. [CrossRef]
8. American Concrete Institute. *ACI 229R*; Controlled Low-Strength Materials; American Concrete Institute: Farmington Hill, M USA, 2013.
9. Fang, X.; Wang, L.; Poon, C.S.; Baek, K.; Tsang, D.C.W.; Kwok, S.K. Transforming waterworks sludge into controlled low-strengt material: Bench-scale optimization and field test validation. *J. Environ. Manag.* **2019**, *232*, 254–263. [CrossRef]
10. Mneina, A.; Soliman, A.M.; Ahmed, A.; El Naggar, M.H. Engineering properties of controlled low-strength materials containin treated oil sand waste. *Constr. Build. Mater.* **2018**, *159*, 277–285. [CrossRef]
11. Ibrahim, M.; Rahman, M.K.; Najamuddin, S.K.; Alhelal, Z.S.; Acero, C.E. A review on utilization of industrial by-products in th production of controlled low strength materials and factors influencing the properties. *Constr. Build. Mater.* **2022**, *325*, 12670 [CrossRef]
12. Wu, J.Y.; Tsai, M. Feasibility study of a soil-based rubberized CLSM. *Waste Manag.* **2009**, *29*, 636–642. [CrossRef] [PubMed]
13. Kuo, W.T.; Gao, Z.C. Engineering properties of controlled low-strength materials containing bottom ash of municipal solid wast incinerator and water filter silt. *Appl. Sci.* **2018**, *8*, 1377. [CrossRef]
14. Yan, D.Y.S.; Tang, I.Y.; Lo, I.M.C. Development of controlled low-strength material derived from beneficial reuse of bottom as and sediment for green construction. *Constr. Build. Mater.* **2014**, *64*, 201–207. [CrossRef]
15. Kim, Y.; Dinh, B.H.; Do, T.M.; Kang, G. Development of thermally enhanced controlled low-strength material incorporatin different types of steel-making slag for ground-source heat pump system. *Renew. Energy* **2020**, *150*, 116–127. [CrossRef]
16. Ghahari, S.; Assi, L.N.; Alsalman, A.; Alyamac, K.E. Fracture properties evaluation of cellulose nanocrystals cement past *Materials* **2020**, *13*, 2507. [CrossRef]
17. Kaliyavaradhan, S.K.; Ling, T.C.; Guo, M.Z. Upcycling of wastes for sustainable controlled low-strength material: A review o strength and excavatability. *Environ. Sci. Pollut. Res.* **2022**, *29*, 16799–16816. [CrossRef]
18. Wu, H.; Huang, B.; Shu, X.; Yin, J. Utilization of solid wastes/byproducts from paper mills in controlled low strength materia (CLSM). *Constr. Build. Mater.* **2016**, *118*, 155–163. [CrossRef]
19. Mohammadinia, A.; Arulrajah, A.; Horpibulsuk, S.; Chinkulkijniwat, A. Effect of fly ash on properties of crushed brick an reclaimed asphalt in pavement base/subbase applications. *J. Hazard. Mater.* **2017**, *321*, 547–556. [CrossRef]
20. Park, J.; Hong, G. Strength characteristics of controlled low-strength materials with waste paper sludge ash (WPSA) for preventio of sewage pipe damage. *Materials* **2020**, *13*, 4238. [CrossRef]
21. Phetchuay, C.; Horpibulsuk, S.; Arulrajah, A.; Suksiripattanapong, C.; Udomchai, A. Strength development in soft marine cla stabilized by fly ash and calcium carbide residue based geopolymer. *Appl. Clay Sci.* **2016**, *127*, 134–142. [CrossRef]

- Xue, S.; Li, X.; Kong, X.; Wu, C.; Li, Y.; Li, M.; Li, C. Alkaline regulation of bauxite residue: A comprehensive review. *Acta Sci. Circumstantiae* **2017**, *37*, 2815–2828.
- Liu, Y.; Su, Y.; Xu, G.; Chen, Y.; You, G. Research progress on controlled low-strength materials: Metallurgical waste slag as cementitious materials. *Materials* **2022**, *15*, 727. [CrossRef] [PubMed]
- Cheng, X.; Long, D.; Zhang, C.; Gao, X.; Yu, Y.; Mei, K.; Zhang, C.; Guo, X.; Chen, Z. Utilization of red mud, slag and waste drilling fluid for the synthesis of slag-red mud cementitious material. *J. Clean. Prod.* **2019**, *238*, 117902. [CrossRef]
- Do, T.M.; Kang, G.; Vu, N.; Kim, Y. Development of a new cementless binder for marine dredged soil stabilization: Strength behavior, hydraulic resistance capacity, microstructural analysis, and environmental impact. *Constr. Build. Mater.* **2018**, *186*, 263–275. [CrossRef]
- Do, T.M.; Kang, G.; Kim, Y. Development of a new cementless binder for controlled low strength material (CLSM) using entirely by-products. *Constr. Build. Mater.* **2019**, *206*, 576–589.
- *JTG 3430-2020;* Test Methods of Soils for Highway Engineering. Ministry of Transport of the People's Republic of China: Beijing, China, 2020.
- *ASTM D6103;* Standard Test Method for Flow Consistency of Controlled Low Strength Material (CLSM). American Society of Testing Materials: West Conshohocken, PA, USA, 2017.
- *JGJ/T 70-2009;* Standard for Test Method of Basic Properties of Construction Mortar. Ministry of Housing and Urban-Rural Development of the People's Republic of China: Beijing, China, 2009.
- Türkel, S. Long-term compressive strength and some other properties of controlled low strength materials made with pozzolanic cement and Class C fly ash. *J. Hazard. Mater.* **2006**, *137*, 261–266. [CrossRef]
- Behera, S.K.; Ghosh, C.N.; Mishra, D.P.; Singh, P.; Mishra, K.; Buragohain, J.; Mandal, P.K. Strength development and microstructural investigation of lead-zinc mill tailings based paste backfill with fly ash as alternative binder. *Cem. Concr. Compos.* **2020**, *109*, 103553. [CrossRef]
- Alizadeh, V.; Helwany, S.; Ghorbanpoor, A.; Sobolev, K. Design and application of controlled low strength materials as a structural fill. *Constr. Build. Mater.* **2014**, *53*, 425–431. [CrossRef]
- Xiao, R.; Polaczyk, P.; Jiang, X.; Zhang, M.; Wang, Y.; Huang, B. Cementless controlled low-strength material (CLSM) based on waste glass powder and hydrated lime: Synthesis, characterization and thermodynamic simulation. *Constr. Build. Mater.* **2021**, *275*, 122157. [CrossRef]
- Ban, C.C.; Ee, T.L.; Ramli, M. The engineering properties and microstructure of sodium carbonate activated fly ash/ slag blended mortars with silica fume. *Compos. Part B Eng.* **2019**, *160*, 558–572.
- Yuan, B.; Yuan, S.; Straub, C.; Chen, W. Activation of binary binder containing fly ash and Portland cement using red mud as alkali source and its application in controlled low-strength materials. *J. Mater. Civ. Eng.* **2020**, *32*, 04019356. [CrossRef]
- Ghalehnovi, M.; Roshan, N.; Hakak, E.; Shamsabadi, E.A.; Brito, J. Effect of red mud (bauxite residue) as cement replacement on the properties of self-compacting concrete incorporating various fillers. *J. Clean. Prod.* **2019**, *240*, 118213. [CrossRef]

coatings

Article

Investigation of the Effect of Filler on Cohesive Bond Strength of Asphalt Mastic Using Binder Bond Strength (BBS) Test

Xiang Ma *, Jinquan Kan, Song Liu, Mingyuan Tu and Dongjia Wang

College of Civil Engineering, Nanjing Forestry University, Nanjing 210037, China
* Correspondence: max@njfu.edu.cn

Abstract: Water damage leads to spalling and loosening of asphalt pavement, which is ultimately due to a reduction in the bond strength of the asphalt mastic. The filler, as an important component of the asphalt mastic, has a great impact on the various properties of the asphalt mastic and even the asphalt mix itself. Therefore, it is important to study the influence of fillers on the performance of asphalt mastic. In this paper, the various properties of twelve types of fillers, including mineral powder, mineral powder partially replaced with cement, slaked lime, and recycled powder, and their effects on the bond strength of asphalt mastic are investigated. Binder bond strength (BBS) was used to evaluate the performance of the different asphalt mastics. The moisture resistance of the asphalt mastic was evaluated by measuring the bond strength of the asphalt mastic after treatment in a water bath. A grey correlation analysis was then carried out to derive the relationship between the filler index and the bond strength of the asphalt mastic. The results show that the addition of slaked lime and cement helped to improve the moisture resistance of the asphalt mastic. The fineness modulus and other indicators have a great influence on the bond strength of asphalt mastic. The interaction between hydrothermal coupling and filler type has a non-negligible effect on the moisture resistance of the asphalt mastic.

Keywords: fillers; asphalt mastic; asphalt mixtures; bond strength; moisture resistance

1. Introduction

Asphalt mixtures are granular materials, and their mechanical strength is generally considered to depend on the friction and embedding forces between the aggregate particles, the cohesion of the asphalt binder, and the bond between the asphalt and the aggregate [1]. Erosion due to water can damage the interface between asphalt and aggregate, reducing the adhesion of the asphalt bond [2–5], resulting in spalling and loosening and reducing the life of the asphalt pavement [6,7].

Many pioneering studies have been carried out on the adhesion between asphalt and aggregate. In general, commonly used tests fall into one of three main categories: energy-based methods (contact angle tests) [8–10], observation-based methods (immersion tests) [11] and mechanical methods (pull-off tests) [12,13]. The binder bond strength (BBS) test, a pull-off test, was developed to obtain mechanical parameters and is widely used to evaluate the adhesion and cohesion of asphalt [14–16]. Canestrari et al. [17] used a modified pull-off head to improve the accuracy of the BBS test. The pull-off tensile strength (POTS) and damage type were obtained from the BBS test and used to evaluate the bond/cohesion properties of the asphalt [18]. Moraes et al. noted that the BBS test results were consistent with surface energy measurements [19]. Wang et al. [1] used the skeleton penetration test, the BBS test and the MBBS test to investigate several aspects of the asphalt–aggregate bond strength, the mastic–aggregate bond strength, the mortar–aggregate bond strength and the cohesive strength of the asphalt, the mastic and the mortar itself. The results show that spalling damage in medium- and low-temperature conditions is dominated by cohesive damage to the mortar; spalling damage due to short-term water damage is dominated by cohesive damage to the mortar, and spalling damage due to long-term water damage is dominated by bonding damage between the asphalt and aggregate.

Studies have shown that asphalt can only form a slurry with mineral fillers in order to adsorb and bond coarse and fine aggregates, forming an asphalt mixture with a certain strength and deformability [20–23]. Therefore, it is necessary to study the influence of the fillers on asphalt mastic and asphalt mixes. Zhu et al. [24] tested six mineral powders for some basic physicochemical parameters (density, hydrophilicity coefficient, methylene blue value) and analyzed the particle size of each mineral powder in detail using a laser particle size analyzer. A slurry was prepared using No. 70 matrix asphalt at different powder-to-mastic ratios and the filler properties were evaluated based on the slurry properties. Li et al. [25] characterized the particle size distribution, chemical composition and microstructure of nine fillers and used the bond strength (BBS) to evaluate the performance of different asphalt mastics.

As can be seen, the influence of fillers on the bond strength of asphalt and asphalt mastic has been of interest to numerous researchers, and many attempts have been made to explore it. However, in regard to the selection of fillers, existing studies have mainly chosen single fillers such as limestone, fly ash, etc., without adding additives to replace the existing fillers in equal amounts and measuring the performance of the newly composed mixed fillers. Therefore, in this study, three grades of limestone mineral powder were prepared by regulating the original grade of limestone mineral powder and adding three additives, cement, slaked lime and recycled powder, to the original grade to replace the mineral powder, and the properties of these 12 fillers were characterized. The asphalt mastic was then prepared for BBS tests, followed by analysis of variance (ANOVA) of the test data to consider the effect of filler type and hydrothermal coupling on the bond strength of the asphalt mastic. Finally, the correlation between the different filler indicators and the bond strength of the asphalt mastic was analyzed using grey correlation analysis.

2. Materials and Methods

2.1. Materials

2.1.1. Fillers

Using limestone mineral powder (powder obtained by crushing mined ore), the filler grade distribution was first analyzed using a laser particle size analyzer (Microtrac S3500, Microtrac Inc., Largo, FL, USA), and the original grade was used as the median grade, on the basis of which coarse and fine grades were adjusted (the adjustment method considered a constant grade, increasing the weight of the coarse or fine particle size). The density and hydrophilicity coefficients of the three particle size gradations were tested, as shown in Table 1.

Table 1. Filler density and hydrophilic coefficient.

Filler	Density (g/cm^3)	Hydrophilic Coefficient
Mineral powder (fine)	2.708	0.509
Mineral powder (medium)	2.669	0.520
Mineral powder (coarse)	2.659	0.565
Mineral powder + 10% recycled powder	2.720	0.582
Mineral powder + 20% recycled powder	2.742	0.632
Mineral powder + 30% recycled powder	2.791	0.658
Mineral powder + 20% cement	2.822	0.714
Mineral powder + 40% cement	2.854	0.738
Mineral powder + 60% cement	2.913	0.792
Mineral powder + 30% slaked lime	2.639	0.623
Mineral powder + 50% slaked lime	2.528	0.651
Mineral powder + 70% slaked lime	2.482	0.684

2.1.2. Additives

Three additives were considered: slaked lime, recycled powder (dust generated during the production by asphalt mixing plants) and cement. The substitution ratios for slaked lime were 30%, 50% and 70%, and for recycled powder 10%, 20% and 30%, while the cement substitution ratios were 20%, 40% and 60%. The densities and hydrophilic coefficients of the nine fillers containing additives were tested, as shown in Table 1.

2.1.3. Asphalt

As the asphalt, 70# base asphalt and finished SBS modified asphalt were selected. Their basic specifications are shown in Tables 2 and 3.

Table 2. Properties of 70# base asphalt.

Items		Measured Value	Test Method [26]
Penetration, 0.1 mm		65.4	T0604
Ductility (5 cm/min, 15 °C), cm		120.4	T0605
Softening Point, °C		48.1	T0606
Thin film oven test (163 °C, 5 h)	Mass variation, %	0.45	T0609
	Residual penetration ratio, %	61.6	T0604
	Ductility (5 °C), cm	79.2	T0605
Dynamic viscosity (60 °C, Pa·s)		290	T0620

Table 3. Properties of SBS modified asphalt.

Items		Measured Value	Test Method [26]
Penetration, 0.1 mm		55.5	T0604
Ductility (5 cm/min, 5 °C), cm		37.9	T0605
Softening Point, °C		86	T0606
Thin film oven test (163 °C, 5 h)	Mass variation, %	−0.15	T0609
	Residual penetration ratio, %	73.2	T0604
	Ductility (5 °C), cm	25	T0605
Dynamic viscosity (60 °C, Pa·s)		4841	T0620

2.1.4. Preparation of Asphalt Mastics

The mass ratio of filler to asphalt was 1, prepared as follows:

A sufficient amount of mineral powder was put in a tin box and placed in an oven at 105 °C for 4 h to eliminate moisture in the mineral powder, kept warm and set aside; a sufficient amount of asphalt was placed in a constant-temperature oven and heated to a flowing state. The heating temperature for SBS modified asphalt was 160 °C, and the heating temperature for 70# base asphalt was 150 °C. Heating time was about 2 h. The asphalt was dispensed into beakers (the mass of bitumen per cup dispensed was recorded). Manual pre-mixing was performed. The divided asphalt was placed on an electric stove to keep the SBS modified asphalt at 160 °C and the 70# base asphalt at 150 °C. The mineral powder was gradually added (powder to mastic ratio of 1) while stirring. The asphalt mastic was put in an insulation jacket and placed under high-speed shear with the rotor immersed in the asphalt mastic. The rotor was turned on and speed was gradually increased to 1500 rpm. The asphalt mastic was stirred for 30 min at a constant stirring temperature. After completion, the mastic was obtained according to the mixture.

2.2. Test and Evaluation Methods

2.2.1. Characterization of Filler Properties

The particle size distribution of fillers reflects their gradation distribution. In this paper, the particle size distribution of 12 fillers was analyzed using a laser particle size analyzer, which applies the Mie scattering theory and tests particle size distribution based on the physical phenomena of diffraction and scattering of light by particles. When the laser beam, which is highly monochromatic and highly directional, is blocked by a particle, a portion of the light is scattered, the angle of the scattered light being inversely proportional to the diameter of the particle. It has been shown that the intensity of the scattered light represents the number of particles of that size and that the scattered light is processed optically. When scattered light at different angles is directed through a Fourier lens onto a multivariate photodetector, the light signals are converted into electrical signals and transmitted to a computer, where they are processed by special software to obtain an accurate particle size distribution [24].

Three parallel analyses were carried out for each filler sample to ensure that the values obtained were reliable. In addition to the particle size distribution curves, indicators such as particle size d (0.1) for 10% passage, d (0.5) for 50% passage, d (0.9) for 90% passage and volume mean particle size D [4,3] were obtained, as shown in Equation (1).

$$\mathrm{D}\,[4,3] = \frac{\sum m_i \times d_i^4}{\sum m_i \times d_i^3} \quad (1)$$

where m_i is the frequency of particles in size class i, having a mean d_i diameter.

In addition, the fineness modulus (FM) is used to reflect the particle size distribution of the filler, as shown in Equation (2) [27].

$$\mathrm{FM} = \frac{\sum(100 - p_i)}{100} \quad (2)$$

where: i ∈ [75,50,30,20,10,5,3,1] μm, and pi is the cumulative passage rate of the ith particle size, %.

2.2.2. Binder Bond Strength Test

The BBS test system consists of an automatic adhesion tester (Positest AT-A, DeFelsko, NY, USA.), an aggregate substrate, a pulling head and an asphalt sample [28]. The AT-A, with its self-aligning system, produces a constant pulling rate, separating the asphalt from the aggregate substrate and allowing the measurement of POTS. There are two types of damage in the BBS test, adhesion damage and cohesion damage, and the corresponding POTS can be divided into adhesion strength (between asphalt and aggregate) and cohesion strength (within the asphalt) [29].

The original pulling head proved to be somewhat problematic in terms of controlling the thickness of the asphalt mastic and the bonding of the asphalt mastic to the pulling head, so in order to improve the accuracy of the BBS test data, some modifications were made to the pulling head. Compared to the original pulling head, the improved pulling head was 20 mm in diameter and 0.2 mm thick around the edge, which provided a lateral restraint for the asphalt mastic during the sample preparation process. The surface of the pulling head was then roughened to improve the adhesion between the pulling head and the asphalt paste. In addition, four cuts were made along the edges to allow excess bitumen to escape when the pulling head was pressed against the aggregate substrate. These modifications allowed the thickness of the asphalt to be controlled and ensured a complete bond between the asphalt and the pulling head [29].

In order to determine the thickness and loading rate of the asphalt tested in the BBS test, Huang et al. [30] recommended a thickness of 0.2 mm and a loading rate of 0.7 MPa/s for the impact test of asphalt. The BBS test followed the following steps:

(1) Slab preparation

Firstly, large pieces of basalt obtained from the quarry were cut into rectangular blocks of 12 cm × 12 cm × 2.5 cm. As the saw teeth used during the cutting process can make the surface of the slabs uneven and of varying roughness, if used directly for the preparation of BBS specimens, this can lead to large errors in the test results, so the surface of the slabs needed to be polished [31]. The stones were sanded with 60 and 180 grit sandpaper for 3 min each and then placed in an ultrasonic cleaner at 60 °C for 30 min to remove the dust from the surface. The stones were then dried in an oven at 160 °C for 2 h. The stones were then removed and set aside.

(2) Specimen preparation

To match the actual asphalt mix, the asphalt mastic was heated at the mixing temperature specified in the specification, while the stone and the pulling head were kept at the same temperature. After the asphalt mastic had softened, a glass rod was dipped into the asphalt mastic and dripped onto the pulling head. The pulling head was then quickly pressed vertically against the surface of the stone and pressure was applied so that the excess asphalt flowed out of the opening, and was held for 10 s. This process must be carried out in an oven to reduce the cooling of the slab, and the drawing head may not be moved again during this period. After the specimen was formed, the slab was removed and the excess asphalt mastic extruded from the bottom was removed using a scraper to prevent it from affecting the drawing results. After completion, the aggregate slab was placed on the drawing head and left at room temperature for 24 h.

(3) Working conditions

The prepared BBS specimens were placed in water at 60 °C, and the effect of three water bath regeneration times of 1 d, 3 d and 5 d on the bond strength of different asphalt mastics was considered.

(4) Testing

The specimens were placed in an oven at 25 °C for 1 h before testing. The bond strength of the asphalt mastic to the aggregate was tested using a Positest AT-A bond tester (Positest AT-A, DeFelsko, NY, USA). The process must be carried out in the oven so as to minimize the effect of temperature variations on the pull-out strength and increase the reliability of the test, as shown in Figure 1. If the bitumen breaks from the interface between the pulling head and the bitumen, this indicates that the bond strength between the bitumen and aggregate is greater than the bond strength between the bitumen and the pulling head. If more than 50% of the final fracture surface slab is adhered to the asphalt, the damage is cohesive; otherwise the damage is adhesive. The data on the display was read and recorded. Each set of tests was repeated three times, and the average of the valid data was taken as the test result.

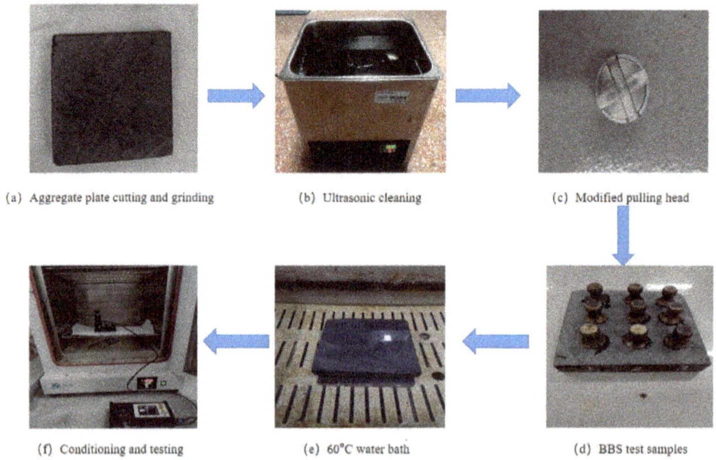

Figure 1. Test flow chart.

2.2.3. Grey Correlation Analysis

The aim of grey correlation analysis is to find important relationships between the factors of a system, which are based on the grey correlation degree. The grey correlation analysis method is a quantitative description and comparison of the developmental trends of a system. The grey correlation is calculated for each comparison series and a reference series, where the greater the correlation with the reference series, the closer the relationship with the reference series. The grey correlation analysis method does not result in a discrepancy between quantitative results and qualitative analysis results. The basic idea is to dimensionlessly process the original observation data of the evaluation indexes, calculate the correlation coefficient and the correlation degree, and rank the indicators to be evaluated according to the magnitude of the correlation degree [1].

3. Results and Discussion

3.1. Characterization of Filler Properties

The data obtained by the laser particle size analyzer was plotted as a particle size distribution graph [32,33], as shown in Figure 2.

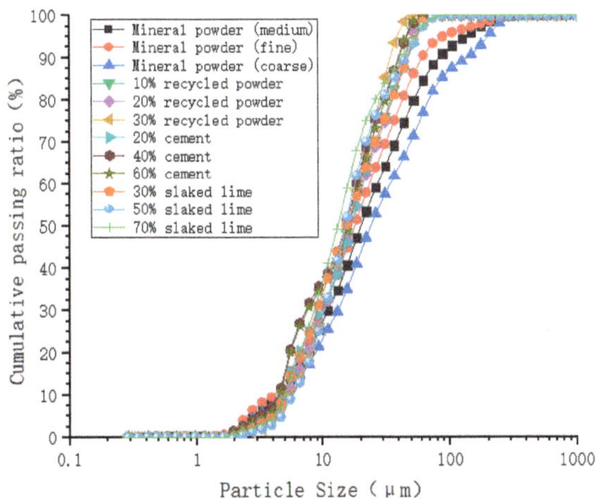

Figure 2. Particle size distributions of different fillers.

As can be clearly seen from the particle size graph of the filler shown in Figure 2, the particle size of the mineral powder (coarse) is the largest, and the filler without additives is somewhat larger than the filler with additives, which indicates that the particle size of additives is generally smaller than that of the mineral powder. The particle sizes of the nine fillers containing additives are relatively similar, requiring the calculation of d (0.1), d (0.5), d (0.9), d [4,3] and FM to quantify the particle size distribution. As shown in Figure 2, the particle size is obviously smaller when replacing mineral powder with additives. In terms of the fineness modulus, 70% slaked lime replacing mineral powder has the smallest particle size, while 50% slaked lime replacing mineral powder is obviously larger, followed by 60% cement replacing mineral powder and 40% cement replacing mineral powder, which shows that the filler particle size not only depends on the type of additives but also on the replacement ratio of additives. The volume average particle size D [4,3] shows that 30% recycled powder replacing mineral powder has the smallest particle size, followed by 60% cement replacing mineral powder and then 40% cement replacing mineral powder, indicating that the cement replacing mineral powder has a significant effect on reducing the particle size of the filler, as shown in Table 4.

3.2. Analysis of Asphalt Mastic Bond Strength and Damage Forms

Data measured with the Posi Test At-A bond tester was recorded and photographed for analysis of the form of damage to the stones and pulling heads, and the average of three parallel specimens under the same working conditions for each sample was taken for analysis to obtain the bond strength of different asphalt mastics under different hydrothermal action times, as shown in Figure 3.

Table 4. Calculated evaluation indicators of different fillers.

Filler	d (0.1)	d (0.5)	d (0.9)	D [4,3]	FM
Mineral powder (fine)	4.54	18.41	61.01	173.56	4.71
Mineral powder (medium)	4.60	19.98	82.79	204.56	4.90
Mineral powder (coarse)	4.93	23.86	137.70	244.53	5.23
10% recycled powder	4.90	16.70	47.87	49.49	4.49
20% recycled powder	5.28	16.84	44.97	49.49	4.43
30% recycled powder	5.27	15.00	34.55	36.82	4.30
20% cement	4.74	16.86	47.59	55.89	4.47
40% cement	4.44	15.19	40.11	44.94	4.28
60% cement	4.55	15.91	39.8	43.67	4.27
30% slaked lime	4.89	15.37	46.85	52.99	4.44
50% slaked lime	5.73	15.01	44.76	58.25	4.41
70% slaked lime	4.96	13.23	42.32	58.39	4.18

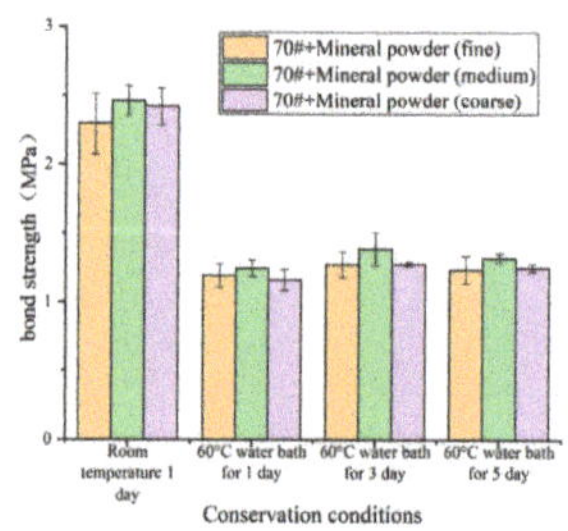

(**a**) 70# base asphalt + mineral powder

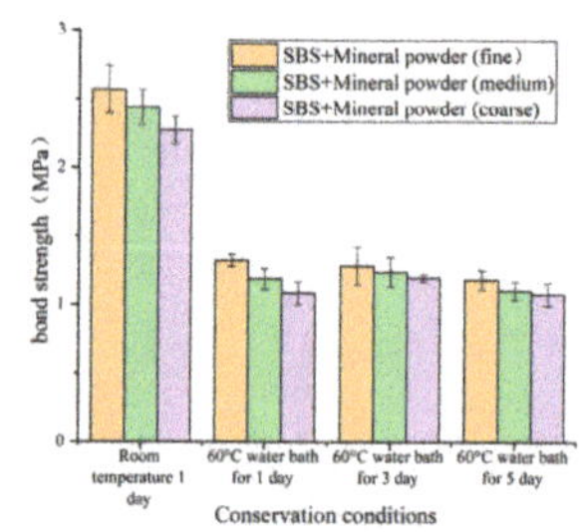

(**b**) SBS modified asphalt + mineral powder

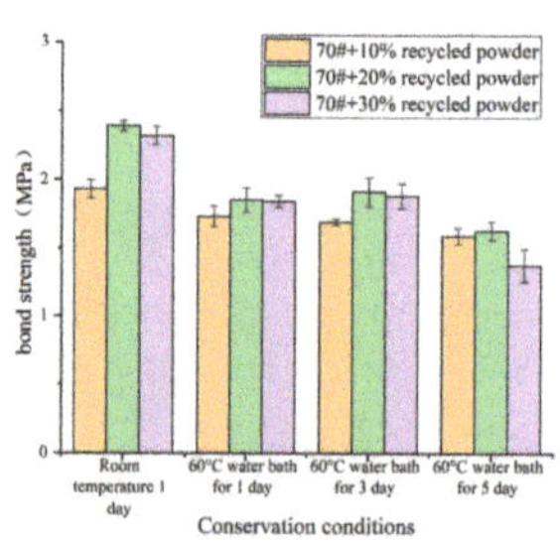

(**c**) 70# base asphalt + recycled powder

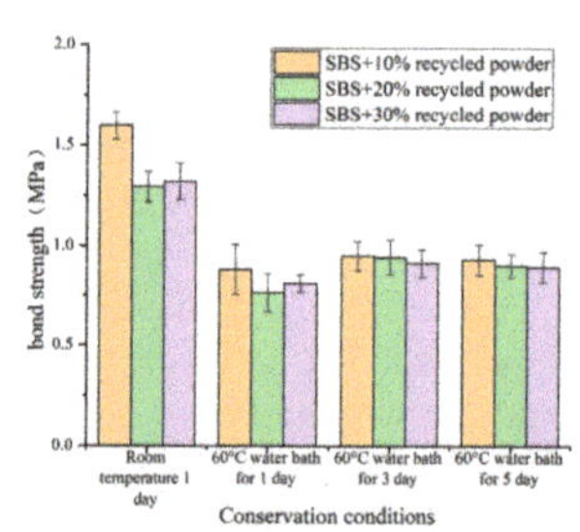

(**d**) SBS modified asphalt + recycled powder

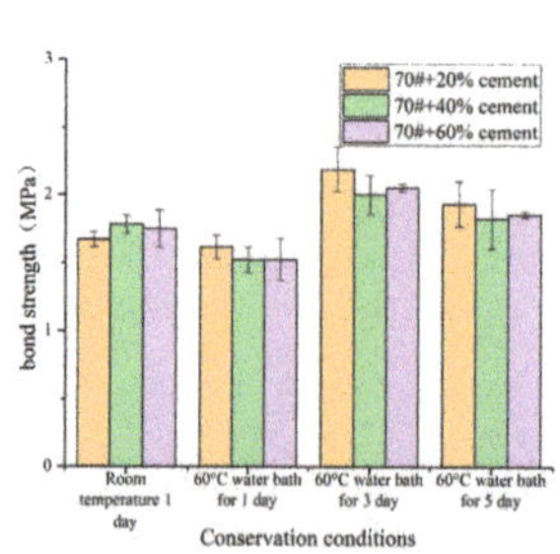

(**e**) 70# base asphalt + cement

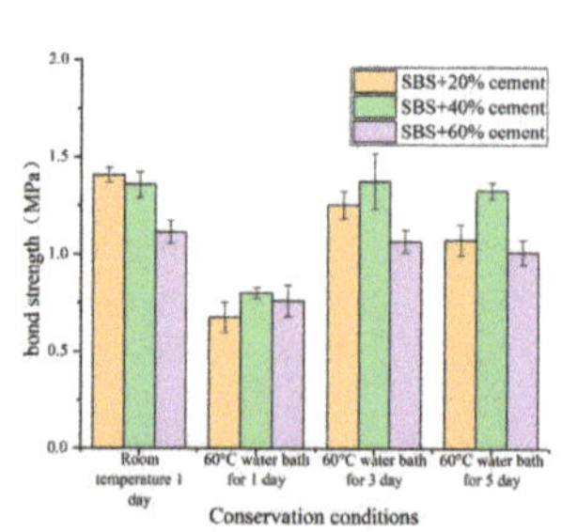

(**f**) SBS modified asphalt + cement

Figure 3. *Cont.*

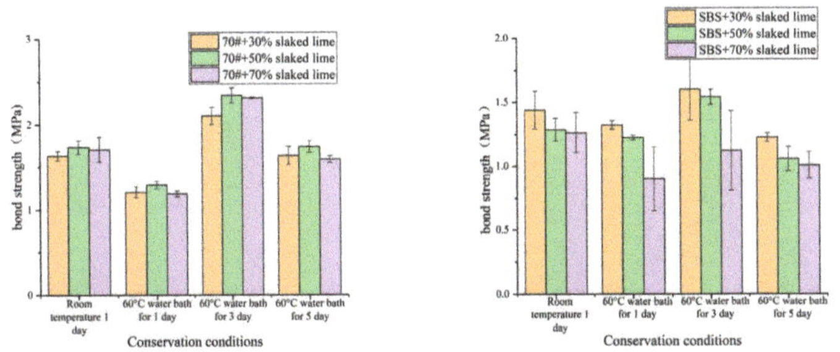

(**g**) 70# base asphalt + slaked lime (**h**) SBS modified asphalt + slaked lime

Figure 3. Asphalt mastics bond strength.

The X-axis represents how the specimen was maintained after preparation, and the time duration.

As can be seen from Table 5, under the action of hydrothermal coupling, the damage form of the 24 asphalt mastic bond strength tests changed gradually from cohesive damage to the asphalt mastic itself to adhesion damage between the asphalt mastic and the aggregate, as shown in Figure 4. When the adhesion damage between the asphalt mastic and the aggregate is changed, it indicates that the bond strength between the asphalt mastic and the aggregate is less than the bond strength of the asphalt mastic itself.

Table 5. Forms of damage in asphalt mastic bond strength tests.

Filler	Forms of Damage							
	Room Temperature 1 Day		60 °C Water Bath for 1 Day		60 °C Water Bath for 3 Day		60 °C Water Bath for 5 Day	
	SBS	70#	SBS	70#	SBS	70#	SBS	70#
Mineral powder (fine)	C	C	C	C	C	C	C	C
Mineral powder (medium)	C	C	C	C	C	C	A	C
Mineral powder (coarse)	C	C	C	C	C	C	A	C
10% recycled powder	C	C	*	C	*	C	*	C
20% recycled powder	C	C	*	C	*	C	*	C
30% recycled powder	C	C	*	C	*	C	*	C
20% cement	C	C	A	A	*	A	*	A
40% cement	C	C	C	A	C	A	C	A
60% cement	C	C	C	C	C	C	C	C
30% slaked lime	C	C	C	C	A	A	*	A
50% slaked lime	C	C	C	C	C	C	A	C
70% slaked lime	C	C	C	C	C	C	C	C

* represents damage between the asphalt and the pulling head; A represents asphalt–aggregate adhesion damage; C represents cohesive damage to the asphalt itself.

3.2.1. Influence of Filler and Hydrothermal Coupling on Bond Strength of Mastic

As can be seen from Figure 3, the bond strength after 24 h in the water bath is significantly lower than that after 24 h at room temperature under hydrothermal coupling, although the bond strength of the asphalt mastic increases again after 72 h in the water bath. The results show that the hydrothermal coupling does not necessarily result in a continuous reduction of the bond strength of the asphalt mastic in the short term, but that the bond strength fluctuates due to the asphalt, the filler and the hydrothermal coupling.

In regard to the mineral powder, the highest bond strength was found in the asphalt mastic made from mineral powder (fine) and SBS modified asphalt, while higher bond strength was found in the mastic made from 70# base asphalt and mineral powder (medium). The results show that the particle size of the mineral powder does not completely determine the bond strength of the asphalt mastic, and that the type of asphalt is also an influencing factor. In terms of the form of damage, the asphalt mastic made using mineral powder as filler changes from cohesive damage to the asphalt mastic itself to adhesion damage between the asphalt mastic and the aggregate under the effect of hydrothermal coupling.

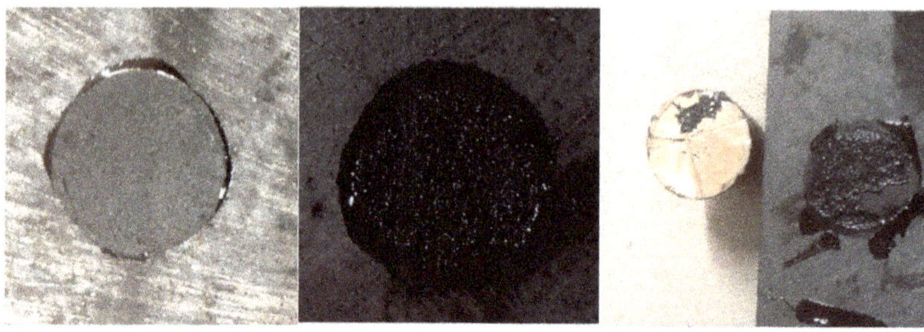

(a) adhesion damage (b) cohesive damage (c) Shedding damage

Figure 4. Forms of damage.

In regard to the recycled powder, the bond strength of the asphalt mastic made from 20% recycled powder replacing mineral powder and 70# base asphalt was the highest, while the bond strength of the mastic made from SBS modified asphalt and filler replaced with recycled powder was higher for the 10% recycled powder replacing mineral powder. The results show that the proportion of recycled powder to replace mineral powder should not be too large, but the mineral powder can be mixed with a small amount of recycled powder. This achieves environmental protection and resource recycling, but does not produce a large change in the performance of the asphalt mastic. In terms of the form of damage, the asphalt mastic made from recycled powder replacing mineral powder as filler with SBS modified bitumen changed from cohesive damage of the asphalt mastic itself to adhesion damage between the asphalt mastic and the aggregate under hydrothermal coupling, and in some specimens to damage between the asphalt mastic and the pulling head, indicating that moisture penetrated between the asphalt mastic and the aggregate plate and between the asphalt mastic and the pulling head for damage, thus indicating that the recycled powder was not as effective as mineral powder as a filler.

In regard to the cement, the bond strength of the asphalt mastic made from 20% cement-replaced mineral powder and 70# base asphalt was the highest, while the bond strength of the slurry made from SBS modified asphalt and filler replaced with cement was higher for the 40% cement-replaced mineral powder. The results indicate that the proportion of cement replacing mineral powder should not be too large, ideally 20% to 40%. The difference is that the bond strength of the 70# base asphalt and the cement-replaced filler is higher when kept in a water bath for 3 days than at room temperature for 1 day, probably due to the reaction of the cement itself with water. When cement meets water, the calcium silicate in it reacts with the water to produce hydrated calcium silicate and calcium hydroxide. The hydrated calcium silicate and calcium hydroxide will rely on hydrogen bonds and intermolecular forces to bind together to form a very high-strength curing material, which has an impact on the bond strength of the asphalt mastic. It is also clear that when cement replaces mineral powder, water does not necessarily have a negative effect on the asphalt mastic, but may also have a beneficial effect on the performance of the asphalt mastic.

In regard to the slaked lime, the bond strength of the asphalt mastic made from 50% slaked lime replaced with mineral powder and 70# base asphalt was highest, while the bond strength of the slurry made from SBS modified asphalt and filler replaced with slaked lime was higher for the mastic made from 30% slaked lime replaced with mineral powder. As with the cement, the bond strength of the 70# base asphalt with filler replaced with slaked lime was higher when kept in a water bath for 3 days than at room temperature for 1 day, which may be related to the effect of slaked lime on the stone. The slaked lime interacts with the acidic stone surface and takes on the nature of an alkaline stone, and the acidic stone surface becomes hydrophobic and lipophilic. When water penetrates the stone to the asphalt interface, the stone becomes more easily bonded to the asphalt and distant from the water and adhesion is improved.

3.2.2. Influence of Type of Asphalt on Bond Strength of Mastic

As can be seen in Figure 3, the bond strength of the mastic varies for the same filler and the same hydrothermal coupling conditions, due to the asphalt. Overall, asphalt mastics made from base asphalt have a greater bond strength than those made from SBS modified asphalt.

From the "mechanical theory" of asphalt adhesion theory, we can learn that there are usually many rough, porous structures on the surface of the aggregate. The asphalt can enter these pores after high temperature flow, and when the asphalt cools and hardens, it forms a "mechanical anchoring force" in the pores. However, the incorporation of SBS causes different degrees of agglomeration within the asphalt, so that the asphalt is not easily adsorbed by the small pores on the surface of the slab, leading to a decrease in the adhesion capacity of the asphalt.

Conversely, SBS modified asphalt has good elastic properties. When evaluating its bond strength, it is necessary to consider not only pulling strength but also the type of pulling process it can withstand. Testing clearly reveals that the changes in the values of the two types of asphalt mastic are different. The pull-out strength of the matrix asphalt drops to zero immediately after reaching its peak, while the pull-out strength of the SBS modified asphalt drops gradually to zero after reaching its peak. This indicates that the mastic specimen made from SBS modified bitumen was still able to withstand a certain tensile load after reaching the pull-out damage value, which indicates that it required more external tensile work than the matrix asphalt mastic specimen to break it completely. In this regard, the bond strength data for 70# base asphalt is more informative.

3.3. *Analysis of Variance for BBS Test Data*

BBS tests were carried out to investigate the effect of filler type and hydrothermal coupling and their interaction with the bond strength of the outgoing asphalt mastic. When testing 70# base asphalt mastic, 12 filler types were used to evaluate the hydrothermal coupling effect. The material was kept at room temperature for 1 day, then placed in a 60 °C water bath for 1 day, 3 days, and 5 days under 4 kinds of working conditions, for a total of 48 tests. Each test was performed 3 times, and the obtained asphalt mastic bond strengths are shown in Table 6.

A two-factor repeated experiment ANOVA was used to analyse the significance of the effect of filler type and hydrothermal action and their interaction on the bond strength of the outgoing asphalt mastic at a significance level of $\alpha = 0.01$ [34].

As can be seen from Table 7, Factor A (filler type), Factor B (hydrothermal coupling effect), and the interaction of Factor A and Factor B all have a significant effect on the bond strength of the asphalt mastic.

In general, the larger the difference between the critical value and the F-value, the greater the influence of the factor corresponding to the F-value on the experimental results. It also indirectly indicates that this factor is more important and requires special attention. The magnitude of the F-value can be used to rank the factors in order of priority: factor B (hydrothermal coupling effect) is the most important, followed by factor A and factor B interaction, and finally factor A (filler type). The interaction is often also a factor influencing the results of experiments performed by the subjects.

Table 6. Asphalt mastics bond strength.

Filler Types	Asphalt Mastics Bond Strength/MPa			
	B1 (Room Temperature 1 Day)	B2 (60 °C Water Bath for 1 Day)	B3 (60 °C Water Bath for 3 Day)	B4 (60 °C Water Bath for 5 Day)
A1 (Fine mineral powder)	1.98	1.12	1.40	1.22
	2.47	1.14	1.19	1.36
	2.42	1.31	1.22	1.12
A2 (Medium mineral powder)	2.61	1.28	1.55	1.31
	2.42	1.29	1.29	1.28
	2.35	1.16	1.31	1.36
A3 (Coarse mineral powder)	2.25	1.18	1.27	1.21
	2.59	1.24	1.29	1.25
	2.41	1.06	1.26	1.28
A4 (20% cement)	1.64	1.52	2.42	1.96
	1.75	1.73	2.03	1.72
	1.63	1.60	2.12	2.12
A5 (40% cement)	1.87	1.65	1.82	1.71
	1.77	1.43	2.02	1.64
	1.71	1.49	2.17	2.13
A6 (60% cement)	1.78	1.57	2.03	1.88
	1.57	1.69	2.09	1.86
	1.90	1.32	2.04	1.83
A7 (30% slaked lime)	1.70	1.29	2.10	1.62
	1.64	1.15	2.22	1.77
	1.57	1.17	1.98	1.51
A8 (50% slaked lime)	1.64	1.26	2.46	1.67
	1.74	1.34	2.26	1.72
	1.83	1.24	2.30	1.83
A9 (70% slaked lime)	1.88	1.19	2.31	1.65
	1.52	1.14	2.30	1.58
	1.72	1.22	2.32	1.55
A10 (10% recycled powder)	1.86	1.65	1.68	1.62
	2.02	1.71	1.72	1.51
	1.91	1.83	1.67	1.64
A11 (20% recycled powder)	2.41	1.98	1.84	1.72
	2.42	1.79	1.83	1.59
	2.34	1.80	2.06	1.57
A12 (30% recycled powder)	2.23	1.82	1.83	1.20
	2.34	1.90	2.01	1.46
	2.38	1.80	1.81	1.46

Table 7. Two-factor repeated trial ANOVA table.

Sources of Variance	Sum of Squared Deviations	Degrees of Freedom	Sum of Mean Deviations Squared	F	Threshold F_{α} (n_1, n_2)	Significance
Factor A (Filler types)	2.5021	11	0.2275	15.9091	$F_{0.01}$ (11, 96) = 2.484	Significant
Factor B (Water-heat coupling)	7.1495	3	2.3832	166.6573	$F_{0.01}$ (3, 96) = 4.022	Significant
Interaction A × B	10.2901	33	0.3118	21.8042	$F_{0.01}$ (33, 96) = 1.526	Significant
Errors E	1.3769	96	0.0143			
Total	21.3186	143				

3.4. *Correlation Analysis of Different Filler Indicators with Bond Strength of Asphalt Mastic*

In this paper, the bond strength of each 70# base asphalt mastic after 1 day at room temperature was selected with reference to the actual measured data for physical and chemical indices of different fillers as a comparative series (see Table 8).

Table 8. Grey correlation analysis of raw data.

Mastic Number	A_1	A_2	A_3	A_4	A_5	A_6	A_7	A_8	A_9	A_{10}	A_{11}	A_{12}
X_0 (Bond strength)	2.29	2.46	2.42	1.67	1.78	1.75	1.64	1.74	1.71	1.93	2.39	2.32
X_1 (Density)	2.71	2.67	2.66	2.72	2.74	2.79	2.82	2.85	2.91	2.64	2.53	2.48
X_2 (hydrophilic coefficient)	0.51	0.52	0.57	0.58	0.63	0.66	0.71	0.74	0.79	0.62	0.65	0.68
X_3 (Fineness modulus)	4.71	4.90	5.23	4.49	4.43	4.30	4.47	4.28	4.27	4.44	4.41	4.18
X_4 (Average particle size)	174	205	245	49.5	49.5	36.8	55.9	44.9	43.7	53.0	58.3	58.4
X_5 [d (0.1)]	4.54	4.60	4.93	4.90	5.28	5.27	4.74	4.44	4.55	4.89	5.73	4.96
X_6 [d (0.5)]	18.4	20.0	23.9	16.7	16.8	15.0	16.9	15.2	15.9	15.4	15.0	13.2
X_7 [d (0.9)]	61.0	82.8	138	47.9	45.0	34.6	47.6	40.1	39.8	46.9	44.8	42.3

Each series is averaged; that is, the individual data in the series is divided by the average of each series, and the averaging process produces a new series. The new series after the averaging process is shown in Table 9.

Table 9. Series generated from correlation analysis.

Mastic Number	A_1	A_2	A_3	A_4	A_5	A_6	A_7	A_8	A_9	A_{10}	A_{11}	A_{12}
X_0 (Bond strength)	1.14	1.23	1.20	0.83	0.89	0.87	0.82	0.87	0.85	0.96	1.19	1.15
X_1 (Density)	1.00	0.99	0.98	1.00	1.01	1.03	1.04	1.05	1.07	0.97	0.93	0.92
X_2 (hydrophilic coefficient)	0.80	0.81	0.89	0.91	0.99	1.03	1.11	1.16	1.24	0.97	1.02	1.07
X_3 (Fineness modulus)	1.04	1.09	1.16	1.00	0.98	0.95	0.99	0.95	0.95	0.98	0.98	0.93
X_4 (Average particle size)	1.94	2.29	2.74	0.55	0.55	0.41	0.62	0.50	0.49	0.59	0.65	0.65
X_5 [d (0.1)]	0.93	0.94	1.01	1.00	1.08	1.07	0.97	0.91	0.93	1.00	1.17	1.01
X_6 [d (0.5)]	1.09	1.19	1.42	0.99	1.00	0.89	1.00	0.90	0.94	0.91	0.89	0.78
X_7 [d (0.9)]	1.09	1.48	2.47	0.86	0.81	0.62	0.85	0.72	0.71	0.84	0.80	0.76

Results of the correlation series:

$\varepsilon_1 = (0.856, 0.771, 0.785, 0.829, 0.873, 0.840, 0.782, 0.814, 0.784, 0.996, 0.758, 0.772)$

$\varepsilon_2 = (0.700, 0.659, 0.720, 0.922, 0.897, 0.836, 0.730, 0.732, 0.672, 1.000, 0.827, 0.908)$

$\varepsilon_3 = (0.900, 0.857, 0.958, 0.836, 0.902, 0.915, 0.823, 0.913, 0.899, 0.983, 0.793, 0.781)$

$\varepsilon_4 = (0.494, 0.423, 0.337, 0.741, 0.704, 0.632, 0.811, 0.687, 0.688, 0.683, 0.594, 0.612)$

$\varepsilon_5 = (0.791, 0.737, 0.804, 0.832, 0.813, 0.800, 0.845, 0.962, 0.919, 0.967, 0.985, 0.854)$

$\varepsilon_6 = (0.951, 0.963, 0.792, 0.841, 0.888, 0.990, 0.814, 0.967, 0.904, 0.953, 0.727, 0.682)$

$\varepsilon_7 = (0.951, 0.759, 0.382, 0.983, 0.913, 0.761, 0.967, 0.849, 0.858, 0.873, 0.671, 0.667)$

Correlation of filler indicators with the bond strength of asphalt mastic:

$\beta_1 = 0.822$, $\beta_2 = 0.800$, $\beta_3 = 0.880$, $\beta_4 = 0.617$, $\beta_5 = 0.859$, $\beta_6 = 0.873$, $\beta_7 = 0.803$

As can be seen in Figure 5, the correlation between filler index and asphalt mastic bond strength, in descending order, is:

$\beta_3 > \beta_6 > \beta_5 > \beta_7 > \beta_1 > \beta_2 > \beta_4$

The correlation between the three indicators of fineness modulus, d (0.5), d (0.1) and the bond strength of asphalt mastic, is greater than 0.85, while the correlation between density, d (0.9) and the hydrophilic coefficient is greater than 0.8. The correlation between the average particle size and the bond strength of asphalt mastic is slightly lower than the others. Therefore, in regard to asphalt mastic bond strength, we should focus on the fineness modulus, d (0.5), d (0.1), to conduct targeted analysis and research.

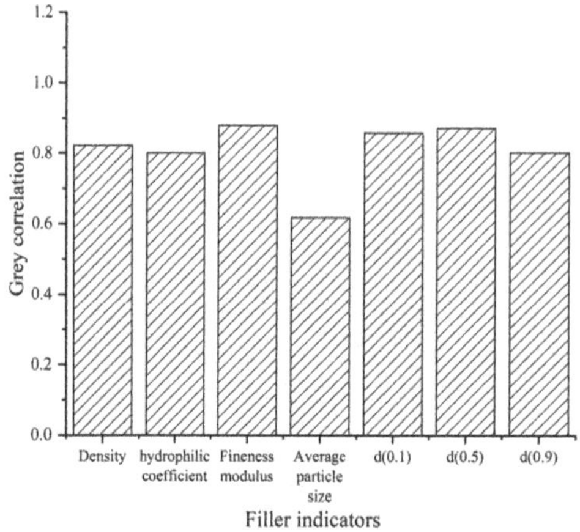

Figure 5. Grey correlation between filler indicators and asphalt mastic bond strength.

4. Conclusions

This paper investigates performance indicators for limestone fines and a total of 12 fillers after replacing the fines with cement, slaked lime and recycled powder. BBS tests were used to evaluate the adhesion and cohesion of the asphalt, while ANOVA data were used to consider the effect of filler type and hydrothermal coupling on the bond strength of the asphalt mastic. Finally, the correlation between different filler indicators and the bond strength of the asphalt mastic was analyzed using grey correlation analysis. The following conclusions can be drawn from this study:

1. Hydrothermal coupling does not necessarily result in a continuous reduction in the bond strength of the asphalt mastic in the short term; rather, the bond strength will fluctuate due to the asphalt, filler and hydrothermal coupling.
2. Filler type, hydrothermal coupling, and the interaction between the two have a significant effect on the bond strength of the asphalt mastic. The order of influence is: hydrothermal coupling > interaction of hydrothermal coupling and filler type > filler type.
3. In regard to asphalt mastic bond strength, the three indicators of fineness modulus, d (0.5) and d (0.1) should be focused on for targeted analysis and research.
4. In regard to mineral fillers, the particle size of the mineral powder should not be too large. The addition of slaked lime and cement helps to improve the moisture resistance of asphalt mastic, but more is not necessarily better.

It should be noted that the asphalt mastic selected for this test was only made with a powder-to-mastic ratio of 1 and a test temperature of 25 °C. The effect of different powder-to-mastic ratios and different test temperatures on the bond strength of the asphalt mastic will be considered in further studies.

Author Contributions: Conceptualization, X.M. and J.K.; methodology, X.M.; software, S.L.; validation, J.K., M.T. and D.W.; formal analysis, J.K.; investigation, S.L..; resources, X.M.; data curation, M.T.; writing—original draft preparation, J.K.; writing—review and editing, X.M.; visualization, S.L.; supervision, X.M.; project administration, J.K.; funding acquisition, X.M. All authors have read and agreed to the published version of the manuscript.

Funding: This research was supported by the Suzhou Science and Technology Research Project (No. SS202032) and the Hebei Transportation Scientific Research Project (No. TH1-202021). The authors gratefully acknowledge the support.

Institutional Review Board Statement: Not applicable.

Informed Consent Statement: Not applicable.

Data Availability Statement: All data that support the findings of this study are included within the article.

Conflicts of Interest: The authors have no conflict of interest regarding this manuscript.

References

1. Wang, X. Strength Mechanism of Porous Asphalt Mixtures and Their Attenuation Process under the Coupling of Multiple Fields. Ph.D. Thesis, Southeast University, Dhaka, Bangladesh, 2018.
2. Ma, X.; Jiang, J.; Zhao, Y.; Wang, H. Characterization of the interconnected pore and its relationship to the directional permeability of porous asphalt mixture. *Constr. Build. Mater.* **2021**, *269*, 121233. [CrossRef]
3. Wang, X.; Hu, X.; Gu, X.; Li, Q.; Li, W.; Chen, B. Microstructure and damage evolution of porous asphalt mixtures under coupled effects of high temperature and moisture. *Int. J. Pavement Eng.* **2021**, 1–15. [CrossRef]
4. Xu, F.; Zhao, Y.; Li, K. Using Waste Plastics as Asphalt Modifier: A Review. *Materials* **2022**, *15*, 110. [CrossRef]
5. Ma, X.; Zhou, P.; Wang, L.; Jiang, J.; Wang, J. Internal structure change of porous asphalt concrete under coupled conditions of load, moisture and temperature. *Constr. Build. Mater.* **2022**, *314*, 125603. [CrossRef]
6. Hu, X.; Wang, X.; Zheng, N.; Li, Q.; Shi, J. Experimental Investigation of Moisture Sensitivity and Damage Evolution of Porous Asphalt Mixtures. *Materials* **2021**, *14*, 7151. [CrossRef]
7. Li, Q.; Zeng, X.; Wang, J.; Luo, S.; Meng, Y.; Gao, L. Aging performance of high viscosity modified asphalt under complex heat-light-water coupled conditions. *Constr. Build. Mater.* **2022**, *325*, 126314. [CrossRef]
8. Cheng, Z.; Zhang, X.; Kong, F.; Guo, P. Analysis of the peeling behavior of asphalt films using surface energy theory and pull-off tests. *Mater. Guide* **2020**, *34*, 1288–1294.
9. Ma, X.; Hu, X.; Wang, L.; Zhong, G. Study on the effect of water-temperature coupling on asphalt binder properties based on surface free energy. *For. Eng.* **2022**, *38*, 140–146.
10. Shi, C.; Cai, X.; Wang, T.; Yi, X.; Liu, S.; Yang, J.; Leng, Z. Energy-based characterization of the fatigue crack density evolution of asphalt binders through controlled-stress fatigue testing. *Constr. Build. Mater.* **2021**, *300*, 124275. [CrossRef]
11. Gronniger, J.; Wistuba, M.P.; Renken, P. Adhesion in bitumen-aggregate-systems: New technique for automated interpretation of rolling bottle tests. *Road Mater. Pavement Des.* **2010**, *11*, 881–898. [CrossRef]
12. Huang, W.; Zhou, L. Evaluation of adhesion properties of modified asphalt binders with use of binder bond strength test. *Transp. Res. Rec.* **2017**, *2632*, 88–98. [CrossRef]
13. Han, Y.; Jiang, J.; Ding, J.; Zhao, Z.; Ma, X.; Ni, F. Investigation of the raveling potential of thin friction course (TFC) under freeze–thaw conditions. *Mater. Struct.* **2021**, *54*, 1–12. [CrossRef]
14. Li, N.; Xu, J.; Xu, T. Preparation, properties and modification mechanism of vulcanized eucommia ulmoides gum modified asphalt. *Constr. Build. Mater.* **2021**, *274*, 121992. [CrossRef]
15. Aguiar-Moya, J.P.; Baldi-Sevilla, A.; Salazar-Delgado, J.; Pacheco-Fallas, J.F.; Loria-Salazar, L.; Reyes-Lizcano, F.; Cely-Leal, N. Adhesive properties of asphalts and aggregates in tropical climates. *Int. J. Pavement Eng.* **2018**, *19*, 738–747. [CrossRef]
16. Zhang, J.; Airey, G.D.; Grenfell, J.; Yao, Z. Laboratory evaluation of Rediset modified bitumen based on rheology and adhesion properties. *Constr. Build. Mater.* **2017**, *152*, 683–692. [CrossRef]
17. Canestrari, F.; Cardone, F.; Graziani, A.; Santagata, F.A.; Bahia, H.U. Adhesive and cohesive properties of asphalt-aggregate systems subjected to moisture damage. *Road Mater. Pavement Des.* **2010**, *11* (Suppl. S1), 11–32. [CrossRef]
18. Huang, W.; Lv, Q. Investigation of critical factors determining the accuracy of binder bond strength test to evaluate adhesion properties of asphalt binders. *J. Test. Eval.* **2017**, *45*, 1270–1279. [CrossRef]
19. Moraes, R.; Velasquez, R.; Bahia, H. Using bond strength and surface energy to estimate moisture resistance of asphalt-aggregate systems. *Constr. Build. Mater.* **2017**, *130*, 156–170. [CrossRef]
20. Zhou, Z. Study on the Effect of Filler on High and Low Temperature Performance of Asphalt Mastic and Prediction of Asphalt Mastic Performance. Master's Thesis, Hunan University, Changsha, China, 2013.
21. Yao, X.; Xu, H.; Xu, T. Void distribution, interfacial adhesion and anti-cracking mechanisms of cold recycled asphalt mixture based on AFM and X-ray CT. *Appl. Surf. Sci.* **2022**, *606*, 155012. [CrossRef]
22. Zhao, Z.; Jiang, J.; Ni, F.; Ma, X. 3D-Reconstruction and characterization of the pore microstructure within the asphalt FAM using the X-ray micro-computed tomography. *Constr. Build. Mater.* **2021**, *272*, 121764. [CrossRef]
23. Li, S.; Ni, F.; Dong, Q.; Zhao, Z.; Ma, X. Effect of filler in asphalt mastic on rheological behaviour and susceptibility to rutting. *Int. J. Pavement Eng.* **2021**, *22*, 87–96. [CrossRef]
24. Zhu, H. Effect of Mineral Fillers on the Performance of Emulsified Asphalt Mastic. Master's Thesis, Shandong University of Construction, Jinan, China, 2015.
25. Li, C.; Liu, Q.; Lei, J.; Zhang, Z.; Fu, C.; Pan, F. Influence of mineral fillers properties on the bonding properties of bitumen mastics. *Constr. Build. Mater.* **2022**, *318*, 126013. [CrossRef]
26. China, Ministry of Transport of the People's Republic of China. *Standard Test Methods of Bitumen and Bituminous Mixtures for Highway Engineering, JTGE20-2011*; China Communications Press Co., Ltd.: Beijing, China, 2011.

27. Can, C.; Yang, H.; Fan, Z. Research on the influence of machine-made sand gradation and fineness modulus on the workabilit and rheological properties of mortar. In *IOP Conference Series: Earth and Environmental Science*; IOP Publishing: Bristol, UK, 202 Volume 760, p. 012031.
28. Yan, C.; Huang, W.; Lv, Q. Study on bond properties between RAP aggregates and virgin asphalt using Binder Bond Strength tes and Fourier Transform Infrared spectroscopy. *Constr. Build. Mater.* **2016**, *124*, 1–10. [CrossRef]
29. Wang, X.; Ren, J.; Gu, X.; Li, N.; Tian, Z.; Chen, H. Investigation of the adhesive and cohesive properties of asphalt, mastic, an mortar in porous asphalt mixtures. *Constr. Build. Mater.* **2021**, *276*, 122255. [CrossRef]
30. Huang, W.; Lv, Q.; Xiao, F. Investigation of using binder bond strength test to evaluate adhesion and self-healing properties o modified asphalt binders. *Constr. Build. Mater.* **2016**, *113*, 49–56. [CrossRef]
31. Meng, W. Evaluation of the Effect of Mineral Powder on the Performance of Asphalt Mastic and the Ability of Asphalt-Minera Powder Interaction. Master's Thesis, Yangzhou University, Yangzhou, China, 2022.
32. Cheng, Y.; Tao, J.; Jiao, Y.; Tan, G.; Guo, Q.; Wang, S.; Ni, P. Influence of the properties of filler on high and medium temperatur performances of asphalt mastic. *Constr. Build. Mater.* **2016**, *118*, 268–275. [CrossRef]
33. Mu, M.; Zhao, H.; Xu, X. Study on the calibration method of multi-parameter in situ laser particle size analyzer. *China Test.* **202** *46*, 95–99.
34. Wu, J.; Li, G.; Ma, L. Analysis of water treatment experimental data using two-factor analysis of variance. *Exp. Technol. Mana* **2015**, *32*, 42–44.

coatings

MDPI

rticle

;tudy on Soil Corrosion Resistance Reinforced with Reactive 'owder Concrete in Chloride Environment

aozhen Wang [1], Xin Cai [1], Xi Peng [2,3,*], Hui Wang [1,*] and Pengqian Wang [1]

1 School of Civil Engineering and Geographical Environment, Ningbo University, Ningbo 315000, China
2 School of Civil Transportation Engineering, Ningbo University of Technology, Ningbo 315211, China
3 Engineering Research Center of Industrial Construction in Civil Engineering of Zhejiang, Ningbo University of Technology, Ningbo 315048, China
* Correspondence: pengxi@nbut.edu.cn (X.P.); wanghui4@nbu.edu.cn (H.W.)

Abstract: The accumulation of residue soil (generally composed of soil, residue, or mud consolidation) is one of the important causes of damage to the environment limiting urban development. At present, the recycling rate of residue soil in developed countries is as high as 90%, while in China it is less than 5%. In marine construction, reinforced concrete often suffers from corrosion, which leads to a decrease in the service life and durability of the structure. Reactive powder concrete (RPC) with high strength and good corrosion resistance can solve these problems. In order to efficiently dispose of residue soil, protect the environment, and promote urbanization development, this study uses residue soil as a raw material to replace some cement in RPC, and studies the corrosion resistance of it (under dry–wet alternations and freeze–thaw cycles). In this study, five types of reinforced RPC with different residue soil contents (0%, 2.5%, 5%, 7.5%, and 10%) are prepared. Firstly, the working performance of blank freshly mixed residue soil RPC slurry is analyzed. Then, the corrosion resistance of residue-soil-reinforced RPC under the dry–wet alternations with 3% NaCl and freeze–thaw cycles is analyzed through parameters such as mass loss rate, electrical resistivity, ultrasonic velocity, AC impedance spectroscopy, and Tafel. The results show that under the dry–wet alternations, when the residue soil content is 10%, the corrosion rate and corrosion depth of the residue-soil-reinforced RPC are the minimum, at 43,744.84 g/m^2h and 640.22 mm/year, respectively. Under the freeze–thaw cycles, the corrosion rate and corrosion depth of the 10% residue soil content group are higher than that of the 5%, being 52,592.87 g/m^2h and 769.71 mm/year, respectivley. Compared to the other groups, the reinforced RPC with 10% residue soil content shows good corrosion resistance in both dry–wet alternations and freeze–thaw cycles. Replacing some of the cement in RPC with residual soil to control the amount of residual soil at 10% of the total mass of RPC can effectively improve the corrosion resistance of residue-soil-reinforced RPC and maximize the consumption of residue soil. This plan provides a feasible method for residue soil treatment in the construction industry, while also providing inspiration for research on the corrosion resistance of concrete in marine buildings.

Keywords: residue soil; RPC; dry–wet alternations; freeze–thaw cycles; corrosion resistance

1. Introduction

Construction residue soil is an important part of construction waste, which is generally composed of soil, residue, or mud consolidation. Research shows that, with the acceleration of the global urbanization process, the construction industry has produced a large amount of residue soil, and billions of tons of residue soil are treated annually [1]. The residue soil will occupy land, pollute the environment, and affect urban development if it is not treated. Furthermore, the recycling rate of residue soil in developed regions, such as the European Union, South Korea, the United States, and Japan, is as high as 90% or above, while in contrast, the comprehensive recycling rate of residue soil in China is no more than 5% [2–4]. The low recycling rate of residue soil in China has become a serious problem. In addition, marine buildings suffer from serious corrosion by the corrosive environment [5–7], and ordinary reinforced concrete will be corroded and destroyed after 10 to 30 years in the ocean [8–10]. These factors can lead to a decrease in the service life and durability of marine buildings, and even their collapse. Ge [11] mentioned that using silicon powder instead of cement in reactive powder concrete (RPC) can improve its strength, impermeability, and corrosion resistance. Kočí [12] proposed that the compressive strength of RPC can usually reach 200 MPa and it can maintain the integrity of the structure for a long time. In this regard, reactive powder concrete (RPC) [13–15], which has the characteristics of ultra-high strength, high toughness, and durability, may become the key to solving these problems.

At present, there is little research on the corrosion resistance of residue soil concrete around the world. Yu and Wang [16] found that the compressive strength and flexural strength of cement mortar will deteriorate when residue soil is added into the cement. Moreover, it shows the highest and lowest strength when the residue soil content is 5% and 15%, respectivley. Wei [17] studied the influence of residue soil content on the performance of C30 concrete. The research showed that the initial slump flow of concrete decreased with the increase in residue soil content.

RPC corrosion resistance is judged by chloride ion permeability. There is little research on the corrosion resistance of RPC doped with residue soil around the world; however, one study [18] measured the diffusion coefficient of chloride ions of RPC with the NEL method as $22.17 \times 10^{14}\ m^2/s$.

For reinforced concrete, there is a lot of research on corrosion resistance around the world. Green [19] found that there is a linear relationship between the reinforcement corrosion rate and the chloride ion content on the reinforcement surface. Feng [7] studied the corrosion propagation behavior of steel reinforcement using a novel custom-designed wire beam electrode (WBE) covered with uncracked and cracked mortar. The study showed that the WBE technique can effectively monitor passivation and pitting corrosion, especially for the propagation of corrosion. To sum up, there is a lot of research on residue soil concrete, including the working and mechanical properties of RPC around the world. However, there are few studies on the corrosion resistance of residue-soil-reinforced RPC. There are also few relevant reports on its corrosion resistance in the marine environment.

The corrosion of steel bars is an electrochemical process [20]. The passive film on the surface of reinforced concrete in marine construction is constantly eroded by chloride ions. This leads to dullness and the corrosion of reinforced concrete. After the failure of passive films, the exposed iron base forms the anode and a large passive area forms the cathode. The anode and cathode finally produce a potential difference. Thus, the corrosion pattern of the small cathode and large anode is formed. At this moment, rust pits will form on the surface of the steel bars, and they will expand rapidly.

The corrosion of reinforced concrete can be reflected through the parameters of electrochemical detection. Among these parameters, the AC impedance spectrum is a more widely used technology. The AC impedance spectrum allows the impedance to vary with the frequency perturbation of the input AC. At different frequencies, the impedance will approach some kind of electrical resistivity or electrical capacitance. Therefore, the core of electrochemical impedance spectrum (EIS) technology can be obtained, in which the entire electrochemistry of the process can be represented by impedance [21–23]. Furthermore, the measured Tafel curve can also indicate the degree to which the electrode potential needs to be changed in order to achieve a given current. Thus, the corrosion degree of reinforcement can be reflected by the ultrasonic velocity, electrical resistivity, AC impedance, and Tafel curve.

In this experiment, five groups of reinforced RPCs are prepared with different residue soil contents (0%, 2.5%, 5%, 7.5%, and 10%). Firstly, the working performance (the setting time and slump flow) of the newly mixed concrete slurry is analyzed. Then, the steel bar is placed into the residue soil RPC to study the parameters of the ultrasonic velocity, the AC electrical resistivity, and the AC impedance spectrum. Finally, the corrosion resistance of RPC in the dry–wet alternations with 3% NaCl and freeze–thaw cycles is analyzed by using these parameters. This study provides a good scheme for the residue soil reclamation treatment and corrosion resistance research of marine construction.

2. Experimental Procedure

2.1. Raw Materials and Preparation Maintenance

2.1.1. Raw Materials

Ordinary Portland cement P·O42.5, produced by Xiangshan Conch Cement Co., Ltd., Ningbo, China, is applied in this study. P·O cement has an initial setting time of 45 min and a final setting time of 390 min. Additionally, P·O cement has a specific surface area of 365 m^2/kg. The residue soil produced by the construction of a subway line in Ningbo is used in this study. Moreover, the residue soil has a particle size of 11 μm and a specific surface area of 190 m^2/kg. The aggregate is quartz sand. The particle sizes are 1~0.71 mm, 0.59~0.35 mm, and 0.15~0.297 mm, and the mass ratio of the sand is 1:1.5:0.8. Polycarboxylic acid superplasticizer (SP) is used in this study as an admixture to improve the fluidity of raw material mixing, and the water-reducing rate can reach 40%. The quartz sand and water-reducing agent were produced by Shanghai Yingshan New Material Technology Co., Ltd., Shanghai, China. Reinforcement of 6.5 mm diameter is used in this study, and the steel mesh is made of stainless steel. Table 1 shows the cumulative raw material passing rate of the cement and residue soil. Table 2 shows the chemical composition of the raw materials.

Table 1. Cumulative pass rate of raw materials (%).

Types \ Particle Size (%)	0.3	0.6	1	4	8	64	360
Cement	0	0.33	2.66	15.01	28.77	93.59	100
Residue soil	31.2	41.2	48.8	82.3	100	100	100

Table 2. Chemical composition of raw materials (%).

Types	Chemical Composition (%)									
	SiO_2	Al_2O_3	Fe_2O_3	MgO	CaO	SO_3	R_2O	P_2O_5	Organic Matter	Loss
OPC	20.86	5.47	3.94	1.73	62.23	2.66	0.48	0	0	2.63
SAC	46.07	13.68	4.81	3.27	12.09	0.26	4.1	0.31	5.2	11.5

2.1.2. Specimen Preparation and Curing

The residue soil contents of five groups of residue-soil-reinforced RPC prepared in this study are 0%, 2.5%, 5%, 7.5%, and 10%, respectively, with six specimens in each group. The specimens are prepared according to GB/T 50081-2019. All weighed raw materials are poured into a UJZ-15 mortar mixer to be stirred for 40s. Then, the time is recorded, and the initial setting time of the specimens is calculated. The mixed concrete is poured into oiled plastic molds, obtaining specimens with sizes of 50 mm × 50 mm × 50 mm when the slump flow test is completed. All specimens are cured at a temperature of 20 ± 2 °C and a relative humidity above 95% (GB50204-2015) after demolding. All specimens are taken out and numbered after 2 days. Figure 1 shows a diagram of the specimens. Table 3 shows the coordination of each group.

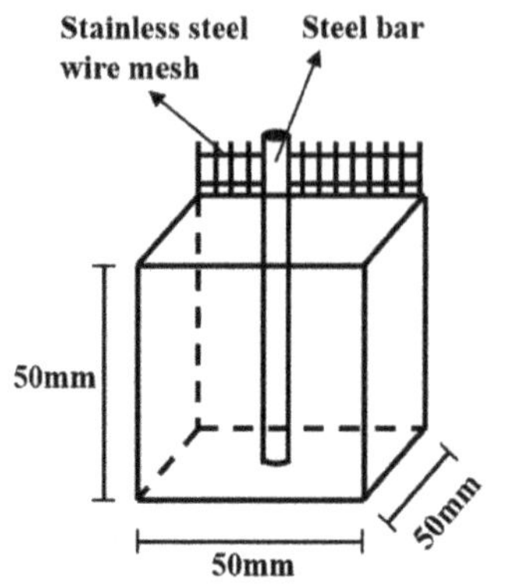

Figure 1. A schematic diagram of the specimen.

Table 3. The mix ratio of concrete.

Group	Water	Cement	Sand	Water Reducer	Residue Soil
1 (0%)	219.9	1466.5	1173.3	47.8	0
2 (2.5%)	219.9	1393.2	1173.3	47.8	73.4
3 (5%)	219.9	1319.9	1173.3	47.8	146.6
4 (7.5%)	219.9	1246.5	1173.3	47.8	219.9
5 (10%)	219.9	1173.3	1173.3	47.8	293.3

Note: The unit is g. The masses of coarse sand, medium sand, and fine sand in each group were 355.5 g, 533.3 g, and 284.4 g, respectively.

2.2. Dry–Wet Alternations and Freeze–Thaw Cycles

The specimens numbered 1, 2 and, 3 in a group are subjected to dry–wet alternations with 3% NaCl and the specimens numbered 4, 5, and 6 are subjected to freeze–thaw cycles with chlorine salt.

To simulate dry–wet alternations with 3% NaCl, specimens are put into the solution (3% NaCl) to soak for 12 h. Subsequently, the specimens are placed into the electric blast drying oven (with the temperature set to 80 °C) for 36 h, with 2 d as a cycle, which is repeated 10 times.

To simulate freeze–thaw cycles of chlorine salts, specimens are put into the freeze–thaw cycle box (temperature range from −18 to 5 °C, with a deviation within 2 °C). One cycle is set for 3 h of freeze–thaw and the cycle repeats 100 times. Furthermore, since the specimen has been wet and saturated during curing before circulation, no additional soaking is required.

2.3. Measurement Method

2.3.1. Measurement of Slump Flow

The slump flow of mixed concrete is measured using the jumping table. The jumping table is produced by Wuxi Prospecting Instrument Factory, Zhejiang Province, and its model number is NLD-3. The power supply voltage of the jumping table is 220 V, the vibration drop distance is 10 mm ± 0.2 mm, the vibration frequency is 1 Hz, and the diameter of the table is 300 mm ± 1 mm. The specific operation process refers to the GB/T 2419-2005 Chinese standard [24].

2.3.2. Measurement of Setting Time

The setting time is determined by the setting time tester in this study. The pressure area of the probe of the setting time tester is 30 mm^2. Moreover, the initial setting time corresponds to pressure 0.7 MPa. According to Equation (1), the pressure corresponding to the initial setting time is 21 N.

$$f_p = \frac{N_p}{A_p} \quad (1)$$

In Equation (1), f_p is the pressure corresponding to the initial setting time, A_p is the pressure area of the probe, and N_p is the pressurized pressure.

The specific operation process refers to the GB/T 1346-2011 Chinese standard [25]. Figure 2 shows the setting time tester and concrete specimens.

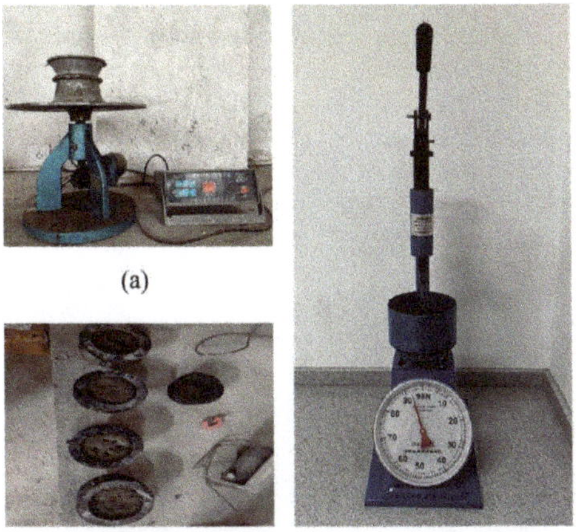

(a) (b) (c)

Figure 2. Slump flow and the setting time measurement. (**a**) Jumping table. (**b**) Concrete specimens. (**c**) Setting time tester.

2.3.3. Measurement of Ultrasonic Velocity

Ultrasonic velocity is measured using a JITAI990 ultrasonic detector in this study. The detector can produce sounds ranging from 1000 m/s to 9999 m/s. Vaseline is applied to both sides of the concrete and the probe is stuck on the axis positions. Ultrasonic velocity signals are collected after all of these steps are completed. The detailed measurement process of ultrasonic velocity can be found in Wang's study [26]. Figure 3 is a picture of the JITAI990 ultrasonic detector.

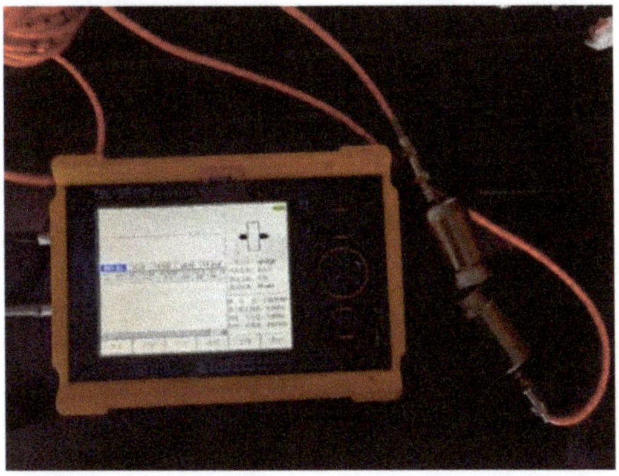

Figure 3. JITAI990 ultrasonic detector.

2.3.4. Measurement of Mass Loss and Electrochemical Parameters

The original rust on the surface of the steel bars is removed by using a sanding process when measuring the mass loss of the specimen. In this study, the original steel bars are then polished with a sand blaster. After polishing, the steel bars are cleaned with citric acid and moved to be dried and weighed. Later, standard cure samples with polished steel bars are applied in the simulated environment. Finally, the mass loss is measured, and the specific measurement process is that of the study of Wang [26].

The AC electrical resistivity is measured using a TH2810D LCR digital bridge. The test voltage is 1 V, and the test frequency is 104 Hz. The steel bar and stainless-steel mesh are used as the two electrodes. Moreover, the digital bridge makes good contact with the electrode. A Princeton electrochemical workstation is used to measure the AC impedance spectrum. The initial AC frequency set in this study is 100,000 Hz, the end AC frequency is 1 Hz, and the voltage is 0.01 V, which meets the conditions of small disturbance. The electrochemical workstation is connected to the electrode of the specimen. The specific measurement process of the AC electrical resistivity and AC impedance spectrum is that of the study of Wang [27]. Figure 4 shows a schematic diagram of the AC electrical resistivity measurement. Figure 5 shows a photo of the electrochemical workstation.

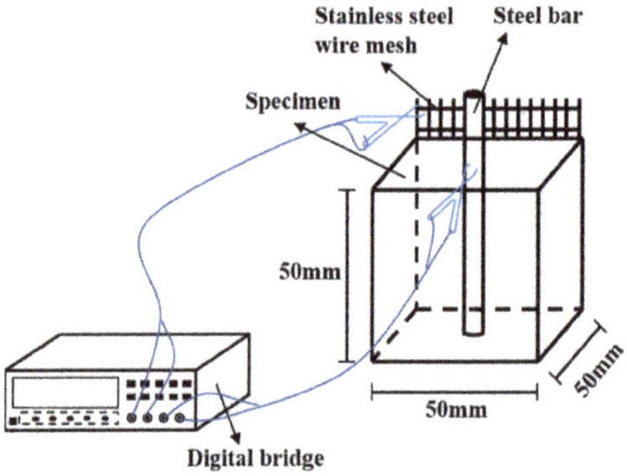

Figure 4. Measurement of the AC electrical resistivity.

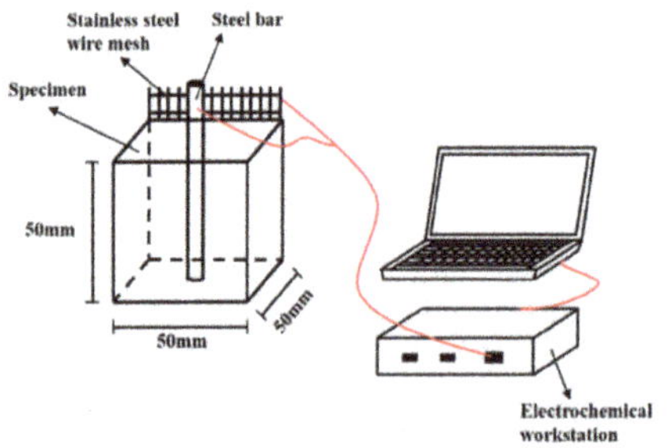

Figure 5. Measurement of the AC impedance spectrum.

3. Results and Discussions

3.1. Working Performance of Residue-Soil-Reinforced RPC

Figure 6 shows the initial setting time (h) and the slump flow (cm) of RPC with different residue soil contents. It can be seen that the initial setting time increases with the increase in residue soil content. The initial setting time changes little when the residue soil content is 0%~5%, and the initial setting time continues to rise when the residue soil content is more than 5%. This is because the residue soil has a much larger surface area than cement in the process of replacing the cement. As a result, the water absorption of the residue soil is stronger than that of cement [28,29], and it affects the hydration process of cement and prolongs the initial setting time of RPC. The initial setting time increases sharply when the residue soil content changes from 5% to 7.5%. This is because the initial setting time is affected by low ambient temperature when the residue soil content is 7.5%.

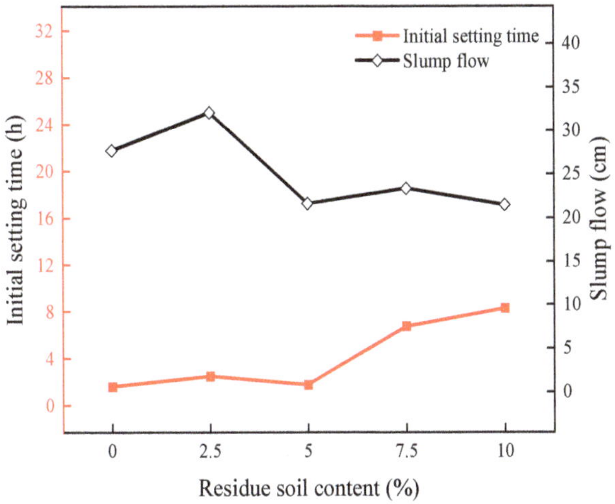

Figure 6. The initial setting time and slump flow with different residue soil contents.

It can also be seen that the slump flow decreased with the increase in residue soil content (from the base of 217.5 mm, slightly increased to 250 mm, and finally dropped to 171 mm). Residue soil has a great influence on the slump flow when its content is less than 5%. Furthermore, when the residue soil content is higher than 5%, the influence is small. This is because the residue soil absorbs the water-reducing agent molecules with the increase in its content [30]. Moreover, water-reducing agent molecules have intercalation in residue soil. The water-reducing agent is consumed by the residue soil, which leads to a reduction in the water-reducing agent acting on the cement and finally results in a reduction in the slump flow.

3.2. *Effect of Dry–Wet Alternations on Corrosion Resistance of Residue-Soil-Reinforced RPC*

3.2.1. Mass Loss Rate of Residue-Soil-Reinforced RPC

Figure 7 shows the mass loss rate of different residue soil contents during dry–wet alternations. As shown in Figure 7, there is no positive or negative correlation between the residue soil content and the mass loss rate. The mass loss rate of RPC with 5% residue soil content is the lowest. Furthermore, the mass loss rate of RPC with residue soil content in each group and the blank control group increases first and then decreases under the dry–wet alternations. Obviously, there is mass loss after five dry–wet alternations. This is because the steel bars are eroded by chloride ions, which leads to the expansion of the steel bars and the peeling of the concrete surface [31]. The mass loss of 10 dry–wet alternations is not as high as that of 5. This indicates that the mass of concrete after 10 dry–wet alternations is greater than that of 5. This is because the mass loss is greater with the influence of temperature stress when the frequency of dry–wet alternations is low.

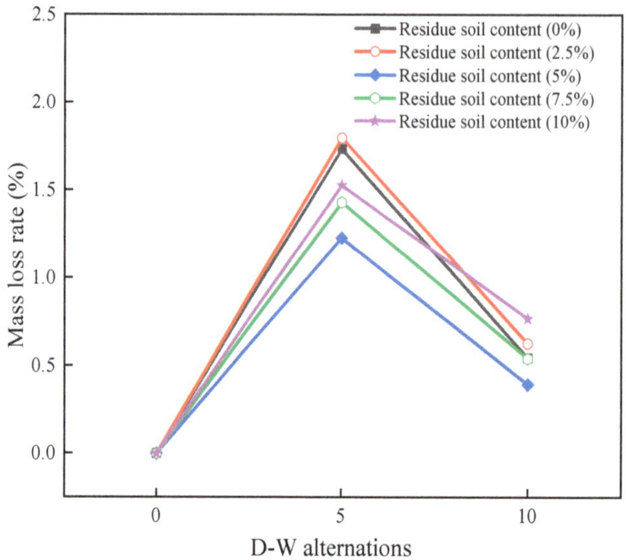

Figure 7. Mass loss rate with different residue soil contents during dry–wet alternations.

3.2.2. Ultrasonic Velocity of Residue-Soil-Reinforced RPC

Figure 8 shows the ultrasonic velocity in different periods. It can be seen from Figure 8 that ultrasonic velocity increases continuously during the curing period. Moreover, ultrasonic velocity reaches the maximum when the curing time is 28 d. In each group, the RPC with 5% residue soil content has the highest ultrasonic velocity, which is 4.45 km/s. The ultrasonic velocity decreases in all of the groups when dry–wet alternation begins. This is because, after repeated immersion in 3% NaCl solution, the chloride ions corrode the steel bars inside of the concrete. This generates rust swelling stress [32] and, because the tensile strength of concrete is less than the rust swelling stress, resulting in cracks appearing inside of the concrete, the ultrasonic velocity decreases [33]. Decky [34] calculates the thermal stress value of cement concrete pavement slabs with a thickness of 12~32 cm and analyzes their sensitivity to different subgrade reaction moduli. The temperature stress level of cement concrete slabs has been determined. When the annual temperature changes from 4.5 to 11.5 °C, a tensile stress in the range of 0.3 to 0.7 MPa will be generated inside of the cement concrete slab. Therefore, temperature stress can also aggravate the cracking of concrete [34,35]. The decrease in ultrasonic velocity before 5 dry–wet alternations is greater than that before 10. This is because the corrosion products of steel bars play a part in filling the micro-cracks in concrete. The reinforced RPC with 5% residue soil content has the largest ultrasonic velocity after 10 dry–wet alternations, which indicates that the reinforced RPC with 5% residue soil content has the largest compaction. The ultrasonic velocity of reinforced RPC with 7.5% residue soil content is the second largest, which is 3.72 km/s. It is close in size to the ultrasonic velocity of reinforced RPC with 10% residue soil content. The ultrasonic velocity of reinforced RPC with 2.5% residue soil content is the smallest, which is 3.52 km/s. This indicates that reinforced RPC with 2.5% residue soil has the smallest compaction. The results are consistent with the mass loss rate chart.

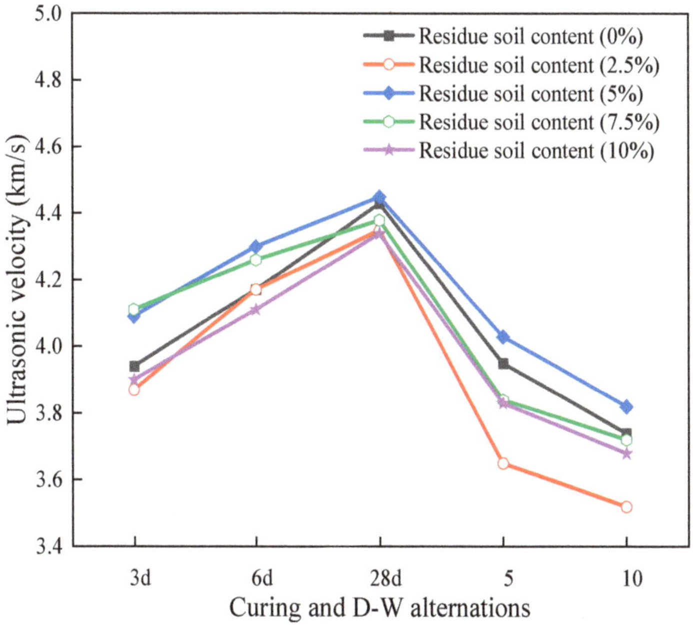

Figure 8. Ultrasonic velocity with different residue soil contents during different periods.

3.2.3. Electrical Resistivity of Residue-Soil-Reinforced RPC

Figure 9 shows the variation of the electrical resistivity in different periods. As shown in Figure 9, the reinforced RPC with 5% and 10% residue soil content have higher electrical resistivity. Furthermore, there is no positive or negative correlation between the residue soil content and the electrical resistivity. The electrical resistivity of each group shows a trend of first decreasing and then increasing again with curing and dry–wet alternations. The electrical resistivity reaches its maximum when the curing time is 28 d. The electrical resistivity of all of the groups decreases after curing for 3 d to 6 d, and the RPC with 5% residue soil content decreases the most. This is because the moisture content in the residue-soil-reinforced RPC increases with the increase in curing time, resulting in the electrical conductivity increasing and the electrical resistivity decreasing. The electrical resistivity of each group increases after curing for 6 d to 28 d. This is because the gradual hydration of the residue-soil-reinforced RPC results in the increase in compactness and the gradual disconnection of pores, which eventually leads to an increase in the electrical resistivity. It can be seen that hydration has a great effect on electrical resistivity. At the beginning of the dry–wet alternations (28 d later), the electrical resistivity begins to decline, which is because the ions in the NaCl solution enter the residue-soil-reinforced RPC and continuously accumulate. At this moment, the electrical resistivity starts to depend on the total amount of charge, and the more the total amount of charge, the lower the electrical resistivity is [36]. The electrical resistivity of all of the groups increases again after five dry–wet alternations. This is because the total charge inside of the residue-soil-reinforced RPC is saturating. Subsequently, hydration is still occurring. At this moment, hydration makes the electrical resistivity increase again.

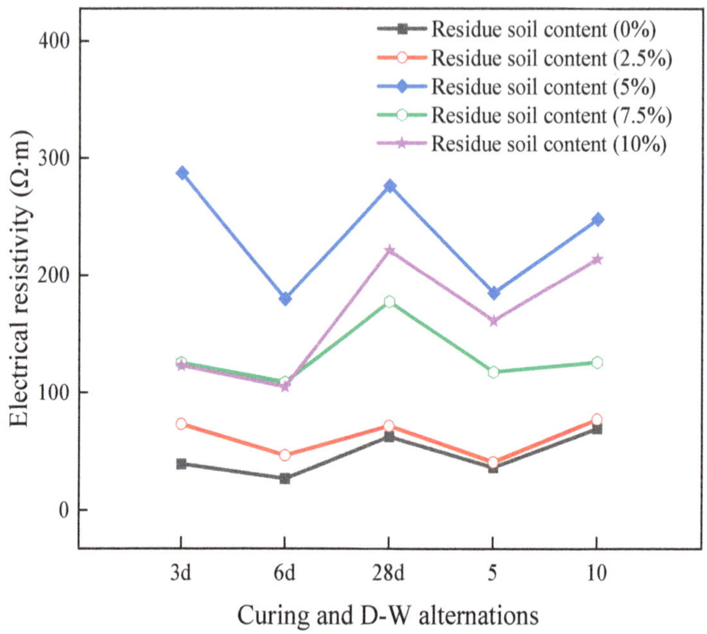

Figure 9. The electrical resistivity of different residue soil contents during different periods.

3.2.4. Electrochemical Performance of Residue-Soil-Reinforced RPC

Figures 10 and 11 show the AC impedance spectrum scatter distribution after 0 and 10 dry–wet alternations, respectively. In Figures 10 and 11, the abscissa Zr (Ω) represents the electrical resistivity, characterizing the corrosion resistance of the residue-soil-reinforced RPC [37], and Zi (Ω) represents the electrical reactance.

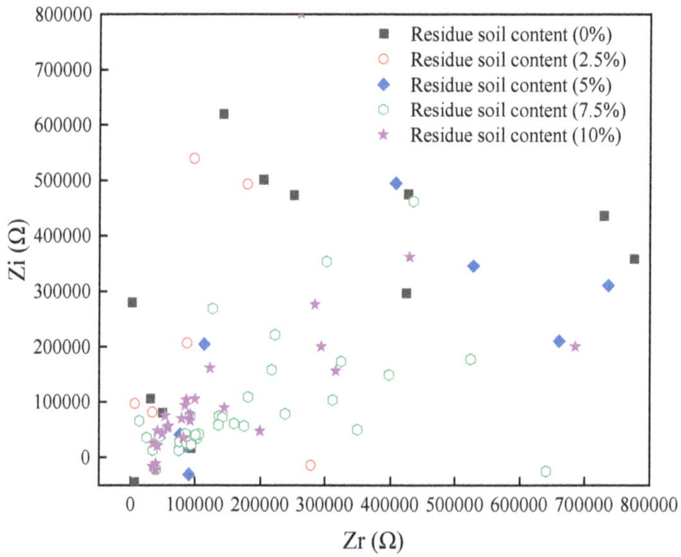

Figure 10. The AC impedance spectrum of 0 dry−wet alternations.

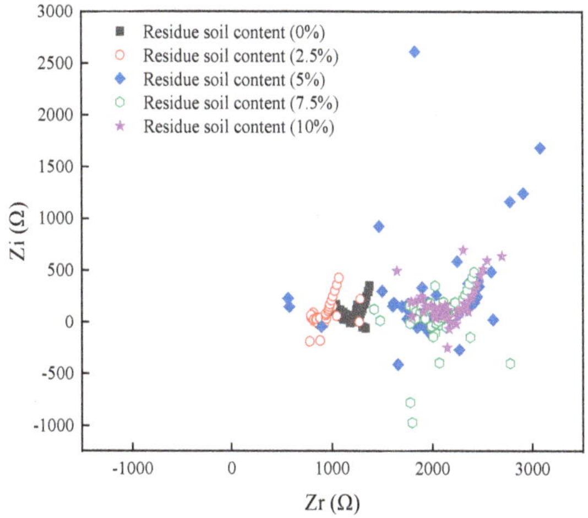

Figure 11. The AC impedance spectrum of 10 dry−wet alternations.

It can be seen from Figure 10 that the impedance values of each group are scattered and irregular after 0 dry–wet alternations. Furthermore, the order of magnitude of the electrical resistivity and the electrical capacitance is extremely large. This indicates that the concrete protective layer of the specimen protects the matrix well and has a strong charge storage ability before the dry–wet alternations [37]. Moreover, the electrical capacitance decreases first and then increases with the increase in the electrical resistivity when the residue soil content is the same. The image moves first to the left and then to the right, and the electrical resistivity value corresponding to the lowest point represents its corrosion resistance with the increase in residue soil content. The higher the electrical resistivity values, the better the corrosion resistance is, which indicates that the residue soil corrosion resistance is better than that of cement. In addition, the corrosion resistance is the best when the residue soil content is 10%. However, the electrical resistivity value of the specimens is smaller than that of the blank control group when the residue soil content is 2.5%. Therefore, concrete with 2.5% residue soil content should be avoided in order to achieve better corrosion resistance in engineering practice.

The corrosion rate of metals can be expressed by the corrosion current density, and the Tafel curve, shown in Figures 12 and 13, can be obtained by converting it with Faraday's law Equations (2) and (3).

$$\text{v} = \frac{10M}{n \times \frac{9.65 \times 10^4}{3.6}} \times I = 3.73 \times 10^{-4}\frac{M}{n}I \quad (2)$$

$$d = \frac{v}{\rho} \quad (3)$$

In Equation (2), v stands for the corrosion velocity, and its unit is g/m^2h. M represents the atomic weight of the metal, and its unit is g. n represents the atomic valence of the metal. I represents the corrosion current density, and its unit is $\mu A/cm^2$. In Equation (3), d indicates the depth of corrosion, and its unit is mm/year. v stands for corrosion velocity, and its unit is g/m^2h. ρ represents the density of the metal, and its unit is g/cm^3.

Figure 12 shows the Tafel curve under 0 dry–wet alternations. Figure 13 shows the Tafel curve under 10 dry–wet alternations. It is not difficult to see that the two branching curves form a complete Tafel curve. The bifurcation point of the curve can reflect the corrosion performance of the specimens [38,39]. It can be seen from Figures 12 and 13 that there is no positive or negative correlation between the residue soil content and the potential of bifurcation points.

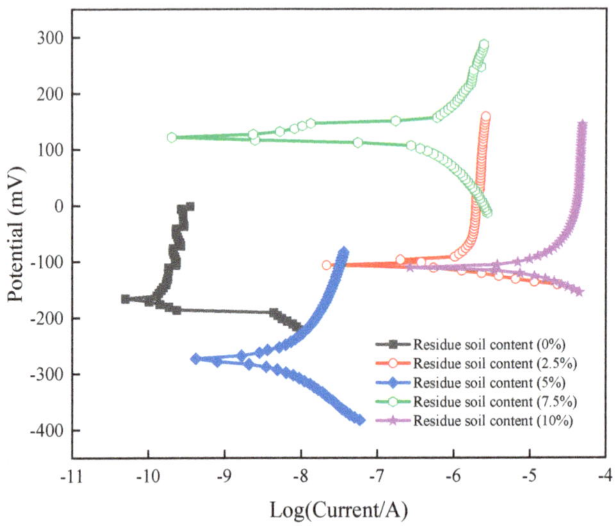

Figure 12. The Tafel curve under 0 dry−wet alternations.

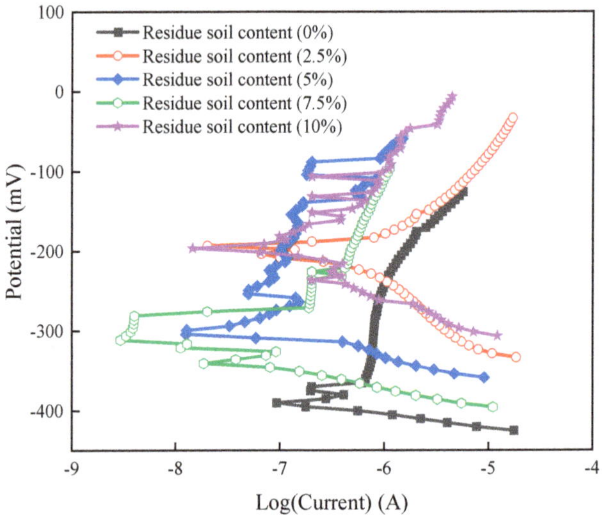

Figure 13. The Tafel curve under 10 dry−wet alternations.

According to the Tafel curve, the voltage and the electrical current of the residue-soil-reinforced RPC in each group are analyzed. Figure 14 shows a bar chart of the corrosion rate and corrosion depth of different residue soil contents under 10 dry–wet alternations. It can be seen from Figure 14 that the corrosion rate and the corrosion depth are not positively correlated with the residue soil content. The corrosion depth of reinforcement increases first and then decreases with the increase in residue soil content. When the residue content is 5%, the maximum corrosion depth is 9043.32 mm/year. When the residue content is 10%, the minimum corrosion depth is 640.21 mm/year. This indicates that residue soil content in a certain range will accelerate the corrosion rate of reinforcement. Moreover, the blank control group and the 10% residue soil content group have better corrosion resistance.

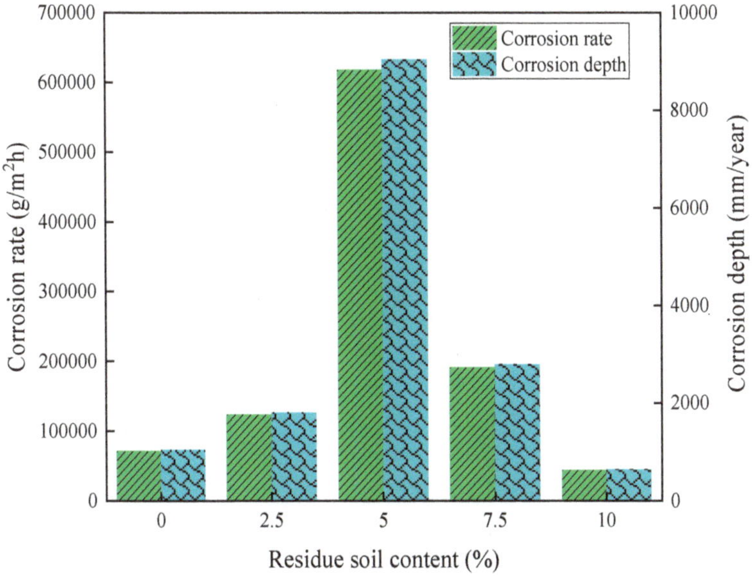

Figure 14. Bar chart of corrosion rate and corrosion depth under 10 dry—wet alternations.

3.3. Effect of Freeze–Thaw Cycles on Corrosion Properties of Residue-Soil-Reinforced RPC

3.3.1. Mass Loss Rate of Residue-Soil-Reinforced RPC

Figure 15 shows the mass loss rate of different residue soil contents during freeze–thaw cycles. It can be seen from Figure 15 that the mass loss rate of each group increases under the freeze–thaw cycle, and when the number of cycles increases from 50 to 100, the mass loss rate increases by an average of 0.72 times. One of the reasons for this is that the NaCl solution inside of the concrete freezes during circulation, resulting in concrete expansion, cracking, and spalling, which finally leads to mass loss [40]. The second reason is that, although the crack is open, the crack also absorbs water instead of part of the mass loss, resulting in the mass loss of 100 cycles being close to, but less than, 1 time. Furthermore, the mass loss rate of reinforced RPC with 10% residue soil content is the smallest. Because the mass loss rate represents the surface erosion of the residue-soil-reinforced RPC and characterizes its corrosion resistance, the reinforced RPC with 10% residue soil content has the best corrosion resistance.

3.3.2. Ultrasonic Velocity of Residue-Soil-Reinforced RPC

Figure 16 shows the ultrasonic velocity during curing and freeze–thaw cycles. It can be seen from Figure 16 that the ultrasonic velocity first increases and then decreases during the curing and freeze–thaw cycles. Moreover, the group with 10% residue soil content and the blank control group have the largest increase. The ultrasonic velocity of each group begins to decrease after the freeze–thaw cycles. The ultrasonic velocity under 100 freeze–thaw cycles is slightly lower than that of 50. This is because the inside of the residue-soil-reinforced RPC is gradually damaged, resulting in micro-cracks, and these micro-cracks grow larger and larger [41]. This is because the ultrasonic propagation velocity in the solution (3% NaCl) with micro-cracks is lower than that in the solid interior of the initial specimen, thus, the ultrasonic velocity decreases [27]. The ultrasonic velocity of the reinforced RPC with 10% residue soil content is the largest, which is 4.24 km/s. Therefore, the RPC with 10% residue soil content has the largest compaction.

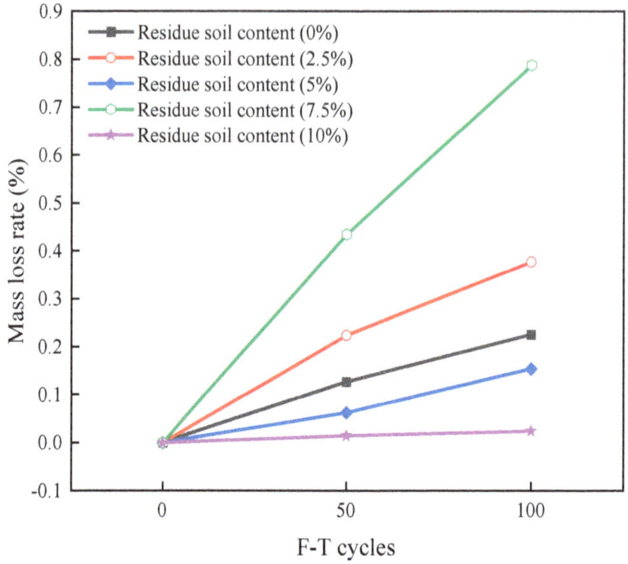

Figure 15. Mass loss rate of different residue soil contents during freeze–thaw cycles.

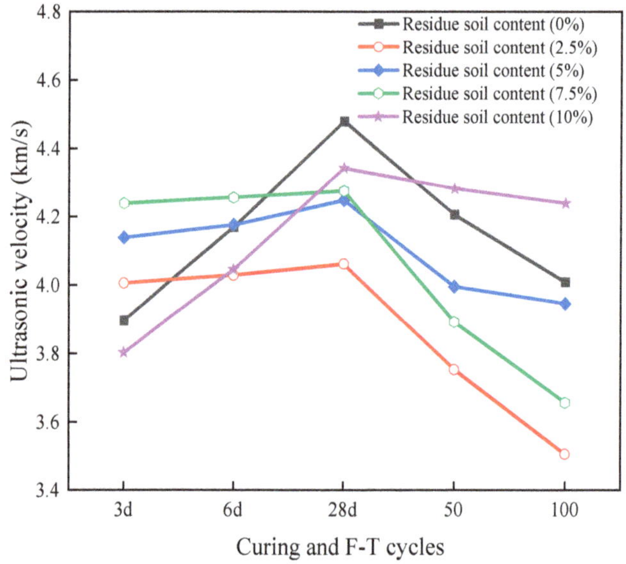

Figure 16. Ultrasonic velocity of different residue soil contents during different periods.

3.3.3. Electrical Resistivity of Residue-Soil-Reinforced RPC

Figure 17 shows the electrical resistivity of each group during different periods. It can be seen from Figure 17 that the electrical resistivity of each group increases with the increase in curing time. Moreover, when comparing 100 freeze–thaw cycles to that of 50, the electrical resistivity growth of each group slows down (5%, 7.5%, and 10%) or even decreases (0% and 2.5%) to different degrees. This is because, although the hydration of concrete is still occurring, the rate of hydration is slower, and the improvement of the concrete compactness is not as significant as it was in the early stage. After 100 freeze–

thaw cycles, the largest electrical resistivity of the group with 7.5% residue soil content is 387.82 Ω·m, followed by that of 10% residue soil content, which is 319.65 Ω·m. This indicates that the reinforced RPC with 7.5% and 10% residue soil contents have the better resistance to chloride ion erosion.

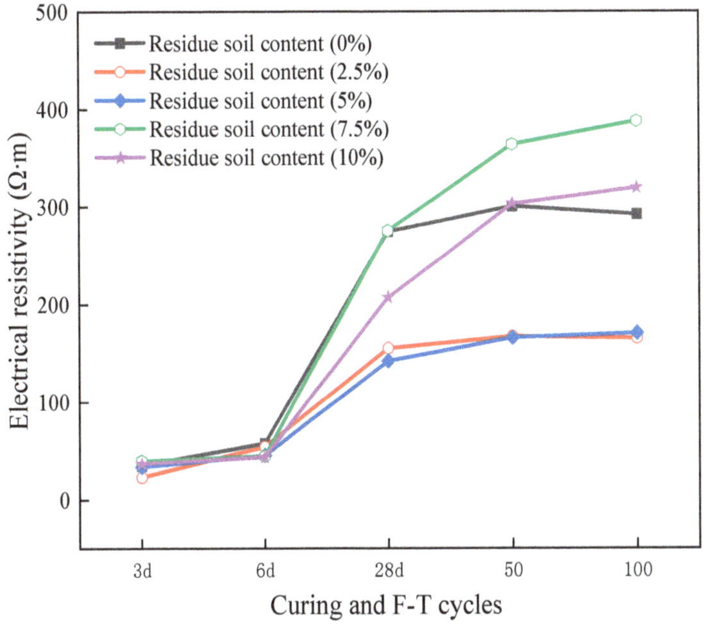

Figure 17. The electrical resistivity of different residue soil contents during different periods.

3.3.4. Electrochemical Performance of Residue-Soil-Reinforced RPC

Figure 18 shows the AC electrical resistivity scatter distribution of 0 freeze–thaw cycles. In the figure, Zr represents the electrical resistance and reflects the degree of corrosion. As with the dry–wet alternations, the AC impedance scatter distribution of 0 freeze–thaw cycles is irregular. This indicates that the concrete protective layer of the test block before the freeze–thaw cycle protects the matrix well and has a strong charge storage ability.

Figure 19 shows the Tafel curve under 0 freeze–thaw cycles. It can be seen from the curve that mixing residue soil into RPC can effectively improve the corrosion resistance of specimens before the freeze–thaw cycle.

Figure 20 shows a bar chart of the corrosion rate and corrosion depth of different residue soil contents under 0 freeze–thaw cycles. It can be seen from Figure 20 that the annual corrosion depth of the RPCs with 0%, 5%, and 10% residue soil contents is less than 1000 mm, which shows good corrosion resistance.

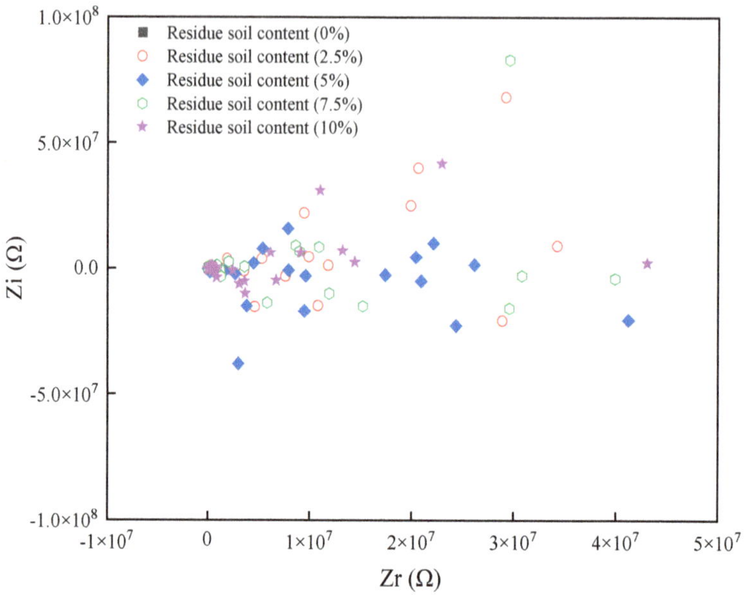

Figure 18. The AC impedance spectrum under 0 freeze–thaw cycles.

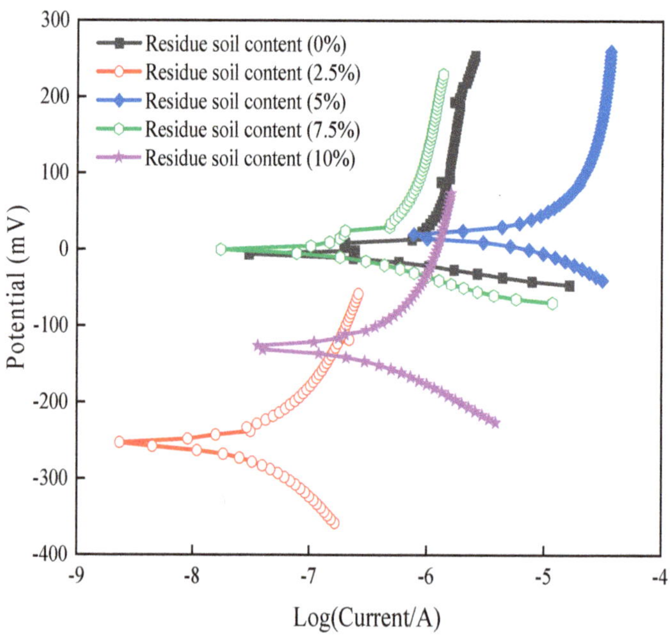

Figure 19. The Tafel curve under 0 freeze–thaw cycles.

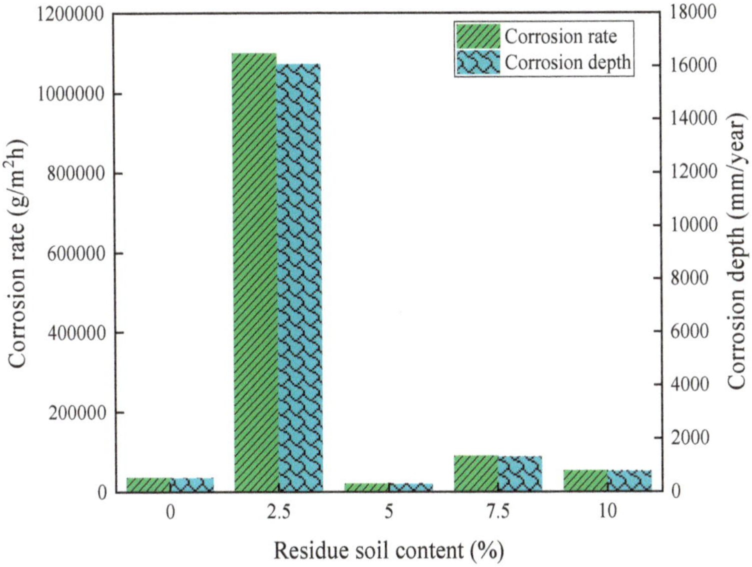

Figure 20. Bar chart of corrosion rate and corrosion depth under 0 freeze−thaw alternations.

4. Conclusions

The multiple properties, including the working performance (slump flow and initial setting time), mass loss rate, and electrochemical parameters (electrical resistivity, ultrasonic velocity, AC impedance spectrum, and Tafel) are analyzed in order to study the corrosion resistance of residue-soil-reinforced RPC. Our research findings can be concluded as follows:

The initial setting time of the residue soil RPC increases with the increase in residue soil content. The slump flow shows a trend of first rising and then falling (from the base of 217.5 mm, then slightly increased to 250 mm, and finally dropped to 171 mm). When the residue soil content is 2.5%, the maximum slump flow is 250 mm, and when the residue soil content is 10%, the minimum slump flow is 171 mm.

The mass loss rate of residue-soil-reinforced RPC increases first and then decreases (when performing dry–wet alternations) and increases (when performing freeze–thaw cycles). The ultrasonic velocity shows the same rule under dry–wet alternations and freeze–thaw cycles. The ultrasonic velocity increases first and then decreases. With the increase in curing age and dry–wet alternations, the electrical resistivity of the residue-soil-reinforced RPC decreases when curing from 3 d to 6 d and then increases when curing from 6 d to 28 d. Moreover, the electrical resistivity decreases (after 28 d) and increases again (after five alternations). Finally, the electrical resistivity reaches its maximum after curing for 28 days. In addition, all of the groups share the same trend. With the increase in curing age and freeze–thaw cycles, the electrical resistivity of the residue-soil-reinforced RPC keeps rising.

The residue-soil-reinforced RPC with 10% residue soil content has good corrosion resistance in both dry–wet alternations and freeze–thaw cycles, which we can see by analyzing the AC impedance spectrum and Tafel curve (a polarization curve that characterizes metals and reflects the relationship between the anodic polarization curve and the current).

Considering the amount of residue soil utilization and corrosion resistance, reinforced RPC with 10% residue soil content is the most suitable for engineering practice due to its high residue soil recycling rate and low corrosion rate. Moreover, reinforced RPC with 2% residue soil content should be avoided in engineering.

Author Contributions: Conceptualization, H.W. (Haozhen Wang), X.P. and H.W. (Hui Wang).; Software, P.W.; Validation, H.W. (Haozhen Wang) and P.W.; Formal analysis, H.W. (Haozhen Wang) and X.C.; Investigation, H.W. (Haozhen Wang), X.C. and P.W.; Resources, X.P. and H.W. (Hui Wang); Data curation, H.W. (Haozhen Wang) and X.C.; Writing—original draft, H.W. (Haozhen Wang); Writing—review and editing, H.W. (Haozhen Wang), X.C., X.P. and H.W. (Hui Wang); Visualization, H.W. (Haozhen Wang); Supervision, X.P. and H.W. (Hui Wang); Project administration, X.P. and H.W. (Hui Wang); Funding acquisition, X.P. and H.W. (Hui Wang). All authors have read and agreed to the published version of the manuscript.

Funding: This research was funded by the Natural Science Foundation of Zhejiang Province, China (No. LY22E080005), Ningbo Natural Science Foundation Project (2023J086), and the National Natural Science Foundation of China (52008215).

Institutional Review Board Statement: Not applicable.

Informed Consent Statement: Not applicable.

Data Availability Statement: The data used to support the findings of this study are available upon request.

Conflicts of Interest: The authors declare no conflict of interest.

References

1. Liu, Y.; Lu, H.; Liu, M.; Cai, L.; Wei, N.; Liu, Y. Microanalytical characterizations, mechanical strength and water resistanc performance of solidified dredged sludge with industrial solid waste and architecture residue soil. *Case Stud. Constr. Mater.* **202** *17*, e01492. [CrossRef]
2. Xu, S.; He, W. Analysis of Resource Utilization of Construction Residue-Take Ningbo City as an Example. *China Resour. Comp Util.* **2022**, *08*, 99–101.
3. Ding, Y.; Zhao, J.; Liu, J.; Zhou, J.; Cheng, L.; Zhao, J.; Shao, Z.; Iris, Ç.; Pan, B.; Li, X.; et al. A review of China's municipal soli waste (MSW) and comparison with international regions: Management and technologies in treatment and resource utilization. *Clean. Prod.* **2021**, *293*, 126144. [CrossRef]
4. Song, Q.; Zhu, Z.; Wang, J.; Long, B.; Luo, W.; Jia, Z.; Wang, M. Status Quo of Resource Utilization of Urban Engineering Dreg *China Resour. Compr. Util.* **2021**, *39*, 90–92.
5. Zhu, P.; Liu, M. Non-uniform Corrosion Mechanism and residual life forecast of marine engineering concrete reinforcement. *Eng. Res.* **2023**, *11*, 100053. [CrossRef]
6. Li, S.; Jin, Z.; Pang, B.; Li, J. Durability performance of an RC beam under real marine all corrosion zones exposure for 7 yea *Case Stud. Constr. Mater.* **2022**, *17*, e01516. [CrossRef]
7. Feng, G.; Jin, Z.; Zhu, D.; Xiong, C.; Li, Z.; Wang, X. Corrosion propagation of steel reinforcement in pre-cracked mortar attacke by seawater using wire beam electrode. *Corros. Sci.* **2022**, *208*, 110655. [CrossRef]
8. Shao, W.; Wang, Y.; Shi, D. Corrosion-fatigue life prediction of reinforced concrete square piles in marine environments. *Eng. Fa Anal.* **2022**, *138*, 106324. [CrossRef]
9. Xu, J.; Wu, J.; Chen, S.; Diao, B.; Zhao, T. Fatigue life prediction of fatigue damaged and chloride corroded RC beams. *J. Buil Struct.* **2022**, *43*, 69–76.
10. Liu, Q.; Hu, Z.; Lu, X.; Yang, J.; Azim, I.; Sun, W. Prediction of Chloride Distribution for Offshore Concrete Based on Statistic Analysis. *Materials* **2020**, *13*, 174. [CrossRef]
11. Ge, W.; Wang, A.; Zhang, Z.; Ge, Y.; Chen, Y.; Li, W.; Jiang, H.; Shuai, H.; Sun, C.; Yao, S.; et al. Study on the workabilit mechanical property and water absorption of reactive powder concrete. *Case Stud. Constr. Mater.* **2023**, *18*, e01777. [CrossRef]
12. Kočí, V.; Vejmelková, E.; Koňáková, D.; Pommer, V.; Grzeszczyk, S.; Matuszek-Chmurowska, A.; Mordak, A.; Černý, R. Bas physical, mechanical, thermal and hygric properties of reactive powder concrete with basalt and polypropylene fibers aft high-temperature exposure. *Constr. Build. Mater.* **2023**, *374*, 130922. [CrossRef]
13. Ambika, D.; Nandhini, V.; Santha Rubini, V.; Poovizhi, G.; Dhinu Priya, S. An exploration on the durability properties of reactiv powder concrete. *Mater. Today Proc.* **2021**, *45*, 529–534. [CrossRef]
14. Yang, Q.; Sun, Y.; Peng, X. Experimental Study on Mechanical Properties of Concrete at Super-Early Age. *Materials* **2022**, *15*, 758 [CrossRef]
15. Peng, X.; Yang, Q.; Cao, H.; Wang, H. Mathematical Model for Early-Aged UHPFRC Compressive Strength Changes. *Coating* **2023**, *13*, 525. [CrossRef]
16. Yu, J.; Zhang, A.; Zhang, L.; Wang, Q.; Li, K.; Wang, H. Study on the performance of low water-binder ratio cement mortar wit excavated soil exposed to NaCl freeze-thaw environment. *Mater. Res. Express* **2021**, *8*, 095511. [CrossRef]
17. Wei, C.; Rong, H.; Song, W.; Qi, Z.; Li, X.; Wang, H. Influence of muck on mechanical properties of C30 concrete. *Concrete* **2016**, 148–150.

8. Yu, Z.; Gao, K.; An, M.; Han, S. Influence of micro-structure on the strength and resistance to chloride ion permeability of reactive powder concrete. *J. Xi'an Univ. Arch. Technol.* **2013**, *45*, 31–37.
9. Green, W. Steel reinforcement corrosion in concrete-an overview of some fundamentals. *Corros. Eng. Sci. Technol.* **2020**, *55*, 289–302. [CrossRef]
0. Sun, S. Research on the whole process of corrosion damage of reinforced concrete based on electrochemical impedance spectroscopy. *ShiJiazhuang TieDao Univ.* **2020**, 4–7. [CrossRef]
1. Ejbouh, A.; Ech-chebab, A.; Hassi, S.; Galai, M.; Benqlilou, H.; Touhami, M. Durability assessment of LC^3-based reinforced concrete under combined chloride-sulfate environment via the EIS technique. *Constr. Build. Mater.* **2023**, *366*, 130194. [CrossRef]
2. Ribeiro, D.; Abrantes, J. Application of electrochemical impedance spectroscopy (EIS) to monitor the corrosion of reinforced concrete: A new approach. *Constr. Build. Mater.* **2016**, *11*, 98–194. [CrossRef]
3. John, D.; Searson, P.; Dawson, J. Use of AC impedance technique in studies on steel in concrete in immersed conditions. *Br. Corros. J.* **2013**, *16*, 102–106. [CrossRef]
4. *GB/T2419-2005*; Test Method for Fluidity of Cement Mortar. General Administration of Quality Supervision, Inspection and Quarantine of the People's Republic of China: Beijing, China, 2005.
5. *GB/T1346-2011*; Test Methods for Water Requirement of Normal Consistency, Setting Time and Soundness of the Portland Cement. General Administration of Quality Supervision, Inspection and Quarantine of the People's Republic of China: Beijing, China, 2011.
6. Wang, H.; Shi, F.; Shen, J.; Zhang, A.; Zhang, L.; Huang, H.; Liu, J.; Jin, K.; Feng, L.; Tang, Z. Research on the self-sensing and mechanical properties of aligned stainless steel fiber reinforced reactive powder concrete. *Cem. Concr. Compos.* **2021**, *119*, 104001. [CrossRef]
7. Sun, H.; Cheng, W.; Xu, H.; Cai, Z.; Yin, M.; Shi, F. Influence of CO_2 Curing on the Alkali-Activated Compound Mineral Admixtures' Corrosion Resistance to NaCl Dry-Wet Alternations. *Coatings* **2023**, *139*, 67. [CrossRef]
8. Yang, H. Influence of Mineral Powder and Fiber on Work and Mechanical Properties of Reactive Powder Concrete. *Multipurp. Util. Miner. Resour.* **2023**, *2*, 197–204.
9. Ge, W.; Zhang, Z.; Ashour, A.; Li, W.; Jiang, H.; Hu, Y.; Shuai, H.; Sun, C.; Li, S.; Liu, Y.; et al. Hydration characteristics, hydration products and microstructure of reactive powder concrete. *J. Build. Eng.* **2023**, *69*, 106306. [CrossRef]
0. Huang, C. Effect of polycarboxylate superplasticizer content on properties of reactive powder concrete. *Adhesion* **2022**, *49*, 116–119.
1. Gong, K.; Wu, M.; Xie, F.; Liu, G.; Sun, D. Effect of dry/wet ratio and pH on the stress corrosion cracking behavior of rusted X100 steel in an alternating dry/wet environment. *Constr. Build. Mater.* **2020**, *260*, 120478. [CrossRef]
2. Li, R.; Yang, J.; Yang, W.; Liu, Y.; Yang, Z. Degradation Model of Compressive Strength of Concrete under the Action of Steel Bar Rust Expansion. *China Concr.* **2022**, *156*, 16–21.
3. Cui, L.; Wang, H. Research on the Mechanical Strengths and the Following Corrosion Resistance of Inner Steel Bars of RPC with Rice Husk Ash and Waste Fly Ash. *Coatings* **2021**, *11*, 1480. [CrossRef]
4. Decky, M.; Papanova, Z.; Juhas, M.; Kudelcikova, M. Evaluation of the Effect of Average Annual Temperatures in Slovakia between 1971 and 2020 on Stresses in Rigid Pavements. *Land* **2022**, *11*, 764. [CrossRef]
5. Liu, Z. Analysis on the Mechanism of Concrete Shrinkage Cracking Under Cooling. *Jiangxi Build. Mater.* **2023**, *1*, 3–8.
6. Wang, H.; Zhang, A.; Zhang, L.; Liu, J.; Han, Y.; Wang, J. Research on the Influence of Carbonation on the Content and State of Chloride Ions and the Following Corrosion Resistance of Steel Bars in Cement Paste. *Coatings* **2020**, *10*, 1071. [CrossRef]
7. Liang, Z.; Peng, X.; Wang, H. The Influence of Aspect Ratio of Steel Fibers on the Conductive and Mechanical Properties of Compound Cement Reactive Powder Concrete. *Coatings* **2023**, *13*, 331. [CrossRef]
8. Mei, K.; He, Z.; Yi, B.; Lin, X.; Wang, J.; Wang, H.; Liu, J. Study on electrochemical characteristics of reinforced concrete corrosion under the action of carbonation and chloride. *Case Stud. Constr. Mater.* **2022**, *17*, e01351. [CrossRef]
9. Zhang, L.; Zhang, A.; Wang, Q.; Han, Y.; Li, K.; Gao, X.; Tang, Z. Corrosion resistance of wollastonite modified magnesium phosphate cement paste exposed to freeze-thaw cycles and acid-base corrosion. *Case Stud. Constr. Mater.* **2020**, *20*, e00421. [CrossRef]
0. Xia, D.; Yu, S.; Yu, J.; Feng, C.; Li, B.; Zheng, Z.; Wu, H. Damage characteristics of hybrid fiber reinforced concrete under the freeze-thaw cycles and compound-salt attack. *Case Stud. Constr. Mater.* **2023**, *18*, e01814. [CrossRef]
1. Chen, D.; Deng, Y.; Shen, J.; Sun, G.; Shi, J. Study on damage rules on concrete under corrosion of freeze-thaw and saline solution. *Constr. Build. Mater.* **2021**, *304*, 124617. [CrossRef]

coatings

MDPI

Article

Evaluation of Curing Effects on Bitumen Emulsion-Based Cold In-Place Recycling Mixture Considering Field-Water Evaporation and Heat-Transfer Conditions

Zili Zhao [1], Fujian Ni [1,*], Junqiu Zheng [2], Zhiqiang Cheng [3,4,*] and Shengjia Xie [3,4]

1 College of Transportation Engineering, Southeast University, Nanjing 211189, China; zili_zhao@seu.edu.cn
2 Engineering & Technology Department, Jiangsu Communications Holding Co., Ltd., Nanjing 210002, China; zjq20140811@sina.com
3 Shanghai Road and Bridge Group Co., Ltd., Shanghai 200433, China
4 Shanghai Engineering Research Center of Green Pavement Materials, Shanghai 200433, China
* Correspondence: 101004746@seu.edu.cn (F.N.); cr1903@tongji.edu.cn (Z.C.)

Abstract: The strength growth of a bitumen emulsion-based cold in-place recycling asphalt mixture (BE-CIR) is time-dependent and time-consuming due to the addition of water. There is a great difference between the curing conditions of specification in the laboratory and the in situ conditions, which often leads to a great discrepancy between the results of lab specimens and the field cores. The main objective of this paper is to evaluate the curing effect on laboratory BE-CIR considering field-water evaporation and heat-transfer conditions. Four different curing methods were designed by using different combinations of waterproof layers, heat insulation layers, and variable temperature modes. The variations in temperature indexes, moisture content, air void, and indirect tensile strength (ITS) with curing time were tested, and the mutual influence of these indicators was analyzed. Furthermore, the results of the laboratory samples were compared with the field cores. Testing results show that the performance of the BE-CIR mixture is significantly different from that with no treatment, which is manifested as higher moisture content and lower air void and ITS under the same curing time. The internal temperature of the mixture is the main factor affecting the variation of moisture content, especially on the first curing day. The air void of the mixture has a strong linear relationship with the moisture content. Moisture content and ITS under different curing methods showed similar trends and could be divided into two stages. Taking the field cores as a benchmark, it can be concluded that the field-water evaporation condition should be considered in the setting of indoor curing methods, while the heat transfer could not.

Keywords: bitumen emulsion; cold in-place recycling; curing method; indirect tensile strength; field cores

1. Introduction

Bitumen Emulsion-based Cold In-place Recycling (BE-CIR) is an environmentally friendly pavement maintenance technology that allows a high percentage of reuse of reclaimed asphalt pavement (RAP), eliminating the requirement for heating and transport during recycling [1–3]. In recent years, it has garnered global recognition and adoption in many countries due to its cost-effectiveness and environmental sustainability [4,5]. It also boasts a range of appealing attributes, including the capability to comprehensively address pavement distresses such as potholes, rutting, irregular cracks, and reflection cracks, and also has the potential to significantly prolong the service life of asphalt pavement, alleviate the burden of landfill disposal, and enhance the overall riding comfort [6–8]. To improve the early strength of BE-CIR mixtures, additives such as cement, fly ash, and lime are often added to the BE-CIR mixture. These additives can facilitate the accelerated breaking of the bitumen emulsion and promote strong bonding between the RAP aggregate and mortar through hydration reactions [9–11].

In contrast to conventional hot-mix asphalt (HMA), the BE-CIR mixture requires the addition of extra water to meet the workability criteria during compaction, ensure a sufficient coating of the bitumen emulsion, and achieve maximum density by lubricating the mixture [12–14]. Due to the presence of moisture and a cementitious co-binder, the evaluation of the performance development of the BE-CIR mixture necessitates taking into account the curing time. Research has revealed that in the early stage of curing time, cement has an important influence on the early strength growth rate of the BE-CIR mixture, while bitumen emulsion plays a decisive role in the initial value of strength and the final in-service strength [15]. During the curing process, a small proportion of water combines with the active minerals (e.g., cement) to form hydration products, while the majority of the free water leaves the system through evaporation [16,17]. This process is greatly affected by external conditions, including temperature, humidity, etc. [18–20], so that the performance of BE-CIR mixtures has significant time-varying characteristics. Because of this, in practical BE-CIR technology projects, it generally needs curing for few days before the construction of the upper structure can continue. This requirement limits the application scope of the technology to a certain extent. Therefore, it is of great significance to study the time-varying performance of BE-CIR mixtures during the curing period to increase the application scenarios of this economically and environmentally beneficial technology.

In order to study the performance change of BE-CIR mixtures in the laboratory, the premise is to simulate the in situ conditions in a laboratory setup. Several curing methods have been compared to obtain the development of strength formation and moisture loss during curing of BE-CIR, such as specimens covered with an impervious material or sealed and cured in a plastic bag [21–23]. Meanwhile, researchers have adopted a combination of varying times, temperatures, and humidity levels to simulate environmental curing conditions [24–26]. However, the aim of most of the current recommended laboratory-curing conditions is to accelerate the actual field-curing process. For example, BE-CIR specimens were recommended to cure in a temperature-controlled oven at 60 °C for 2 days after fabrication [27]. Whereas this may be beneficial for evaluating the structural contribution of the BE-CIR mixture when it is fully cured, it may not be capable of providing conditions to study time-varying characteristics of mixtures. Additionally, the use of high temperatures to accelerate the curing process may change the properties of the mixture.

However, under field-curing conditions, the internal temperature of the pavement is constantly fluctuating due to ambient temperature [28,29], and only the upper surface is in contact with the external environment until the cover layer is constructed. Since few studies have considered the moisture and heat conditions, like field, in laboratory simulations, results of the mechanical performance's development during curing could hardly correspond to the field-cored samples. To fill this gap, this study aims to establish a practical laboratory-curing method simulating in situ curing conditions as much as possible. The main objectives of this research are: (1) investigate the properties, including temperature indexes, volume parameters, moisture content, and indirect tensile strength (ITS), during the curing time considering field-water evaporation and heat-transfer conditions; (2) explore the mutual influence between properties during the curing process; (3) evaluate the designed curing method based on the results of the field-core samples.

2. Materials and Experimental Program

2.1. Mix Design

This work was based on a highway maintenance project using BE-CIR technology in Lianyungang, China. The structure design for this project was to regenerate the original 10 cm surface layer into an entire BE-CIR layer, followed by a 5 cm overlay layer on top of it. The materials and mixture design were consistent with the project [30]. Therefore, the laboratory samples were prepared using pure RAP, and the gradation of RAP material from this project fell within the medium gradation limitations specified by the Jiangsu Province agency [31], as demonstrated by the black curve in Figure 1. One cationic slow-setting emulsified bitumen was used and its basic properties are listed in Table 1 based on Chinese standards [32].

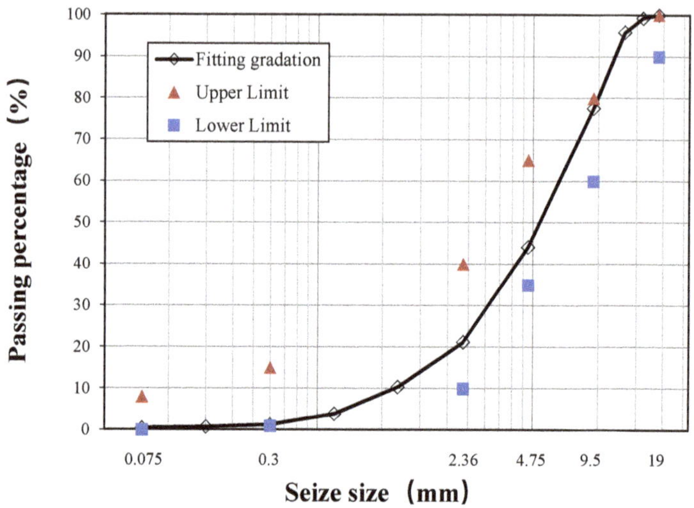

Figure 1. Aggregate gradation of CIR mixtures.

Table 1. Properties of emulsified bitumen.

Characteristics of the Emulsified Bitumen	Result
Residual content (%)	64
Penetration (25 °C; 0.1 mm)	59.1
Softening point (°C)	59.9
Ductility (15 °C; cm)	110
Sieve residue (1.18 mm; %)	0.03
Storage stability (5 d, 25 °C; %)	1.47
Storage stability (1 d, 25 °C; %)	0.42

An ordinary Portland cement (PO. 42.5) from Anhui Conch Cement Co., Ltd, (Hong Kong, China) was selected as an additive in this project. It is important to highlight that since the BE-CIR mixture in this project was used in the asphalt layer, and the CIR layer was only topped by a 5 cm overlay. In order to solve the problems of the lack of high-temperature performance and the weak early strength, a relatively higher cement content of 2.2% was determined based on previous project experiences. The optimum emulsified bitumen content was determined to be 3.3% by weight of the RAP based on the results from Marshall Stability and Indirect Tensile Strength. The optimal moisture content was obtained using the maximum dry density method as 4.01% by weight of the RAP [33]. The composition of the optimum moisture content included the water in the emulsified bitumen as well as the extra water. Since the residue content by evaporation of emulsified bitumen was 64%, the water in the emulsified bitumen was 1.188% (3.3×0.36), and the content of extra water was therefore calculated to be 2.8%. Table 2 shows the mixture design parameters based on the Jiangsu Province agency [31]. In this process, the specimens were compacted using a Marshall hammer with 75 blows per side, then cured in a temperature-controlled oven at 60 °C for 48 h and finally cooled at room temperature for 12 h for subsequent tests.

Table 2. Properties of mixture design.

Properties of Mixture Design	Results
Optimal emulsified bitumen content (%)	3.3
Cement content (%)	2.2
Optimal moisture content (%)	4.01
Bulk density (g/cm^3)	2.184
Target air voids (%)	10.0

2.2. Curing Method Design

In most of the field projects in China, there are usually several days of curing depending on the external environment for the BE-CIR layer before the construction of the cover layer. It is difficult to determine the exact curing time required in the actual field condition using the methods in the specification. Therefore, in an attempt to provide a better approach to solving this problem, this paper only considers the development of the mixture properties during curing prior to overlay. Based on this, compared with the actual project condition, two main aspects of water evaporation and heat transfer are not considered in curing conditions of the current specification. In order to solve these two problems, two targeted specimen preparation methods are proposed in this paper. The first is to simulate the actual moisture evaporation by adding a waterproof layer to the specimen. As shown in Figure 2a, after the mixture is fabricated, a fully enclosed tarpaulin is wrapped around its sides and bottom surface. This method leaves only the upper surface of the mixture in contact with the external environment, so that moisture inside the mixture can only evaporate from the upper surface. The second is to simulate the heat transfer by applying an insulation treatment to the specimens, as shown in Figure 2b. A thermal insulation pad is wrapped around the specimen, which leads to heat transfer from the upper surface only. One important assumption of this simulating method in this paper was ignoring the effect of the material on the moisture migration in the substructure of BE-CIR layer as well as on the sides. In the project for this paper, the substructure was a densely graded asphalt layer with almost zero percolation rate whose impact could be neglected, while for other different substructures this method could be improved.

(a) (b)

Figure 2. Simulation of curing schematic (**a**) waterproofing (**b**) heat insulation.

For the temperature conditions during the curing period, two different temperature control modes are considered. One is constant temperature mode, which means during the curing time, the temperature remains the same, and this mode is similar to the specification. Another is variable temperature mode, which is formed by a programmable environmental chamber to achieve temperature conditions close to those in the field. The settings of the curing temperature were mainly based on the BE-CIR project of Fenguan highway (G15) in June 2021. Based on the results of the temperature measurement of the road surface on site, it is known that the temperature of the road surface can reach 50 °C at the highest temperature at noon and drop to around 20 °C at the lowest temperature at night. More details on field temperature can be found in our previous study [30]. Therefore, for constant temperature mode, the constant temperature was set to the highest and lowest average temperature of 35 °C. For variable temperature mode, due to the difficulty of simulating the real temperature change of each day, a segmented temperature control method was applied in this study. Since the temperature of road surface is influenced by the solar radiation, according to the observation in the field, road surface temperature is relatively high during 8:00–18:00 each day. Therefore, for variable temperature mode, 10 h high temperature +14 h low temperature was set. In order to be closer to the actual project, 45 °C was used for high temperature and 25 °C was used for low temperature. The relative humidity during curing was set as 45% controlled by the environmental chamber.

Combining different specimen-preparation methods and temperature-control modes, four different curing methods are designed in this paper. As shown in Table 3, for the convenience of expression, "N", "W", and "WT" represent different specimen-preparation methods, namely no treatment, waterproofing, and waterproofing + heat insulation, respectively. "C" and "V" stand for constant temperature mode and variable temperature mode, respectively.

Table 3. Design of curing methods.

Curing Method	Specimen-Preparation Method	Temperature-Control Mode
N + C	No Treatment	Constant 35 °C
W + C	Waterproofing	Constant 35 °C
W + V	Waterproofing	Variable 10 h (45 °C) + 14 h (25 °C)
WT + V	Waterproofing + Heat Insulation	Variable 10 h (45 °C) + 14 h (25 °C)

2.3. Test Program

Previous studies have shown that the Superpave gyratory compaction method is more in line with the actual situation of pavement construction than the Marshall compaction method. Therefore, the Superpave gyratory compaction method was chosen to prepare BE-CIR mixture in this paper. In order to match the actual project, specimens with a diameter of 100 mm and a height of 100 mm were compacted. The gyratory count was set to 30 to achieve the same air void as the in situ project. All specimens were compacted by the same number of gyrations using the same mass.

Since one of the significant objectives of this study was to observe the performance development of a cold recycled mixture during the curing period under different curing conditions, five different curing days, namely 1, 3, 7, 14, and 28 d, were selected. Three replicate specimens were compacted for each day, so that fifteen replicate specimens were prepared for each curing condition. Considering four different conditions, a total of 60 specimens were produced; part of them are shown in Figure 3a. Two programmable environmental chambers were applied for different temperature-control modes. In addition, temperature sensors were buried in the specimens to obtain the temperature variation inside the mixture under different curing conditions. Considering the N + C method has no preparation treatment with the constant permaculture temperature, the internal temperature of the mix would keep consistent with the environmental temperature. Therefore, a sample was selected from each of the other three curing methods to monitor internal temperature changes, considering that if the temperature sensor was installed in the core of specimens, the installation process would inevitably cause an eventual damage and mass loss. The temperature sensors were stuck to the sample surface before the waterproofing, so the measured temperature was only the surface temperature of the specimen, which may be slightly different from that inside the sample. This can be seen in Figure 3b; two temperature sensors were installed at 25 mm and 75 mm from top to bottom, respectively.

(**a**) Specimen for different curing method (**b**) Sensor embedment

Figure 3. Specimens preparing and testing.

In this study, several material properties, including the water loss, moisture content, bulk volume relative density (γ_f), air void (AV), and indirect tensile strength (ITS) of BE-CIR mixture during curing period, were tested. The daily mass of each sample was recorded to calculate the water loss. The moisture content of each sample was also calculated based on the water content in the initial state. After curing 1, 3, 7, 14, and 28 days, three replicate specimens of each curing method were taken out. The bulk volume relative density (γ_f) of each specimen was tested and the air void (AV) could be therefore calculated according to AASHTO T 269 [34]. The ITS carried out at 15 °C was measured after testing γ_f. Before ITS tests, the upper part of the specimen with a height of 63.5 cm (measured down from the side in contact with the environment) was cut out for testing due to the dimensional requirements of the indirect tensile strength test. In addition, field-cored samples with a diameter of 100 mm were collected from this project at different times of curing to test ITS the same as laboratory specimens.

3. Results Analysis

3.1. Analysis of Temperature Variation

To evaluate the temperature variation of specimens under different curing methods, one temperature cycle (24 h) was selected for analysis, as shown in Figure 4. It was found that there was a great difference in the temperature variation between different curing methods. For the method W + C, the specimen rose rapidly to the set temperature and then remained stable. However, the two curing methods with variable temperature modes had two phases in each cycle. Specimens from the W + V method were similar to those from the W + C, rising rapidly to the maximum temperature of 45 °C and falling rapidly to 25 °C when cooling down. Because of the presence of insulation, the internal temperature of the WT + C method both rose and fell more slowly. The maximum temperature of WT + C did not reach the environmental temperature of 45 °C and the minimum temperature was also slightly less than 25 °C. Comparing results in the two different positions at the top and bottom, the results under different curing methods were similar, which means temperature in the top part is higher than that at the bottom part in the high-temperature phase, while lower in the low-temperature phase. This is mainly due to the high heat capacity and low thermal conductivity of BE-CIR, which leads to a slower temperature change inside than on the surface, and the phenomenon is also consistent with the data measured in the field [30].

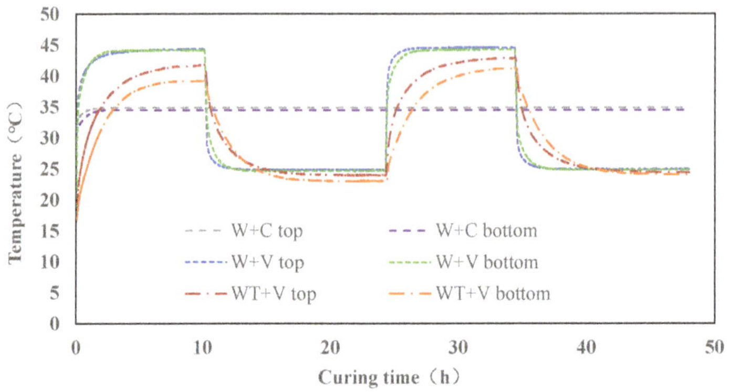

Figure 4. Temperature change curve for different curing methods.

In order to quantitatively analyze the curve of temperature-curing time, some indices were extracted from the curve, as shown in Figure 5. Firstly, the temperature-curing time curve of one cycle was divided into two phases—the high-temperature phase and the low-temperature phase. Each phase had three indexes, including maximum or minimum temperature, heating or cooling time, and Gt. The maximum and minimum temperatures were recorded as T-max and T-min, respectively. The time required to reach the maximum or minimum temperature was defined as the heating time or cooling time. The area enclosed by the time-temperature curve was defined as Gt, which can be interpreted as the temperature energy. It can be inferred that higher Gt indicates the specimens that absorbed or dissipated more heat at this stage. The Gt for the high and low-temperature phases was denoted as Gt-h and Gt-l, respectively. Two indicators were chosen for the whole cycle, the average temperature (T-a) and the total Gt (Gt-a). The results of these indicators are shown in Table 4.

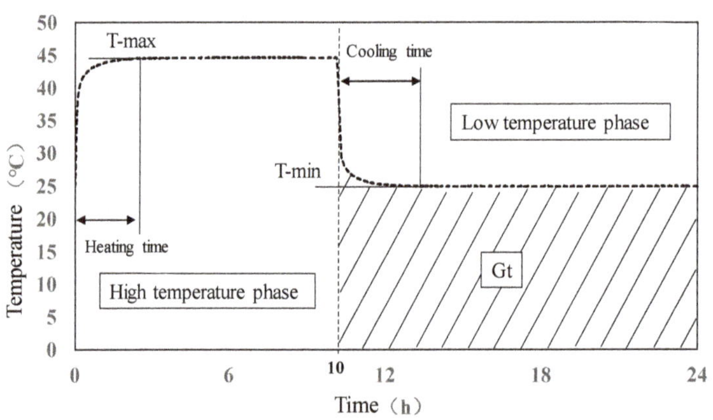

Figure 5. Quantitative analysis of temperature-curing time curve.

From Table 4, considering the full cycle, the T-a and Gt-a of W + C were the largest of the four curing methods, and the T-a was close to the set temperature of 35 °C. However, the T-a and Gt-a of the another two variable temperature modes had great differences. The T-a of WT + V was larger than that of W + V, while the Gt-a of WT + V was smaller. This is mainly due to the fact that the curve for WT + V rose and fell more slowly, which led to a concentration of temperature, resulting in a larger T-a. W + V remained at its T-max for a longer time, resulting in a larger Gt-a under the whole cycle. Comparing different positions of the mixture, both T-a and Gt-a were greater at the top than the bottom of the different methods.

Table 4. Quantitative analysis index of temperature for different curing methods.

Property	W + C-Top	W + C-Bottom	W + V-Top	W + V-Bottom	WT + V-Top	WT + V-Bottom
			High-temperature phase (0–10 h)			
T-max (°C)	34.9	34.5	44.7	44.6	43	41.3
Heating time (h)	2	2.1	2.5	3.3	7.9	8.5
Gt-h (°C·h)	348.07	343.28	442.88	436.95	405.44	377.50
			Low-temperature phase (11–24 h)			
T-min (°C)	34.9	34.5	24.9	24.8	24.4	24.1
Cooling time (h)	-	-	3.2	3.9	10.6	11
Gt-l (°C·h)	488.21	482.62	355.29	356.20	368.13	374.01
			Full cycle			
T-a (°C)	34.84	34.41	32.86	32.65	33.35	32.34
Gt-a (°C·h)	836.28	825.90	798.17	793.15	773.58	751.51

For different phases, in the high-temperature phase, the heating time of W + C method was only 2 h, and the three indexes of the upper and bottom were essentially the same. Specifically, the heating time of W + V was longer than that of W + C due to the difference in T-max. Meanwhile, the heating time for the top and bottom was also different, with the bottom being 3.3 h, 0.8 h longer than the 2.5 h in the top position. Additionally, the Gt-h at the top of W + V was the largest among the different methods. WT + V had a slightly lower T-max compared with W + V, and the top and bottom positions were 43 °C and 41.3 °C, respectively, which were 1.7 °C and 2.3 °C lower than the corresponding top and bottom position of W + V. In addition, the heating time of WT + V was considerably longer, with the top and bottom position being 7.9 h and 8.5 h, respectively, which lasted almost the entire high-temperature phase. Also, WT + V presented a smaller Gt-h compared with W + V; the order of Gt-h was W + V > WT + V > W + C. In the low-temperature phase, the cooling time of W + V was longer than the heating time, with the top and bottom parts being 3.2 h and 3.9 h, respectively, and the corresponding Gt-l was the smallest among the three methods. Similarly, comparing WT + V to W + V, WT + V showed slightly lower T-min, longer cooling time, and greater Gt-l. Therefore, the order of Gt-l became W + C > W + V > WT + V.

It can be concluded from the comparison that the average temperature of the variable temperature mode was lower than that of the constant temperature mode. When the insulation was added, the temperature change became slower. However, from the difference in heating and cooling time between the top and bottom position in W + C and W + V, it could also indicate that the waterproofing layer plays a certain role in blocking heat transfer. This may be also because of the fact that water in the specimens with waterproofing only diffuses in one direction, resulting in a difference in water content in thickness direction, which also affects the temperature transmission.

3.2. Moisture Variation Analysis

The results of moisture content over curing time are shown in Figure 6. The initial value for moisture content was 4%. It can be seen from the figure that under different curing methods, the water content decreased rapidly in the early stage, and remained stable as curing time increased. However, there are significant differences between the four curing methods. The moisture content of the N + C method was obviously lower than the other three methods, only 0.29% at twenty-eight days, and decreased rapidly in the first few days, especially from 4% to 1.547% on the first day. After curing for 7 days, the moisture content dropped to less than 1% and stayed stable earlier than the other methods, mainly because the water inside can diffuse freely from all sides of the specimen. For the other three curing methods, the curves of moisture content over curing time are relatively similar due to the specimen preparation of the waterproofing. The moisture content was still more than 2% after 5 days of curing, and it tended to be stable in about 12 days. Between the three curing methods, the moisture content ranked W + C < W + V < WT + V from 0 to 10 days, while after 10 curing days the moisture content of W + V gradually converged with that of W + C. Eventually, the water content of W + V and W + C kept almost the same after curing for 28 days (0.974% and 0.943%, respectively), which were smaller than that of WT + V (1.18%), while obviously greater than N + C (0.29%), indicating that there was still a significant amount of residual water within the mixture and this water may continue to dissipate with increasing curing.

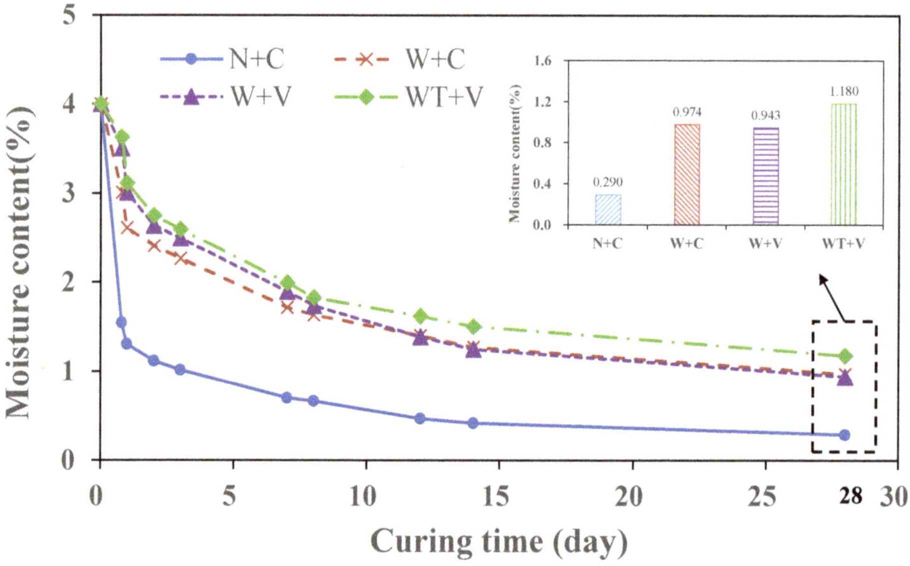

Figure 6. The variation of moisture content with curing time.

It is well-known that the changes in moisture content are mainly influenced by temperature, and the analysis of temperature shows that the rank of T-a is W + C > WT + V > W + V, while that for Gt-a is W + C > W + V > WT + V, indicating that Gt-a is consistent with the regularity exhibited by water content. Furthermore, considering the high- and low-temperature phases, Gt-h is W + V > WT + V in the high-temperature phase, while it is the opposite for Gt-l in the low-temperature phase, which infers that the high-temperature phase has a greater effect on the moisture content of the specimen.

3.3. *Analysis of the Changing Air Void Content*

In Figure 7, it can be found that the air void of the samples under different curing methods increases with the increase in curing time. This phenomenon is reasonable because after the BE-CIR mixture is formed, the internal voids are filled with free water. During the curing time, the free water in the voids is partially hydrated with the cement, and the rest of the water undergoes phase changes and evaporates into the environment, resulting in a continuous increase in the air void of the mixture. As can be seen from Figure 7, the air void for the N + C method is much greater than that for the other three methods and increases rapidly in the early stages, reaching 9.88% at 7 days, and remains essentially stable after curing for 7 days. The air void of the other three methods increases gradually with the curing time and does not reach a similar period of stability to N + C. The increase between 14 and 28 days is still 0.2%–0.3% for the three methods except N + C. Comparing those three curing methods, the air void of different methods did not show a consistent regularity under different curing days, and remained essentially the same when curing time was longer than 14 days. However, it is theoretically clear that the variety in air void was mainly caused by the dissipation of internal moisture, since the rank of moisture content was W + C < W + V < WT + V, and specimens with lower moisture content should have a greater air void. It can also be found that the variability in air voids is relatively large, and the maximum difference between different samples can reach 0.5%. When compared with N + C, the air void is well-discriminated due to differences between 0.8% and 1.6%, while comparisons between W + C, W + V, and WT + V show only 0.1%–0.3% differences in air voids between them, and this difference is even smaller than the variability in the test methods.

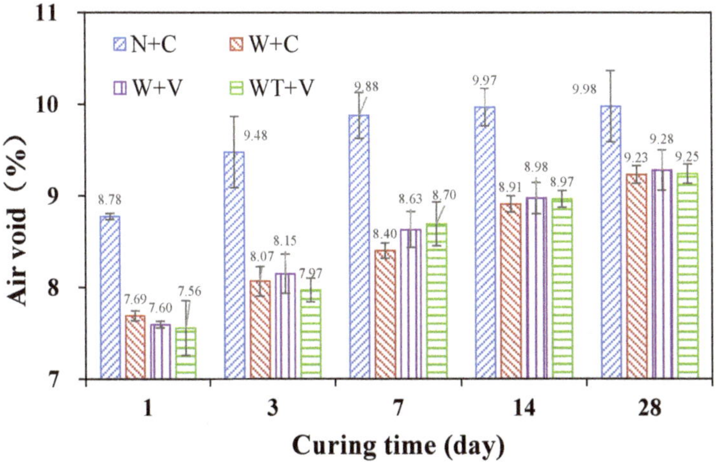

Figure 7. Variation in air void with curing time.

3.4. Analysis of the ITS

The variation in ITS with curing time for the different curing methods is shown in Figure 8. ITS increased over curing time but showed great differences between curing methods. Similar to moisture content and air void, the ITS of N + C was significantly greater than that of the other three curing methods since the optimum water content for cement was about 1/3 of its mass, i.e., about 0.8% water (as opposed to 2.2% cement). However, during the curing process, according to the results in Section 3.2, the moisture content of N + C was about 1% after 3 days of curing, whereas the other three methods were all greater than 2%, which was greater than the optimum water content for cement. The higher water content may, thus, inhibit the hydration of the cement, contributing to the lower ITS of the BE-CIR. In addition, the ITS of N + C increased from 0.74 MPa (14 d) to 0.963 MPa (28 d) due to continuous hydration of the cement, which is close to the increase from days 1 to 14 (0.253 MPa).

The other three curing methods had relatively low ITSs of less than 0.28 MPa on day 1, with 0.2 MPa smaller than N + C. From day 1 to day 3, the ITS increased rapidly, and the gap between N + C was only less than 0.12 MPa after curing for 14 days. The rank of these three methods was W + C > W + V > WT + V, which was consistent with the rank of moisture content, suggesting that higher moisture content inhibits cement hydration even more. Moreover, it could be found that there was great difference for the three curing methods with N + C at 28 days, similar to moisture content and air void, indicating that the curing may be still in progress after curing for 28 days at least for the sealed specimens. This is mainly due to the fact that the cement is not fully hydrated in this condition because of the high moisture content. It is interesting whether or when these indexes converge with different curing methods and different curing conditions; more studies will be conducted in the future.

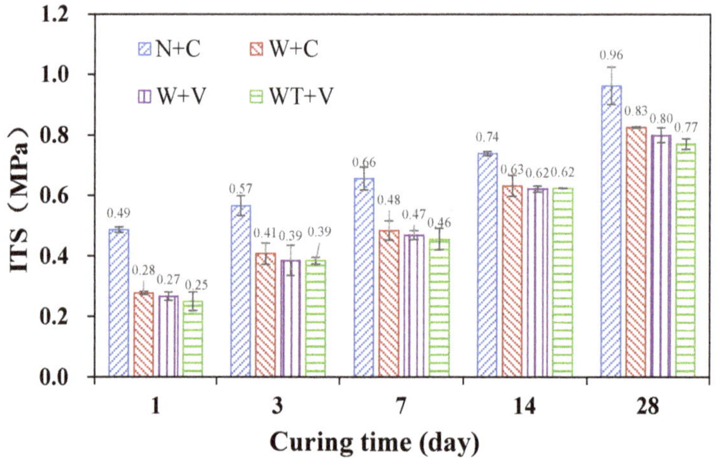

Figure 8. Results of ITS over curing time.

4. Mutual Influence of Indicators

In Section 3, four indicators of temperature, moisture content, air void, and ITS were analyzed over curing time, and clearly there was a strong connection between these four indicators. Theoretically, the temperature changes inside specimens under different curing methods directly affect the moisture content, and since free water mainly exists in the voids, the decrease in water content will directly cause the change in the air void. The ITS of the mixture is the result of the comprehensive influence of the curing conditions. Therefore, this section's analysis of the mutual influence of different indicators includes temperature indexes with moisture content, moisture content with air void, and moisture content/air void with ITS. The average value of these indicators was used in this section.

4.1. Effect of Temperature Indexes on Moisture Content

Since specimens of the N + C method were not treated with a waterproofing layer, although it had the same temperature conditions as the W + C method, the moisture content decreased faster due to how it could be dissipated from all sides of the specimens. It proved that the waterproofing layer had a significant effect on changes in the moisture content of the specimen. Therefore, results of the other three curing methods were detailed and analyzed.

From Section 3.2, the moisture content of different curing methods displays as W + C < W + V < WT + V, which corresponds to the rank of Gt-a. Figure 9a illustrates the moisture content and Gt-a. The rank of Gt-a was always W + C > W + V > WT + V since Gt-a is accumulated continuously with curing time. However, the rank of moisture content was different. During the first 7 curing days, the moisture content of W + C was the smallest and that of WT + V was the largest. When the curing period exceeded 14 days, the moisture content of W + C and W + V were closer, while that for WT + V was still the largest, which exhibited that for the specimens with only the treatment of waterproofing, the effect of different temperature modes on the moisture content decreased with time, and there was almost no effect after curing for 14 days. In addition, from the comparison of the W + V and WT + V methods, it can be seen that the influence of the thermal insulation on the moisture content was continuous during curing for 28 days, and the moisture content of the samples with the thermal insulation layer was significantly larger.

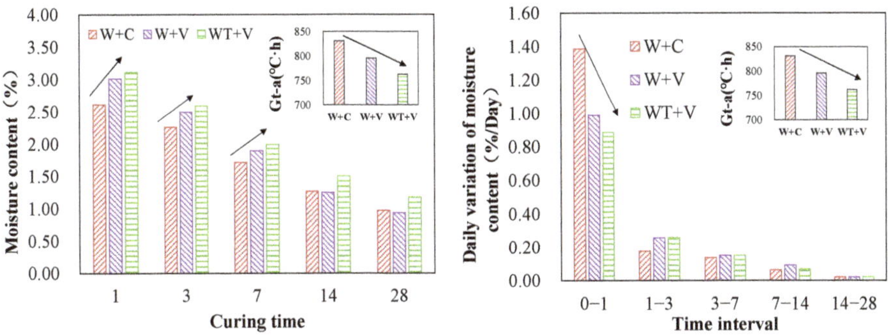

(**a**) Moisture content with Gt-a (**b**) Daily variation of moisture content with Gt-a

Figure 9. Effect of Gt-a on moisture content.

In order to further analyze the influence of the temperature index on the change in moisture content, the daily variation in moisture content was calculated, that is, the difference in moisture content divided by the number of curing days, and the results are shown in Figure 9b. It can be found that the effect of temperature was mainly on the first curing day. The moisture content of W + C decreased 1.4% on the first day, and the variation for W + V was slightly larger than that for WT + V. Subsequently, the daily variation in moisture content of W + C was the smallest among the three methods, which was opposite to its Gt-a. Similarly, the daily variation in moisture content of W + V and WT + V was not the same rank as that of Gt-a except for the first curing day. Therefore, it can be concluded that the effect of different temperature modes on moisture content is mainly reflected in the initial stage of curing, especially the first day.

4.2. Relationship between Moisture Content and Air Void

Figure 10 illustrates the relationship between moisture content and air void during the curing period. A clear linear relationship between moisture content and air void can be observed regardless of the curing method, and the greater the moisture content, the smaller the air void. This indicates that after compaction of the BE-CIR, residual water fills the void, and the free water continues to evaporate, migrate, or be absorbed by cement during curing, resulting in the increase in air void. Meanwhile, there is a maximum value of the air void, which is about 10% when the water content is close to 0%, indicating that if the internal water inside the mixture is completely dissipated, the real air void inside the mixture is about 10%.

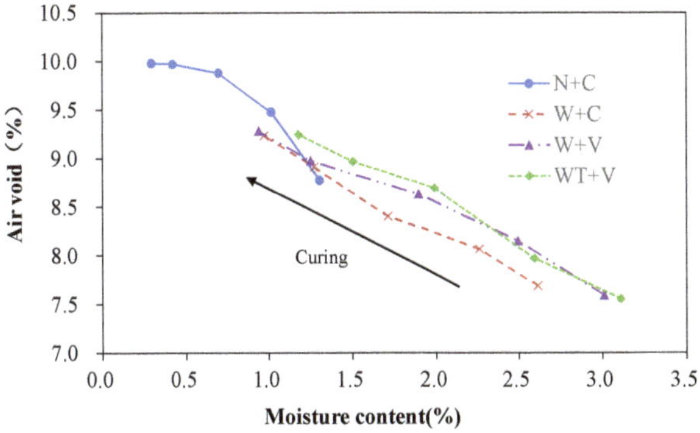

Figure 10. Relationship between moisture content and air void.

4.3. Effect of Moisture Content on ITS

Since there was a certain systematic error in the air void and a strong linear relationship between moisture content and air void, only the effect of moisture content on ITS is analyzed in this section, as shown in Figure 11. From the curve of N + C, it can be found that the relationship between the ITS and moisture content obviously presented two stages. During the first 14 curing days, the moisture content and ITS show a linear relationship, while from 14 to 28 days, ITS increased significantly with only a slight decrease in moisture content. Meanwhile, while the curves of W + C and W + V had similar characteristics, development of ITS with moisture content before and after 14 days of curing was not the same. Therefore, it can be inferred that the strength growth of the BE-CIR during the curing period can be divided into two stages. In the first stage, the growth of ITS is mainly affected by the internal moisture dissipation, high residual water inhibits the hydration of the cement, and prevents the formation of asphalt film. With the continuous loss of moisture during the curing process, the inhibiting effect of water on the hydration of the cement diminishes, leading to a gradual increase in ITS. In the second stage, it can be considered that the free water in the void has basically evaporated. In this stage, more hydration products were formed and better bonded to the bitumen film, resulting in the strength growth of the BE-CIR mixture.

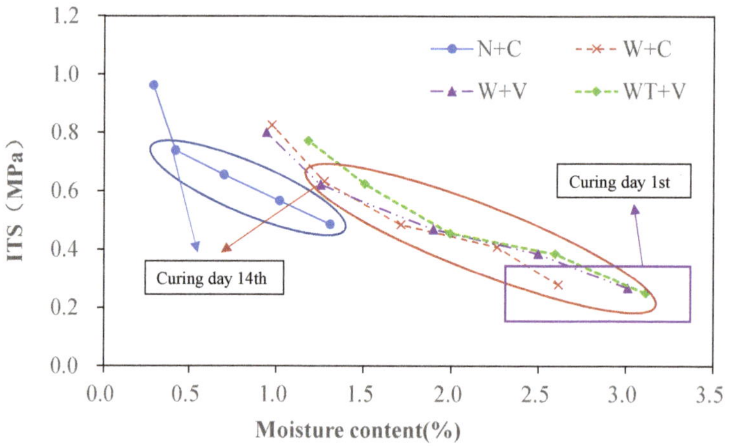

Figure 11. Relationship between moisture content and ITS.

In addition, by comparing the moisture content and ITS of the three curing methods except N + C on the first day of curing, it can be found that the moisture content of the two methods under variable temperature modes is closer. W + C with constant temperature mode had a smaller moisture content, while ITS was basically the same as that of the other two. For the moisture content and ITS after curing for 28 days, the moisture content of WT + V was relatively large and the strength was slightly smaller.

5. Comparison with Field Cores

The field cores were drilled at four curing times, 2, 7, 14, and 28 days, and two samples were taken at each time for testing. Due to the inevitable presence of large amounts of water during the coring process, the moisture content of the field cores could not be tested precisely. Also, this makes it difficult to obtain the actual air void at a special curing time. Therefore, only the ITS of the field cores was tested and the comparison of laboratory samples and field cores is shown in Figure 12. It is also worth noting that the field conditions were more complex than the laboratory conditions, with the temperature and relative humidity varying cyclically every day. The average ambient temperature during the curing was about 25–30 °C and there were few days of rain. In order to match with the indoor data, the overlay was not constructed during the 28 days of curing in the field-test section.

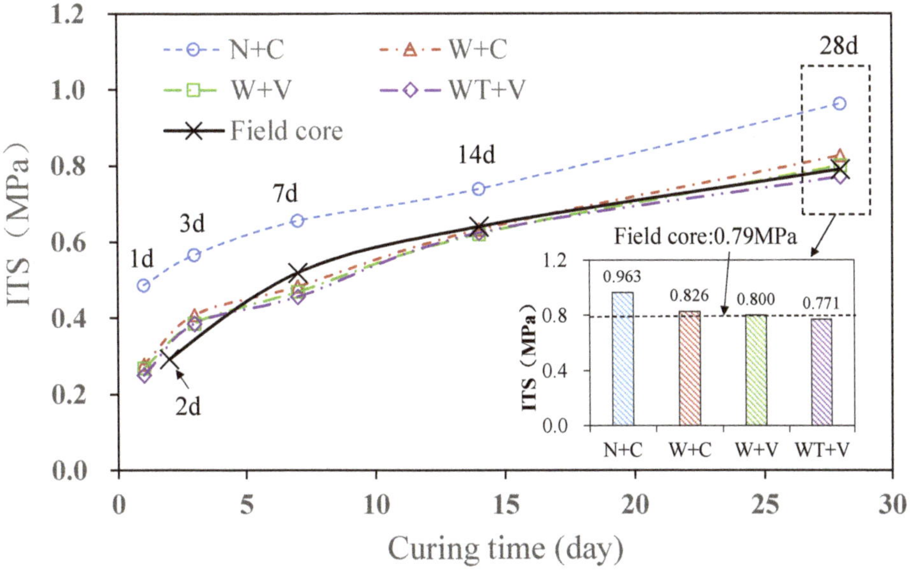

Figure 12. Comparison of ITS between laboratory samples and field cores.

It can be found that the ITS obtained from the curing method of N + C was significantly larger than that of the field cores, indicating that the curing method without any treatment is inconsistent with the actual situation. For the other three curing methods, the increase in ITS over curing time was close to that of the field cores, suggesting that the indoor curing method should consider field-water evaporation. In terms of ITS at a curing time of 28 days, W + V was the closest to the field cores with a difference of only 0.01 MPa, W + C was slightly larger, and WT + V was slightly smaller. Therefore, results of specimens under the curing method of W + V, that is, waterproofing + variable temperature mode, were the closest to those in the actual field curing conditions.

However, considering that the variable temperature mode is relatively complicated and difficult to be realized in the lab, the errors of results between W + C and core samples are also small. The only difference between W + C and W + V is the moisture content in the first few curing days. Therefore, the curing method of W + C, which is the specimen with waterproofing + constant temperature mode, is the most suitable indoor curing method for the BE-CIR, is closer to the real condition, and easier to popularize. To sum up, the indoor curing method for BE-CIR should consider field-water evaporation, while the heat transfer does not have to be considered in the indoor settings.

6. Findings and Conclusions

Four different curing methods considering water evaporation and heat transfer were designed. The variations in temperature indexes, moisture content, air void, and ITS with curing time were tested, and the mutual influence of these indicators was analyzed. Furthermore, results of the laboratory samples after curing for 28 days were compared with the field cores. Several findings and conclusions can be summarized as follows:

(1) The performance development of the BE-CIR mixture cured in the method considering the field-water evaporation condition had significant differences from that with no

treatment (N + C), which was manifested as higher moisture content, lower air void, and ITS under the same curing time. This indicates that the higher water content within the mix may inhibit the hydration of the cement and, thus, lead to a lower strength.

(2) Comparing curing methods considering different heat-transfer conditions, the moisture content showed a rank of W + C < W + V < WT + V, and the ITS had similar regularity. The temperature energy (Gt-a) was better aligned with this ranking than the average temperature. In addition, the differences between the three methods were much smaller than the differences considering water evaporation.

(3) The internal temperature of the mixture was the main factor affecting the variation in moisture content, especially on the first curing day. The air void of the mixture had a strong linear relationship with the moisture content. The relationship between water content and ITS can be divided into two stages. In the first stage, the ITS increased linearly with the decrease in moisture content, indicating that the water dispersion had a significant influence on the strength growth of the BE-CIR mixture. In the second stage, the moisture content decreased slightly and ITS increased rapidly, implying that the free water basically evaporated, and the development of strength was mainly caused by the continuous hydration of cement.

(4) Compared with the results of field cores, N + C had the largest difference in both moisture content and ITS due to the quick water loss rate. The waterproofing and variable temperature mode can better simulate the in situ environmental conditions and delay the water-loss rate of the specimen to a reasonable value.

To sum up, in the laboratory curing setup of BE-CIR, it is important to consider field-water evaporation while the heat-transfer conditions have less influence. Based on this, the waterproofing plus constant temperature mode was considered as the optimal curing method due to its simplicity and a good match. However, at present, only the results of field cores from one actual pavement project were used for the verification of our method. Therefore, more field cores from different sites with different environmental conditions are needed to further verify the adaptability of this curing method.

Although limited in adaptability, the curing method of W + C provides an option to better simulate in situ curing conditions in the laboratory. Different from the purposes of most of the current recommended laboratory curing conditions, the curing method promoted by this paper may not be usually used in the material design of BE-CIR mixtures since it costs more curing time; it is more suitable as a supplementary laboratory test for QC/QA. From the results of this method, pavement engineers and practitioners could achieve a model of mechanical property over curing time to reliably predict the recommended time for the placing of overlay on CIR layers under specific curing conditions.

Author Contributions: Conceptualization, F.N. and Z.C.; Methodology, F.N. and S.X.; Software, Z.Z.; Validation, J.Z.; Formal analysis, Z.Z.; Data curation, Z.Z.; Writing—original draft, Z.Z.; Writing—review & editing, F.N. and J.Z.; Visualization, Z.Z.; Project administration, J.Z.; Funding acquisition, F.N. All authors have read and agreed to the published version of the manuscript.

Funding: This research was funded by the Fundamental Research Funds for the Central Universities (Grant No. 3221002139D) and Shanghai municipal administration commission of housing and urban-rural development (2022-002-002).

Institutional Review Board Statement: Not applicable.

Informed Consent Statement: Not applicable.

Data Availability Statement: The data presented in this study are available on request from the corresponding author.

Conflicts of Interest: The authors declare no conflict of interest.

eferences

. Gu, F.; Ma, W.Y.; West, R.C.; Taylor, A.J.; Zhang, Y.Q. Structural performance and sustainability assessment of cold central-plant and in-place recycled asphalt pavements: A case study. *J. Clean. Prod.* **2019**, *208*, 1513–1523. [CrossRef]

Modarres, A.; Rahimzadeh, M.; Zarrabi, M. Field investigation of pavement rehabilitation utilizing cold in-place recycling. *Resour. Conserv. Recycl.* **2014**, *83*, 112–120. [CrossRef]

. Hussein, Z.; Yaacob, H.; Idham, M.; Abdulrahman, S.; Choy, L.; Jaya, R. Rejuvenation of hot mix asphalt incorporating high RAP content: Issues to consider. In *IOP Conference Series: Earth and Environmental Science*; IOP Publishing: Bristol, UK, 2020; p. 012009.

Turk, J.; Pranjic, A.M.; Mladenovic, A.; Cotic, Z.; Jurjavcic, P. Environmental comparison of two alternative road pavement rehabilitation techniques: Cold-in-place-recycling versus traditional reconstruction. *J. Clean. Prod.* **2016**, *121*, 45–55. [CrossRef]

. Alkins, A.E.; Lane, B.; Kazmierowski, T.; Environmental, S.P. Economic, and Social Benefits of In Situ Pavement Recycling. *Transp. Res. Rec.* **2008**, *2084*, 100–103. [CrossRef]

. Xiao, F.; Yao, S.; Wang, J.; Li, X.; Amirkhanian, S. A literature review on cold recycling technology of asphalt pavement. *Constr. Build. Mater.* **2018**, *180*, 579–604. [CrossRef]

. Wang, Z.C.; Wang, Q.; Jia, C.X.; Bai, J.R. Thermal evolution of chemical structure and mechanism of oil sands bitumen. *Energy* **2022**, *244*, 123190. [CrossRef]

. Al-Saffar, Z.H.; Yaacob, H.; Katman, H.Y.; Satar, M.K.I.M.; Bilema, M.; Jaya, R.P.; Eltwati, A.S.; Radeef, H.R.J.S. A review on the durability of recycled asphalt mixtures embraced with rejuvenators. *Sustainability* **2021**, *13*, 8970. [CrossRef]

. Mohammad, L.N.; Abu-Farsakh, M.Y.; Wu, Z.; Abadie, C. Louisiana experience with foamed recycled asphalt pavement base materials. *Transp. Res. Rec.* **2003**, *1832*, 17–24. [CrossRef]

0. Kuna, K.; Airey, G.; Thom, N. Mix design considerations of foamed bitumen mixtures with reclaimed asphalt pavement material. *Int. J. Pavement Eng.* **2017**, *18*, 902–915. [CrossRef]

1. Konitufe, C.; Abubakar, A.; Baba, A.S.J.C. Influence of Aggregate Size and Shape on the Compressive Strength of Concrete. *Construction* **2023**, *3*, 15–22. [CrossRef]

2. Flores, G.; Gallego, J.; Miranda, L.; Marcobal, J.R. Cold asphalt mix with emulsion and 100% rap: Compaction energy and influence of emulsion and cement content. *Constr. Build. Mater.* **2020**, *250*, 118804. [CrossRef]

3. Deb, P.; Singh, K.L. Mix design, durability and strength enhancement of cold mix asphalt: A state-of-the-art review. *Innov. Infrastruct. Solut.* **2021**, *7*, 61. [CrossRef]

4. Du, S.W. The Optimum Pre-mixing Water Content in Asphalt Emulsion Mixture with Cement. *J. Test. Eval.* **2021**, *49*, 4560–4575. [CrossRef]

5. Lin, J.; Wei, T.; Hong, J.; Zhao, Y.; Liu, J. Research on development mechanism of early-stage strength for cold recycled asphalt mixture using emulsion asphalt. *Constr. Build. Mater.* **2015**, *99*, 137–142. [CrossRef]

6. Golian, M.; Katibeh, H.; Singh, V.P.; Ostad-Ali-Askari, K.; Rostami, H.T. Prediction of tunnelling impact on flow rates of adjacent extraction water wells. *Q. J. Eng. Geol. Hydrogeol.* **2020**, *53*, 236–251. [CrossRef]

7. Yang, Y.H.; Wang, H.B.; Yang, Y.; Zhang, H.Z. Evaluation of the evolution of the structure of cold recycled mixture subjected to wheel tracking using digital image processing. *Constr. Build. Mater.* **2021**, *304*, 124680. [CrossRef]

8. Cardone, F.; Grilli, A.; Bocci, M.; Graziani, A. Curing and temperature sensitivity of cement-bitumen treated materials. *Int. J. Pavement Eng.* **2015**, *16*, 868–880. [CrossRef]

9. Graziani, A.; Godenzoni, C.; Cardone, F.; Bocci, M. Effect of curing on the physical and mechanical properties of cold-recycled bituminous mixtures. *Mater. Des.* **2016**, *95*, 358–369. [CrossRef]

0. Graziani, A.; Iafelice, C.; Raschia, S.; Perraton, D.; Carter, A. A procedure for characterizing the curing process of cold recycled bitumen emulsion mixtures. *Constr. Build. Mater.* **2018**, *173*, 754–762. [CrossRef]

1. Fu, P.C.; Jones, D.; Harvey, J.T.; Halles, F.A. Investigation of the Curing Mechanism of Foamed Asphalt Mixes Based on Micromechanics Principles. *J. Mater. Civ. Eng.* **2010**, *22*, 29–38. [CrossRef]

2. Lee, H.; Im, S. *Examination of Curing Criteria for Cold In-Place Recycling*; Technical Report for Iowa Highway Research Board, University of Iowa: Iowa City, IA, USA, 2008.

3. Pasetto, M.; Pasquini, E.; Baliello, A.; Raschia, S.; Rahmanbeiki, A.; Carter, A.; Perraton, D.; Preti, F.; Gouveia, B.C.S.; Tebaldi, G. Influence of curing on the mechanical properties of cement-bitumen treated materials using foamed bitumen: An interlaboratory test program. In Proceedings of the 9th International Conference on Maintenance and Rehabilitation of Pavements—Mairepav9, Zurich, Switzerland, 1–3 July 2020; Springer: Berlin/Heidelberg, Germany, 2020; pp. 55–65.

4. Kim, Y.; Im, S.; Lee, H.D. Impacts of curing time and moisture content on engineering properties of cold in-place recycling mixtures using foamed or emulsified asphalt. *J. Mater. Civ. Eng.* **2011**, *23*, 542–553. [CrossRef]

5. Tebaldi, G.; Dave, E.V.; Marsac, P.; Muraya, P.; Hugener, M.; Pasetto, M.; Graziani, A.; Grilli, A.; Bocci, M.; Marradi, A.J.R.M.; et al. Synthesis of standards and procedures for specimen preparation and in-field evaluation of cold-recycled asphalt mixtures. *J. Mater. Civ. Eng.* **2014**, *15*, 272–299. [CrossRef]

6. Kuna, K.; Airey, G.; Thom, N.J.C.; Materials, B. Development of a tool to assess in-situ curing of Foamed Bitumen Mixtures. *Constr. Build. Mater.* **2016**, *124*, 55–68. [CrossRef]

7. Yang, W.; Ouyang, J.; Meng, Y.; Han, B.; Sha, Y.J.C.; Materials, B. Effect of curing and compaction on volumetric and mechanical properties of cold-recycled mixture with asphalt emulsion under different cement contents. *Constr. Build. Mater.* **2021**, *297*, 123699. [CrossRef]

28. Ma, X.Y.; Dong, Z.J.; Quan, W.W.; Dong, Y.K.; Tan, Y.Q. Real-time assessment of asphalt pavement moduli and traffic loads usin monitoring data from Built-in Sensors: Optimal sensor placement and identification algorithm. *Mech. Syst. Signal Process.* **202** *187*, 109930. [CrossRef]
29. Wang, H.; Zhang, X.; Jiang, S.C. A Laboratory and Field Universal Estimation Method for Tire-Pavement Interaction Noise (TPIN Based on 3D Image Technology. *Sustainability* **2022**, *14*, 12066. [CrossRef]
30. Zhao, Z.L.; Jiang, J.W.; Chen, Z.W.; Ni, F.J. Moisture migration of bitumen emulsion-based cold in-place recycling pavement afte compaction: Real-time field measurement and laboratory investigation. *J. Clean. Prod.* **2022**, *360*, 132213. [CrossRef]
31. DOT, J.P. *Applied Manual for Asphalt Emulsion Cold In-Place Recycling Technology*; Department of Transport of Jiangsu provinc Nanjing, China, 2010. (In Chinese)
32. *JTG E20-2011*; Standard Test Methods of Bitumen and Bituminous Mixtures for Highway Engineering. Ministry of Transport c the People's Republic of China: Beijing, China, 2011. (In Chinese)
33. *JTG/T 5521-2019*; Technical Specifications for Highway Asphalt Pavement Recycling. Ministry of Transport of the People' Republic of China: Beijing, China, 2019. (In Chinese)
34. *AASHTO T 269*; Standard Method of Test for Percent Air Voids in Compacted Dense and Open Asphalt Mixtures. America Association of State Highway and Transportation Officials: Washington, DC, USA, 2014.

coatings

MDPI

Article

An Evaluation of Asphalt Mixture Crack Resistance and Identification of Influential Factors

Liheng Shu [1], Fujian Ni [1,*], Hui Du [2] and Yajin Han [1]

[1] School of Transportation, Southeast University, No.2 SEU Road, Nanjing 211189, China; shuliheng@seu.edu.cn (L.S.); yajinhan@seu.edu.cn (Y.H.)

[2] School of Civil Engineering, Fujian University of Technology, No.3 Xueyuan Road University Town, Fuzhou 350118, China; duhui1105@163.com

* Correspondence: 101004746@seu.edu.cn

Abstract: The crack resistance of asphalt pavement mixtures directly impact pavement service condition and pavement distress. And characterizing the crack resistance of a pavement mixture can reflect the crack resistance potential of asphalt pavement. This study analyzes several representative highway sections based on time, material, and service conditions to identify the mixture type of three layers. Semi-circular bending tests are conducted at 15 °C, and load–displacement curves are recorded. Factor independence analysis is performed, and combinations showcasing the cracking performance of the surface layer, middle layer, and bottom layer are selected. Analysis of variance (ANOVA) evaluating the indices versus selected factors for the three layers identifies significant influencing factors, and the crack resistance is analyzed based on these significant factors. The crack resistance of the middle layer with the highest truck loads is significantly lower than the two other lanes and the shoulder. Transverse crack spacing (TCS) can be used to directly evaluate the crack resistance of the mixture. The Factor dots upper rate (FUDR) and absolute Factor dots upper rate (absFUDR) indices are introduced to quantify the percentage deviation of a factor specimen from the average crack resistance index–fracture energy ratio, indicating whether the crack curve becomes sharper or flatter. The factor dots upper rate index is then applied to characterize the factors, and the results are reasonable. It is found only on the surface and middle layers that the service age has significant impacts on crack resistance, the Transverse crack spacing has significant impacts on crack resistance index, and the Factor dots upper rate can identify the brittleness of mixtures with different factors.

Keywords: crack resistance; cores; SCB; service age; asphalt aggregate ratio; load displacement curve; TCS

1. Introduction

After asphalt pavement construction, the asphalt mixture in pavement undergoes a complex process of distress until the end of its service life. The main types of distress are rutting, transverse and longitude crack, stripping and potholes [1]. Among these forms of distress, crack-resistance performance plays an important role in pavement condition, as pavement service often leads to obvious crack distress, including fracture and fatigue cracks [2–4]. Crack resistance distress in pavements is mainly caused by environmental and loading factors such as traffic volume, oxidation aging, and in the case of pavements with a semi-rigid base, the presence of many transverse cracks. Additionally, the materials or structural factors of the pavement, such as mixture type and structural combination, can also affect the pavement crack resistance distress [5]. To further investigate the difference in crack resistance and the factors causing this difference, the field service, which is complex and affected by numerous factors, was analyzed. Four important factors were examined: accumulated traffic loads (ESALs), pavement service age (years), pavement section area

(area), and materials used in different layers (mixture type). The significant factors were selected for pavement crack evaluation [6–8].

Previous research has found a good relationship between the crack resistance of the mixture and weather conditions, such as low temperature or intermediate temperature, allowing crack-resistance performance to be used to evaluate pavement crack condition [9,10]. Insufficient crack resistance may result in small or longitudinal cracks. Field core testing is a direct and reasonable approach used to illustrate actual pavement condition [3]. The semi-circular bending (SCB) test is an effective method used to evaluate mixture fracture performance compared to other crack tests, such as the indirect tensile test (IDT), bending beam test, thermal stress restrained sample test (TSRST) or overlayer test (OT) [9,11,12]. The SCB test offers advantages such as material and time saving, and accuracy of results [13]. And the un-notched SCB could be a promising alternative to distinguish asphalt mixtures with regard to crack resistance, which is more convenient in mass specimens testing without increasing testing variance [14,15] Thus, all the semi-circular specimens tested in this study are un-notched SCB specimens at an intermediate temperature of 15 °C. The cores were prepared as SCB specimens and tested at intermediate temperature to evaluate fracture and fatigue performance [16]. Three indices, namely tensile strength, fracture energy and crack resistance index (CRI) [17], calculated based on the load–displacement curve, were used for all the tested specimens. Each index has its own advantages and disadvantages [15].

The aged and reclaimed asphalt pavement (RAP) containing mixture specimens exhibits different fracture types, which can be identified using these crack indices [18,19]. Unmodified aged asphalt is harder than unaged asphalt, and unmodified aged asphalt mixture has a high modulus and is more likely to experience brittle fracture. Modified aged asphalt and asphalt mixture may not be harder than modified aged asphalt and asphalt mixture, but they also tend to experience brittle fracture in the intermediate-temperature crack test. This means that when the load–displacement curves of aged specimens' SCB tests are plotted, they are likely to have a short or nonexistent post-peak curve [20–22]. The fracture energy of aged specimens is smaller than that of unaged specimens of the same mixture type, and the crack indices of aged specimens also differ [16]. The tensile strength of aged mixture specimens is likely to be higher than that of unaged mixture for both modified and unmodified asphalt mixture over service age. The fracture energy of aged mixture specimens is likely to be lower than that of new mixture specimens [23]. The CRI of aged mixture specimens does not follow a consistent pattern. The indices study of crack resistance mostly focusses on lab-aged mixtures and mixtures containing RAP, and the indices study of crack resistance on field cores focuses on the effects of low temperature [10]. Studies investigating the crack resistance of field cores at intermediate temperature are needed to improve our understanding of the crack resistance change on asphalt pavement through service.

The objective of this research is to identify the factors that independently and significantly influence the crack resistance of pavement specimens and to conduct a statistical analysis of crack curve classification. A schematic diagram of the study is shown in Figure 1. Firstly, the independent significant influence is explored via factors independent analysis to determine whether these factors are significant to the crack resistance of a certain layer. Secondly, an analysis of variance (ANOVA) and Tukey–Kramer comparison analysis of factors on crack resistance is conducted to determine whether a factor significantly affects the crack resistance of specimens. Thirdly, the crack curve statistical analysis aims to distinguish between brittle fractures and ductile fractures on mixture specimens with different factor conditions.

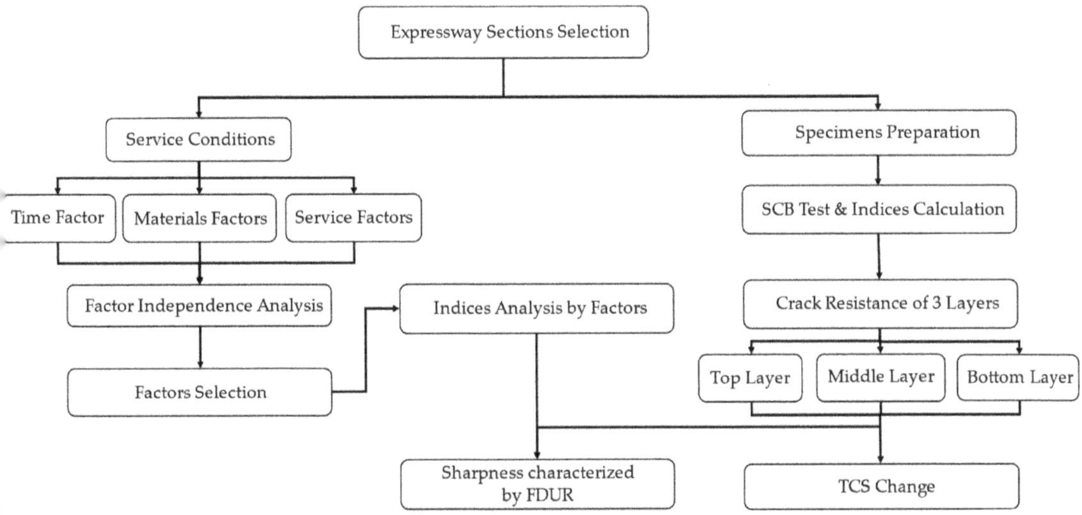

Figure 1. Schematic diagram of this study.

2. Materials Preparation and Test Method

2.1. Pavement Section Selection and Factors Indentification

To evaluate mixture performance in pavement, several sections without any maintenance were selected. These sections were collected from 9 highways, as shown in Table 1. After the year 2000, all newly built and maintained projects used a modified asphalt mixture for the surface layer. Similarly, after the year 2007, all newly constructed and maintained middle layers also used modified asphalt mixture. Therefore, the surface and middle layer mixtures with a service age of less than 15 years are all modified.

The crack resistance of pavement in service is affected by various factors, which can be categorized as the time factor, materials factors, and service factors. The time factor only consists of one factor: service age. The materials factors are influenced by the original materials properties, such as air void, mixture type, and asphalt aggregate ratio. The service factors describe the service loads and environment, including the climate, average temperature and equivalent single-axle loads (ESALs). Most expressways have a service age of less than 20 years. If a pavement's service age is greater than 20 years, maintenance is conducted.

Table 1. The selected pavement section information.

Section ID	Service Age (year)	Structure	Lane	TCS^2 (m)
1 RYDQ	17	4 cm SMA13 + 6 cm AC20 + 8 cm AC25 + 36 cm CTB + 20 cm LFS	1, 2, 3	44
2 TZDQ	5	4 cm SMA13 + 6 cm AC20 + 8 cm AC25 + 38 cm CTB + 20 cm CTB	2, shoulder	200
3 TZDQ	2 surface layers 9 others	4 cm SMA13 + 6 cm AC20 + 8 cm AC25 + 38 cm CTB + 20 cm CTB	3, shoulder	21
4 XY1	4	4 cm SMA13 + 6 cm SUP20 + 8 cm SUP25 + 34 cm CTB + 20 cm LFTB	2	30
5 XY2	17	4 cm AK13 + 6 cm AC20 + 8 cm AC25 + 34 cm CTB + 20 cm LFTB	2	13
6 LL	9	5 cm SMA13 + 6 cm SUP20 + 10 cm SUP25 + 38 cm CTB + 20 cm LFS	2	5.4

Table 1. *Cont.*

Section ID	Service Age (year)	Structure	Lane	TCS[2] (m)
7 JT	14	4 cm SMA13 + 5 cm SUP20 + 7 cm SUP25 + 36 cm CTB + 20 cm LFS	1, 2	31.3
8 YH1	15	4 cm SMA13 + 6 cm AC20 + 8 cm AC25 + 36 cm CTB + 20 cm LFS	1	25
9 YH2	15	4 cm AK13 + 6 cm AC20 + 8 cm AC25 + 36 cm CTB + 20 cm LFS	1, 3	8.3
10 QY	17	4 cm AK13 + 6 cm AC20 + 8 cm AC25 + 36 cm CTB + 20 cm LFS	1	25
11 FG	17	4 cm AK13 + 6 cm AC20 + 8 cm AC25 + 38 cm CTB + 20 cm LFS	1	3.9
13 LX2	18	4 cm AK13 + 6 cm AC20 + 8 cm AC25 + 36 cm LFTB + 20 cm LFS	1, 2	16.8

Note 1: AC or SUP stands for different HMA design criterions; 13 represents the nominal maximum aggregate size of mixture; CTB = cement-treated base; CSM = cement stabilized macadam; LFTB = lime fly ash-treated base; LFS = lime-fly ash soil. Note 2: TCS = transverse crack spacing (transversal cracks count per 100 m).

2.2. Specimens Preparation

Sections within a 300 m-long drilling zone from the selected sections were chosen. At least 30 m was maintained between each drilled core. The cores were examined to ensure that there was no obvious crack on the core surface and were then labeled with the expressway name, position, and lane. These cores drilled from different pavement sections were then marked and cut into specimens [10]. Since the optimal width for the SCB test is 50 mm and common surface-layer thickness in Jiangsu is 40 mm, the specimen thickness of the surface, middle and bottom layer is 40 mm, 50 mm and 50 mm, respectively. The radius of the SCB specimen was 75 mm. A total of 144 specimens were prepared, with 48 specimens for each layer. Each selected section layer mixture had two parallel specimens to minimize deviation. The specimen preparation is shown in Figure 2.

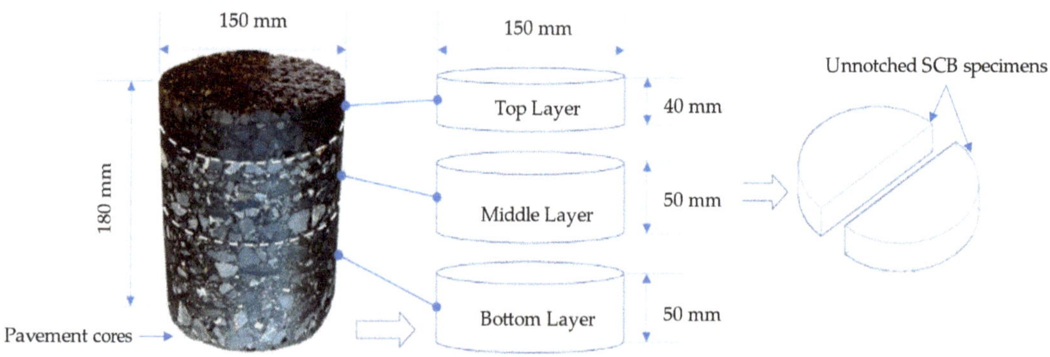

Figure 2. Schematic of SCB specimens' preparation [15].

2.3. SCB Test and Indices

The un-notched SCB test components are shown in Figure 3. The test specimen has a diameter (R) of 75 mm and is placed on two cylinder supports with a width of 0.8 diameter (120 mm). The actuator is placed on top of the specimen and applies a vertical load. To ensure uniform temperature in specimens, the specimens were placed in the environmental chamber at the test temperature for 4 h prior to testing. The test temperature is 15 °C, and the loading speed is 50 mm/min [24]. In a typical SCB test, the load–displacement curve appears similar to that in Figure 4, showing a single peak curve. This peak can be divided into the before peak zone and the post-peak zone, with the dividing point being the maximum force point. For the intermediate temperature crack test on asphalt mixture, the load–displacement curve shows a continuous ascending curve before peak and a smooth continuous descending curve post peak.

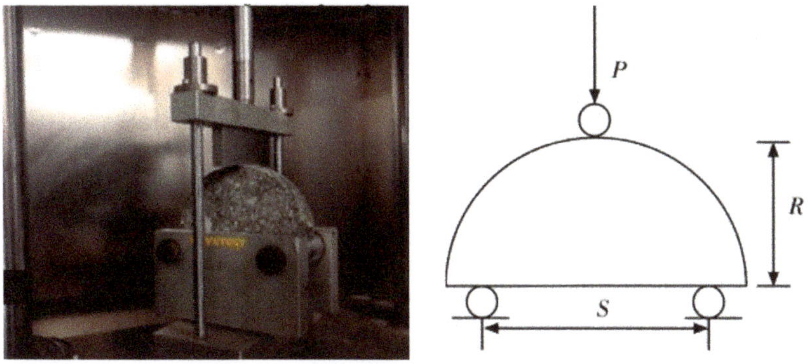

Figure 3. Un-notched SCB test component and load graph.

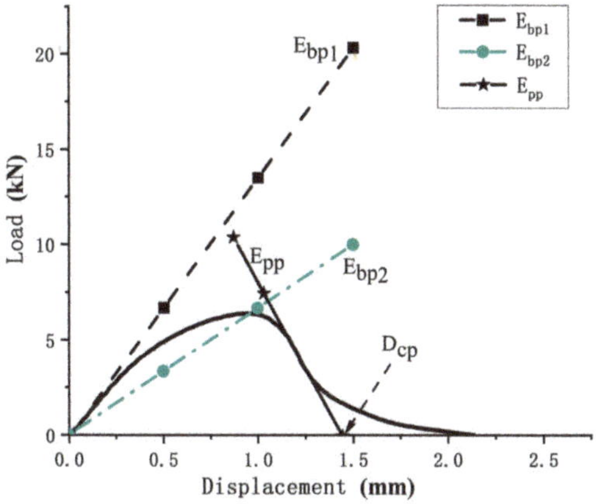

Figure 4. SCB test component and load graph.

After loading is completed, the curve is plotted, and three indices are calculated to evaluate the crack resistance of the mixture. These indices are tensile strength, fracture energy and CRI (crack index), as shown in Equations (1)–(3) below [8,25–27]:

$$\sigma_t = \frac{4.8F}{BD} \tag{1}$$

where σ_t = SCB tensile strength (MPa); F = peak load (kN); B = specimen thickness (mm); D = specimen diameter (mm).

$$G_f = \frac{W_f}{A_{lig}} \tag{2}$$

where G_f = fracture energy (J/m^2); W_f = work of fracture (J); $W_f = \int Pdu$; P = applied load (kN); u = vertical displacement (mm); A_{lig} = ligament area (m^2); $A_{lig} = (1/2)D \times B$.

$$CRI = \frac{G_f}{F} \tag{3}$$

where CRI = crack index (1/m)

The tensile strength indicates the ability of the mixture to resist strength and is used to measure the load capacity of the specimen. A higher tensile strength indicates a better load resistance for the tested specimen. The CRI is a combined index that divides the fracture energy by the peak load [28–30]. Because crack progress could be divided into three phases: uncrack working, crack work and failure, a higher CRI, at the same fracture or tensile strength, indicates a better overall fracture resistance performance for the tested specimen.

2.4. Significance Factors Analysis

In terms of SCB test data, they were collected from different pavements under service conditions, resulting in a range of data variation. To address this, it is necessary to identify and remove possible outliers from the original test data. The JMP software platform was utilized for this purpose to reduce bias, and all test data were entered into JMP as numerical variables. ANOVA was employed as a statistical significance analysis model to determine the relationship between dependent factors and one (or more) independent factors [31,32].

To illustrate the distribution of the selected data, means diamonds and Tukey–Kramer comparison circles were plotted. In the diamond graph, the top and bottom points represent the $(\alpha)*100$ confidence interval for each group, and a confidence interval $\alpha = 95\%$ was selected. The mean lines across the center of the diamond represent the group mean. Overlap marks appear as lines above and below the group mean, and they are calculated as group mean $\pm\left(\sqrt{2}\right)/2 \times \mathrm{CI}/2$, where CI represents the confidence interval. For data groups of equal sizes, overlap marks indicate that the group means are not significantly different at the given confidence level. Similar to the mean diamonds, the Tukey–Kramer comparison circle was used to compare the significance between two variables. The Tukey–Kramer comparison circle is depicted in Figure 5 on the inner-right side of the mean diamond, with the circle radius equal to corresponding various CI. When observing the comparison circles of two variables, if the circles do not intersect or only have a small intersection, this indicates that the two variables are significantly different.

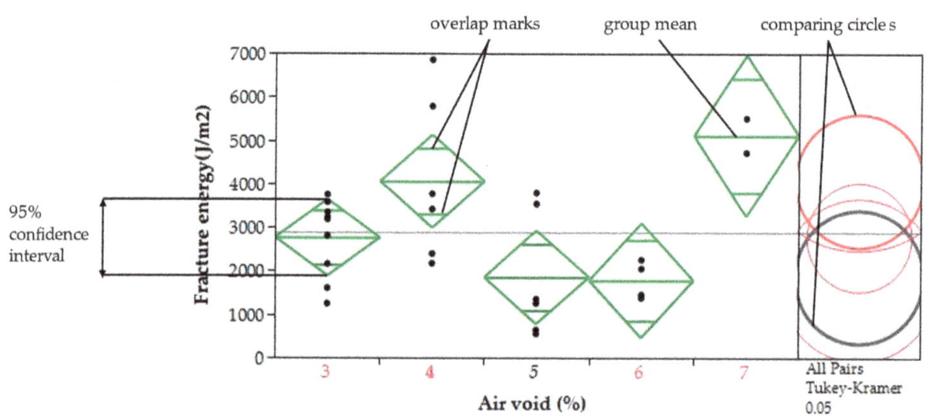

Figure 5. Means diamond of 95% confidence interval and Tukey–Kramer circles.

In the graph showing the Tukey–Kramer comparison circles, test results were used to reveal the sensitivities of variables such as mixture type, service age (year), asphalt aggregate ratio and so on. The circles were plotted on the right side of the diamond graph, with each circle corresponding to a diamond on the left. The center of the circle was set to the same value as the mean value, and the diameter was equal to the confidence interval. By analyzing the intersection of any circles, the significance of the parameters could be determined. If any two circles did not intersect or had only a small intersection area, this indicated that the two corresponding parameters were significantly different. When a circle was selected, it was highlighted in red, while the circles representing significant parameters were colored gray and bold [33].

3. Results and Discussion

3.1. Factor Independence Analysis

Since the asphalt mixtures for each layer have different structure, function and mechanism requirements, in this section, the results of three layers are discussed separately. Firstly, the factor independence analysis was conducted on all selected factors, because the performance of pavement mixture is complex and influenced by multiple factors [30]. To avoid the correlated factors, several factors of each layer were selected via factor independence analysis for further analysis. The stepwise regression method was used to test if one factor was significantly different from another. The stepwise regression used minimum Bayesian information criterions (BIC) as the stopping rule and forward direction. This regression can handle both numeric and nominal types of factors, and BIC can control the complexity of model at an appropriate level. The correlation of selected factors by three layers is listed in Tables 2–4. The asterisk in the table means that the row and column of the two factors are not independent. Then, the independent factors of the three layers were selected; the independent factors of the surface layer were service age, mixture type and air void, while the independent factors of the middle layer were service age, lane and air void, and the independent factors of the bottom layer were service age, air void and asphalt aggregate ratio. The total selected factors of three layers are shown in Table 5.

Table 2. Factors independence analysis of surface-layer specimens.

Factor	ESALs	Service Age (year)	Area	Mixture Type	Lane	Air Void (%)	Asphalt Aggregate Ratio (%)	Factor Selected
ESALs	\	*		*				
Service age (year)		\			*		*	√
Area			\	*	*	*	*	
Mixture type				\			*	√
Lane					\			
Air Void (%)						\		√
Asphalt aggregate ratio (%)							\	√

\ represents same factor. * represents the two factors are not independent. √ in the last column represent this factor is selected from this table.

Table 3. Factors independence analysis of middle-layer specimens.

Factor	ESALs	Service Age (year)	Area	Mixture Type	Lane	Air Void (%)	Asphalt Aggregate Ratio (%)	Factor Selected
ESALs	\	*	*		*	*	*	
Service age (year)		\	*	*				√
Area			\	*	*			
Mixture type				\		*	*	
Lane					\		*	√
Air Void (%)						\	*	√
Asphalt aggregate ratio (%)							\	

\ represents same factor. * represents the two factors are not independent. √ in the last column represent this factor is selected from this table.

Table 4. Factors independence analysis of bottom-layer specimens.

Factor	ESALs	Service Age (year)	Area	Mixture Type	Lane	Air Void (%)	Asphalt Aggregate Ratio (%)	Factor Selected
ESALs	\	*	*		*		*	
Service age (year)		\	*	*	*			√
Area			\	*	*	*		
Mixture type				\			*	
Lane					\			
Air Void (%)						\		√
Asphalt aggregate ratio (%)							\	√

\ represents same factor. * represents the two factors are not independent. √ in the last column represent this factor is selected from this table.

Table 5. Factor selection results of three layers.

Layer	Factor Selection			
Surface	Service age	Mixture type	Air void	Asphalt aggregate ratio
Middle	Service age	Lane	Air void	
Bottom	Service age		Air void	Asphalt aggregate ratio

In the analysis of factor significance, it was found that some independent factors, such as service age and air void, were selected for all three layers. In addition to these common factors, the surface layer had also had the mixture type and asphalt aggregate ratio as independent factors. This might be the result of two different mixture types, AC13 and SMA13. The middle layer had an independent factor lane, which meant loads on lanes had an impact on the crack resistance of middle-layer specimens. Since trucks on the expressway were recommended to drive in lane 3, the shear loads had the most impact on the middle layer. This might result in crack resistance in the middle layer. The independent factors of all three layers were mostly materials factor and service age; the performance of materials and service age were the most important factors for selected expressway crack-resistance performance.

3.2. Indices Analysis Results by Factors

To clarify the influence of each factor on the crack-resistance performance, ANOVA tests were conducted on three crack indices versus selected factors of three layers. The ANOVA tests were conducted with an alpha value of 0.05 at the 95% confidence level. The results of ANOVA Tests of crack-resistance performance indices versus selected factors of three layers were shown in Tables 6–8, the annotations were attached behind Table 6. The p-value was used to evaluate the significance of factors. A small p-value (<0.05) indicated that the selected factor has a statistically significant impact on the crack index. The smaller p-value (<0.01 or <0.001) indicated that the selected factor has a relatively higher significant impact.

From Table 6, we could see that the tensile strength was significant to the mixture type and asphalt aggregate ratio, the fracture energy and CRI were more significant to service age and the tensile strength was significant with regard to the mixture and asphalt aggregate ratio. Crack indices of surface layer specimens did not have a significant relationship with air voids. Table 7 showed that tensile strength and fracture energy are both significant to service age; fracture energy was more significant to service age, tensile strength and CRI. And tensile strength was significant to lane. Table 8 shows tensile strength, fracture energy and CRI were significant to service age, and all three indices were significant at the highest level. The fracture energy and CRI were also significant to air voids.

From Tables 6–8, we learn the significant factors are service age and one internal material factor, such as air voids or the asphalt aggregate ratio. The environmental factors only have a significant impact on the crack resistance of middle-layer specimens.

Table 6. ANOVA results of crack-resistance performance of surface layer.

Index	Factor	DF [a]	SS [b]	MS [c]	F-Value [d]	*p*-Value [e]	Significant Effect [f]
Tensile Strength	Service age	1	0.65	0.65	0.2250	0.6373	-
	Mixture type	1	22.88	22.88	9.26	0.0037	**
	Air void	1	1.93	1.93	0.64	0.4273	-
	Asphalt aggregate ratio	1	15.11	15.11	5.76	0.0200	*
Fracture Energy	Service age	1	3.22×10^7	3.22×10^7	8.62	0.0049	**
	Mixture type	1	1.85×10^6	1.85×10^6	0.4295	0.5151	-
	Air void	1	3.75×10^6	3.75×10^6	0.9398	0.3376	-
	Asphalt aggregate ratio	1	1.53×10^7	1.53×107	3.7627	0.0578	-
CRI	Service age	1	1.82×10^5	1.82×10^5	7.3128	0.0092	**
	Mixture type	1	3.31×10^4	3.31×10^4	1.1918	0.2800	-
	Air void	1	5.52×10^4	5.52×10^4	2.2667	0.1393	-
	Asphalt aggregate ratio	1	2.22×10^4	2.22×10^4	0.7916	0.3777	-

[a] Degree of freedom. [b] Sum of squares. [c] Ratio of mean square. [d] F-value. [e] *p*-value. [f] Significant with * (*p*-value < 0.05), ** (*p*-value < 0.01).

Table 7. ANOVA results of crack-resistance performance of middle layer.

Index	Factor	DF [a]	SS [b]	MS [c]	F-Value [d]	*p*-Value [e]	Significant Effect [f]
Tensile Strength	Service age	1	29.02	29.02	6.59	0.0133	*
	Lane	3	51.10	17.03	4.12	0.0111	*
	Air void	1	9.12	9.12	1.6465	0.2063	-
Fracture Energy	Service age	1	3.43×10^7	3.43×10^7	7.17	0.0099	**
	Lane	3	1.45×10^7	4.82×10^6	0.90	0.4504	-
	Air void	1	1.89×10^7	1.89×10^7	3.41	0.0715	*
CRI	Service age	1	1.08×10^5	1.08×10^5	5.44	0.0237	*
	Lane	3	2.52×10^4	8.40×10^3	0.37	0.7716	-
	Air void	1	3.38×10^4	3.38×10^4	1.68	0.2019	-

[a] Degree of freedom. [b] Sum of squares. [c] Ratio of mean square. [d] F-value. [e] *p*-value. [f] Significant with * (*p*-value < 0.05), ** (*p*-value < 0.01).

Table 8. ANOVA results of crack-resistance performance of bottom layer.

Index	Factor	DF [a]	SS [b]	MS [c]	F-Value [d]	*p*-Value [e]	Significant Effect [f]
Tensile Strength	Service age	1	27.53	27.53	17.19	0.0001	***
	Air void	1	1.68	1.68	0.83	0.37	-
	Asphalt aggregate ratio	1	3.12	3.12	1.48	0.2299	-
Fracture Energy	Service age grade	1	3.58×10^7	3.58×10^7	23.58	<0.0001	***
	Air void	1	5.31×10^6	5.31×10^6	2.95	0.0934	-
	Asphalt aggregate ratio	1	1.04×10^7	1.04×10^7	5.07	0.0290	*
CRI	Service age grade	1	9.57×10^4	9.57×10^4	18.71	<0.0001	***
	Air void	1	1.16×10^4	1.16×10^4	2.04	0.16	-
	Asphalt aggregate ratio	1	3.58×10^4	3.58×10^4	5.63	0.0217	**

[a] Degree of freedom. [b] Sum of squares. [c] Ratio of mean square. [d] F-value. [e] *p*-value. [f] Significant with * (*p*-value < 0.05), ** (*p*-value < 0.01), *** (*p*-value < 0.001).

3.3. *Crack-resistance performance on Three-Layer Specimens*

3.3.1. Crack-resistance performance Change on Surface Layer Specimens

From Table 6, we learn that tensile strength is significant to the mixture type and asphalt aggregate ratio, while fracture energy and CRI are significant to service age. The ANOVA and Tukey–Kramer results of surface layer were shown in Figures 6–8. The means and 95% confidence intervals were shown in green diamonds. And the Tukey–Kramer circles were shown in the right side; when selecting a parameter, the circles of parameters with significant differences to the selected parameter were painted gray and others were painted red. Thus, the impact of service age, mixture type and air void was introduced when considering the impact of mixture type and air void. Figure 6 shows the tensile versus mixture type; Figure 7 shows the tensile strength versus air voids. From Figure 8, we learn that surface-layer specimens' tensile strength was significantly different in terms of mixture type, while the tensile strength of AC13M was significantly larger for SMA13Ms. And the circles in Tukey–Kramer circles did not interact.

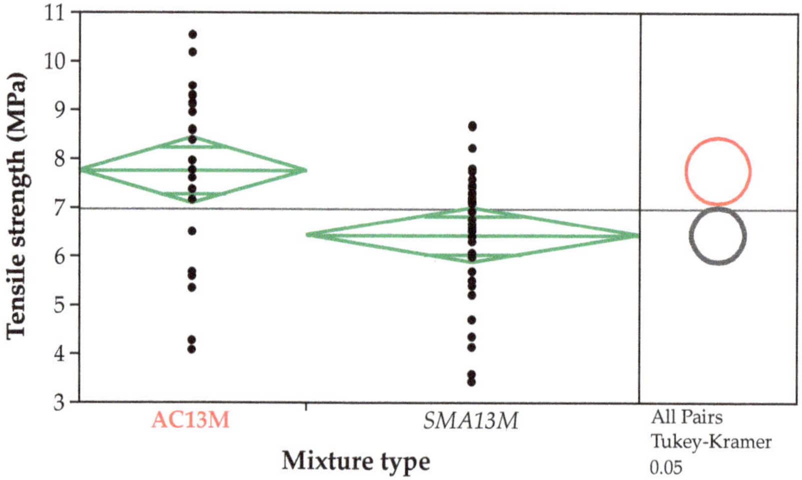

Figure 6. ANOVA and Tukey–Kramer comparison of tensile strength of surface-layer specimens by mixture type.

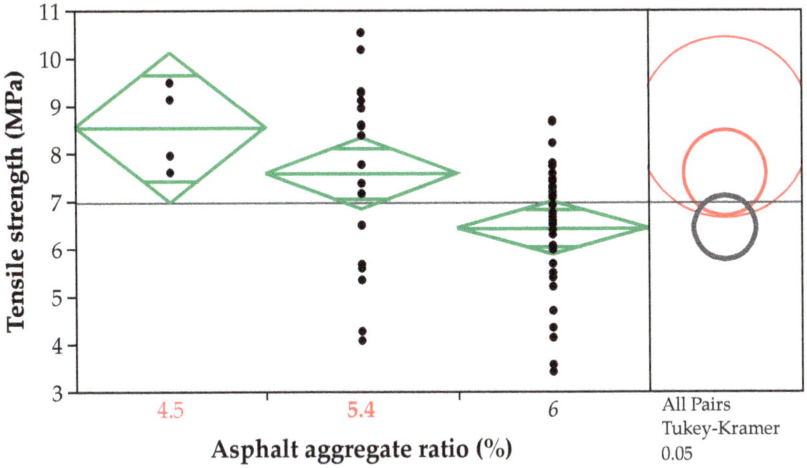

Figure 7. ANOVA and Tukey–Kramer comparison of tensile strength of surface-layer specimens based on the asphalt aggregate ratio.

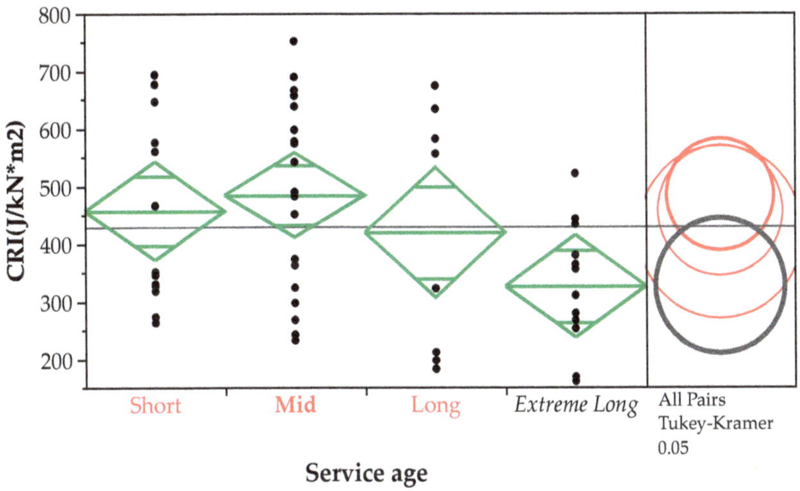

Figure 8. ANOVA and Tukey–Kramer comparison of CRI of surface-layer specimens by service age.

From Figure 6, we learn that surface-layer specimens' tensile strength differed significantly based on the asphalt aggregate ratio. The higher the asphalt aggregate ratio was, the lower the tensile strength was. And the tensile strengths of 4.5% asphalt aggregate ratio and 5.4% specimens were significantly different to the tensile strength of the 6% asphalt aggregate ratio. This result also indicates that the AC13M's specimens had a higher tensile strength because Figure 7 showed that the lower the asphalt aggregate ratio was, the higher the tensile strength was.

From Figure 8, we learn that surface-layer specimens' CRI was significantly different based on service age. The mid service age surface-layer specimens had the highest CRI and extremely long service age surface-layer specimens had the lowest CRI. The average CRI increased from short to mid service age and then decreased from mid to extremely long. The CRI of mid service age and extremely long service age had significant differences. And this crack resistance difference was significant at extremely long service age. This result indicated the crack-resistance performance did not decrease for pavement which had a service age of 0 years; the surface-layer specimens' crack resistance reinforced over the first few years, reached the highest level at mid service age and then decreased with the accumulating service age from mid service age. This decreasing tendency led to significant differences in the crack resistance of surface-layer specimens when their service age reached about 15 years.

3.3.2. Crack-Resistance Performance Change in Middle-Layer Specimens

From Table 7, we learn that tensile strength was significant to service age and lane, fracture energy and CRI were also significant to service age. Figure 9 showed the average fracture energy of specimens of mid service age was significant to that of the short, long and extremely long service age. When the middle layers were paved, the mixtures of the middle layer exhibited a fracture energy decrease in the first 5 years and then increased from 5 years to 20 years. This tendency was the opposite to the tendency of surface-layer specimens.

Figure 9. ANOVA and Tukey–Kramer comparison of fracture energy of middle-layer specimens by service age.

The middle-layer specimens from Figure 10 showed that different lanes exhibited significant difference in tensile strength. Lane 3 was arranged as the lane for trucks and vehicles from China's regular traffic, and lane 1 and 2 were arranged as the lanes for vehicles.

Thus, greater loads were applied to lane 3. And more trucks led to fracture of aggregates which generated small cracks and degraded the tensile strength. Lane 2 also had a small number of trucks, so the tensile strength of lane 2 specimens was smaller than the tensile strength of lane 1 specimens and larger than the tensile strength of lane 3 specimens.

From Figure 11 we can learn the fracture energy of middle-layer specimens first decreased with air voids at a percentage lower than 6%, then increased with the presence of air voids of more than 6%.

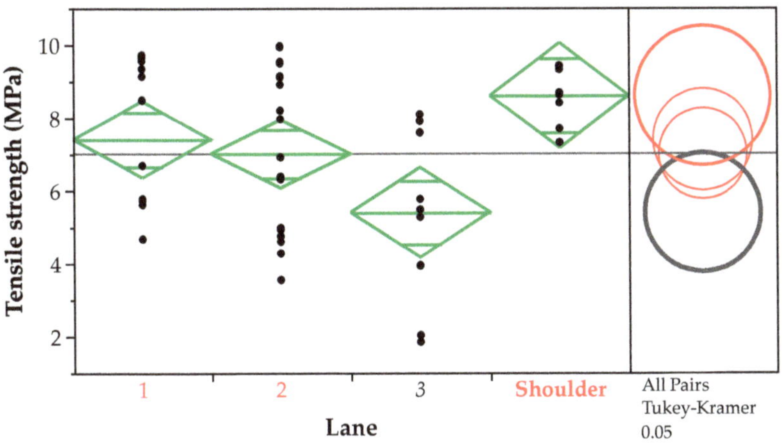

Figure 10. ANOVA and Tukey–Kramer comparison of tensile strength of middle-layer specimens by lane.

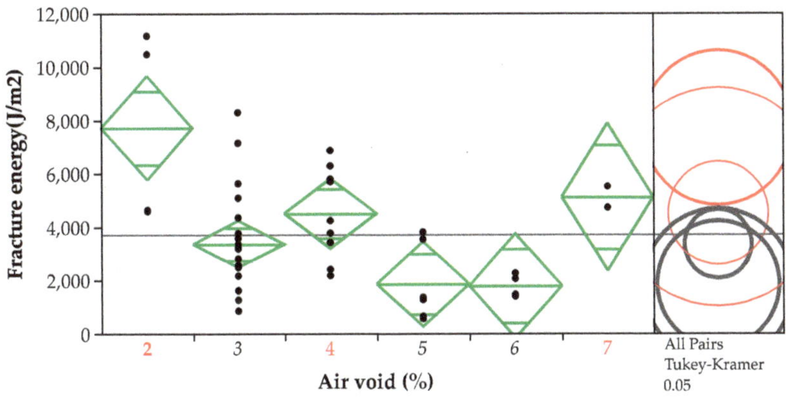

Figure 11. ANOVA and Tukey–Kramer comparison of fracture energy of middle-layer specimens by air void.

3.3.3. Crack Resistance Change on Bottom-Layer Specimens

From Table 8, we learn that the tensile strength, fracture energy and CIR were significant to service age and asphalt aggregate ratio, fracture energy and CRI were significant to the asphalt aggregate ratio. The bottom-layer specimens' crack resistance indices were significant to service age and the asphalt aggregate ratio. All the indices were not significant to the air void of bottom-layer specimens. The CRI of bottom-layer specimens was selected based on the service age and asphalt aggregate ratio. From Figure 12, we learn that the CRI of extremely long service age was more significant than CRI of short, mid and long service age. The crack resistance of bottom-layer specimens experienced significant distress at about 15 to 20 years, like the crack resistance of surface-layer specimens.

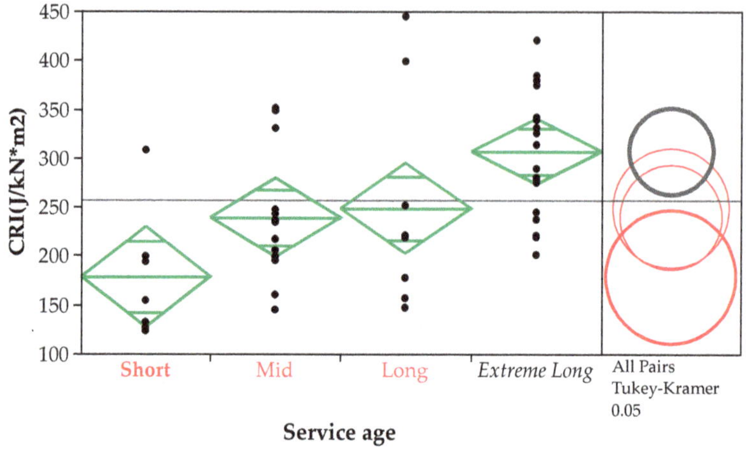

Figure 12. ANOVA and Tukey–Kramer comparison of CRI of bottom-layer specimens by service age.

From Figure 13, we learn that the CRI of specimens with a 4.3% asphalt aggregate ratio was significantly larger compared to the CRI of specimens with a 4.1% asphalt aggregate ratio. This indicates the asphalt content of the original plant mixture had a long-term effect on the bottom-layer mixture.

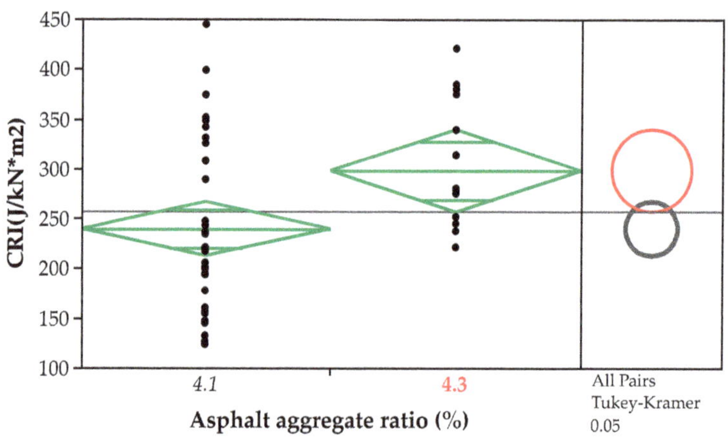

Figure 13. ANOVA and Tukey–Kramer comparison of CRI of bottom-layer specimens based on the asphalt aggregate ratio.

3.3.4. TCS Indicated by Fracture Indices of Mixture

The TCS is an index representing the density of transverse cracks. A larger value means a better pavement condition, with larger average spacing between transverse cracks, and a smaller value means a worse pavement condition with smaller average spacing between transverse cracks. The TCS grades in the local standard of Jiangsu are recommended as follows: TCS $\geq$ 250 as excellent, 100 $\leq$ TCS $\leq$ 250 as good, 60 $\leq$ TCS $\leq$ 100 as medium, 30 $\leq$ TCS $\leq$ 60 as inferior and TCS $\leq$ 30 as bad. TCS and TCS grade was used for evaluating good and bad conditions of a section over a continuous distance. Thus, it also reduced by time and the TCS-service age is shown in Figure 14 the TCS grade was significant to service age. The short service age pavement had relatively high TCS and mid, long and extremely long service age pavements had similar TCS.

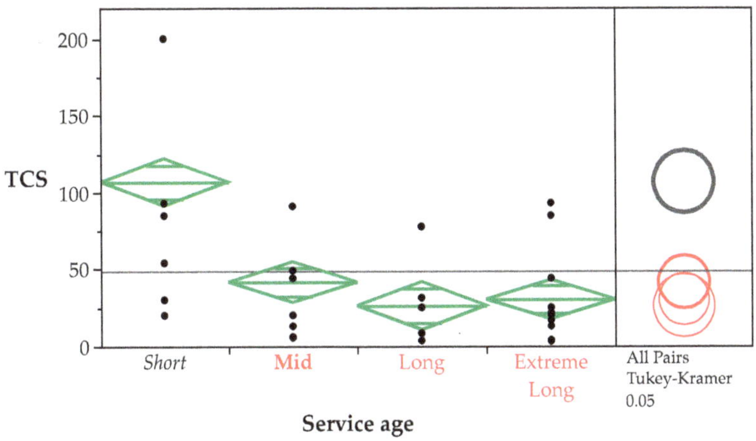

Figure 14. ANOVA and Tukey–Kramer comparison of TCS by section service age.

The relationships between TCS grade and mixture crack indices were also examined and the results showed only the CRI of the bottom layer is significant to TCS, while other indices were not significant to the TCS grade. This also shows the pavement sections with many transverse cracks might not have low crack-resistance mixtures and the crack resistance of selected sections could not evaluated via TCS. Figure 15 shows the ANOVA results of bottom CRI and TCS grade; it is shown that the bottom-layer specimens had lower CRI of the section with TCS at good condition, but the bottom CRI increased while TCS decreased and the TCS grade became worse. Since TCS decreased with service age, as Figure 14 shows, and the bottom CRI increased by service age, as Figure 12 shows, the crack resistance of bottom-layer specimens increased but the number of transverse cracks also increased. This showed the TCS or TCS grade of the pavement section had no direct effect on the crack resistance of this section.

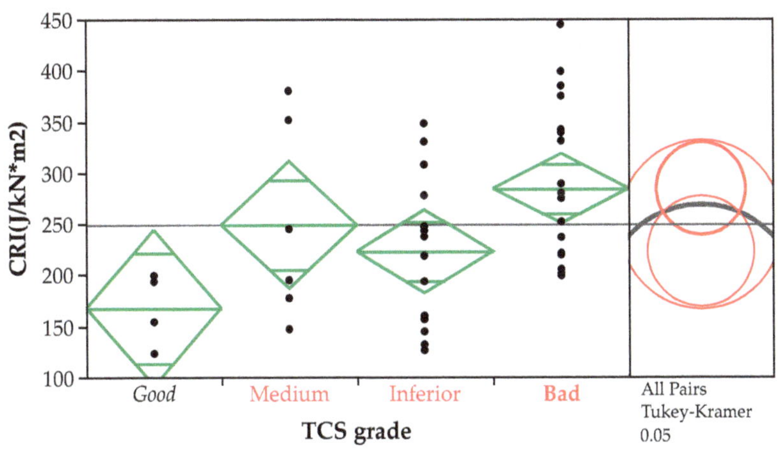

Figure 15. ANOVA and Tukey–Kramer comparison of CRI of bottom-layer sections' TCS grade.

3.3.5. Correlation between CRI with Fracture Energy and Index FUDR Analysis

Since all the indices reflect some aspects of mixture crack resistance, some of them may have a linear relationship [11]. The results of factor analysis in Tables 2–4 show that some indices have the same distinguishment on one factor, which means the two indices,

CRI and fracture energy have same significance to the factors and they could be replaced by one another. Some researchers found fracture energy could not characterize whole fracture resistance performance, and the mixture specimens which have the same fracture energy may have totally different fracture phenomena. They found CRI is sensitive to mixture type and can be used to identify the mixture containing RAP. The mixtures containing RAP have greater hardness compared to mixtures containing no RAP and mixtures containing RAP have smaller CRI compared to mixtures containing no RAP when they have the same fracture energy. Based on a similar thought, the CRI combined fracture energy can be used to distinguish SCB specimens with different factors like service area, service layer, asphalt aggregate ratio or air void. In this section, all the specimens were analyzed to reveal the general rule, including the impact of the service layer. The correlation between CRI and fracture energy of all specimens was plotted in Figure 16, which shows a linear relationship between CRI and fracture energy, and the R^2 is 0.712.

Figure 16 features colored data dots that indicate factor categories. The factors are service age, asphalt aggregate ratio and air void. Since CRI is calculated by fracture energy divided by peak load, data dots in the upper lines zone mean the peak load is smaller than that on the fit line and the curve is flatter than that on the line. Data dots in the lower line zone mean that the peak load is bigger than that on the fit line and the curve is higher than that on the line. And the fit line is the average CRI versus fracture energy, which is the average peak load. The surface layers dots in Figure 15 are mostly in the upper zone; the bottom layer dots are mostly in the lower zone and the middle layer does not have obvious deviation, which indicates surface-layer specimens' curves are higher than average and bottom-layer specimens' curves are flatter.

The index factor dots upper rate (FDUR) was promoted to evaluate the percentage of factor data dots set on the upper side of the fit line. Most previous research focused on specimens and the difference between various asphalt mixtures was characterized by means, variance or correlation of fracture indices. This index, FDUR, tried to characterize a large number of specimens with different factor conditions. Since the best fitting factor has factor dots with an upper rate equaling 50%, −50% was added for normalization. The larger the absolute FDUR ($absFDUR_{factor}$) one factor was, the greater the impact the factor had on the fracture pattern. FDUR and absFDUR were calculated in Equation (4) and (5).

$$FDUR_{factor} = \frac{N_{upper-fitline}}{N_{total-factor}} \times 100\% - 50\% \quad (4)$$

$$abs\left(FDUR_{factor}\right) = \left|\frac{N_{upper-fitline}}{N_{total-factor}} \times 100\% - 50\%\right| \quad (5)$$

The CRI-fracture energy graph classified by factors is shown in Figure 15 and the FDUR is shown in Table 9.

The FDUR and absFDUR data were used to study curve form. The absFDUR indicated the impact level one factor has. The larger absFDUR one factor type was, the more bias this factor type caused. The positive or negative sign of FDUR indicated the curve was sharp or flat. If one specimen was FDUR positive, the load–displacement curve was flatter comparing to the average level; otherwise, it was sharper. For further analysis, the FDUR of three layers was plotted and shown in Table 9 and Figure 16.

In Figure 17, the positive value is plotted as a solid column, the negative value is plotted as a transverse line column and the 0 value is plotted as a hollow column. Since each layer mixture has its unique mixture type with a specific aggregate–asphalt ratio, the mixture type was not introduced for the whole CRI-fracture energy curve analysis.

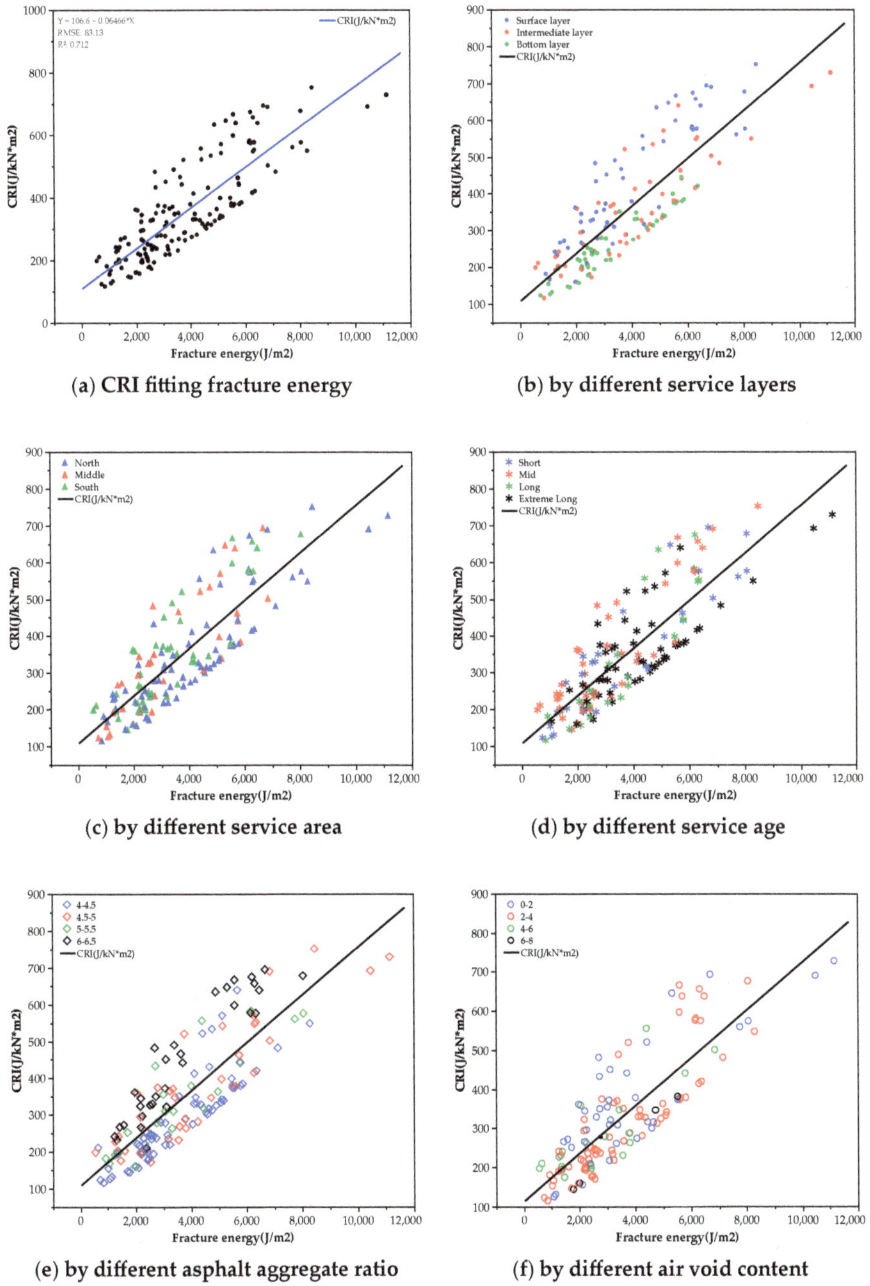

Figure 16. CRI fitting fracture energy based on different factors.

Table 9. FDUR value versus different factors.

Factor	Factor Type	Factor Percent (%)	Factor Count	FDUR (%)	absFDUR (%)
Layers	Surface	34.4	54	31.5	31.5
	Middle	33.8	53	33.0	33.0
	Bottom	31.8	50	−46	46
Area	North	49.7	78	−16.7	16.7
	Middle	21.0	33	−1.5	1.5
	South	29.3	46	4.4	4.4
Service age (years)	Short	18.5	29	−13.3	13.3
	Mid	29.3	46	15.0	15.0
	Long	17.8	28	42.9	42.9
	Extremely long	34.4	54	−18.5	18.5
Asphalt aggregate ratio (%)	4–4.5	42.7	67	−38.1	38.1
	4.5–5	26.1	41	−1.2	−1.2
	5–5.5	11.4	18	0	0
	5.5–6	0	0	-	-
	6–6.5	19.8	31	46.8	46.8
Air void (%)	0–2	26.7	36	5.6	5.6
	2–4	55.5	75	−19.3	19.3
	4–6	14.8	20	0.0	0.0
	6–8	3.0	4	−50	50

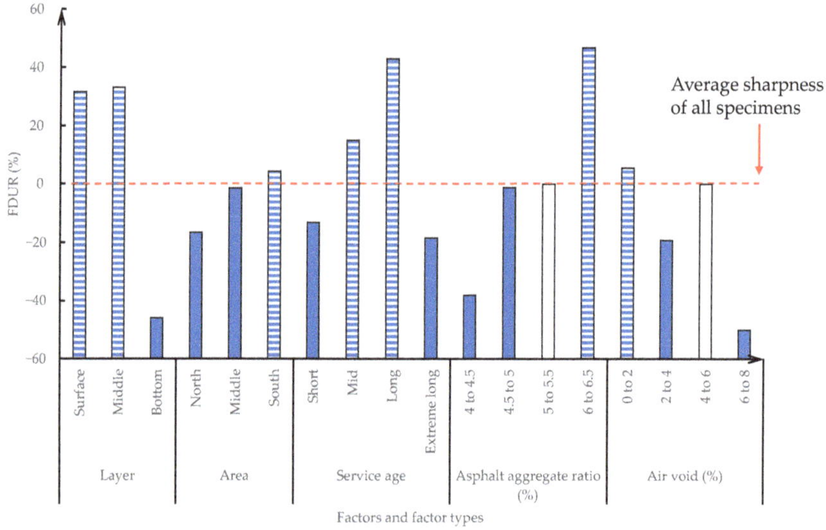

Figure 17. FDUR versus factors and factor types.

From Figure 16b and Table 9, it is shown that the factor "layer" has largest curve form impact. The FUDR of the surface and middle specimens is 31.5% and 33%, respectively. And the FUDR of the bottom specimens is −16%. The surface-layer mixture AC13 and SMA13 are entirely composed of modified asphalt mixtures and the middle-layer AC20 and SUP20 are mostly composed of modified asphalt mixture. This indicates the surface and middle layer mixture have flatter fracture curves compared to the bottom-layer mixture.

From Figure 16c and Table 9, it is shown the factor "area" specimens have no significant difference area. The north area specimens have a relatively large absFUDR. The north and middle area specimens are tougher than others with −16.7% and −1.2% FUDR. South area specimens are softer than others with 4.4% FUDR. The north and middle area specimens are sharper compared to others and south area specimens are flatter. And the north area specimens have more bias curve phenomena.

From Figure 16d and Table 9, it is shown the factor "service age" specimens have significant difference; the short and extremely long service age specimens have negative FDUR and mid and long service age specimens have positive FDUR. The specimens of long service age have largest absFUDR 42.9%. This indicates short and extremely long service age specimens are harder than others and mid and long service age specimens are softer than others. The long service age specimens have the most data above the CRI-fracture energy fit line.

From Figure 16e and Table 9, it is shown the factor "asphalt aggregate ratio" also refers to the mixture types, the 4% to 4.5% asphalt aggregate ratio often suits the bottom layer of the AC25 and SUP25 mixture, the 5% to 5.5% asphalt aggregate ratio often suits the middle layer of the AC20 and SUP20 mixture, the 5% to 5.5% asphalt aggregate ratio often suits the surface-layer AC13 mixture and the 6% to 6.5% asphalt aggregate ratio often suits the surface-layer SMA mixture. The 4% to 4.5% and 6% to 6.5% asphalt aggregate ratios have relatively large absFUDR, which indicates the fracture phenomena of the bottom-layer mixture and surface-layer SMA13 are different in terms of their average curve phenomena. The 4% to 4.5% asphalt aggregate ratio AC25 and SUP25 specimens are harder than others with −38.1% FUDR, and 6% to 6.5% asphalt aggregate ratio SMA13 specimens are softer than others with 48.6% FUDR.

From Figure 16f and Table 9, it is shown the factor "air void" also has a curve form impact. The 4% to 6% air void specimens have 0 FUDR, so their data have average distribution on both sides of the CRI-fracture energy curve. The 0% to 2% air void specimens are slightly softer than others with 5.6% FUDR; correspondingly, the 2% to 4% and 6% to 8% specimens are tougher than others with −19.3% and −50% FUDR.

Figure 17 gives a total review of FUDR. As shown by the analysis above, the high absFUDR values in terms of the layer, service age and asphalt aggregate ratios are as follows: the factor layer with surface, middle and bottom types; the factor service age with long and extremely long types; the factor asphalt aggregate ratio with 4% to 4.5% and 6% to 6.5% types and the factor air void with 6% to 8% types. These types of factors can obviously affect the fracture curve form.

4. Conclusions

1. Several representative expressway sections were selected and the SCB intermediate temperature fracture test was conducted. The factors independence analysis and different indices for different layers were applied. Service age and air void are independent factors for all three layers. The mixture type is a significant factor in relation to tensile strength on surface-layer specimens, lane is a significant factor to tensile strength on middle layers and the asphalt aggregate ratio is a significant factor to CRI on bottom layers.
2. ANOVA and Tukey–Kramer comparison analysis of three layers was conducted and the indices of each layer and factors were analyzed. The most important factors affecting mixture crack resistance were one time factor (service age) and one material factor like air voids or the asphalt aggregate ratio. Only middle-layer specimens have another environmental factor lane, and lane 3 with the most truck loads has the lowest crack resistance of all lanes.
3. For the surface layer, AC13M's crack resistance is significantly larger than SMA13M's, the crack resistance of 4.5% and 5% asphalt aggregate ratio specimens is significantly larger than that of the 6% ratio specimens, the crack resistance of extremely long service age specimens is significantly smaller than short, mid and long service age specimens. For the middle layer, the crack resistance of mid service age specimens is significantly smaller than the crack resistance of short, long and extremely long service age specimens. For the bottom layer, the crack resistance of 4.1% asphalt aggregate ratio specimens is significantly smaller than the crack resistance of 4.3%

asphalt aggregate ratio specimens, while the crack resistance of extremely long service age specimens is significantly smaller than the crack resistance of short, mid and long service age specimens.

4. Only the CRI index of the bottom-layer mixture has a significant relationship with TCS of the section from which the mixture was cored, while other mixture indices cannot represent the pavement transverse crack spacing of the section. The pavement sections with many transverse cracks may not have low crack-resistance mixtures.
5. The index FUDR and its absolute value absFUDR are promoted as an index for determining if a mixture of a certain factor type is softer or tougher than average on a statistical level. And this index is sensitive to many factors. The specimens with surface layer, middle layer, long service age or a 6% to 6.5% asphalt aggregate ratio are obviously softer than others and have an obviously flatter load–displacement curve on a statistical level.

Author Contributions: Conceptualization, L.S. and F.N.; Formal analysis, L.S., H.D. and Y.H.; Investigation, H.D.; Methodology, L.S., F.N. and H.D.; Supervision, F.N.; Writing—original draft, L.S.; Writing—review and editing, Y.H. All authors have read and agreed to the published version of the manuscript.

Funding: This research received no external funding.

Institutional Review Board Statement: Not applicable.

Informed Consent Statement: Not applicable.

Data Availability Statement: The data presented in this study are available in the article.

Conflicts of Interest: The authors declare no conflict of interest.

References

1. Wang, H.P.; Guo, Y.X.; Wu, M.Y.; Xiang, K.; Sun, S.R. Review on Structural Damage Rehabilitation and Performance Assessment of Asphalt Pavements. *Rev. Adv. Mater. Sci.* **2021**, *60*, 438–449. [CrossRef]
2. Idris, I.I.; Sadek, H.; Hassan, M. State-of-the-Art Review of the Evaluation of Asphalt Mixtures' Resistance to Reflective Cracking in Laboratory. *J. Mater. Civ. Eng.* **2020**, *32*, 03120004. [CrossRef]
3. Gong, M.; Xiong, Z.; Chen, H.; Deng, C.; Chen, X.; Yang, J.; Zhu, H.; Hong, J. Evaluation on the Cracking Resistance of Semi-Flexible Pavement Mixture by Laboratory Research and Field Validation. *Constr. Build. Mater.* **2019**, *207*, 387–395. [CrossRef]
4. Wu, S.; Wen, H.; Zhang, W.; Shen, S.; Mohammad, L.N.; Faheem, A.; Muhunthan, B. Field Performance of Top-down Fatigue Cracking for Warm Mix Asphalt Pavements. *Int. J. Pavement Eng.* **2019**, *20*, 33–43. [CrossRef]
5. Zhou, L.; Ni, F.; Zhao, Y. Evaluation Method for Transverse Cracking in Asphalt Pavements on Freeways. *Transp. Res. Rec.* **2010**, *2153*, 97–105. [CrossRef]
6. Canestrari, F.; Ingrassia, L.P.; Virgili, A. A Semi-Empirical Model for Top-down Cracking Depth Evolution in Thick Asphalt Pavements with Open-Graded Friction Courses. *J. Traffic Transp. Eng. (Engl. Ed.)* **2022**, *9*, 244–260. [CrossRef]
7. Willis, J.R.; Taylor, A.J.; Nash, T.M. Laboratory and Field Evaluation of Florida Mixtures at the 2012 National Center for Asphalt Technology Pavement Test Track. *Transp. Res. Rec.* **2016**, *2590*, 65–73. [CrossRef]
8. Saha, G.; Biligiri, K.P. Fracture Properties of Asphalt Mixtures Using Semi-Circular Bending Test: A State-of-the-Art Review and Future Research. *Constr. Build. Mater.* **2016**, *105*, 103–112. [CrossRef]
9. Zaumanis, M.; Valters, A. Comparison of Two Low-Temperature Cracking Tests for Use in Performance-Based Asphalt Mixture Design. *Int. J. Pavement Eng.* **2020**, *21*, 1461–1469. [CrossRef]
10. Du, H.; Ni, F.; Zhao, Y. Characterization of the Influence Factors of Low-Temperature Fracture Properties for Serving Asphalt Pavements Using Layer Core Samples. *J. Transp. Eng. Part B Pavements* **2021**, *147*, 04021016. [CrossRef]
11. Seitllari, A.; Boz, I.; Habbouche, J.; Diefenderfer, S.D. Assessment of Cracking Performance Indices of Asphalt Mixtures at Intermediate Temperatures. *Int. J. Pavement Eng.* **2022**, *23*, 70–79. [CrossRef]
12. Budziński, B.; Ratajczak, M.; Majer, S.; Wilmański, A. Influence of Bitumen Grade and Air Voids on Low-Temperature Cracking of Asphalt. *Case Stud. Constr. Mater.* **2023**, *19*, e02255. [CrossRef]
13. Zhou, Z.; Gu, X.; Dong, Q.; Ni, F.; Jiang, Y. Low- and Intermediate-Temperature Behaviour of Polymer-Modified Asphalt Binders, Mastics, Fine Aggregate Matrices, and Mixtures with Reclaimed Asphalt Pavement Material. *Road Mater. Pavement Des.* **2020**, *21*, 1872–1901. [CrossRef]
14. Yan, C.; Zhang, Y.; Bahia, H.U. Comparison between SCB-IFIT, Un-Notched SCB-IFIT and IDEAL-CT for Measuring Cracking Resistance of Asphalt Mixtures. *Constr. Build. Mater.* **2020**, *252*, 119060. [CrossRef]

5. Xu, D.; Ni, F.; Du, H.; Zhao, Z.; Wang, J.; Chen, S. Investigation of Factors Affecting the Intermediate-Temperature Cracking Resistance of In-Situ Asphalt Mixtures Based on Semi-Circular Bending Test. *Coatings* **2023**, *13*, 384. [CrossRef]
6. Zhou, F.; Im, S.; Hu, S.; Newcomb, D.; Scullion, T. Selection and Preliminary Evaluation of Laboratory Cracking Tests for Routine Asphalt Mix Designs. *Road Mater. Pavement Des.* **2017**, *18*, 62–86. [CrossRef]
7. Batioja-Alvarez, D.; Lee, J.; Haddock, J.E. Understanding the Illinois Flexibility Index Test (I-FIT) Using Indiana Asphalt Mixtures. *Transp. Res. Rec.* **2019**, *2673*, 337–346. [CrossRef]
8. Zhou, Z.; Gu, X.; Ni, F.; Li, Q.; Ma, X. Cracking Resistance Characterization of Asphalt Concrete Containing Reclaimed Asphalt Pavement at Intermediate Temperatures. *Transp. Res. Rec.* **2017**, *2633*, 46–57. [CrossRef]
9. Gao, L.; Li, H.; Xie, J.; Yu, Z.; Charmot, S. Evaluation of Pavement Performance for Reclaimed Asphalt Materials in Different Layers. *Constr. Build. Mater.* **2018**, *159*, 561–566. [CrossRef]
0. Ding, H.; Gyasi, J.; Hesp, S.A.M.; Marks, P.; Nie, Y.; Somuah, M.; Tabib, S.; Tetteh, N.; Ubaid, I. Performance Grading of Extracted and Recovered Asphalt Cements. *Constr. Build. Mater.* **2018**, *187*, 996–1003. [CrossRef]
1. Jaskula, P.; Szydlowski, C.; Stienss, M. Influence of Bitumen Type on Cracking Resistance of Asphalt Mixtures Used in Pavement Overlays. *IOP Conf. Ser. Mater. Sci. Eng.* **2018**, *356*, 012010. [CrossRef]
2. Saeidi, H.; Aghayan, I. Investigating the Effects of Aging and Loading Rate on Low-Temperature Cracking Resistance of Core-Based Asphalt Samples Using Semi-Circular Bending Test. *Constr. Build. Mater.* **2016**, *126*, 682–690. [CrossRef]
3. Dave, E.V.; Hoplin, C.; Helmer, B.; Dailey, J.; Van Deusen, D.; Geib, J.; Dai, S.; Johanneck, L. Effects of Mix Design and Fracture Energy on Transverse Cracking Performance of Asphalt Pavements in Minnesota. *Transp. Res. Rec.* **2016**, *2576*, 40–50. [CrossRef]
4. Jiang, J.; Dong, Q.; Ni, F.; Zhao, Y. Effects of Loading Rate and Temperature on Cracking Resistance Characteristics of Asphalt Mixtures Using Nonnotched Semicircular Bending Tests. *J. Test Eval.* **2019**, *47*, 2649–2663. [CrossRef]
5. Mirzaiyanrajeh, D.; Dave, E.V.; Sias, J.E.; Voels, J.; Dai, S. Effects of Laboratory Compacted Asphalt Mixtures Air-Void Variations on Fracture Properties at Low Temperatures. *Adv. Civ. Eng. Mater.* **2021**, *10*, 262–275. [CrossRef]
6. Radeef, H.R.; Abdul Hassan, N.; Zainal Abidin, A.R.; Mahmud, M.Z.H.; Mohd Satar, M.K.I.; Mohd Warid, M.N.; Al Saffar, Z.H. Determining Fracture Energy in Asphalt Mixture: A Review. *IOP Conf. Ser. Earth Env. Sci.* **2021**, *682*, 012069. [CrossRef]
7. Kaseer, F.; Yin, F.; Arámbula-Mercado, E.; Epps Martin, A.; Daniel, J.S.; Salari, S. Development of an Index to Evaluate the Cracking Potential of Asphalt Mixtures Using the Semi-Circular Bending Test. *Constr. Build. Mater.* **2018**, *167*, 286–298. [CrossRef]
8. Das, B.P.; Siddagangaiah, A.K. Identification of Parameter to Assess Cracking Resistance of Asphalt Mixtures Subjected to Aging and Moisture Conditioning. *J. Traffic Transp. Eng. (Engl. Ed.)* **2022**, *9*, 293–304. [CrossRef]
9. Haslett, K.E. *Evaluation of Cracking Indices for Asphalt Mixtures Using SCB Tests at Different Temperatures and Loading Rates Evaluation of Cracking Indices for Asphalt Mixtures Using SCB Tests At Senior Honors Theses of Bachelor of Science*; University of New Hampshire: Durham, UK, 2018; Available online: https://scholars.unh.edu/cgi/viewcontent.cgi?article=1385&context=honors (accessed on 3 July 2023).
0. Zhou, F.; Newcomb, D.; Gurganus, C.; Banihashemrad, S.; Sakhaeifar, M.; Park, E.S.; Lytton, R.L. National Academies of Sciences, Engineering, and Medicine. In *Field Validation of Laboratory Tests to Assess Cracking Resistance of Asphalt Mixtures: An Experimental Design*; The National Academies Press: Washington, DC, USA, 2016. [CrossRef]
1. Guo, R.; Zhou, F.; Nian, T. Indices Relation and Statistical Probability Analysis of Physical and Mechanical Performance of Asphalt Mixtures. *Case Stud. Constr. Mater.* **2022**, *16*, e01091. [CrossRef]
2. Yang, Q.; Deng, Y. Evaluation of Cracking in Asphalt Pavement with Stabilized Base Course Based on Statistical Pattern Recognition. *Int. J. Pavement Eng.* **2019**, *20*, 417–424. [CrossRef]
3. Du, H.; Ni, F.; Ma, X. Crack Resistance Evaluation for In-Service Asphalt Pavements by Using SCB Tests of Layer-Core Samples. *J. Mater. Civ. Eng.* **2021**, *33*, 04020418. [CrossRef]

coatings

MDPI

Article

Modification of the Crumb Rubber Asphalt by Eucommia Ulmoides Gum under a High-Temperature Mixing Process

Yueyang Shi [1], Zongliang Tong [1], Qingbo Ren [1], Zhigang Li [2,*] and Jingling Wang [2]

[1] China Communications Fourth Navigation Bureau Seventh Engineering Co., Ltd., Guangzhou 511466, China; y06021425@163.com (Y.S.); tzl1987@outlook.com (Z.T.); rqingbo@cccc4.com (Q.R.)

[2] School of Transportation, Southeast University, Nanjing 211189, China; jingling_wang@seu.edu.cn

* Correspondence: 103200039@seu.edu.cn

Abstract: The crumb rubber (CR) asphalt has some defects of high viscosity and poor storage stability, which brings great challenge to the high-quality construction of the CR asphalt pavement. To improve the interaction between the CR and base binder, the Eucommia ulmoides gum (EUG) with double-bond structure similar to trans-polyoctenamer rubber (TOR) was used to modify the CR asphalt. However, the original EUG double bond is basically inactive at room temperature and cannot form the effect of TOR. Open double bonds of EUG with asphalt and rubber powder form a network structure similar to TOR-modified rubber asphalt by high-temperature mixing with EUG in a torque rheometer. The effects of modified CR on rubber asphalt were analyzed by macro- and micro-experiments such as rotational viscosity tests, segregation tests, FTIR tests, and PG tests. It was found that the high-temperature mixing process works in both physical and chemical ways to mix the CR and EUG into an inseparable substance. The modified CR has higher chemical activity after desulfurization and degradation, which allows it to form a more effective chemical connection with asphalt. EUG can build a stable spatial crosslinking structure in CR asphalt due to the sulfurization reaction, which significantly improves the construction workability and system stability of the CR asphalt.

Keywords: rubber asphalt; Eucommia ulmoides gum; high-temperature mixing process; asphalt modification; sulfurization reaction; desulfurization and degradation

1. Introduction

Crumb rubber (CR), which is made of waste tires, can be used in the field of modified asphalt, and it brings some benefits of saving resources and protecting the environment. Rubber asphalt has unique advantages, such as the properties of anti-rutting deformation at high temperature and anti-cracking at low temperature [1,2]. However, the liquid–solid two-phase system and high viscosity of rubber asphalt result in the poor storage stability and construction workability. Therefore, it is necessary to further modify the performance of rubber asphalt [1,3,4]. In order to improve the compatibility between CR and asphalt and to establish an effective chemical connection between the CR and asphalt, the scholars have used trans-polyoctenamer rubber (TOR) to modify the rubber asphalt, which has achieved positive results [5]. TOR is a kind of polymer with a double-bond structure that can crosslink the sulfur in the asphalt with the sulfur on the surface of the CR to form a network structure, thus establishing a chemical connection between the CR and asphalt [6,7]. Although TOR has a good modification effect on the rubber asphalt, it is too expensive to be used widely in the construction of highways.

Through extensive investigations, one kind of polymer called Eucommia ulmoides gum (EUG) was found to be useful. EUG has double-bond structure, is a unique renewable resource in China, and can be used to modify the rubber asphalt with the similar effect of TOR. The EUG is a kind of natural high polymer material, and it has the same chemical composition as natural rubber, but the chemical constitution of them is different. EUG is the trans-polyisoprene, but the natural rubber is the cis-polyisoprene. The cis-polyisoprene is orderly and easy to gather and crystal. So it is a kind of flinty solid instead of an elastomer such as natural rubber. The EUG molecular chain has a lot of double bonds, and damaging its crystallization through the methods of cross-joining or grafting, which can damage its double bonds or orderly structure, can make its melting point drop, and even can make it become a kind of completely amorphous material such as natural rubber so that it can be used in some engineering occasion. To not depend on imports promoting efficient reactions between rubber crumb and asphalt and express the technical advantages of a dry-process rubber bitumen mixture with domestic natural rubber, the Eucommia ulmoides gum (EUG) was used to create chemical links between rubber crumb and asphalt. After rubber crumb was mixed with EUG, EUG's own double bond structure can be vulcanized with the sulfur of rubber crumb, founded by the microscopic analysis and macroscopic representation of the mixture. EUG can also be grafted with maleic anhydride (MA) reacting with the asphalt amino-group, so chemical links were built between the EUG and asphalt. The EUG can also greatly improve the high- and low-temperature performance and system stability of rubber asphalt [8]. However, the original EUG double bond is basically inactive at room temperature and cannot form the effect of TOR. Through the preliminary test [9], it is found that when EUG is mixed with rubber powder at high temperature, the double bond can be opened by the sulfur atom of rubber powder, and open double bonds of EUG with asphalt and rubber powder form a network structure similar to TOR-modified rubber asphalt. However, existing studies are limited to the direct modification of EUG to rubber asphalt, which does not involve the modification of CR under high-temperature mechanical action. In this paper, a new process with high temperature was investigated. Firstly, quantitative CR and EUG were mixed in the torque rheometer at high temperature to form a new material (EUG-modified CR), then the EUG-modified CR was used to modify the asphalt binder. Finally, the effects of the high-temperature mixing process and EUG on the properties of CR are studied by laboratory tests.

2. Mechanism Analysis

2.1. Modification Mechanism of the High-Temperature Mixing Process on EUG and CR

The CR contains sulfur, while the double bonds on the molecular chains of different EUG can be crosslinked by sulfur during the sulfurization process. Gradually a crosslinking network with the increase of chemical crosslinking points was formed in the sulfurization process [10,11], as shown in Figure 1. Therefore, these properties can be used to combine the EUG with the CR organically.

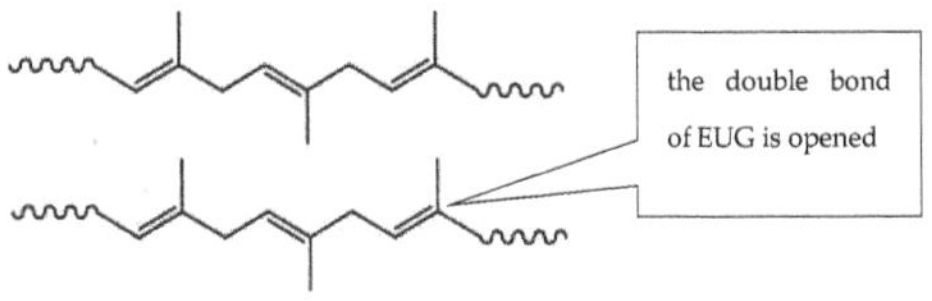

Figure 1. The sulfurization mechanism of EUG.

Under the high-temperature condition and the intense mechanical action with the torque rheometer, the sulfurization rate and the sulfurization degree of EUG are both improved significantly. Through the actions of thermal energy and the shearing force, the CR will undergo the desulfurization reaction and the degradation reaction, which are conducive to improving the compatibility between the CR and the asphalt. With the existence of the "wrapping effect" of EUG, the CR cannot be excessively degraded, thus the EUG has a positive impact on the CR.

In the high-temperature mixing process, the sulfurization reaction can form a tight chemical crosslinking network between the EUG and the CR. And the intense extrusion of the torque rheometer's rotor can integrate the EUG with the CR at the physical level. Finally, the two materials can be closely combined into a new composite material.

2.2. Modification Mechanism of EUG on Rubber Asphalt

Due to the poor compatibility between the CR and the asphalt, the rubber asphalt has some disadvantages, such as the high viscosity and the poor storage stability. The CR is prone to excessive swelling in the asphalt, so the light components of the asphalt can be excessively absorbed. However, the rubber particles tend to aggregate with each other, which leads to the high viscosity of the rubber asphalt and internal segregation phenomenon in the rubber asphalt.

The density of the CR is larger than that of asphalt, which leads to the deposition of CR in the asphalt. While the density of the EUG is smaller than that of asphalt, the composite material's density is close to the density of the asphalt. According to the "equal density theory" [12,13], the composite of CR and EUG can be stably suspended in the asphalt.

After the desulfurization of CR, the broken sulfur bond on the surface of the rubber particle will react chemically with the asphalt, which can improve the compatibility between the CR and the asphalt. Moreover, the degraded rubber particles will be smaller, which can prevent them from aggregating into agglomerates and excessively swell in the asphalt. The "wrapping effect" of the EUG on the CR also has a similar effect as above. The double bond of the EUG can not only react with the sulfur in the CR but can also combine with the sulfur in the asphalt. Thus, the connection network among the EUG, the CR, and the asphalt can be formed through the above methods, which changes the interior of the rubber asphalt into a stable system. The modification mechanism of the EUG on the rubber asphalt is shown in Figure 2.

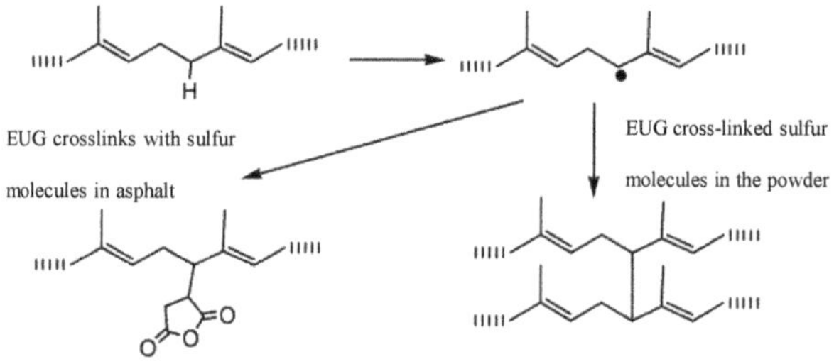

Figure 2. Modification mechanism of EUG on rubber asphalt.

3. Materials and Experimental Plan

3.1. Materials

1. EUG

EUG is a kind of natural high polymer material, and it has the same chemical composition as natural rubber, but the chemical constitution of them is different. EUG is the trans-polyisoprene, but the natural rubber is the cis-polyisoprene. The cis-polyisoprene is orderly and easy to gather and crystal. So it is a kind of flinty solid instead of an elastomer such as natural rubber. The main properties of the EUG used in this research are shown in Table 1.

Table 1. Properties of EUG.

Specific Gravity	Hardness Shore A	Melting Point/°C	Softening Point/°C	Coefficient of Cubical Expansion/(/°C)	Thermal Conductivity/ (cal/ (cm·s·°C))
0.96~0.99	98	65	55	0.0008	3.1×10^4

2. CR

The size of the CR used in this research is 30 mesh, and the main physical and chemical properties of CR are shown in Table 2.

Table 2. Physical and chemical properties of CR.

Item	Relative Density	ω (Water)/%	ω (Metal)/%	ω (Fiber)/%	ω (Ash Content)/%	ω (Acetone Extract)/%	ω (Carbon Black)/%	ω (Rubber Hydrocarbon)/%
Standard	1.10–1.30	<1.00	<0.05	<1.00	≤8.00	≤22.00	≥28.00	≥42.00
Test result	1.18	0.04	0.03	0.55	7.00	7.50	29.00	50.00

3. Asphalt

The asphalt used in this research is one neat petroleum asphalt with penetration of 60/70, and the main technical properties of the neat asphalt are shown in Table 3.

Table 3. Technical properties of asphalt.

Item	Penetration (25 °C, 100 g, 5 s)/(0.1 mm)	Ductility (5 cm/min, 15 °C)/cm	Softening Point/°C	RTFOT (163 °C, 5 h)		
				Mass Loss/%	Penetration Ratio /%	10 °C Ductility/cm
Standard	60~80	≥100	≥46	−0.8~0.8	≥61	≥6
Test result	70	>150	47.0	0.15	67	7

3.2. Experimental Indicators (Test Method: Refer to the JTGE20-2011 Test Code for Asphalt and Asphalt Mixtures in Highway Engineering)

1. The rotational viscosity test: The rotational viscosity shows the workability of the asphalt in construction. The instrument used in this test is an NDJ-1C rotational viscometer. The rotor type is 27#, and the rotational speed is 50 r/min. The rotational viscosity test was conducted at the temperature of 180 °C.
2. Storage stability test of polymer-modified asphalt: The storage stability test was based on the results obtained from the softening point test. The smaller the difference in softening point is, the better the storage stability of the modified asphalt is.
3. The Fourier-transform infrared spectroscopy (FTIR) test: Infrared spectroscopy can accurately detect the structure and structural changes of polymers in the asphalt, providing plenty of information to study the modification mechanism of the polymers on the asphalt. The instrument used in this test is the Nicolet 5700 Fourier infrared spectrometer.
4. The scanning electron microscope (SEM) test: The morphological characteristics of polymers in the asphalt can be directly reflected by the microscopic images. The instrument used in this test is the SEM-3200M scanning electron microscope.
5. The fluorescence microscope (FM) test: The distribution of the polymers in the asphalt can be observed by the fluorescence microscope. The instrument used in this test is the Eclipse 80i fluorescence microscope.

3.3. Experimental Scheme

Under the intense mechanical action in the high-temperature environment of the torque rheometer, CR will be desulfurized and degraded [9], but there are few references about the influence of the high-temperature mixing process on the CR in the presence of the EUG. Therefore, the relevant research was carried out according to the formula of the experimental materials in Table 4. In this table, the ratio of EUG and rubber powder is 1:10.

Table 4. The formulae of experimental materials.

Formula	Different Rubber Asphalt Corresponding Modifier
Y-1	CR (unprocessed)
Y-2	Mixing of CR and raw EUG
Y-3	High temperature mixing of CR
Y-4	High temperature mixing of raw EUG and CR

According to the principle of the physical blending method, the EUG-modified CR was prepared by the XSS-300 torque rheometer (as shown in Figure 3). There are three process parameters in the use of the torque rheometer, which are temperature, time, and rotational speed. On the basis of previous studies, the following mixing process parameters of Y3 and Y4 were adopted in this research, which are 150 °C (mixing temperature), 15 min (mixing time), and 50 r/min (rotational speed). The speed and time of Y1 and Y2 are the same as Y4, and the temperature is 25 °C at room temperature.

Figure 3. (**a**) Untreated CR; (**b**) CR modified by EUG in a torque rheometer.

4. Analysis of Experimental Results

4.1. The Tests of Rotational Viscosity and Softening Point Difference

The results of the rotational viscosity test and softening point difference test of the rubber asphalt with different formulas are shown in Figures 4 and 5, respectively.

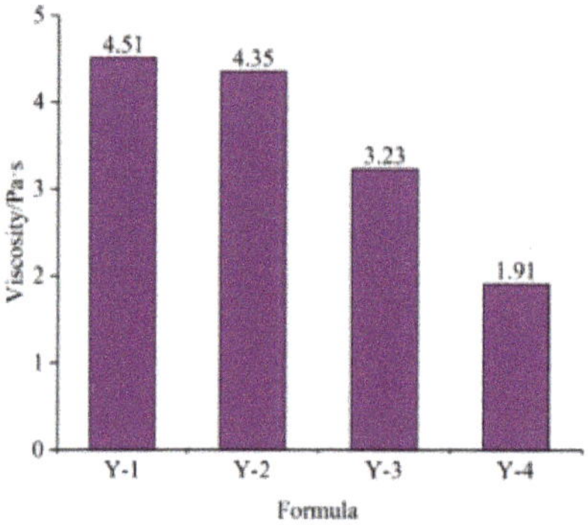

Figure 4. The results of the rotational viscosity tests.

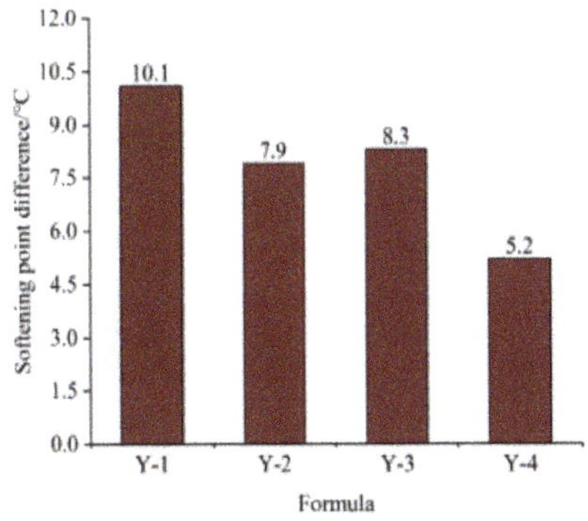

Figure 5. The results of the softening point difference tests.

It can be seen from Figure 4 that the viscosity of the ordinary rubber asphalt (Y-1) is the highest among the four modified asphalt binders, indicating that the unmodified CR has undergone an excessive swelling reaction in the asphalt, which makes the viscosity of rubber asphalt increase sharply, thus affecting the binder's workability in construction. Meanwhile, the viscosity of the mixed CR-modified asphalt (Y-3) is much lower than that of the ordinary rubber asphalt, which means that the mixed CR has undergone an intense desulfurization and degradation reaction. The rubber macromolecules are decomposed into a large number of small molecules, so the phenomenon of excessive swelling of the CR in the asphalt is changed. Furthermore, the viscosity of the raw EUG-modified rubber asphalt (Y-2) is almost unchanged compared with the ordinary rubber asphalt (Y-1), so it can be concluded that the raw EUG has a slight effect on the workability of the rubber asphalt in construction.

Figure 5 demonstrates that the subsidence phenomenon of the CR in the ordinary rubber asphalt (Y-1) is the most serious among the four modified asphalt binders. The system stability of the mixed CR-modified asphalt (Y-3) is superior to that of the ordinary rubber asphalt, indicating that the high-temperature mixing process could result in the depolymerization and the sulfur bond's fracture of the CR, which not only decreases the binder's molecular weight but also enhances the binder's chemical activity [9]. At the same time, it is effective for the dispersion of the rubber particles in the asphalt and the formation of the crosslinking network between the CR and the asphalt; thereby, the storage stability of the modified asphalt is improved. In the same case of adding the EUG, the segregation degree of the modified asphalt without mixed CR (Y-2) is also larger than that of the modified asphalt with mixed CR (Y-4).

In Figure 4, the modification effect of the CR on the asphalt is further enhanced under the joint actions of the EUG and the high-temperature mixing process (Y-4). The EUG can crosslink the sulfur in the CR and the sulfur in the asphalt to form a network structure, and the network structure can separate the entangled rubber particles and reduce the molecular weight of the polymer. In addition, low melting point is a key characteristic of the EUG. The comprehensive effects of these factors have led to a significant decrease in the viscosity of the rubber asphalt.

As Figure 5 shows, the separation degree of the rubber asphalt under the joint actions of the EUG and the high-temperature mixing process (Y-4) is the smallest among the four modified asphalt binders, which shows that the EUG can promote the dispersion of the rubber particles in the asphalt and improve the compatibility between the CR and the asphalt. Therefore, the suspension stability of the CR in the asphalt is greatly improved.

4.2. *Microscopic Examinations [14]*

Figure 6a is a SEM image of the ordinary rubber asphalt (Y-1). It can be seen that the asphalt surface is uneven and the sizes of different rubber particles are very inconsistent.

Moreover, the rubber particles are irregularly dispersed in the asphalt, and the aggregation of the rubber particles is very obvious. The above phenomena show that the swelling and diffusion reactions occur after the blending between the CR and the asphalt, but the rubber particles are still dispersed in the solid phase. Due to the poor compatibility between the CR and the asphalt, the ordinary rubber asphalt results in high viscosity and bad storage stability.

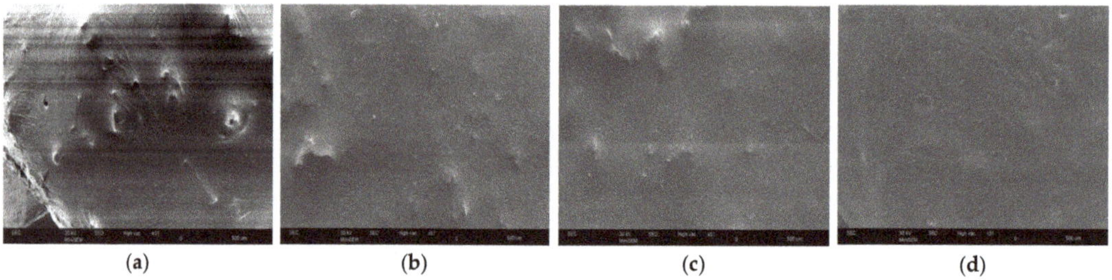

Figure 6. SEM images of rubber asphalt with different formulas. (a) SEM image of Y-1; (b) SEM image of Y-2; (c) SEM image of Y-3; and (d) SEM image of Y-4.

Figure 6c is a SEM image of the mixed CR-modified asphalt (Y-3). Compared with Figure 7a, it can be seen that the bulges on the asphalt surface are reduced and the agglomeration of rubber particles is improved. The above phenomena indicate that the molecular weight of the CR is reduced under the effect of the high-temperature mixing process. The fracture of the crosslinking bonds among rubber molecules leads to their inability to agglomerate, which improves the compatibility between the CR and the asphalt; therefore, the segregation degree of the modified asphalt during the storage process is decreased.

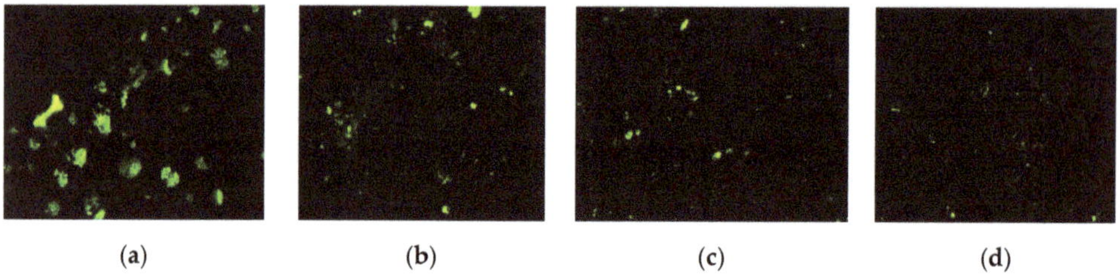

Figure 7. FM images of rubber asphalt with different formulas. (a) FM image of Y-1; (b) FM image of Y-2; (c) FM image of Y-3; and (d) FM image of Y-4.

By comparing Figures 6d and 6b, it can be found that the rubber particles in the mixed EUG and CR composite modified asphalt (Y-4) are smaller and more evenly distributed than those in the raw EUG-modified rubber asphalt (Y-2), which indirectly indicates that the high-temperature mixing process can improve the system stability of the rubber asphalt.

The intrinsic fluorescence properties of the CR and the asphalt are different, which show different colors under the blue light of FM. Therefore, the modifier shows an obvious yellow color, while the asphalt shows a dark one. Due to the viscoelasticity of the CR and the complex flow field in the blending process, the formation of the morphological structure of the rubber asphalt is a complex process. Moreover, the morphological structure is multi-layered, so there will be some granular, fibrous, or banded structures in the rubber asphalt.

Figure 7a is a FM image of the ordinary rubber asphalt (Y-1). It can be seen that the sizes of different rubber particles after excessive swelling are inconsistent, and the distribution of the rubber particles is uneven. It can be seen from Figure 7b that the distribution of the rubber particles in the mixed CR-modified asphalt (Y-3) is relatively regular, and the sizes of different rubber particles are relatively uniform, which indicates that CR can undergo a proper swelling reaction with the asphalt after the action of the high-temperature mixing process, thus the thermal storage stability of the mixed CR-modified asphalt (Y-3) is better than that of the ordinary rubber asphalt (Y-1).

Under the condition of adding the EUG, the microstructure of the mixed EUG and CR composite-modified asphalt (Y-4) is more orderly, and almost none of the big particles can be seen compared with the raw EUG-modified rubber asphalt (Y-2), which reflects the importance of the high-temperature mixing process.

4.2.1. The Effect of EUG on the Properties of Rubber Asphalt

Figure 6 shows that the surface of the mixed EUG and CR composite modified asphalt (Y-4) is very smooth and basically free of bulge, and the agglomeration phenomenon of the rubber particles disappears. It means that the compatibility between the CR and the asphalt is greatly improved by the joint actions of the EUG and the high-temperature mixing process, and the rubber particles are evenly dispersed in the asphalt. However, the separate action of the EUG and the high-temperature mixing process only has a limited modification effect on the CR. Therefore, it can be seen from Figure 6 that the microstate of the rubber particles in the raw EUG-modified rubber asphalt (Y-2) and the mixed CR-modified asphalt (Y-3) are both worse than that in the mixed EUG and CR composite modified asphalt (Y-4).

It can be seen from Figure 7 that the distribution characteristics of the rubber particles in the four modified asphalts under the observation of FM are basically consistent with the observation results of SEM. The distribution of the rubber particles in the raw EUG-modified rubber asphalt (Y-2) is more reasonable, and the sizes of the particles are smaller compared with the ordinary rubber asphalt (Y-1). Moreover, the microstructure of the mixed EUG and CR composite-modified asphalt (Y-4) is also superior to the mixed CR-modified asphalt (Y-3). These phenomena show that the EUG has a positive modification effect on the rubber asphalt.

4.2.2. A Summary of Microscopic Examinations

According to the three microscopic examinations, the fusion and dispersion of the rubber particles in the mixed EUG and CR composite modified asphalt (Y-4) are obviously better than the other three modified asphalt binders. It means that both the addition of the EUG and the high-temperature mixing process are helpful to improve the performance of the rubber asphalt. Therefore, the results of macroperformance tests are verified at the micro level.

4.3. Performance Grade Test (AASHTO M320: Standard Specification for Performance-Graded Asphalt Binder)

SBS-modified asphalt is a kind of high-performance modified asphalt binder, which is often used in road engineering. In order to fully prove the effectiveness of the EUG and the high-temperature mixing process, three kinds of asphalt binders were selected in comparison with the SBS-modified asphalt through the PG test. The results of the test are shown in Table 5.

According to the high-temperature indicator ($G^*/\sin\delta$), it can be found that the raw EUG-modified rubber asphalt (Y-2) is one grade higher than the ordinary rubber asphalt (Y-1), and the mixed EUG and CR composite-modified asphalt (Y-4) is one grade higher than Y-2 and the same as SBS-modified asphalt. It means that the EUG and the high-temperature mixing process are both effective to improve the high-temperature performance of the rubber asphalt.

Table 5. Performance grade test results.

Test	Item		Temperature/°C	Y-1	Y-2	Y-4	SBS-Modified Asphalt	Requireme
DSR	Virgin test	(G*/sinδ)/kPa	64	7.729	8.235	13.318	14.752	≥1.0
			70	2.121	4.511	9.261	11.355	
			76	1.653	2.375	4.524	5.349	
			82	1.014	1.210	2.101	2.408	
			88	0.360	0.312	0.862	0.992.	
	RTFOT test		64	11.995	14.422	21.213	21.726	≥2.2
			70	5.804	7.517	14.670	15.440	
			76	2.106	2.976	8.467	8.847	
			82	—	1.724	4.596	5.020	
			88	—	—	2.140	2.193	
	PAV test	(G*·sinδ)/kPa	22	3345	2580	2029	2310	≤5000
			19	4921	4022	2840	3402	
			16	5309	4910	3560	4023	
			13	—	5340	4510	5122	
			10	—	—	5063	—	
BBR	PAV test	Creep stiffness/MPa	−12	182	165	124	130	≤300
			−18	361	324	225	244	
			-24	—	—	343	371	
		Creep rate	−12	0.320	0.373	0.422	0.392	≥0.3
			−18	0.282	0.291	0.375	0.304	
			−24	—	—	0.293	0.280	
Performance grade				PG70-22	PG76-22	PG82-28	PG82-28	

According to the intermediate-temperature indicator (G*·sinδ), it can be found that the raw EUG-modified rubber asphalt (Y-2) is one grade higher than the ordinary rubber asphalt (Y-1) and the same as SBS-modified asphalt. The mixed EUG and CR composite modified asphalt (Y-4) is one grade higher than SBS-modified asphalt, which shows that the EUG and the high-temperature mixing process are both effective to enhance the fatigue cracking performance of the rubber asphalt.

According to the low-temperature indicator (creep stiffness/creep rate), the raw EUG has little effect on the low-temperature performance of the rubber asphalt, and the joint action of the EUG and the high-temperature mixing process can improve the low-temperature performance of the rubber asphalt to the same level as the SBS-modified asphalt.

The results of the above test show that the performance of the rubber asphalt can be significantly enhanced under the joint action of the EUG and the high-temperature mixing process.

5. Conclusions

The following conclusions and findings can be drawn:

1. The strong mechanical action in the high-temperature condition leads to the desulfurization and degradation of the rubber particles. The desulfurized and degraded rubber particles can react more intensely with the asphalt, which can improve the compatibility between the CR and the asphalt.
2. The EUG can effectively prevent the occurrence of excessive swelling phenomena and the agglomeration phenomenon of the CR in asphalt. Moreover, a tight crosslinking network can be formed among the EUG, the CR, and the asphalt by the sulfurization reaction, and finally a stable spatial crosslinking structure can be established inside the rubber asphalt. Therefore, the workability of construction and the system stability of the mixed EUG and CR composite-modified asphalt are much better than that of the ordinary rubber asphalt.
3. After the action of the high-temperature mixing process, a better chemical connection can be built between the CR and asphalt. The EUG can also improve the degree and

the efficiency of the chemical reaction between the CR and the asphalt. Only under the joint action of these factors can the performance of the rubber asphalt be significantly enhanced.

Author Contributions: Conceptualization, Z.L. and Y.S.; methodology, Y.S. and Z.T.; formal analysis, Q.R.; investigation, Y.S.; resources, Z.L.; writing—original draft preparation, Y.S.; writing—review and editing, J.W.; supervision, Z.L.; funding acquisition, Z.L. All authors have read and agreed to the published version of the manuscript.

Funding: This research was sponsored by the National Natural Science Foundation of China (NSFC) (Project No. 51778621).

Institutional Review Board Statement: Not applicable.

Informed Consent Statement: Not applicable.

Data Availability Statement: Data are contained within the article.

Conflicts of Interest: Authors Yueyang Shi, Zongliang Tong, Qingbo Ren were employed by the company China Communications Fourth Navigation Bureau Seventh Engineering Co., Ltd. The remaining authors declare that the research was conducted in the absence of any commercial or financial relationships that could be construed as a potential conflict of interest.

References

1. Presti, D.L. Recycled Tyre Rubber Modified Bitumens for road asphalt mixtures: A literature review. *Constr. Build. Mater.* **2013**, *49*, 863–881. [CrossRef]
2. Milad, A.; Ahmeda, A.G.; Taib, A.M.; Rahmad, S.; Solla, M.; Yusoff, N.I.M. A review of the feasibility of using crumb rubber derivedfrom end of life tire as asphalt binder modifier. *J. Rubber Res.* **2020**, *23*, 203–216. [CrossRef]
3. Jeong, K.D.; Lee, S.J.; Amirkhanian, S.N.; Kim, K.W. Interaction effects of CR modified asphalt binders. *Constr. Build. Mater.* **2010**, *24*, 824–831. [CrossRef]
4. Egami, K.; Kimura, M.; Hamaguchi, T. Effect of CR modifier dissolution on storage stability of CR-modified asphalt. *Transp. Res. Rec. J. Transp. Res. Board* **2013**, *2370*, 109–115.
5. Katman, H.Y.; Ibrahim, M.R.; Karim, M.R.; Salim Mashaan, N.; Koting, S. Evaluation of permanentdeformation of unmodified and rubber-reinforced SMA asphalt mixtures using dynamic creep test. *Adv. Mater. Sci. Eng.* **2015**, *2015*, 247149. [CrossRef]
6. Liu, H.; Chen, Z.; Wang, W.; Wang, H.; Hao, P. Investigation of the rheological modification mechanism of CR modified asphalt (CRMA) containing TOR additive. *Constr. Build. Mater.* **2014**, *67*, 225–233. [CrossRef]
7. Liang, M.; Xin, X.; Fan, W.; Ren, S.; Shi, J.; Luo, H. Thermo-stability and aging performance of modified asphalt with CR activated by microwave and TOR. *Mater. Des.* **2017**, *127*, 84–96. [CrossRef]
8. Li, Z.; Li, C.; Cai, M. Modification mechanism and effect of grafted eucommia ulmoides gum on dry-process rubber bitumen mixture. *J. Southeast Univ. (Nat. Sci.)* **2014**, *44*, 845–848.
9. Li, Z.; Deng, S.; Tan, M. Analysis on temperature and time for mixing preparation of crumb rubber modified by eucommia ulmoides gum. *J. Southeast Univ. (Nat. Sci. Ed.)* **2020**, *50*, 1109–1114.
10. Fang, Q.; Jin, X.; Yang, F.; Ma, C.; Gao, Y.; Wang, N. Preparation and characterizations of eucommia ulmoides gum/polypropylene blend. *Polym. Bull.* **2016**, *73*, 357–367. [CrossRef]
11. Mu, Y.X.; Yang, F.; Kang, H.L.; Wang, N.; Fang, Q.H. Effects of mixing process on the properties of eucommia ulmoides gum/natural rubber compound. *Polym. Mater. Sci. Eng.* **2018**, *34*, 108–114+119.
12. Ren, R.B.; Geng, L.T.; Xu, Q. Preparation of reclaimed rubber-plastic alloying agent and its modification mechanism on matrix asphalt. *J. Build. Mater.* **2016**, *19*, 528–533.
13. Li, J.; Wang, S.; Zhang, Y.; Zhang, Y. Preparation of CR modified asphalt with high temperature storage stability by mechanochemical equal density method. *J. China Foreign Highw.* **2011**, *31*, 270–274.
14. Tan, Y.; Li, G.; Dan, L.; Lyu, H.; Meng, A. Research progress of bitumen microstructures and components. *J. Traffic Transp. Eng.* **2020**, *20*, 1–17.

coatings MDPI

Article

Effect of Chloride Salt Erosion on the Properties of Straw Fiber Reactive Powder Concrete

Hangyang Wang [1,2], Kaiwei Gong [1,2], Bingling Cheng [3], Xi Peng [2,4,*], Hui Wang [1,*] and Bin Xu [4]

1 School of Civil Engineering and Geographical Environment, Ningbo University, Ningbo 315000, China; 1105390194@aliyun.com (H.W.); 1095957844@aliyun.com (K.G.)
2 School of Civil Transportation Engineering, Ningbo University of Technology, Ningbo 315211, China
3 Ningbo Yonghuan Yuan Environmental Protection Engineering Technology Co., Ltd., Ningbo 315000, China; 263454376@aliyun.com
4 Ningbo Roaby Technology Industrial Group Co., Ltd., Ningbo 315800, China; 5632374376@aliyun.com
* Correspondence: pengxi@nbut.edu.cn (X.P.); huiwang123@aliyun.com (H.W.)

Abstract: Straw fibers are renowned for their cost-effectiveness, sustainability, and durability. They represent a promising natural reinforcement option for reactive powder concrete (RPC). This paper investigated the impact of straw fibers on RPC's workability, mechanical performance (mechanical strength and flexural toughness), and electrical properties (electrical resistance and AC impedance spectroscopy curves). The straw fiber volumes ranged from 1% to 4.0% of the total RPC volume. Specimens were cured under standard curing conditions for 3, 7, 14, and 28 days. Mechanical and electrical properties of the specimens were tested before chloride salt erosion. The mass loss and ultrasonic velocity loss of the samples were measured under NaCl freeze–thaw cycles (F-Cs). The mass loss, ultrasonic velocity loss, and mechanical strengths loss of the samples were measured under NaCl dry–wet alternations (D-As). The findings indicated that incorporating straw fibers enhanced RPC's flexural strength, compressive strength, and flexural toughness by 21.3% to 45.76%, −7.16% to 11.62%, and 2.4% to 32.7%, respectively, following a 28-day curing period. The addition of straw fibers could augment the AC electrical resistance of the RPC by 10.17% to 58.1%. The electrical characteristics of the RPC adhered to series conduction models. A power function relationship existed between the electrical resistance and mechanical strengths of the RPC. After 10 NaCl D-As, the mass loss rate, ultrasonic velocity loss rate, flexural strength, and compressive strength loss rates of the RPC decreased by 0.42% to 1.68%, 2.69% to 6.73%, 9.6% to 35.65%, and 5.41% to 34.88%, respectively, compared to blank samples. After undergoing 200 NaCl F-Cs, the rates of mass loss and ultrasonic velocity loss of the RPC decreased by 0.89% to 1.01% and 6.68% to 8.9%, respectively.

Keywords: straw fibers; reactive powder concrete; chloride salt erosion; mechanical performance; electrical properties

1. Introduction

Ordinary concrete is a fundamental material in civil engineering, appreciated for its cost-effectiveness and ease of construction. It is widely used in coastal infrastructure projects, including bridges, tunnels, harbors, and waterways [1]. However, under challenging service conditions and heightened standards for specialized projects [2–4], conventional concrete often fails to meet project requirements [5–7]. It is imperative to find a new concrete material.

Reactive powder concrete (RPC), a novel cementitious composite, offers exceptional mechanical properties and durability [8–13]. Originating in the 1990s through Bouygues Corporation's innovation [14], RPC is designed to maximize particle packing density. Precast RPC components have been extensively utilized in engineering projects [15–18], enhancing the longevity of offshore concrete structures. Scholars, including Blunt [19], advocate for fiber reinforcement as an optimal solution to enhance crack control in concrete structures exposed to harsh erosive environments. The commonly used fibers are steel fibers, synthetic fibers, and basalt fibers. For instance, studies by P.R. et al. showed that RPC with 2% fiber content exhibits ductility 5.21 times greater than that of fiber-less specimens. Steel fibers enhance both ductility and residual load-carrying capacity of the RPC [20]. A study by Hiremath et al. indicated that RPC with 0.1% polypropylene and polyester fibers demonstrates good spalling resistance, while a 0.5% fiber dosage enhances residual mechanical and durability properties at elevated temperatures [21]. Wang et al. demonstrated that incorporating 1.2% oriented steel fibers with an aspect ratio of 150 significantly enhances flexural and compressive strengths, as well as bending toughness [22]. Branston et al. demonstrated that incorporating BF enhanced the initial crack resistance of concrete under flexural loading [23]. However, the high cost of steel fibers represents a significant barrier to their widespread adoption in construction projects. Synthetic fibers have a serious effect on concrete flow and steel fiber corrosion. Basalt fibers have strict requirements on the composition of the raw ore [24].

As global environmental awareness grows, the engineering field faces increasingly stringent requirements regarding carbon emissions. Straw fiber is one of the abundant natural fibers and a major by-product of agriculture. In China, nearly 900 million tons of straw are produced annually, with the majority being disposed of by burning. This practice not only pollutes the environment but also results in a significant waste of resources [25,26]. Straw fibers offer several advantages as a natural material: they are lightweight, renewable, extremely tough, and easy to manufacture [27,28]. The use of straw fibers has been shown to improve the mechanical strength of concrete [29]. It also helps to reduce the overall dead weight of the concrete [30]. For example, Wu et al. showed that incorporating CSF (corn straw fiber) moderately enhances the compressive strength, flexural strength, and cracking toughness of EPS concrete [31]. Additionally, these fibers promote the formation of denser hydration products, enhancing crack repair and increasing permeability [32]. This approach not only improves the structural integrity of concrete, but also its durability under various conditions [33]. At present, a considerable number of scholars have conducted in-depth studies on the performance mechanisms of various types of straw fiber, such as rice straw fiber, corn straw fiber, and highland barley straw fiber, in different concrete materials [34–36]. However, the performance of straw fiber RPC under conditions of chloride salt attack remains an area that has not been sufficiently explored.

This study focused on investigating the effects of chloride salt erosion on the performance of straw fiber RPC. Before chloride salt erosion, the feasibility and excellence of incorporating straw fibers into RPC were validated through tests of workability, mechanical properties, and electrochemical performance. After chloride salt erosion, the defects within RPC and changes in its performance were detected through mass, ultrasonic, and mechanical strength loss experiments. In addition, the mechanical strength of the straw fiber RPC was characterized by electrical performance. The results of this study will provide a benchmark for the use of straw fiber RPC in chloride salt erosion environments and will also aid in developing specialized methods to predict future performance.

2. Experimental

2.1. Materials and Sample Preparation

2.1.1. Materials

The research utilized ordinary Portland cement (OPC) obtained from Jiangsu Huaxi Cement Manufacturing Co., Ltd., based in Wuxi, China. This OPC demonstrates a specific surface area of 350 m^2/kg and a density of 3.0 g/cm^3. The OPC displayed initial and final setting times of 121 min and 233 min, respectively. Silica fume (SF), boasting a density of 2.2 g/cm^3 and a specific surface area measuring 15 m^2/g, and containing over 98% SiO_2, was utilized as an additional cementitious component. Quartz sand (QS) was employed as the aggregate, consisting of three particle sizes: 1~0.71 mm, 0.59~0.35 mm, and 0.15~0.297 mm. The mass ratio of coarse, medium, and fine sands was 1:1.5:0.8. QS and SF were manufactured at Lingshou Xinhui Mining Processing Factory, Shijiazhuang, China.

We utilized a polycarboxylate-based superplasticizer (SP) manufactured by Shanghai Yingshan New Material Technology Co., Ltd., Shanghai, China. The tawny microemulsion solution exhibited a 40% water-reduction capability. Additionally, we employed a defoamer known as DF-04. The defoamer was a polyether-based surfactant comprising a mixture of dry powder carriers, which are white or light grey in color, with a density of 300–450 g/L and a pH of 7.0–7.5. The two additives in question were manufactured by Shanghai Yingshan New Materials Technology Co., Ltd., Shanghai, China. The straw fibers, with a density of 27 kg/m^3, were obtained from Jiangsu Fangyuan Tong Biotechnology Co., Ltd., Sucheng, China. The fiber lengths exhibited a range of 1 mm to 20 mm. The primary chemical constituents of straw fiber were quite similar, predominantly including cellulose, hemicellulose, and lignin. The physical diagram of straw fibers is shown in Figure 1. The chemical compositions of OPC, QS, and SF are presented in Table 1 for reference.

Figure 1. The physical diagram of straw fibers.

Table 1. Chemical composition of raw materials (%).

Type	Chemical Composition/%							
	SiO_2	Al_2O_3	Fe_2O_3	MgO	CaO	SO_3	R_2O	SiO_2
OPC	20.86	5.47	3.94	1.73	62.23	2.66	0.48	20.86
QS	34.06	14.74	0.83	9.73	35.93	0.23	3.51	34.06
SF	90	0.8	0.6	0.8	0.4	0	7.4	90

2.1.2. Sample Preparation

The experiment was conducted using a 3 L volume to prepare the mixture proportion, as illustrated in Table 2. All materials were weighed using an electronic balance. The particulate-sized quartz sand, silica fume, and Portland cement were successively introduced into a pre-wetted UJZ-15 mortar mixer. The mixing time was set at eight minutes, after which the mixer was initiated. After mixing for 30 s, a blend of water, a water-reducing agent, and a defoamer was gradually introduced. The duration was then noted to determine the setting time of the sample. One minute after commencing the mixing process, the addition of straw fibers was undertaken in order to achieve uniform mixing. Upon completion of the mixing process, the machine automatically ceased operation, and the resulting concrete slurry exhibited satisfactory flowability, indicating that it was ready for casting specimens. The RPC mixture from the mixer was poured into prism-shaped specimens of dimensions 40 × 40 × 160 mm^3 and 50 × 50 × 50 mm^3, with stainless steel mesh inserted on both sides (size 45 × 60 mm^2) and covered with plastic film. Each group was labeled and placed in a SHBY-40B standard constant temperature and humidity curing chamber at a temperature of 20 ± 2 °C and humidity above 95% for a period of 24 h. Following the curing period, the molds were removed and each specimen was correctly labeled. The production process is illustrated in Figure 2.

Table 2. Mass (kg/m^3) and mass ratio of proportioned fresh cement paste.

Materials	Water	OPC	SF	QS	Water Reducer	Defoamer	Straw Fibers
1	339.6	1018.9	339.6	679.2	26.4	1.32	0
2	339.6	1018.9	339.6	679.2	26.4	1.32	27
3	339.6	1018.9	339.6	679.2	26.4	1.32	54
4	339.6	1018.9	339.6	679.2	26.4	1.32	81
5	339.6	1018.9	339.6	679.2	26.4	1.32	108
mass ratio	14.12%	42.36%	14.12%	28.24%	1.09%	0.05%	

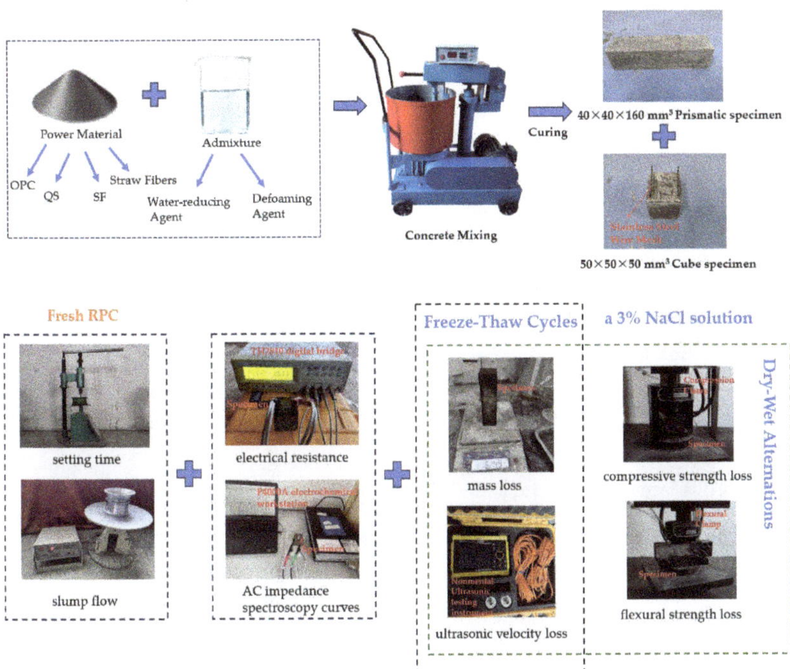

Figure 2. Preparation and experimental procedure of straw fiber RPC.

2.2. *Chloride Salt Environments*

Concrete test blocks were immersed in a 3% NaCl solution for 4 days after a standard curing period of 24 days. Following this, samples were placed into a rapid freeze–thaw machine from Tianjin Gangyuan Test Instrument Factory (Tianjin, China) for conducting the rapid freeze–thaw test. Concrete test blocks were exposed to temperatures spanning from −15 °C to 8 °C within plastic sleeves filled with a 3% NaCl solution during the experiment. Each freeze–thaw cycle extended over a period of 3 h, comprising 2 h of freezing followed by 1 h of thawing. A total of 200 cycles were conducted on the samples during the F-Cs test, with measurements of the mass and ultrasonic velocity taken at 50-cycle intervals. The NaCl D-As were assessed using a cyclic salt spray tester produced by Jiangsu Jiuyi Power Equipment Co., Ltd., Yangzhou, China. Following each five-alternation interval, the mass, ultrasonic velocity, and strength of the specimens were measured. Each cycle of D-As spanned 48 h, commencing with drying specimens free of surface moisture at 80 °C for 36 h, succeeded by a 2 h cooling phase. Subsequently, the cooled specimens underwent a 10 h immersion in NaCl solution prior to measurements, maintaining consistent conditions throughout the brief measurement procedure.

2.3. Measurement Method

2.3.1. Slump Flow and Setting Time Experiments

According to the Chinese standard GB/T 2419-2005 for measuring the fresh RPC slump flow [37], the procedure entailed cleaning the tabletop, tamping bar, and molds with a wet cloth. Subsequently, the fresh RPC was poured in two layers into the testing molds, with each layer tamped 15 times with a rod. Subsequently, the mold was elevated vertically and subjected to 25 vibrations using an electric vibrating table. Subsequently, the maximum and perpendicular diameters were measured and averaged to determine the slump flow. The setting time was determined using a ZKS-100 pointer device. This device was produced by Hebei Xin Test Machine Manufacturing Co., Ltd., Cangzhou, China, and the procedures followed were by the ASTM C191-19 [38]. The cement paste was poured into the mold. The Vicat needle was positioned above the surface and slowly inserted vertically into the paste. The time at which the needle first contacted the surface was recorded as the initial setting time. For the final measurement, the needle was inserted until it could no longer penetrate the paste, and this time was recorded as the final setting time.

2.3.2. Mechanical Property Experiments

The assessment of mechanical strength was conducted using the YAW-300D microcomputer automatic cement compression and folding testing machine. This machine is produced by Jinan Liling Testing Machine Co., Ltd., Jinan, China. Specimens measuring $40 \times 40 \times 160$ mm^3 were subjected to testing for flexural strength, compressive strength, and flexural toughness in accordance with the Chinese standard GB/T17671-1999 [39]. The loading rates employed were 0.05 kN/s for flexural strength and 2.4 kN/s for compressive strength. The displacement at the mid-bottom of the specimen was recorded using the CMT 5205 microcomputer-controlled electronic universal testing machine produced by Zhengzhou Muchen Automation Technology Co., Ltd., Zhengzhou, China, along with a linear variable displacement transducer (LVDT). The loading rate employed was 0.05 mm/min for the flexural toughness. The flexural toughness was determined by integrating the initial segment of the bending force-displacement curve from the peak load point during flexural strength testing. The precise measurement procedures were outlined in Ref. [40]. The methodology employed for the assessment of flexural toughness is illustrated in Figure 3.

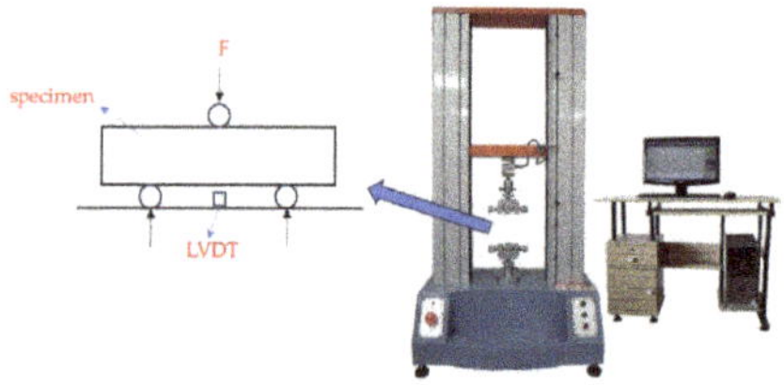

Figure 3. Flexural toughness test of the RPC.

2.3.3. Compactness Experiments

This study utilized the ZBL-U5100 ultrasonic testing instrument. It was manufactured by Beijing Zhibolian Technology Co., Ltd., Beijing, China. According to ASTM C597-1, both sides of the test block were first cleaned, and petroleum jelly was uniformly applied to both sides of the sample to ensure good ultrasonic transmission [41]. Two metal probes were then positioned close to the axis on both sides of the sample, and the propagation time of ultrasonic waves in the concrete was recorded. After all steps were completed, the ultrasonic velocity was calculated based on the measured propagation time.

2.3.4. Electrical Performance Test

According to ASTM C57, the electrical resistance of the sample was tested using a two-electrode method. Two electrodes, composed of 316 L stainless steel mesh with an aperture of 4 mm and spaced 40 mm apart, were connected to the surface of the concrete specimen [42]. The electrical resistance was measured utilizing a TH2810 digital bridge (Changzhou Tonghui Co., Ltd., Changzhou, China) at a constant frequency of 10 kHz and a voltage of 1 V. The AC impedance spectrum was determined using the P4000A electrochemical workstation provided by AMETEK Trading (Shanghai) Co., Ltd., Shanghai, China, which operated between frequencies of 10^5 Hz and 1 Hz and within a voltage range of −10 mV to 10 mV [43,44].

3. Results and Discussions

3.1. Slump Flow and Setting Time Analysis

The impact of straw fibers on the slump flow and setting time of the RPC is illustrated in Figures 4 and 5. As the proportion of straw fiber content increased, the slump flow of the RPC initially decreased, then increased, and subsequently gradually decreased. Compared to the blank specimens, the RPC showed a slump flow reduction of 9.8%~13.1%. This phenomenon can be explained by two main factors. First, with a constant water–cement ratio, the significant water absorption of straw fibers reduces the water available to the cementitious material, thereby decreasing the fluidity of the RPC and, consequently, the slump flow. Second, adding straw fibers results in the formation of a spatial mesh structure within the RPC. The mechanical interlocking of straw fibers and matrix increases the friction within the RPC, reducing its fluidity and consequently the slump flow. However, silica fume forms agglomerates during the mixing process. Following the addition of straw fibers, these agglomerates disintegrate into fine particles, increasing the fluidity of the concrete [45].

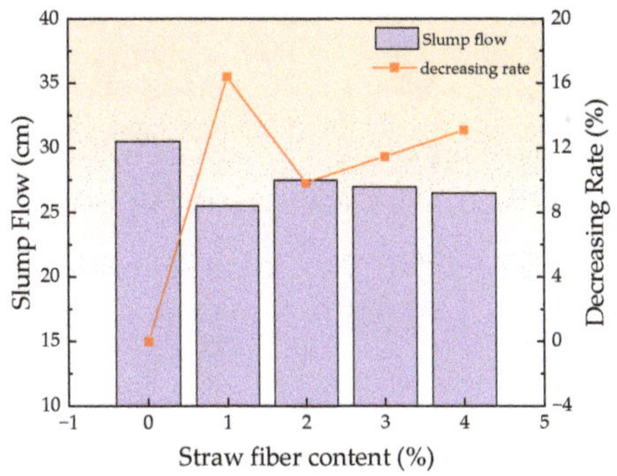

Figure 4. The impact of straw fibers on the slump flow of the RPC.

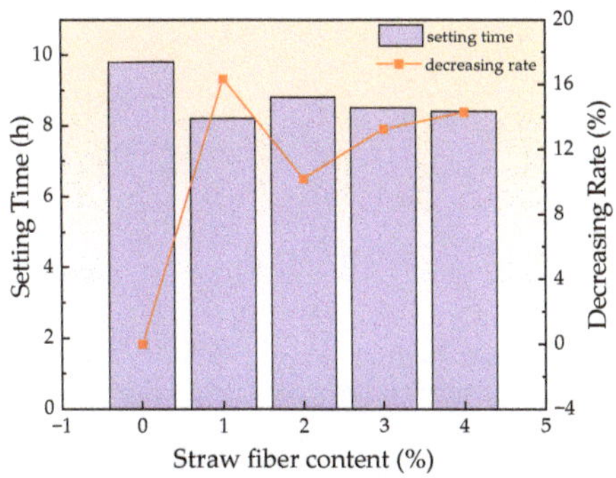

Figure 5. The impact of straw fibers on the setting time of the RPC.

Additionally, incorporating straw fibers has been shown to promote the setting time of the RPC. With an increase in straw fiber content, the setting time initially decreases, then increases, and finally stabilizes. Compared to the blank specimens, the setting time of the RPC decreased by 10.2%~16.3%. This phenomenon can be explained by two main factors. First, the straw fibers have developed a relatively cohesive spatial network structure within the RPC, characterized by substantial mechanical interlocking and frictional forces. This has notably accelerated the initial setting rate of the RPC. Second, the dissolution of hemicellulose and other substances present in straw fibers in alkaline environments prevented the initial setting of cement [46].

3.2. Mass Loss Analysis

Figure 6 depicts the mass loss rates of RPC following various cycles of NaCl D-A and F-C. In the RPC specimens containing 1%, and 3% straw fibers, a gradual increase in the mass loss rate was observed during the dry–wet alternations. Conversely, the RPC specimens doped with 2% and 4% straw fibers initially showed an increase followed by a decrease in mass loss rates during the alternating dry–wet alternations. The concrete spalling observed on the specimen surface was a consequence of the contraction and expansion of the specimen during the alternating wet and dry cycles. Moreover, the initial products were observed to be encapsulated on the surface of the cement particles, and the water did not react sufficiently with the cement particles. This consequently affects the hydration of the concrete. Consequently, the mass of the product is considerably less than that of the spalled material, resulting in a reduction in the overall mass of the specimen. The immersion of the specimen in brine over an extended period of time results in a chemical reaction between the salt and substances such as tricalcium chlorate, leading to the formation of calcium chloroaluminate polycrystalline materials and an increase in specimen mass [47]. During F-Cs, the reduction in specimen mass is attributable to stress-induced fissures caused by moisture freezing, while the increase may be linked to secondary hydration of the cementitious material and crystal formation through chloride salt erosion [48]. After undergoing 10 cycles of D-A and 200 F-Cs using an NaCl solution, the incorporation of straw fibers resulted in decreased mass loss rates of the RPC, ranging approximately from 0.42% to 1.68% and 0.89% to 1.01%.

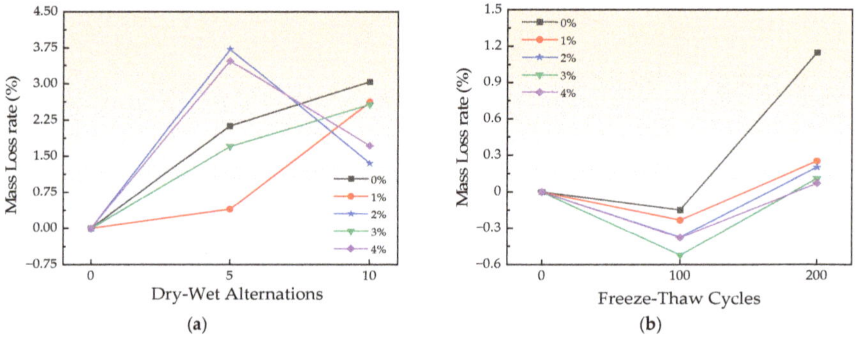

Figure 6. The mass loss rate after different numbers of NaCl D-As and F-Cs: (**a**) D-As; (**b**) F-Cs.

3.3. Ultrasonic Velocity Loss Analysis

Figure 7 depicts the ultrasonic velocity and attenuation rate of the RPC following various cycles of NaCl D-As and F-Cs. Ultrasonic velocity showed a declining trend as fiber content increased during the initial NaCl D-As. This could be attributed to the hollow and porous structure of straw fibers. The inclusion of straw fibers in RPC samples increases the internal porosity, which in turn reduces the density of the samples [49]. In the initial five cycles of NaCl D-As, the RPC specimens without straw fibers exhibited significantly greater ultrasonic velocity loss compared to those with straw fibers. This occurs because even minimal quantities of straw fibers can serve as fillers. Strong chemical bonds and mechanical interlocking are formed by the straw fibers with the cement matrix, thereby enhancing the density of the samples [50]. Between cycles 5 to 10 of D-As, the erosion damage to the specimens tends to stabilize. As the number of F-Cs increased, the ultrasonic velocity of the specimens without straw fibers exhibited a notable initial increase, followed by a subsequent decrease. This pattern mirrors the mass loss under freeze–thaw conditions discussed in Section 3.2. The average velocity trend remains stable across increasing F-Cs when straw content is between 1% and 4%. After 10 NaCl D-As and 200 NaCl F-Cs, the addition of straw fibers led to a reduction in ultrasonic velocity loss rates of the RPC, ranging from 2.69% to 6.73% and 6.68% to 8.9%, respectively. At 2% straw fiber content, the specimens are least affected by chloride salt erosion, and changes in internal density are minimal.

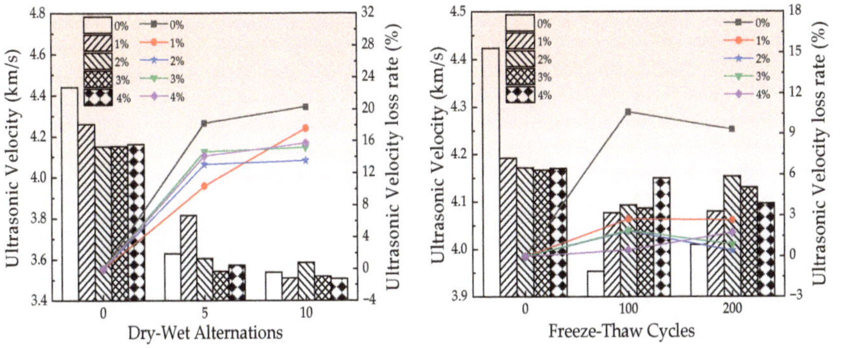

Figure 7. The ultrasonic velocity after different numbers of NaCl D-As and F-Cs: (**a**) D-As; (**b**) F-Cs.

3.4. Mechanical Strength Analysis

Figure 8 presents the RPC samples' mechanical strengths across varying curing times. The figures show that incorporation of an optimal quantity of straw fibers has been demonstrated to enhance the flexural and compressive strengths of the RPC at both the early and late stages of the material's curing. After 7 days of curing, the incorporation of straw fibers resulted in a 7.2% to 26.7% increase in flexural strength of the RPC and a −6.17% to 8.43% increase in compressive strength of the RPC. The addition of straw fibers increased the flexural and compressive strengths of the RPC by 21.3% to 45.76% and −7.16% to 11.62%, respectively, after curing for 28 days. A 2% straw fiber content represents the optimal ratio for achieving maximum flexural and compressive strengths of the RPC. However, increasing the straw fiber content to 3% and 4% led to a reduction in compressive strength of the specimens compared to specimens without straw fibers. This result can be attributed to two factors: Firstly, straw fibers, characterized by a hollow and porous structure, can cause aggregation when present in excessive amounts in concrete. These issues may lead to internal defects in RPC, thereby creating points of weakness [29]. Secondly, straw fibers, with their high water absorption capacity, can absorb excessive free water during the concrete's hydration process. This interference with the concrete's normal hydration can subsequently reduce the compressive strength of the samples [51].

Figure 9 illustrates the cumulative loss of flexural and compressive strengths in RPC specimens with varying straw fiber contents under D-A conditions. Both straw fiber content and the number of D-As considerably influence RPC's strength. As the number of D-A cycles increases, RPC's strength loss initially increases and then stabilizes. After 10 NaCl D-As, adding straw fibers reduced the loss rates of flexural and compressive strengths of the RPC by 9.6%~35.65% and 5.41%~34.88%, respectively, compared to blank samples.

This phenomenon is attributable to two factors: Under D-A conditions, incorporating straw fibers into RPC leads to secondary hydration of the material, resulting in more new cementitious gel formation. Additionally, stresses generated by the alternating dry and wet conditions, due to drying shrinkage and temperature variations, lead to the enlargement of internal pores within the samples. These horizontal stresses are effectively counteracted by straw fibers, which improve the structural stability and significantly reinforce the samples. According to M.S. Ammari et al., straw displays remarkable ductility, which further enhances the mechanical properties of concrete [52]. Following 10 NaCl D-As, RPC specimens with 2% straw fibers showed the lowest levels of flexural and compressive strength loss. Figure 10 displays the images of specimens both before and after exposure to chloride salt erosion under experimental loads, which led to their failure. The failure mode in the RPC specimens experiencing flexural failure was characterized by complete fracture. Compared to specimens containing 2% fiber, those without fibers exhibited significantly larger crack widths. This observation clearly demonstrates that the incorporation of straw fibers inhibits concrete fracturing and spalling.

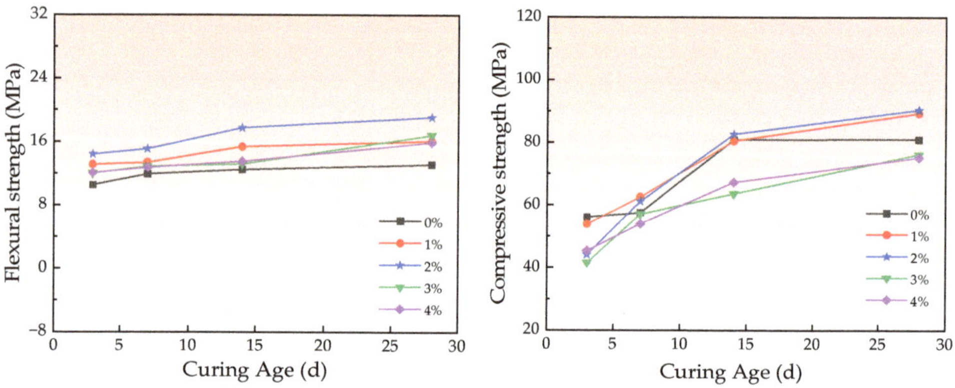

(a) (b)

Figure 8. (**a**) Flexural strength of the RPC over different curing periods; (**b**) compressive strength of the RPC contents over different curing periods.

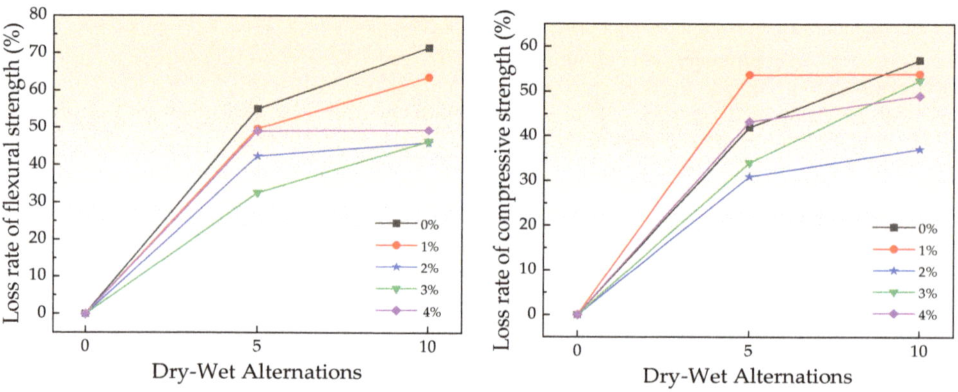

(a) (b)

Figure 9. (**a**) The cumulative loss of flexural strength of the RPC under D-As; (**b**) the cumulative loss of compressive strength of the RPC under D-As.

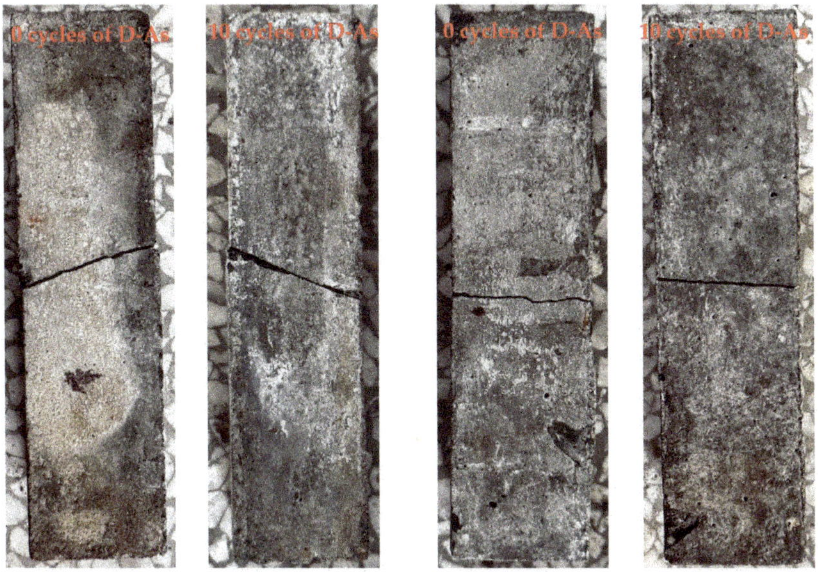

Figure 10. Images of specimens failing under experimental loading before and after chloride salt attack.

3.5. Flexural Toughness Analysis

Figure 11 illustrates the flexural load and displacement curves of the RPC at different curing ages. The graph shows that during the RPC flexural strength test, the force–displacement curve initially developed horizontally. As displacement increased, the force on the RPC specimen gradually increased until reaching the maximum flexural failure load. Ultimately, the specimen abruptly fractured and the force dropped to zero. Although the basic trend of each group's test curve was similar, the curves exhibited fluctuations. This phenomenon can be attributed to the rough edges of the RPC specimens and incomplete contact with the loading machine, resulting in uneven force distribution during the testing process, resembling peak-like patterns [53].

Figure 12 illustrates the impact of curing age on the flexural toughness of the RPC. The graph shows that flexural toughness initially increased and then decreased as straw fiber content increased. After 7 days of curing, straw fibers increased flexural toughness of the RPC by 3.36%~13.6%. Following 28 days of curing, the flexural toughness of the test blocks reached its peak at a 2% fiber content, exhibiting a 32.7% increase compared to the control group. Straw fibers' bridging effect in concrete can enhance the material's toughness and tensile strength. Furthermore, straw fibers can absorb and disperse stress around cracks, delaying crack propagation and enhancing crack resistance. However, increasing the straw fibers may cause excessive internal voids and defects in the concrete, resulting in lower compressive strength of the concrete. This can result in early deterioration of the concrete under external loads, leading to decreased concrete toughness [54].

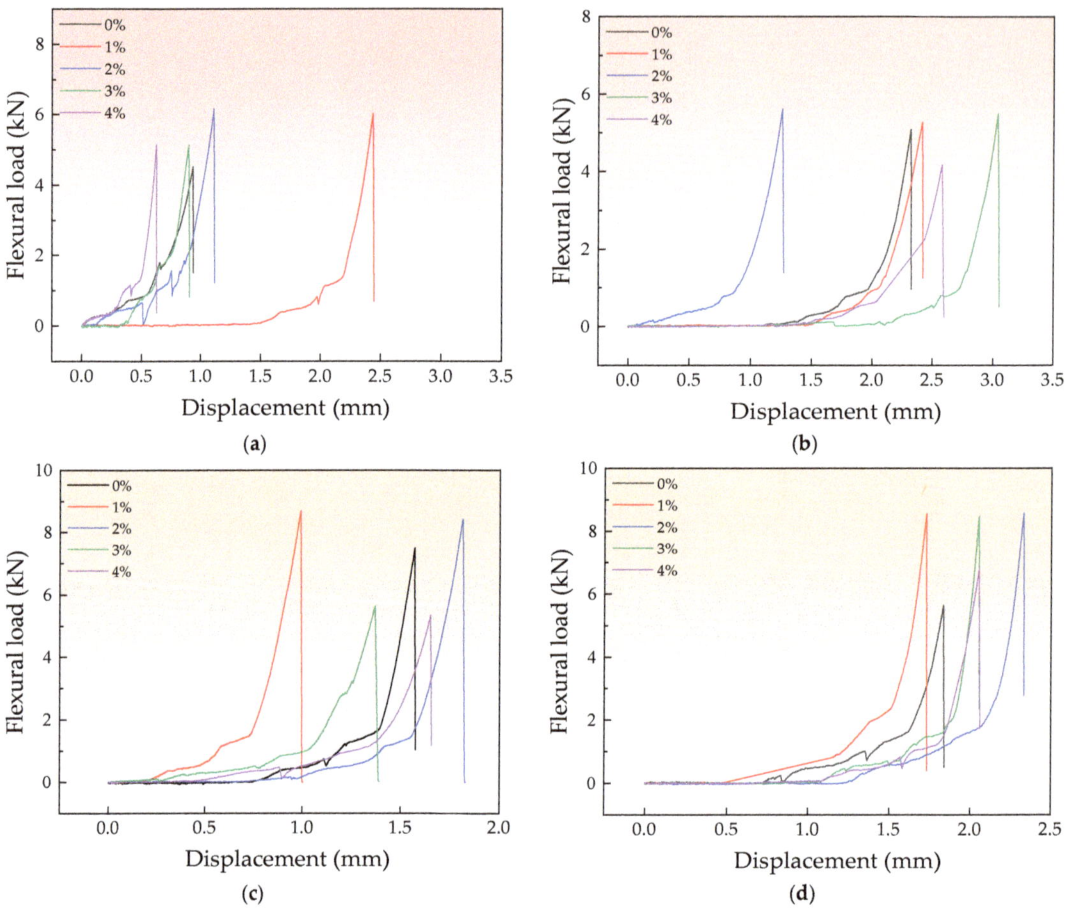

Figure 11. Flexural load and displacement of the RPC at different curing ages: (**a**) 3 d; (**b**) 7 d; (**c**) 14 d; (**d**) 28 d.

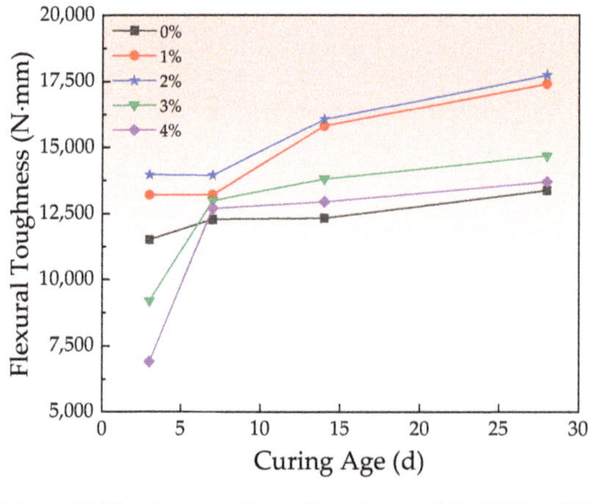

Figure 12. The change in flexural toughness of the RPC at different curing ages.

3.6. Electrochemical Performance Analysis

Figure 13 displays the electrochemical impedance spectroscopy of the RPC cured for 28 days with varying straw fiber contents. The horizontal axis represents the real part, Zr, which represents the AC electrical resistance; the vertical axis represents the imaginary part, -Zi, which represents the capacitance [55]. The AC impedance spectral curve for the specimen's imaginary part first increases, then decreases, aligning with the real part—typical behavior for the EIS of gelled composites. Each curve exhibits an inflection point, where the AC electrical resistance value can, to some extent, indicate the erosion state of the RPC. Additionally, scattered points on the curves arise from polarization and electrophoresis phenomena [56]. The results indicate that the addition of straw fibers could increase the AC electrical resistance of the RPC by rates of 10.17%~58.1%. The maximum AC electrical resistance was observed at a fiber content of 4%.

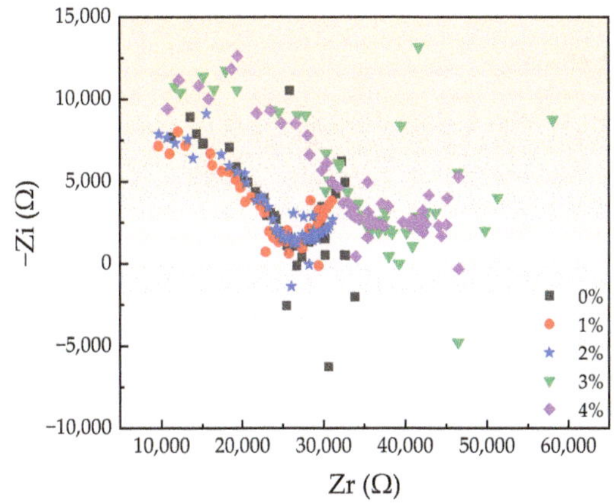

Figure 13. Electrochemical AC impedance spectra of the RPC.

Figure 14 displays the equivalent circuit diagram of the RPC specimen. As depicted, the electrical characteristics of the samples adhered to series conduction models. These models encompassed parallel electrical resistance and capacitance across the pore solution, RPC matrix, and straw fibers, interconnected through electrode contact resistance with the specimens. A comprehensive examination of equivalent circuit diagrams can be found in the studies conducted by Wang et al. [57].

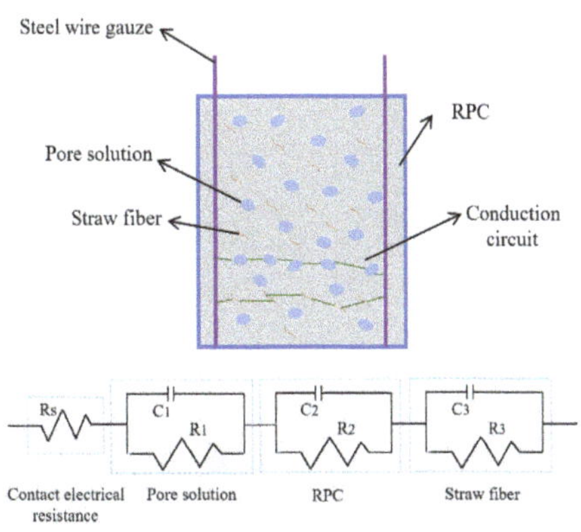

Figure 14. The equivalent circuit diagram for electrochemical properties in the RPC.

Figure 15 depicts the fitting curves of electrical resistance in the RPC with varying straw fiber content throughout the curing period. The electrical resistance of the RPC increased exponentially with curing age, provided the straw fiber content remained constant. The functional relationship can be derived by fitting the data, as demonstrated in Table 3. This phenomenon is attributable to the consumption of pore solution and ions during cement hydration, reducing RPC's conductivity. Cement hydration products filling RPC pores lead to reduced pore size and connectivity, further diminishing conductivity. Only the 1% straw fiber specimens had higher resistance values than the blank specimens. This phenomenon arises from the water absorption characteristics of straw fibers, which decrease the pore solution volume and thereby augment electrical resistance. Moreover, the ion-rich water in straw fibers can create a conductive path, further reducing RPC's electrical resistance [58].

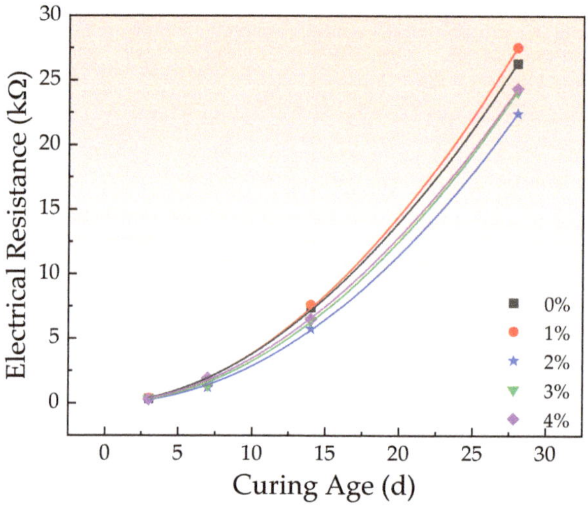

Figure 15. The fitting curve of the RPC electrical resistance with curing age.

Table 3. The fitting function of the RPC electrical resistance with curing age.

Fitting Function	$R = a \cdot t^b$				
	0%	**1%**	**2%**	**3%**	**4%**
A	0.05	0.05	0.03	0.04	0.05
B	1.87	1.91	1.99	1.95	1.88
R^2	0.99	0.99	0.99	0.99	0.99

Figure 16 shows the fitted curves for the mechanical strength and electrical resistance of the RPC. From the figure, it is evident that the flexural and compressive strengths of the RPC rose with the augmentation of electrical resistance. This phenomenon can be attributed to two primary factors: Initially, as RPC hydrates and undergoes curing, the density and bonding strength between the straw fibers and the matrix enhance, consequently boosting the specimen's strength and ductility. In addition, RPC electrical resistance is positively correlated with curing age. This leads to an increase in electrical resistance and an increase in mechanical strengths. Second, the addition of suitable fibers improves the homogeneity of the RPC, increases the electrical conductivity of the material, and enhances fiber bridging and mechanical properties. However, this can result in decreased mechanical strengths of concrete and a rise in the electrical resistance of concrete [59]. The functional relationship can be derived by fitting the data, as demonstrated in Tables 4 and 5.

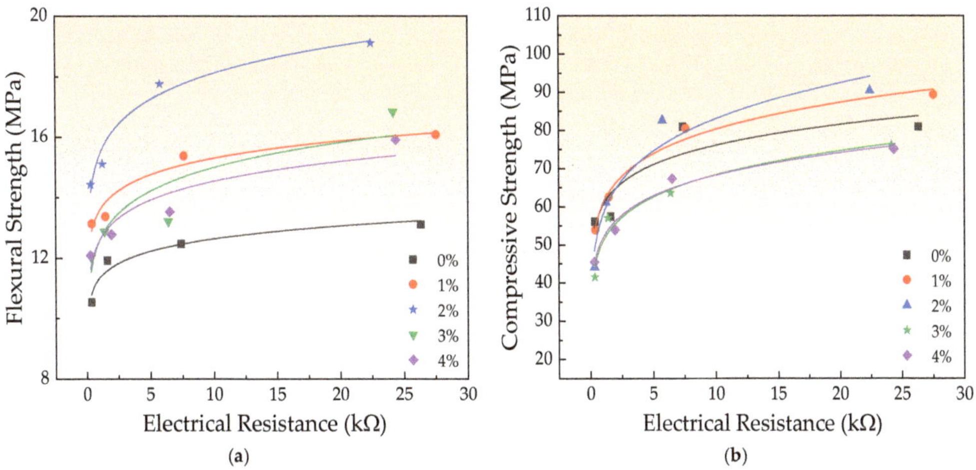

Figure 16. (**a**) The fitting curve of flexural strength of the RPC with electrical resistance; (**b**) the fitting curve of the compressive strength of the RPC with electrical resistance.

Table 4. The fitting function of flexural strength of the RPC with electrical resistance.

Fitting Function	$f_b = a \cdot R^b$				
	0%	1%	2%	3%	4%
a	11.32	13.56	15.39	12.59	12.58
b	0.047	0.05	0.07	0.08	0.06
R^2	0.94	0.94	0.95	0.81	0.88

Table 5. The fitting function of the compressive strength of the RPC with electrical resistance.

Fitting Function	$f_c = a \cdot R^b$				
	0%	1%	2%	3%	4%
a	60.44	60.85	58.01	50.92	51.89
b	0.10	0.12	0.15	0.12	0.12
R^2	0.83	0.98	0.94	0.96	0.98

4. Conclusions

In this experiment, the macroscopic properties of straw fiber RPC before and after chloride salt erosion were investigated. The primary findings of the investigation are outlined below:

Incorporating straw fibers led to decreases in both the slump flow and setting time of fresh RPC by 9.8% to 13.1% and 10.2% to 16.3%, respectively.

The electrical characteristics of straw fiber RPC were simulated by employing series conduction models, which include parallel electrical resistance and capacitance within the pore solution, RPC matrix, and straw fibers. These components were interconnected via contact resistance between the electrodes and the specimens. The addition of straw fibers could augment the AC electrical resistance of the RPC by 10.17%~58.1%. A power function relationship was identified between the electrical resistance of straw fiber RPC and its flexural and compressive strengths.

After 7 days of curing, the incorporation of straw fibers increased flexural strength, compressive strength, and flexural toughness of the RPC by 7.2%~26.7%, −6.17%~8.43%, and 3.36%~13.6%. After 28 days of curing, the incorporation of straw fibers increased flexural strength, compressive strength, and flexural toughness of the RPC by 21.3%~45.76%,

−7.16%~11.62%, and 2.4%~32.7%. The best mechanical properties were obtained with a fiber content of 1% and 2%. Following 10 NaCl dry–wet cycles, the incorporation of straw fibers reduced the loss rates of flexural and compressive strength of the RPC by 9.6%~35.65% and 5.41%~34.88%, respectively, compared to the blank samples.

After subjecting the specimens to 10 cycles of D-A and 200 F-Cs in an NaCl solution, the incorporation of straw fibers resulted in decreased mass loss rates of the RPC by 0.42%~1.68% and 0.89%~1.01%. Similarly, the addition of straw fibers led to reduced rates of ultrasonic velocity loss of the RPC after the same exposure conditions by 2.69%~6.73% and 6.68%~8.9%.

According to the above experimental results, the optimum straw fiber incorporation content in RPC is 1%–2% in a chloride salt erosion environment. This study establishes a characterization relationship between electrical resistance and mechanical strength, which will also contribute to the development of specialized methods for predicting future performance.

Fuzzy logic excels at handling ambiguity and imprecise information; however, it may encounter challenges when faced with extreme conditions or high-dimensional data. Therefore, prior to the experiments, we established hypotheses regarding issues such as consistency of materials, loading conditions, and age-related factors to design our experimental plan. Similarly, while Bayesian networks are powerful in inferring complex dependencies, their effectiveness is highly contingent on the choice of prior probabilities and the quality of the data. Discrepancies between the samples produced in the laboratory and the complexities present in actual engineering scenarios may lead to inaccuracies in the experimental results.

Author Contributions: Conceptualization, H.W. (Hui Wang); methodology, H.W. (Hangyang Wang); software, H.W. (Hangyang Wang); validation, B.C. and K.G.; formal analysis, H.W. (Hangyang Wang); investigation, H.W. (Hangyang Wang) and H.W. (Hui Wang); resources, B.C. and B.X.; data curation, H.W. (Hui Wang); writing—original draft preparation, H.W. (Hangyang Wang); writing—review and editing, H.W. (Hangyang Wang), X.P., K.G. and H.W. (Hui Wang); visualization, H.W. (Hangyang Wang) and H.W. (Hui Wang); supervision, K.G., X.P. and H.W. (Hui Wang); project administration, K.G., X.P. and H.W. (Hui Wang); funding acquisition, H.W. (Hui Wang) and X.P. All authors have read and agreed to the published version of the manuscript.

Funding: This research was supported by Zhejiang Provincial Natural Science Foundation of China (No. LY22E080005, NO. LY24E080010).

Institutional Review Board Statement: Not applicable.

Informed Consent Statement: Not applicable.

Data Availability Statement: The data used to support the findings of this study are available on request.

Conflicts of Interest: Author Bingling Cheng was employed by Ningbo Yonghuan Yuan Environmental Protection Engineering Technology Co., Ltd. Authors Xi Peng and Bin Xu were employed by Ningbo Roaby Technology Industrial Group Co., Ltd. The remaining authors declare that the research was conducted in the absence of any commercial or financial relationships that could be construed as a potential conflict of interest.

References

1. Yusuke, K.; Masaki, M.; Ryuji, N. Verification test of restoration effect of offshore concrete structure using permanent formwor panels and underwater inseparable mortar. *Int. J. Civ. Eng.* **2021**, *19*, 1111–1124.
2. Li, B.; Chen, Z.; Wang, S.; Xu, L. A Review on the Damage Behavior and Constitutive Model of Fiber Reinforced Concrete a Ambient Temperature. *Constr. Build. Mater.* **2024**, *412*, 134919.
3. Wang, Y.; Qiao, P.; Sun, J.; Chen, A. Influence of Fibers on Tensile Behavior of Ultra-High Performance Concrete: A Review. *Cons Build. Mater.* **2024**, *430*, 136432.
4. Cao, Z.; Wang, K.; Peng, X.; Wang, H.; Huang, R. Influence of NaCl Solution External Erosion on Corrosion Resistance of RP Reinforced with Straw Fiber. *Coatings* **2023**, *13*, 1308. [CrossRef]
5. Grzymski, F.; Musiał, M.; Trapko, T. Mechanical Properties of Fibre Reinforced Concrete with Recycled Fibres. *Constr. Buil Mater.* **2019**, *198*, 323–331.

6. Esaker, M.; Thermou, G.E.; Neves, L. Impact Resistance of Concrete and Fibre-Reinforced Concrete: A Review. *Int. J. Impact Eng.* **2023**, *180*, 104722.
7. Allam, H.; Duplan, F.; Amziane, S.; Burtschell, Y. Assessment of Manufacturing Process Efficiency in the Dispersion of Carbon Fibers in Smart Concrete by Measuring AC Impedance. *Cem. Concr. Compos.* **2022**, *127*, 104394.
8. Cao, X.; Quan, Y.; Ren, Y.; Fu, F.; Jin, Q.; He, D.; Zheng, Y. Experiment study on reactive powder concrete beams using spirals reinforcement under torsion. *Eng. Struct.* **2023**, *290*, 116361.
9. Ge, W.; Zhang, Z.; Ashour, A.; Li, W.; Jiang, H.; Hu, Y.; Shuai, H.; Sun, C.; Li, S.; Liu, Y. Hydration characteristics, hydration products and microstructure of reactive powder concrete. *J. Build. Eng.* **2023**, *69*, 106306.
10. Liang, Z.; Peng, X.; Wang, H. The Influence of Aspect Ratio of Steel Fibers on the Conductive and Mechanical Properties of Compound Cement Reactive Powder Concrete. *Coatings* **2023**, *13*, 331. [CrossRef]
11. Alharbi, Y.R.; Abadel, A.A.; Salah, A.A.; Mayhoub, O.A.; Kohail, M. Engineering properties of alkali-activated materials reactive powder concrete. *Constr. Build. Mater.* **2021**, *271*, 121550.
12. Hou, X.; Wang, N.; He, T.; Chen, C. Compressive Stress-Strain Relationship of Steam Free Reactive Powder Concrete at Ultra-Low Temperatures. *Cem. Concr. Compos.* **2024**, *152*, 105655.
13. Mayhoub, O.A.; Nasr, E.S.A.; Ali, Y.A.; Kohail, M. The influence of ingredients on the properties of reactive powder concrete: A review. *Ain Shams Eng. J.* **2021**, *12*, 145–158.
14. Richard, P.; Cheyrezy, M. Composition of reactive powder concretes. *Cem. Concr. Res.* **1995**, *25*, 1501–1511.
15. Shafieifar, M.; Farzad, M.; Azizinamini, A. New connection detail to connect precast column to cap beam using ultra-high-performance concrete in accelerated bridge construction applications. *Transp. Res. Rec.* **2018**, *2672*, 207–220.
16. Duan, Y.; Lei, H.; Zhou, Y.; Jin, S. Experimental Study on Bearing Capacity of Compression Members of Space Grid Structures Reinforced by RPC. *Appl. Sci.* **2022**, *12*, 7809. [CrossRef]
17. Ren, P.; Hou, X.; Kodur, V.K.R.; Ge, C.; Zhao, Y.; Zhou, W. Modeling the fire response of reactive powder concrete beams with due consideration to explosive spalling. *Constr. Build. Mater.* **2021**, *301*, 124094.
18. Lai, Z.; Yao, P.; Huang, W.; Chen, B.; Ying, Z. Reactive powder concrete-filled steel tube (RPCFT) members subjected to axial tension: Experimental study and design. *Structures* **2020**, *28*, 933–942.
19. Blunt, J.; Jen, G.; Ostertag, C.P. Enhancing corrosion resistance of reinforced concrete structures with hybrid fiber reinforced concrete. *Corros. Sci.* **2015**, *92*, 182–191.
20. PR, K.R.; Mathangi, D.P.; Sudha, C.; Neelamegam, M. Experimental investigation of reactive powder concrete exposed to elevated temperatures. *Constr. Build. Mater.* **2020**, *261*, 119593.
21. Hiremath, P.; Yaragal, S.C. Performance of polypropelene and polyester fibres-reinforced reactive powder concretes at elevated temperatures. *Constr. Build. Mater.* **2023**, *373*, 130862.
22. Wang, H.; Shi, F.; Shen, J.; Zhang, A.; Zhang, L.; Huang, H.; Liu, J.; Jin, K.; Feng, L.; Tang, Z. Research on the Self-Sensing and Mechanical Properties of Aligned Stainless Steel Fiber-Reinforced Reactive Powder Concrete. *Cem. Concr. Compos.* **2021**, *119*, 104001.
23. Branston, J.; Das, S.; Kenno, S.Y.; Taylor, C. Mechanical Behaviour of Basalt Fibre Reinforced Concrete. *Constr. Build. Mater.* **2016**, *124*, 878–886.
24. Zhang, G.Z.; Liu, C.; Cheng, P.F.; Li, Z.; Han, Y.; Wang, X.Y. Enhancing the Interfacial Compatibility and Self-Healing Performance of Microbial Mortars by Nano-SiO_2-Modified Basalt Fibers. *Cem. Concr. Compos.* **2024**, *152*, 105650.
25. Le, D.L.; Salomone, R.; Nguyen, Q.T. Circular Bio-Based Building Materials: A Literature Review of Case Studies and Sustainability Assessment Methods. *Build. Environ.* **2023**, *244*, 110774.
26. Le, D.L.; Salomone, R.; Nguyen, Q.T.; Versele, A.; Piccardo, C. Status and Barriers to Circular Bio-Based Building Material Adoption in Developed Economies: The Case of Flanders, Belgium. *J. Environ. Manag.* **2024**, *367*, 121965.
27. Singh, A.K.; Bedi, R.; Khajuria, A. A review of composite materials based on rice straw and future trends for sustainable composites. *J. Clean. Prod.* **2024**, *457*, 142417.
28. Wu, B.; Zhang, S.; Zhao, Y. Dynamic compressive properties of rubberized mortar reinforced with straw-fiber additive. *Constr. Build. Mater.* **2023**, *408*, 133679.
29. Basta, A.H.; Lotfy, V.F.; Shafik, E.S. Synergistic Valorization of Rice Straw and Red Brick Demolition in development performance of Lightweight Cement Mortars. *J. Build. Eng.* **2024**, *92*, 109769.
30. Ahmad, J.; Mohammed, J.; Tayyab, N.M. Improvement in the strength of concrete reinforced with agriculture fibers: Assessment on mechanical properties and microstructure analysis. *J. Eng. Fibers Fabr.* **2024**, *19*, 1226480.
31. Wu, Z.; Wang, X.; Chen, Z. Experimental study on Corn Straw Fiber (CSF) toughening EPS concrete. *Constr. Build. Mater.* **2024**, *429*, 136325.
32. Fu, Z.; Wang, X.; Yang, W.; Ren, J.; Dong, B.; Xing, F. Enhanced Repetitive Self-Healing of Cementitious Materials through Modified Biochar and Optimised Superabsorbent Polymer. *Constr. Build. Mater.* **2024**, *443*, 137716.
33. Alioui, A.; Kaitouni, S.I.; Azalam, Y.; Bendada, E.M.; Mabrouki, M. Effect of Straw Fibers Addition on Hygrothermal and Mechanical Properties of Carbon-Free Adobe Bricks: From Material to Building Scale in a Semi-Arid Climate. *Build. Environ.* **2024**, *255*, 111380.
34. Nian, T.; Li, P.; Ge, J.; Song, J.; Wang, M. Green Environmental Protection and Sustainable Utilization of Straw: Investigation for Molecular Dynamics of Highland Barley Straw Fiber to Enhancing Road Performance of Asphalt. *J. Clean. Prod.* **2024**, *452*, 141940.

35. Jin, Z.; Mao, S.; Zheng, Y.; Liang, K. Pre-Treated Corn Straw Fiber for Fiber-Reinforced Concrete Preparation with High Resistanc to Chloride Ions Corrosion. *Case Stud. Constr. Mater.* **2023**, *19*, e02368.
36. Wu, H.; Shen, A.; Cheng, Q.; Cai, Y.; Ren, G.; Pan, H.; Deng, S. A review of recent developments in application of plant fibers a reinforcements in concrete. *J. Clean. Prod.* **2023**, *419*, 138265.
37. Zhang, L.; Huang, M.; Yang, F.; Zhang, W. A Novel Hydrophilic Modification Method of EPS Particles: Conception Design an Performances in Concrete. *Cem. Concr. Comp.* **2023**, *142*, 105199.
38. Ma, J.; Yuan, H.; Zhang, J.; Zhang, P. Enhancing Concrete Performance: A Comprehensive Review of Hybrid Fiber Reinforce Concrete. *Structures* **2024**, *64*, 106560.
39. Luo, J.; Quan, Z.; Shao, X.; Li, F.; He, S. Mechanical Performance of RPC and Steel–RPC Composite Structure with Different Fibe Parameters: Experimental and Theoretical Research. *Polymers* **2022**, *14*, 1933. [CrossRef] [PubMed]
40. Shi, X.; Ning, B.; Wang, J.; Cui, T.; Zhong, M. Improving Flexural Toughness of Foamed Concrete by Mixing Polyvinyl Alcoho Polypropylene Fibers: An Experimental Study. *Constr. Build. Mater.* **2023**, *400*, 132689.
41. Bortoletto, M.; Sanches, A.O.; Santos, J.A.; da Silva, R.G.; Tashima, M.M.; Payá, J.; Soriano, L.; Borrachero, M.V.; Malmonge, J.A Akasaki, J.L. New Insights on Understanding the Portland Cement Hydration Using Electrical Impedance Spectroscopy. *Const Build. Mater.* **2023**, *407*, 133566.
42. Meng, Z.; Liu, Q.; She, W.; Cai, Y.; Yang, J.; Farjad Iqbal, M. Electrochemical Deposition Method for Load-Induced Crack Repair o Reinforced Concrete Structures: A Numerical Study. *Eng. Struct.* **2021**, *246*, 112903.
43. Ojala, T.; Ahmed, H.; Kuusela, P.; Seppänen, A.; Punkki, J. Monitoring of Concrete Segregation Using AC Impedance Spectroscop *Constr. Build. Mater.* **2023**, *384*, 131453.
44. de Almeida, E.F., Jr.; Martini, S. Measurements of Electrical Impedance in Cementitious Mortars: Influence of Electrodes an Physical Dimensions of Specimens. *Case Stud. Constr. Mater.* **2023**, *18*, e01880.
45. Gong, K.; Liang, Z.; Peng, X.; Wang, H. Research into Preparation and Performance of Fast-Hardening RPC Mixed with Strav *Materials* **2023**, *16*, 5310. [CrossRef] [PubMed]
46. Liu, J.; Xie, X.; Li, L. Experimental Study on Mechanical Properties and Durability of Grafted Nano-SiO_2 Modified Rice Strav Fiber Reinforced Concrete. *Constr. Build. Mater.* **2022**, *347*, 128575.
47. Tai, Y.; Yang, L.; Gao, D.; Kang, K.; Cao, Z.; Zhao, P. Properties Evolution and Deterioration Mechanism of Steel Fiber Reinforce Concrete (SFRC) under the Coupling Effect of Carbonation and Chloride Attack. *J. Build. Eng.* **2024**, *95*, 110275.
48. Tu, Y.; Liu, D.; Yuan, L.; Zhang, Y. Corrosion Resistance of Concrete Strengthened with Fibre-Reinforced Polymer Sheets. *Ma Concr. Res.* **2022**, *74*, 54–69.
49. Wang, H.; Jin, K.; Zhang, A.; Zhang, L.; Han, Y.; Liu, J.; Shi, F.; Feng, L. External Erosion of Sodium Chloride on the Degradatio of Self-Sensing and Mechanical Properties of Aligned Stainless Steel Fiber Reinforced Reactive Powder Concrete. *Constr. Buil Mater.* **2021**, *287*, 123028.
50. Marcos-Meson, V.; Michel, A.; Solgaard, A.; Fischer, G.; Edvardsen, C.; Skovhus, T.L. Corrosion Resistance of Steel Fibr Reinforced Concrete-A Literature Review. *Cem. Concr. Res.* **2018**, *103*, 1–20.
51. Abolhasani, A.; Nazarpour, H.; Dehestani, M. The Fracture Behavior and Microstructure of Calcium Aluminate Cement Concret with Various Water-Cement Ratios. *Theor. Appl. Fract. Mech.* **2020**, *109*, 102690.
52. Ammari, M.S.; Belhadj, B.; Bederina, M.; Ferhat, A.; Quéneudec, M. Contribution of Hybrid Fibers on the Improvement of San Concrete Properties: Barley Straws Treated with Hot Water and Steel Fibers. *Constr. Build. Mater.* **2020**, *233*, 117374.
53. Su, Q.; Xu, J. Mechanical Properties of Concrete Containing Glass Sand and Rice Husk Ash. *Constr. Build. Mater.* **2023**, *39* 132053.
54. Zhang, Y.; Wang, J.; Wang, J.; Qian, X. Preparation, Mechanics and Self-Sensing Performance of Sprayed Reactive Powde Concrete. *Sci. Rep.* **2022**, *12*, 7787.
55. Miyazaki, Y.; Nakayama, R.; Yasuo, N.; Watanabe, Y.; Shimizu, R.; Packwood, D.M.; Nishio, K.; Ando, Y.; Sekijima, M.; Hitosug T. Bayesian Statistics-Based Analysis of AC Impedance Spectra. *AIP Adv.* **2020**, *10*, 045231.
56. Valcuende, M.; Lliso-Ferrando, J.R.; Ramón-Zamora, J.E.; Soto, J. Corrosion Resistance of Ultra-High Performance Fibre Reinforced Concrete. *Constr. Build. Mater.* **2021**, *306*, 124914.
57. Wang, H.; Du, T.; Zhang, A.; Cao, P.; Zhang, L.; Gao, X.; Liu, J.; Shi, F.; He, Z. Relationship between Electrical Resistance an Rheological Parameters of Fresh Cement Slurry. *Constr. Build. Mater.* **2020**, *256*, 119479.
58. Xu, Z.; Cao, P.; Wang, D.; Wang, H. The Corrosion Resistance of Reinforced Magnesium Phosphate Cement Reactive Powde Concrete. *Materials* **2022**, *15*, 5692. [CrossRef] [PubMed]
59. Wang, H.; Cai, X.; Rao, C.; Wang, K.; Wang, J. Mechanical and Electrical Properties of Rapid-Strength Reactive Powder Concret with Assembly Unit of Sulphoaluminate Cement and Ordinary Portland Cement. *Materials* **2022**, *15*, 3371. [CrossRef]

coatings

MDPI

rticle

The Relationship between the Fresh Sludge Ceramsite Concrete's Fluidity and the Sludge Ceramsite's Dispersion

ehan Yu [1,2], Bing Xiao [3], Zihao Cao [1,2], Bingling Cheng [4], Xi Peng [2,5,*] and Hui Wang [1,*]

1 School of Civil Engineering and Geographical Environment, Ningbo University, Ningbo 315000, China; yuyehan0127@163.com (Y.Y.); 489471783@aliyun.com (Z.C.)
2 School of Civil Transportation Engineering, Ningbo University of Technology, Ningbo 315211, China
3 Road & Bridge International Co., Ltd., Beijing 100084, China; 170057398@aliyun.com
4 Ningbo Yonghuan Yuan Environmental Protection Engineering Technology Co., Ltd., Ningbo 315000, China; 263454376@aliyun.com
5 Ningbo Roaby Technology Industrial Group Co., Ltd., Ningbo 315800, China
* Correspondence: pengxi@nbut.edu.cn (X.P.); huiwang123@aliyun.com (H.W.)

Abstract: Sludge ceramsite (SC) can be utilized as a lightweight aggregate in concrete, especially in external wall materials, due to the increasing volume of polluted sludge, which contributes to water system deterioration and poses greater threats to human health. The influence of the fresh mortar's slump flow on the dispersion of ceramsite was studied. The ultrasonic sound velocity, capillary water absorption rate, compressive strength, and coefficient of variation (CV) were measured in this study. Thermogravimetric (TG) analysis, ultra depth-of-field microscope scanning, X-ray diffraction (XRD), scanning electron microscopy (SEM), and energy dispersive spectrometry (EDS) were used to analyze the performance mechanism of the ceramsite concrete. The results indicated that adding SC could reduce the fluidity of the fresh concrete, with a reduction by rates of up to 2.04%. The addition of WRA could improve the fluidity by rates of up to 60.77%. The relationship between the ultrasonic sound speed and the increasing fluidity could be deduced as a negative correlation. The water absorption was negatively correlated with the compressive strength. The concrete with a slump flow of 12.35 and 12.5 cm reached the maximum compressive strength, which had the lowest water absorption, and demonstrated internal homogeneity. The optimum slump flow was 12.35 and 12.5 cm. With the slump flow of 12.5 cm, the corresponding CV was the lowest, showing the optimum SC's dispersion. Through TG, XRD, and SEM analyses, it was verified that the addition of 0.6% WRA promoted the hydration of cement. In addition, SC increased the hydration products.

Keywords: sludge ceramsite; dispersion; slump flow; compressive strength; X-ray diffraction

1. Introduction

Due to industrial discharges, the amounts of industrial pollutants and contaminated river sludge are increasing [1,2]. In the whole world, the annual municipal sludge production has been expected to reach 100 million tons by 2024 [3,4]. Sludge accumulation in rivers severely affects the entire water system [5]. It can be used as a raw material to produce industrial ceramsite [6,7]. In this way, it can provide a solution to the problem of waste disposal and also offer economic benefits, achieving low-cost resource utilization of sludge [8]. Through technological innovation and market development, the production of sludge ceramsite (SC) has great economic potential [9].

SC can serve as lightweight aggregate in concrete, offering benefits such as reduced self-weight, low thermal conductivity, good hard film properties, and convenient construction [10,11]. It can be widely used in roof slopes, thermal insulation layers, non-load-bearing wall sound insulation, and so on [12]. It should be mentioned that this ceramsite is different from lightweight expanded clay aggregate (LECA). LECA is usually made by high-temperature roasting and swelling, while the production process for sludge ceramsite includes raw material pretreatment, batching, molding, pre-burning, roasting, cooling, screening and other steps [13]. Compared with LECA, it has a denser structure and more active properties [14].

At home and abroad, scholars have performed a lot of experiments on the preparation and performance of ceramsite concrete using waste materials such as sludge, paper and waste gangue. Xie et al. [15] used papermaking sludge to prepare expanded ceramsite and invented a green self-insulation wall block with ceramsite. The wall block shows a thermal conductivity of 0.39 W/(M^2·K). Wang et al. [16] found that when the gangue ceramsite was used as coarse aggregate with a 40% substitution rate, the compressive strength could be up to 50.8 MPa, which was only 1.7 MPa lower than ordinary concrete [17,18]. A study by Li et al. [19] reported that the lightweight ceramsite concrete made from paper sludge could enhance the compressive strength by 53%. The cement hydration products can enter the interior of ceramsite, which promotes the close combination of ceramsite and cementitious materials [20].

At present, the use of ceramsite concrete still faces several challenges. Firstly, the density difference between ceramsite and cement mortar causes the ceramsite to float, resulting in the separation of the mortar layers [21]. Secondly, during the mixing process, the ceramsite absorbs water, causing a decrease in the slump flow [22]. Ultimately, it affects the bearing capacity, fluidity, and durability [23]. Bu et al. [24] found that with the increase in the rubber ceramsite content, the compressive strength of rubber ceramsite concrete increased first and then decreased. The water absorption was negatively correlated with the compressive strength. To solve this problem, Wang et al. [25] discovered that the segregation degree of sludge ceramsite, which was crushed, coated with ash, and pre-wetted for 1.0–2.0 h before pouring, could be reduced to approximately 2%. Wang et al. [26,27] showed that the coefficient of variation could be used to characterize the dispersion of the internal filler. This has certain reference significance for our research on the dispersion of sludge ceramsite. Although there has been some research on sludge ceramsite, the corresponding mechanism of fresh mixed ceramsite concrete mortar has not been reported. In particular, the dispersion of sludge ceramsite at different slump flows has not been investigated.

This study aims to evaluate the influence of the water-reducing agent (WRA) content (0%~1.2% by the mass of cement) and SC content (10% and 20% by the total mass of all materials) on the fluidity of ceramsite concrete. In this paper, the influence of fluidity on ultrasonic velocity, capillary water absorption, and compressive strength was studied. The ceramsite dispersion was characterized by the coefficient of variation (CV). Moreover, thermogravimetric (TG), ultra depth-of-field microscope scanning, X-ray diffraction (XRD), scanning electron microscopy (SEM), and energy dispersive spectrometry (EDS) were used to reveal the mechanism of performance change. This study provides a new approach to the treatment of fresh SC concrete.

2. Experimental

2.1. Raw Materials

Ordinary Portland cement (OPC) was used in this study. The initial setting time and the final setting time of the Portland cement were 155 min and 305 min, respectively, with a specific surface area lower than 350 m^2/kg. In addition, quartz sand (particles with a size range of 0.35 mm to 0.59 mm) and an efficient WRA (polyarboxylic acid system, brown microemulsion solution, water reduction rate of 40%) were prepared in this study. Ceramsite, a material produced by Anhui Taotianxia Environmental Protection Technology Co., Ltd., Huainan, China, was used. The length radius was 5–10 mm. The apparent density was 620 kg/m^3, with a 24 h water absorption of 18.1%. Table 1 shows the chemical composition of the raw materials. The appearance of SC is shown in Figure 1.

Table 1. Chemical composition of the raw materials (%).

Types	Chemical Composition (%)							
	SiO_2	Al_2O_3	Fe_2O_3	MgO	CaO	SO_3	R_2O	Loss (%)
OPC	21.18	5.19	3.84	3.91	62.51	2.90	0.47	1.55
Medium sand	99.66	0.12	-	-	0.11	-	0.11	-
SC	54.99	20.55	14.26	3.42	2.8	3.98	-	23.6

Figure 1. Appearance of the SC.

2.2. Specimen Preparation

The water–cement ratio of 0.4 was maintained, and the total mass of the medium sand and SC remained unchanged. Ten groups of ceramsite concrete were manufactured with the proportions shown in Table 2.

Table 2. Mix proportions of ceramsite concrete per one cubic meter (kg).

Group	Water	Cement	Medium Sand	SC	WRA
C1	144	360	792	144	0.00
C2	144	360	648	288	0.00
C3	144	360	792	144	1.44
C4	144	360	648	288	1.44
C5	144	360	792	144	2.16
C6	144	360	648	288	2.16
C7	144	360	792	144	2.88
C8	144	360	648	288	2.88
C9	144	360	792	144	4.32
C10	144	360	648	288	4.32

The preparation of each specimen was in accordance with the JGJ/T 12-2019 and GB/T 17431.2-2010 [28,29]. Firstly, the cement, quartz sand and SC were poured into the UJZ-15 mortar mixer. After dry mixing for 30 s and stirring evenly, the solution of the WRA and water was slowly poured and mixed. Once the materials had been stirred for 480 s, the mixing pot was removed from the turntable. Then, some of the fresh mortar was removed to be used for the slump flow test. The remaining fresh concrete was poured into the mold to form 6 specimens with dimensions of 100 mm × 100 mm × 100 mm. After 24 h, when the specimens had reached a certain strength, the specimens were removed from the molds. Then, all the specimens were placed in the standard curing environment (temperature of 20 °C ± 2 °C, humidity of more than 95%). The specimens were maintained for 28 d. The procedure for the preparation is shown in Figure 2.

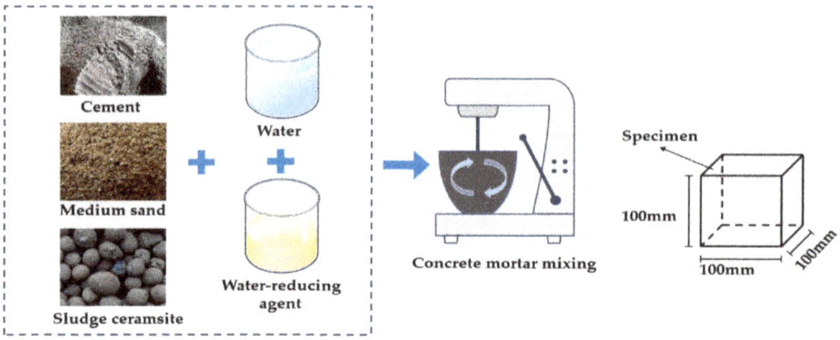

Figure 2. The procedure for the preparation.

2.3. *Measurement Methods*

2.3.1. Slump Flow Test

The slump flow test was in accordance with the GB/T 2419-2005 method for the flow determination of cement mortar [30]. Take out part of the fresh concrete and pour it into the test mold in the center of the jump table until basically level with the mold opening. Start the instrument immediately after removing the mold. Stop the instrument after jumping the table 25 times. Use a ruler to measure the maximum vertical diameter of the bottom surface of the concrete and calculate the arithmetic average of the two vertical diameters as the slump flow of the group. The 10 groups of specimens were tested according to the above experimental steps 6 times.

2.3.2. Ultrasonic Testing

The ultrasonic testing was carried out using the ZBL-U5100 ultrasonic testing instrument provided by Times Shuangjie Instrument Technology Co., Ltd., Tianjin, China. It was conducted after curing the samples for 7 days, 14 days, and 28 days. For the specific experimental methods, refer to the article written by Chen Bo and Ghani Razaqpur [31]. Since the speed of propagation of ultrasonic velocity in the solid phase is greater than that in the liquid and gas phases, concrete with higher ultrasonic velocity will have fewer cracks. In this experiment, there were 6 blocks in each group and 2 test points were sampled. The average value and coefficient of variation were taken in the final analysis.

2.3.3. Capillary Water Absorption Test

The methods for the capillary water absorption test referred to the article by Zhang C. et al. [32]. In this paper, only the water absorption at the bottom of the block was tested. Put the test block in the high-temperature drying oven. When the block is completely dry, first measure the length at the bottom of the block and calculate the area at the bottom of the block. Then, use epoxy resin glue from the bottom of the test block to 3–4 cm up the top (AB glue was mainly used in this paper) without a space. Then, weigh the mass of the sample. The depth of the water should not exceed 3–4 cm. Put the bottom of the test block down into the water, and remove the test block every two minutes and weigh the test block. When the quality of the block remains unchanged, the test of the particular test block can be ended. Six samples in each group were tested. The average value and coefficient of variation were taken in the final analysis.

2.3.4. Compressive Strength Test

After the completion of all the experiments above, the compressive strength test could commence [33]. The measuring process followed the Chinese standard GBJ 81–85 [34]. The strength measured from the 6 specimens in each group was eventually averaged for the final analysis.

2.3.5. Experiments on Microscopic Properties

The specimens were dried in an electric constant temperature blast-drying oven at (60 ± 5) °C for 24 h until they reached a constant weight. Each specimen was taken out and cooled in an indoor environment at (20 ± 5) °C. Then, the samples were subjected to microscopic mechanism tests. The TG-209F3 produced by the company NETZSCH, Bavaria, Germany was used in the TG analysis experiment. The samples were finely ground through a 0.1 mm sieve. Then, 100 mg was weighed on a thermal balance. The heating speed was controlled through a heating device, and the sensor monitored the weight changes of the sample in real time. The temperature ranged from 30 °C to 900 °C, with a heating rate of 10 °C/min and a nitrogen gas flow rate of 150 mL/min. By recording the sample quality with the temperature changes through a data collection and processing system, a graph of the relationship between the sample quality and the temperature changes could be obtained, and its characteristics could be analyzed. The Olympus DSX-500 provided by Shimai Technology Co., Ltd., Wuxi, China was used for the ultra depth-of-field microscope scanning experiment. To obtain a flat and undamaged sample surface, the sample of the cross-section was polished. The sample was placed under the microscope to observe and analyze the microstructure, surface texture, and distribution of the sludge particles. The samples were pulverized and used for XRD. It was carried out using a D8 Discover X-ray diffractometer from Brooke, Berlin, Germany. The scanning method was continuous slow scanning, with a working voltage of 40 kV, a current of 40 mA, a scanning speed of 0.6°/min, and a scanning range of 5° to 90°. After the experiment was completed, the XRD patterns were obtained. The crystal phase, crystal structure, and other information about its existence could be determined by this analysis. For the SEM-EDS experiment, a sample of a flat, rice-sized portion from inside the concrete was taken, and a layer of gold-plated film was sprayed on the surface before scanning. The sample was then placed in a vacuum environment for observation. In this experiment, a Nova Nano SEM scanning electron microscope, produced by FEI, Hillsboro, OR, USA, was used to observe the microstructure, with a resolution of 3.5 nm, an acceleration voltage of 500–30,000 V, a magnification of 18–30,000, a sample stage diameter of 30 mm, and a Brooke Quantum EDS spectrometer manufactured by Brooke, Berlin, Germany was used for elemental analysis in the micro range. The experimental procedure is shown in Figure 3.

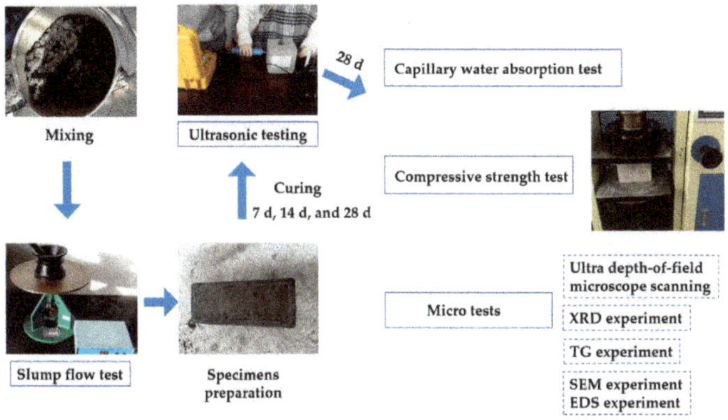

Figure 3. The experimental procedure.

3. Results and Discussion

3.1. Slump Flow Test Results

Figure 4 shows the slump flow of the fresh concrete with different WRA and different SC content. Figure 4 demonstrates that as the WRA content increased from 0% to 1.2%, the slump flow of the mortar exhibited an upward trend. The increasing rate of the slump flow with 10% SC varied from 0% to 60.77%. This was attributed to WRA's ability to enhance the fluidity of the cement mixture [35]. However, the slump flow with a larger content of SC was relatively small, with a reduction by rates of up to 2.04%, indicating that increased SC particles lead to higher water absorption and reduced fluidity. Therefore, both the SC and WRA content affect the slump flow, although WRA has a more pronounced effect compared to SC.

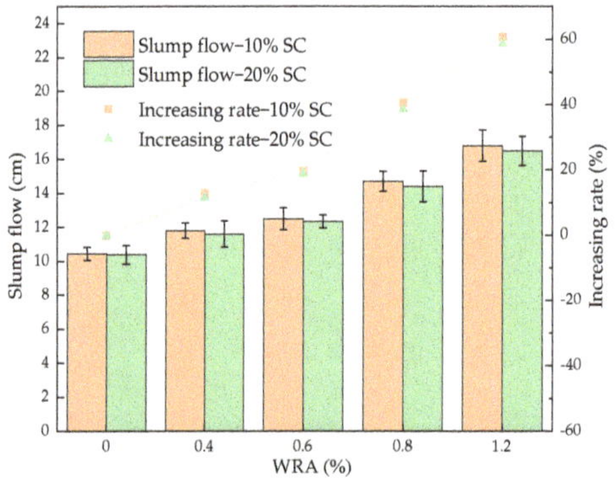

Figure 4. Slump flow of SC fresh concrete.

3.2. *Ultrasonic Testing Results*

Figure 5a illustrates the correlation between the slump flow and the ultrasonic velocity. With increased fluidity, the ultrasonic sound speed of the hardened specimens decreased by 4.1~8.5%, indicating a deteriorated distribution of SC particles and increased porosity. Moreover, as shown in Figure 5a, the height of the curves shows that the ultrasonic velocity in each group increased continuously with age. It can be observed that the ultrasonic velocity curve of the groups with 10% SC was always above the curve with 20% SC, indicating that the internal compactness decreased with the increasing addition of SC. The ultrasonic velocity in some groups with lower WRA was higher than in other groups, indicating that excessive WRA can negatively impact concrete [36,37]. Figure 5b displays the coefficient of variation (CV) in the slump flow. A smaller CV indicated a positive effect on the concrete dispersion. The lowest CV was observed at a slump flow of 12.35 cm, indicating an optimal dispersion.

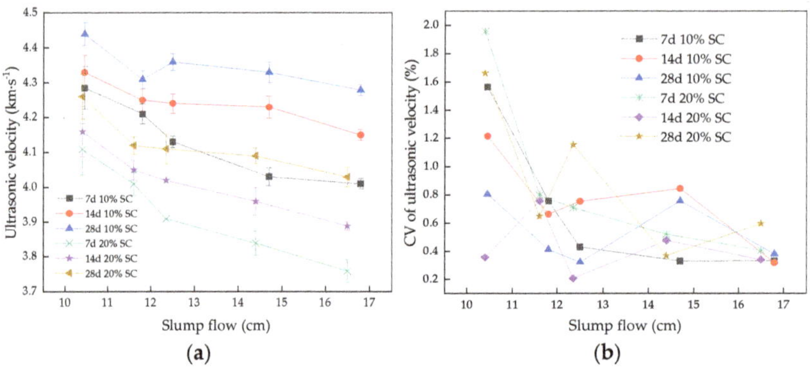

Figure 5. Curves of the relation between the ultrasonic velocity and the slump flow, as well as the relation between CV of ultrasonic velocity and slump flow: (**a**) ultrasonic velocity; and (**b**) CV of the ultrasonic velocity.

3.3. Capillary Water Absorption Test Results

Figure 6 shows the capillary water absorption of the different slump flows with the soaking time. Table 3 presents the fitting results for the soaking time (T) and water absorption (W). Initially, all the groups exhibited rapid increases, followed by stabilization. The rapid water absorption rate was due to the dry block, facilitating water absorption into the voids. As depicted in Figure 6, C9 and C10 with 1.2% WRA reached a maximum water absorption of 0.11%, indicating that the dispersion was the lowest [38]. This could be attributed to several factors. Firstly, the excessive WRA content may lead to overly fluid fresh concrete, causing bleeding in the poured test block. Additionally, the lower density of ceramsite compared to fresh mortar, after thinning, may result in segregation within fresh concrete [39,40]. The lowest final water absorption of C5 and C6 confirms the relatively symmetrical concrete interiors, with slump flows of 12.35 cm and 12.5 cm. It can be concluded that WRA can improve the condition of fresh concrete within a certain range, but exceeding that range results in adverse effects.

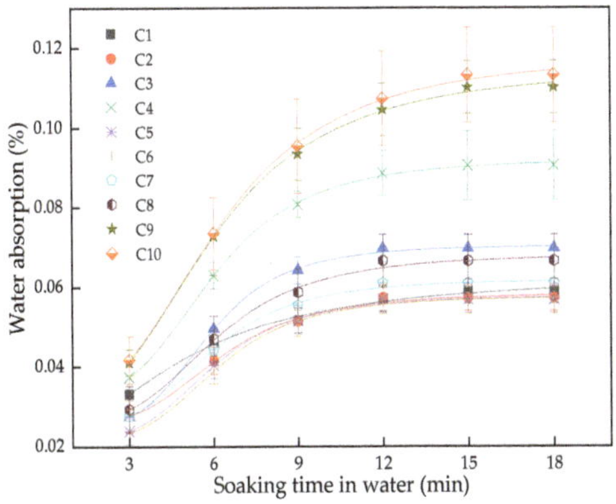

Figure 6. The change in water absorption with the soaking time.

Table 3. The fitting results of the soaking time in water (T) and water absorption (W).

Equation	Types	a	xc	k	R^2
$W = a/(1 + e^{(-k(T-xc))})$	C1	0.060	2.20	0.30	0.99
	C2	0.058	3.35	0.36	0.99
	C3	0.071	4.02	0.45	0.99
	C4	0.092	3.96	0.40	0.99
	C5	0.058	3.88	0.41	0.99
	C6	0.058	4.01	0.40	0.99
	C7	0.062	3.57	0.40	0.99
	C8	0.068	3.76	0.37	0.99
	C9	0.11	4.39	0.37	0.99
	C10	0.11	4.49	0.37	0.99

3.4. Compressive Strength Test Results

Figure 7 shows the compressive strength variation with the different SC and WRA. In Figure 7, the increase in fluidity caused the compressive strength of the concrete to first increase and then decrease. At a slump flow of 12.5 cm, the specimen achieved a maximum compressive strength of 15.51 MPa. The compressive strength of the concrete with the same WRA increased as the content of SC ranged from 10% to 20%. This was attributed to the porous nature of SC, enabling it to absorb free water in the concrete, thereby reducing the water–cement ratio and enhancing the strength [41–43]. The addition of WRA can enhance concrete's compressive strength within a specific range. However, exceeding this range can lead to bleeding and other adverse conditions, thereby affecting the concrete strength [44]. To summarize, it was found that when the slump flow was 16.5 cm and 16.8 cm, the concrete had high water absorption and low compressive strength, showing poor concrete properties. By contrast, when the flow degree was around 12.5 cm, the water absorption was low, along with high compressive strength. The relationship between water absorption and compressive strength could be deduced as a negative correlation. As can be seen from the figure, the ultrasonic velocity was not necessarily related to the other two parameters. The CV was lower for slump flows of 12.5 cm and 12.45 cm. This indicates the good dispersion of concrete with 0.6% WRA. The CV with a 11.6 cm slump flow was the highest, indicating the poor dispersion of concrete with 20% SC and 0.4% WRA.

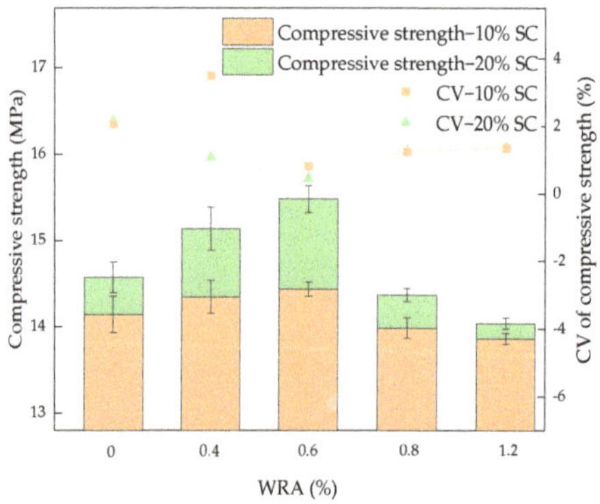

Figure 7. The relation between the compressive strength and the slump flow, as well as the relation between the CV of compressive strength and the slump flow.

3.5. Interpretation of TG Curves

To quantify the hydration degree of the concrete under different fluidity, TG/DTG tests were carried out at a temperature ranging from 30 °C to 900 °C. Figure 8 shows the TG and DTG curves of the concrete. Based on the DTG curves, the mass loss of the sample can be mainly divided into three steps [45]. The first heat absorption peak (70–400 °C) was caused by the decomposition of C-S-H, AFt, etc.; the second heat absorption peak (400–500 °C) was caused by the dehydration reaction of calcium hydroxide (CH); and the third heat absorption peak (500–800 °C) was caused by the decomposition of calcium carbonate ($CaCO_3$) [46]. Figure 8a presents the differences between the concrete with 10% SC and 20% SC when the addition of WRA was 0%. The decomposition of organic matter occurred before the decomposition of CH. It could be seen that the organic matter of 20% SC was higher than 10% SC. The decomposition content of CH in 20% SC was less than 10% SC, which may be because the organic matter in SC prevented the formation of CH. As shown in Figure 8b–e, the total mass loss increased and then decreased with the increase in fluidity. When the addition of WRA was 0.6%, the mass loss was the largest, indicating the highest hydration degree.

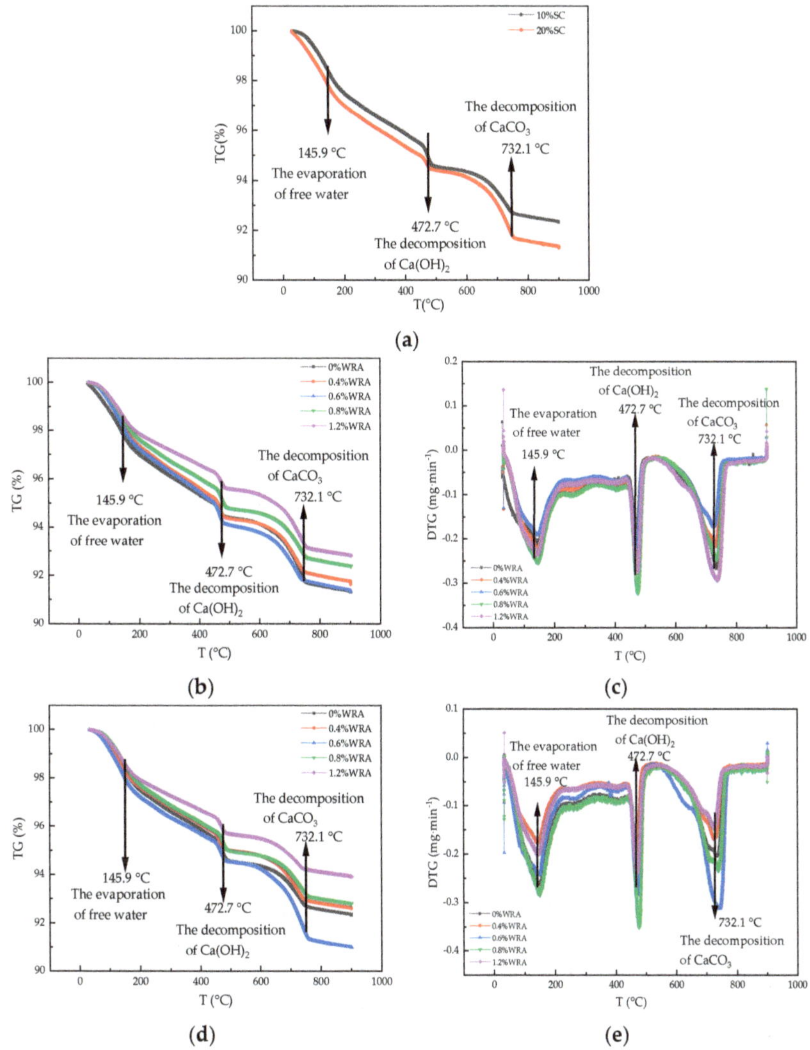

Figure 8. TG analysis curves: (**a**) TG curve of 0% WRA; (**b**) TG curve of 10% SC; (**c**) DTG curve of 10% SC; (**d**) TG curve of 20% SC; and (**e**) DTG curve of 20% SC.

3.6. Ultra Depth-of-Field Microscope Results

The concrete was scanned in section. The results are shown in Figure 9. Due to the instrument, the figures in the upper left corner were displayed in Chinese. It can be seen that the dispersion of ceramsite changed from uniform to dispersed with the increase in fluidity. The samples with 0% WRA cannot show obvious ceramsite. With the increasing addition of SC, there was an area distribution of ceramsite and the interface could be seen more easily. When the addition of WRA was 0.6%, it can be seen that the ceramsite and other materials were relatively bonded, with no obvious gaps. In contrast, the distribution of ceramsite in the 0.8% and 1.2% WRA samples was more dispersed. The concrete surface was relatively uneven [47]. In particular, as can be seen in Figure 9j, the addition of WRA was prone to the stratification of aggregate and hydration products.

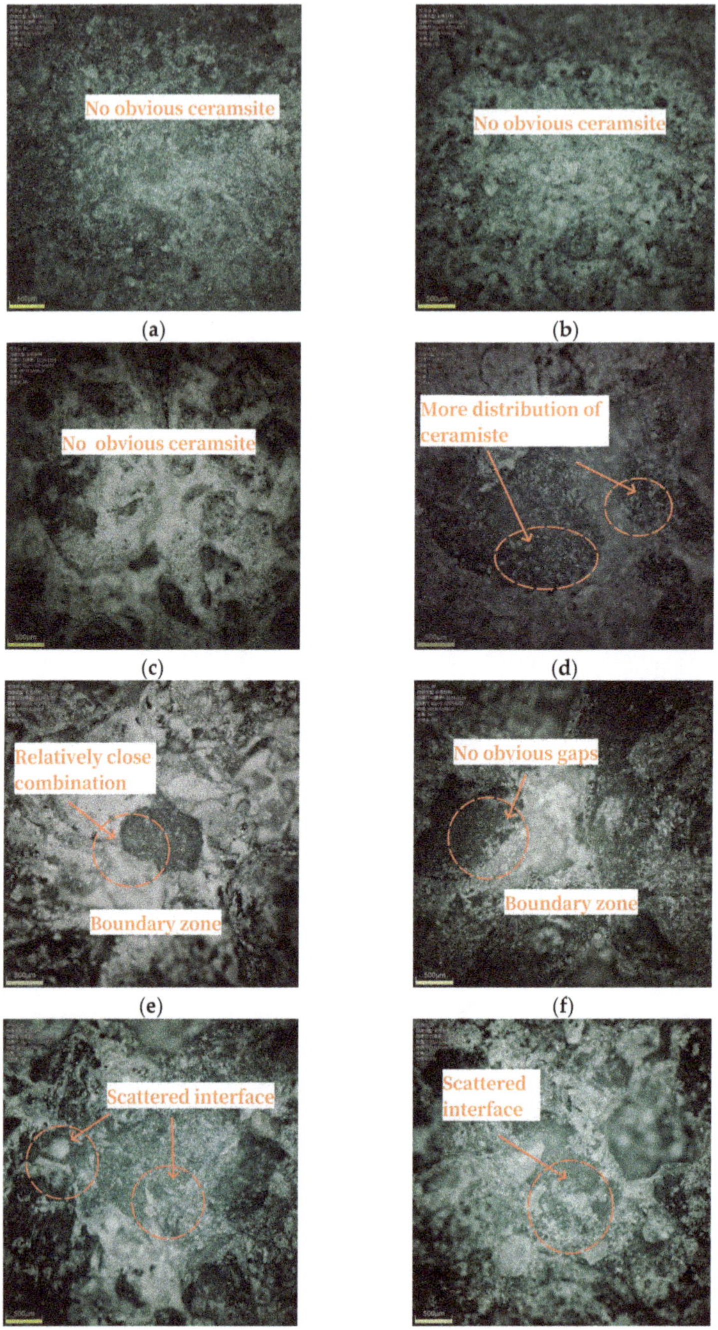

Figure 9. *Cont.*

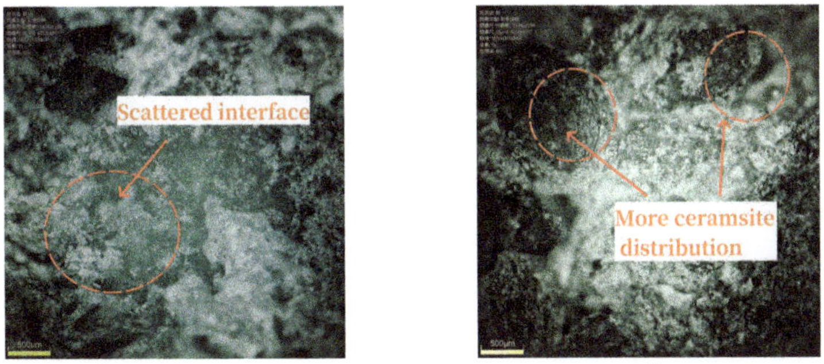

(i) (j)

Figure 9. Ultra depth-of-field microscope scanning of specimens: (**a**) 10% SC 0% WRA; (**b**) 20% SC 0% WRA; (**c**) 10% SC 0.4% WRA; (**d**) 20% SC 0.4% WRA; (**e**) 10% SC 0.6% WRA; (**f**) 20% SC 0.6% WRA; (**g**) 10% SC 0.8% WRA; (**h**) 20% SC 0.8% WRA; (**i**) 10% SC 1.2% WRA; and (**j**) 20% SC 1.2% WRA.

3.7. XRD Analysis

Figure 10 shows XRD images of the specimens. From Figure 10, the diffraction peaks of CH and SiO_2 mainly appeared, in addition to the presence of calcium feldspar (anorthite -$CaAl_2Si_2O_8$) and hydration products such as calcium silicate hydrate (C-S-H). In the 10% and 20% SC concrete, the SiO_2 diffraction peaks of the samples with WRA of 0.6% and 0.8% were lower, indicating that more SiO_2 was involved in the hydration-hardening reaction to change the crystalline form. However, the sample of 1.2% WRA had greater SiO_2 content, which was because more water caused by segregation could not react completely. Furthermore, the SiO_2 reduced with the increasing addition of SC in the hydration reaction, indicating that the participation of the components was enhanced [48]. In the 20% SC concrete, the samples with 0.4% and 0.8% WRA had a higher calcium feldspar peak intensity. The presence of this crystalline state could provide some mechanical strength to the SC, resulting in higher compressive strength.

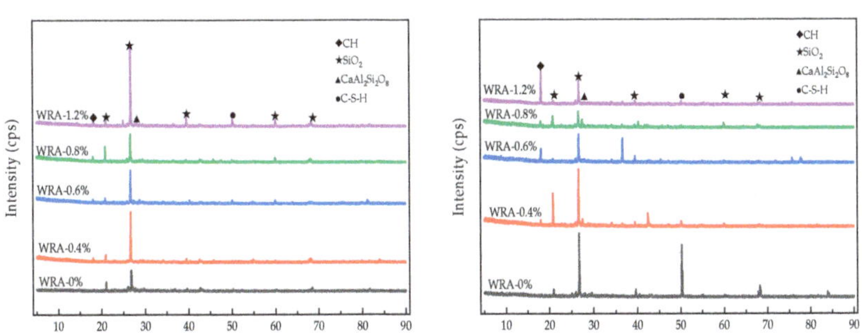

(**a**) (**b**)

Figure 10. X-ray diffraction patterns of the specimens: (**a**) concrete with 10% SC; and (**b**) concrete with 20% SC.

3.8. SEM-EDS Analysis

SEM images of each group are shown in Figure 11. Table 4 shows the element distribution obtained by EDS. The interfacial transition zone (ITZ) was the primary focus of this study. As observed in these images, the amount of calcium hydroxide (CH) first decreased and then increased with the increasing fluidity. Concurrently, as the amount of SC increased, the quantity of hydration products also increased. The ITZ of the sample with the addition of 10% SC and 0% WRA exhibited a greater number of particles and more pronounced pores. The concrete interface appeared rougher, with a reduced presence of nearby hydration products. In the samples with the addition of 20% SC and 0% WRA, distinct macropores were observed. Near the ITZ, the distribution of the hydration products was uneven. Additionally, more obvious clustered products were observed in the ITZ of the samples with 0.4% WRA. In the ITZ of samples with the addition of 10% SC and 0.6% WRA, the C-S-H gel was more tightly wrapped, forming a more continuous bond [49]. In the samples with the addition of 20% SC and 0.6% WRA, a higher number of flocculent products were observed in the left region, along with dense hydration products and increased densification in the ITZ [50,51]. This finding reinforced the notion that the concrete with 0.6% WRA exhibited a superior internal structure. The ITZ of the samples with the addition of 0.8% WRA and 1.2% WRA showed an interior with no obvious hydration products. Particularly in C9, there was a small amount of flocculent products generated, and the overall connection was not tight enough. Similarly, it was verified that the concrete strength was lower at this WRA content.

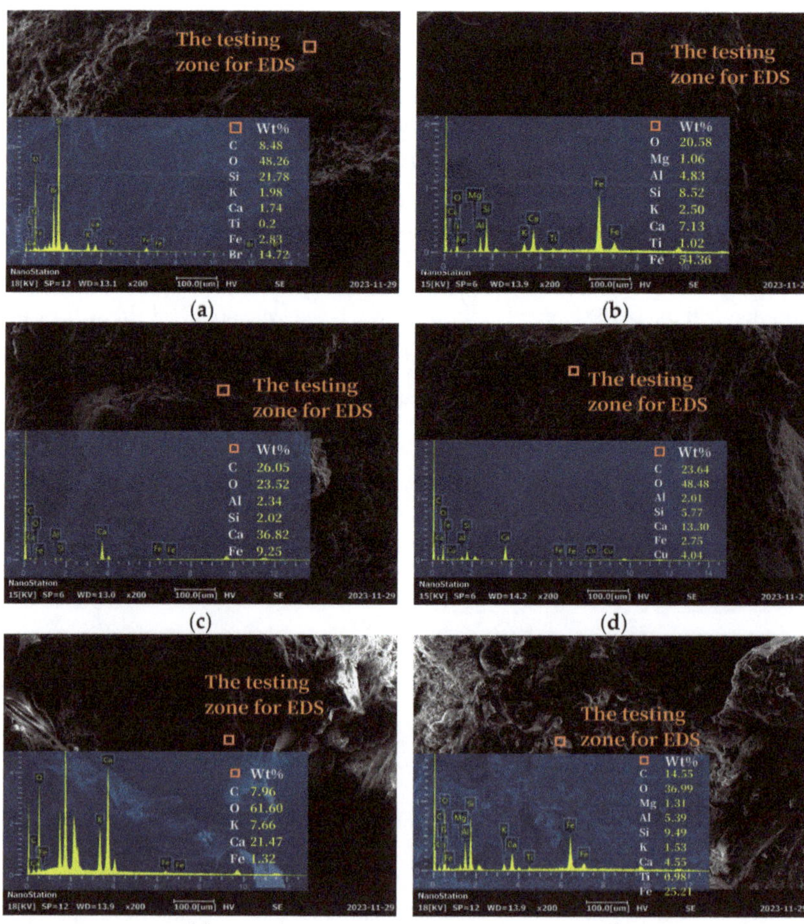

Figure 11. *Cont.*

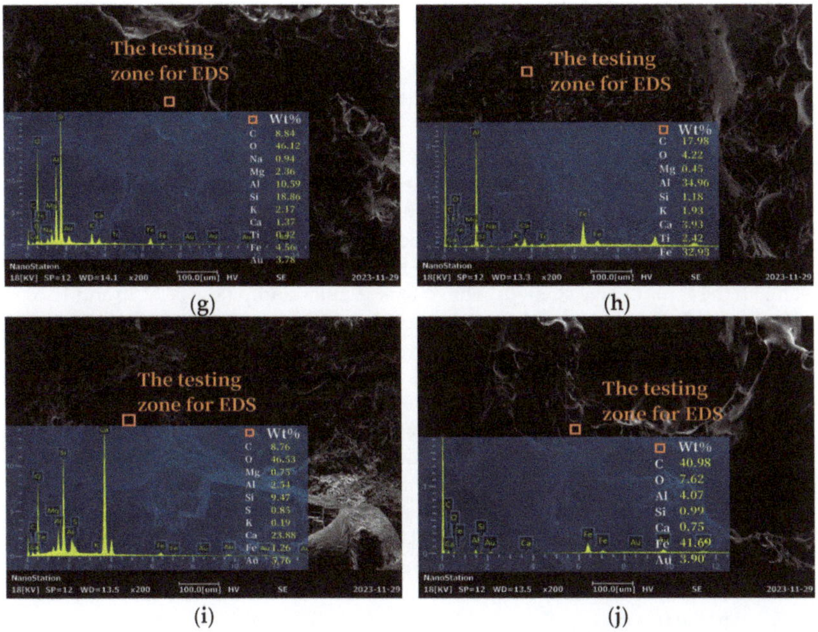

Figure 11. SEM-EDS micrographs of the specimens: (**a**) 10% SC 0% WRA; (**b**) 20% SC 0% WRA; (**c**) 10% SC 0.4% WRA; (**d**) 20% SC 0.4% WRA; (**e**) 10% SC 0.6% WRA; (**f**) 20% SC 0.6% WRA; (**g**) 10% SC 0.8% WRA; (**h**) 20% SC 0.8% WRA; (**i**) 10% SC 1.2% WRA; and (**j**) 20% SC 1.2% WRA.

Table 4. The element distribution obtained by EDS (%).

Group	C	O	Si	K	Ca
C1	8.48	48.26	21.78	1.98	1.74
C2	-	20.58	8.52	2.50	7.13
C3	26.05	23.52	2.02	-	36.82
C4	23.64	48.48	5.77	-	13.30
C5	7.96	61.60	-	7.66	21.47
C6	14.55	36.99	9.49	1.53	4.55
C7	8.84	46.12	18.86	2.17	1.37
C8	17.98	4.22	1.18	1.93	3.93
C9	8.76	46.53	9.47	0.19	23.88
C10	40.98	7.62	0.99	-	0.75

Table 5 shows the Ca/Si ratio obtained by EDS. The Ca/Si ratios of C1, C5, C6 and C7 were relatively low. This may be due to the SiO_2 in the aggregate. The low Ca/Si ratio of the C-S-H gels was beneficial for the gels and concrete, because the average length of the silicate chain of C-S-H gels increased with the low Ca/Si ratio, the degree of polymerization increased, and the performance was better.

Table 5. The Ca/Si ratio obtained by EDS (%).

Group	C1	C2	C3	C4	C5	C6	C7	C8	C9	C10
Ca/Si ratio	0.08	0.84	18.23	2.31	0	0.47	0.07	3.33	2.52	0.76

4. Conclusions

In this study, the influence of fluidity on ultrasonic velocity, capillary water absorption, and compressive strength was studied. The ceramsite dispersion was characterized by CV. The optimum slump flow was 12.35 and 12.5 cm, which was also verified by microscopic tests. The conclusions are as follows:

(1) The addition of WRA improved the fluidity. Conversely, the addition of SC reduced the fluidity. WRA caused a much greater change in the concrete fluidity.

(2) The relationship between the ultrasonic sound speed and the increasing fluidity could be deduced as a negative correlation. The addition of SC increased the internal defects and compressive strength. The water absorption was negatively correlated with the compressive strength. The concrete with a slump flow of 12.35 and 12.5 cm reaching the maximum compressive strength, which had the lowest water absorption, demonstrated internal homogeneity.

(3) By analyzing of the ultrasonic velocity's CV and the compressive strength's CV, the fresh cement concrete with a slump flow of 12.5 cm shows the lowest CV, indicating the optimum SC dispersion. The optimum slump flow was 12.35 and 12.5 cm. That is the concrete with 0.6% WRA. The TG analysis revealed that the addition of 0.6% WRA promoted the hydration of cement. Combined with the XRD and SEM results, the amount of calcium hydroxide (CH) was the least with the concrete of 0.6% WRA addition. Furthermore, the organic substances in SC increased the hydration products. These phenomena indicated that a more homogeneous dispersion of SC and better concrete properties were obtained at a slump flow of 12.35 and 12.5 cm, which can be used in the future.

Author Contributions: Conceptualization, B.X. and B.C.; methodology, Y.Y.; software, Y.Y.; validation, B.X. and Z.C.; formal analysis, Y.Y.; investigation, Y.Y. and H.W.; resources, B.C.; data curation, H.W.; writing—original draft preparation, Y.Y.; writing—review and editing, Y.Y., X.P., Z.C. and H.W.; visualization, Y.Y. and H.W.; supervision, Z.C., X.P. and H.W.; project administration, B.X., X.P. and H.W.; funding acquisition, H.W. and X.P. All authors have read and agreed to the published version of the manuscript.

Funding: This research was supported by Zhejiang Provincial Natural Science Foundation of China (No. LY22E080005, NO. LY24E080010).

Institutional Review Board Statement: Not applicable.

Informed Consent Statement: Not applicable.

Data Availability Statement: The data used to support the findings of this study are available on request.

Conflicts of Interest: Author Bing Xiao was employed by Road & Bridge International Co., Ltd., Author Bingling Cheng was employed by Ningbo Yonghuan Yuan Environmental Protection Engineering Technology Co., Ltd. Author Xi Peng was employed by Ningbo Roaby Technology Industrial Group Co., Ltd. The remaining authors declare that the research was conducted in the absence of any commercial or financial relationships that could be construed as a potential conflict of interest.

References

1. Bandieira, M.; Zat, T.; Schuster, S.L.; Justen, L.H.; Weide, H.; Rodríguez, E.D. Water treatment sludge in the production c red-ceramic bricks: Effects on the physico-mechanical properties. *Mater. Struct.* **2021**, *54*, 168. [CrossRef]
2. Schumacher, K.; Samannshausen, N.; Pritzel, C.; Trettin, R. Lightweight aggregate concrete with an open structure and a porou matrix with an improved ratio of compressive strength to dry density. *Constr. Build. Mater.* **2020**, *264*, 120167. [CrossRef]
3. Alderete, N.M.; Joseph, A.M.; Heede, P.V.D.; Matthys, S.; Belie, N.D. Effective and sustainable use of municipal solid was incineration bottom ash in concrete regarding strength and durability. *Resour. Conserv. Recycl.* **2021**, *167*, 105356. [CrossRef]
4. Yang, Z.; Ji, R.; Liu, L.; Wang, X.; Zhang, Z. Recycling of municipal solid waste incineration by-product for cement composite preparation. *Constr. Build. Mater.* **2018**, *162*, 794–801. [CrossRef]
5. Wang, H.; Xu, J.; Sheng, L. Preparation of ceramsite from municipal sludge and its application in water treatment: A review. *Environ. Manag.* **2021**, *287*, 11237. [CrossRef] [PubMed]

6. Lim, Y.C.; Shih, Y.-J.; Tsai, K.-C.; Yang, W.-D.; Chen, C.-W.; Dong, C.-D. Recycling dredged harbor sediment to construction materials by sintering with steel slag and waste glass: Characteristics, alkali-silica reactivity and metals stability. *J. Environ. Manag.* **2020**, *270*, 110869. [CrossRef] [PubMed]
7. Kuandykova, A.; Taimasov, B.; Potapova, E.; Sarsenbaev, B.; Kolesnikov, A.; Begentayev, M.; Kuldeyev, E.; Dauletiyarov, M.; Zhanikulov, N.; Amiraliyev, B. Production of Composite Cement Clinker Based on Industrial Waste. *J. Compos. Sci.* **2024**, *8*, 257. [CrossRef]
8. Hu, K.; Zhang, Z.; Zhou, Q.; Fang, Y.; Li, W.; Zhang, W.; Zhang, W.; Fan, S. Experimental study on the preparation of lightweight ceramsite with engineering muck. *New Build. Mater.* **2022**, *49*, 6–9.
9. Zeng, Y.; Zhou, X.; Tang, A. Shear performance of fibers-reinforced lightweight aggregate concrete produced with industrial waste ceramsite-lytag after freeze-thaw action. *J. Clean. Prod.* **2021**, *328*, 129626. [CrossRef]
10. Zhu, H.; Xiao, Y.; Li, X.; Wang, Y.; Wen, S. Study on Flexural Strength of Interface between Full Lightweight Ceramsite Concrete and Ordinary Concrete. *Coatings* **2023**, *13*, 1383. [CrossRef]
11. Vishwajith, A.G.H.; Mahanama, K.R.R.; Wijesinghe, L.P.R.J. Investigation on the effective disposal of sludge from a water treatment plant. *Water Pract. Technol.* **2023**, *18*, 130–139. [CrossRef]
12. Jia, G.; Wang, Y.; Yang, F.; Ma, Z. Preparation of CFB fly ash/sewage sludge ceramsite and the morphological transformation and release properties of sulfur. *Constr. Build. Mater.* **2023**, *373*, 130864. [CrossRef]
13. Rashad, A. Lightweight expanded clay aggregate as a building material–An overview. *Constr. Build. Mater.* **2018**, *170*, 757–775. [CrossRef]
14. Wang, J.; Wang, S.; Wang, H.; He, Z. Influence of Ceramsite with Assembly Unit of Sludge and Excavated Soil on the Properties of Cement Concrete. *Materials* **2022**, *15*, 3164. [CrossRef]
15. Xie, J.; Liu, J.; Liu, F.; Wang, J.; Huang, P. Investigation of a new lightweight green concrete containing sludge ceramsite and recycled fine aggregates. *J. Clean. Prod.* **2019**, *235*, 1240–1254. [CrossRef]
16. Wang, C.-Q.; Duan, D.-Y.; Huang, D.-M.; Chen, Q.; Tu, M.-J.; Wu, K.; Wang, D. Lightweight ceramsite made of recycled waste coal gangue & municipal sludge: Particular heavy metals, physical performance and human health. *J. Clean. Prod.* **2022**, *376*, 134309.
17. Khazaleh, M.A.; Kumar, P.K.; Mohamed, M.; Kandasamy, A. Influence of coarse coal gangue aggregates on properties of structural concrete with nano silica. *Mater. Today Proc.* **2023**, *72*, 2089–2095. [CrossRef]
18. Luo, Z.; Guo, J.; Liu, X.; Mu, Y.; Zhang, M.; Zhang, M. Preparation of ceramsite from lead-zinc tailings and coal gangue: Physical properties and solidification of heavy metals. *Constr. Build. Mater.* **2023**, *368*, 130426. [CrossRef]
19. Li, J.; Gao, G. Preparation of ceramsite concrete from paper mill sludge and its application ingreen self-insulation wall material. *New Build. Mater.* **2021**, *48*, 154.
20. Fan, L.; Zhang, Z.; Yu, Y.; Li, P.; Cosgrove, T. Effect of elevated curing temperature on ceramsite concrete performance. *Constr. Build. Mater.* **2017**, *153*, 423–429. [CrossRef]
21. Bu, C.; Yang, H.; Liu, L.; Zhu, D.; Sun, Y.; Yu, L.; Ouyang, Y.; Cao, X.; Wei, Q. Quantification of Ceramsite Granules in Lightweight Concrete Panels through an Image Analysis Technique. *Materials* **2022**, *15*, 1603. [CrossRef] [PubMed]
22. Yakovlev, G.; Saidova, Z.; Gordina, A.; Kuzmina, N.; Ginchitskaya, Y.; Knyazeva, S.; Elrefai, A. Application of ceramsite dust as an active pozzolan additive in the cement-based compositions. *IOP Conf. Ser. Mater. Sci. Eng.* **2021**, *1203*, 032022. [CrossRef]
23. Wei, H.; Song, B.; Huan, Q.; Song, C.; Wang, S.; Song, M. Preparation of iron tailings-based porous ceramsite and its application to lead adsorption: Characteristic and mechanism. *Sep. Purif. Technol.* **2024**, *342*, 126839. [CrossRef]
24. Bu, C.; Zhu, D.; Liu, L.; Lu, X.; Sun, Y.; Yan, Z.; Yu, L.; Wei, Q. A Study on the Mechanical Properties and Microcosmic Mechanism of Basalt Fiber Modified Rubber Ceramsite Concrete. *Buildings* **2022**, *12*, 103. [CrossRef]
25. Wang, P.; Zhang, J.; Duan, K. Experimental Study on Influencing Factors of the Separation of Silt Ceramsite Concrete. *J. Jinling Inst. Technol.* **2019**, *35*, 47–51.
26. Wang, H.; Gao, X.; Wang, R. The influence of rheological parameters of cement paste on the dispersion of carbon nanofibers and self-sensing performance. *Constr. Build. Mater.* **2017**, *134*, 673–683. [CrossRef]
27. Wang, Z.; Gao, J.; Ai, T.; Jiang, W.; Zhao, P. Quantitative evaluation of carbon fiber dispersion in cement based composites. *Constr. Build. Mater.* **2014**, *68*, 26–30. [CrossRef]
28. *JGJ/T 12-2019*; Technical Standard for Application of Lightweight Aggregate Concrele. Ministry of Housing and Urban-Rural Development of the People's Republic of China: Beijing, China, 2019.
29. *GB/T 17431.2-2010*; Lightweight Aggregates and Its Test Methods-Part 2: Test Methods for Lightweight Aggregates. The State Bureau of Quality and Technical Supervision: Beijing, China, 2010.
30. *GB/T 2419-2005*; Test Method for Fluidity of Cement Mortar. The State Bureau of Quality and Technical Supervision: Beijing, China, 2005.
31. Chen, B.; Razaqpur, A.G. Effect of air-entrainment and phase transition on chloride diffusion in partially frozen concrete. *Cold Reg. Sci. Technol.* **2022**, *196*, 103513. [CrossRef]
32. Zhang, C.; Zhang, S.; Yu, J.; Kong, X. Water absorption behavior of hydrophobized concrete using silane emulsion as admixture. *Cem. Concr. Res.* **2022**, *154*, 106738. [CrossRef]
33. Amer, I.; Abdelkhalik, A.; Mayhoub, O.A.; Kohail, M. Development of sustainable slag-based geopolymer concrete using different types of chemical admixtures. *Int. J. Concr. Struct. Mater.* **2024**, *18*, 27. [CrossRef]

34. *GBJ81-85*; Test Method for Mechanical Properties of Ordinary Concrete. National Standards of people's Republic of China: Beijing, China, 1986.
35. Nahhab, A.H.; Ketab, A.K. Influence of content and maximum size of light expanded clay aggregate on the fresh, strength, and durability properties of self-compacting lightweight concrete reinforced with micro steel fibers. *Constr. Build. Mater.* **2020**, *23*, 117922. [CrossRef]
36. Xu, F.; Chang, R.; Zhang, D.; Liang, Z.; Wang, K.; Wang, H. Improvement of CO_2-cured sludge ceramsite on the mechanical performances and corrosion resistance of cement concrete. *Materials* **2022**, *15*, 5758. [CrossRef] [PubMed]
37. Peng, X.; Shi, F.; Yang, J.; Yang, Q.; Wang, H.; Zhang, J. Modification of construction waste derived recycled aggregate via CO curing to enhance corrosive freeze-thaw durability of concrete. *J. Clean. Prod.* **2023**, *405*, 137016. [CrossRef]
38. Zhuang, Y.Z.; Chen, C.Y.; Ji, T. Effect of shale ceramsite type on the tensile creep of lightweight aggregate concrete-sciencedirec *Constr. Build. Mater.* **2013**, *46*, 13–18. [CrossRef]
39. Tong, L.; Ji, J.; Yang, J.; Qian, X.; Li, X.; Wang, H.; Zhou, S.; Wu, Y.; Zhao, Y.; Yuan, X. Sludge-based ceramsite for environmental remediation and architecture ingredients. *J. Clean. Prod.* **2024**, *448*, 141556. [CrossRef]
40. Zang, J.; Pan, C.; Hu, Y.; Qu, S.; Li, M.E. Preparation of Ceramsite Using Dehydrated Silt Soil and Its Performance on Compressive Strength of Ceramsite Concrete Block. *Sustainability* **2023**, *15*, 9134. [CrossRef]
41. Reza, S.M.; Mohammad, M.; Mohammad, A.H. Impact of fine lightweight aggregates and coal waste on structural lightweight concrete: Experimental study and gene expression programming. *Structures* **2024**, *63*, 106397.
42. Paul, A.; Murgadas, S.; Delpiano, J.; Moreno-Casas, P.A.; Walczak, M.; Lopez, M. The role of moisture transport mechanisms on the performance of lightweight aggregates in internal curing. *Constr. Build. Mater.* **2021**, *268*, 121191. [CrossRef]
43. Hasan, M.; Saidi, T.; Afifuddin, M. Mechanical properties and absorption of lightweight concrete using lightweight aggregate from diatomaceous earth. *Constr. Build. Mater.* **2021**, *277*, 122324. [CrossRef]
44. Alghrairi, N.; Aziz, F.N.A.; Rashid, S.A.; Mohamed, M.Z.B.; Ibrahim, A.M. Impact of nano-silica on the mechanical properties of lightweight concrete. *Earth Environ. Sci.* **2024**, *1369*, 012033. [CrossRef]
45. Ma, C.; Bao, S.; Zhang, Y.; Luo, Y.; Gui, Y.; Ren, Y. Preparation of non-sintered sewage sludge based ceramsite by alkali-thermal activation and hydration mechanism. *Ceram. Int.* **2022**, *48*, 31606–31613. [CrossRef]
46. Yue, D.; Yue, Q.; Gao, B. Preparation and bloating mechanism of porous ultra-lightweight ceramsite by dehydrated sewage sludge and Yellow River sediments. *J. Wuhan Univ. Technol.-Mater. Sci. Ed.* **2014**, *29*, 1129–1135. [CrossRef]
47. Liu, L.; Yang, J.; She, Y.; Lv, S.; Yang, Z.; Hu, P. Thermal and mechanical properties of coal gasification slag based foam concrete. *Environ. Sci. Pollut. Res. Int.* **2023**, *30*, 49905–49916. [CrossRef] [PubMed]
48. Shi, Y.; Guo, W.; Jia, Y.; Xue, C.; Qiu, Y.; Zhao, Q. Preparation of non-sintered lightweight aggregate ceramsite based on red mud-carbide slag-fly ash: Strength and curing method optimization. *J. Clean. Prod.* **2022**, *372*, 133788. [CrossRef]
49. Guan, H.; Yu, J.; Kibugenza, A.S.U.; Sun, Q. Preparation of coal gangue ceramsite high-strength concrete and investigation of its physico-mechanical properties. *Sci. Rep.* **2022**, *12*, 16369. [CrossRef]
50. Zhang, G.Z.; Liu, C.; Cheng, P.F.; Li, Z.; Han, Y.; Wang, X.Y. Enhancing the Interfacial Compatibility and Self-Healing Performance of Microbial Mortars by Nano-SiO_2-Modified Basalt Fibers. *Cem. Concr. Compos.* **2024**, *152*, 105650.
51. Zhang, Q.; Xu, Y.; Zhang, H.; Gu, X.; Ni, S.; Xu, Y.; Liu, R. Preparation and performance influencing factors of sludge-based non-sintering ceramsite foam concrete block. *Constr. Build. Mater.* **2024**, *426*, 136240. [CrossRef]

coatings

MDPI

Article

Enhancing Interface Performance Through Self-Assembly Mechanisms of APTES on Surface-Modified Tuff Aggregates

ingxin Lai [1], Xiaoying Gao [2], Lin Kong [3,4,*], Lizong Chen [5], Guoan Gan [3,4], Haixing Lin [6], Jiakang Zhang [3,4], en Zhang [1], Yueling Lin [7], Hongming Zhu [1] and Xinping Zhang [1]

1 Fujian First Highway Engineering Group Co., Ltd., Quanzhou 362122, China
2 Fujian Expressway Group Co., Ltd., Fuzhou 350001, China
3 School of Civil Engineering, Southwest Jiaotong University, Chengdu 610031, China
4 Key Laboratory of Road Engineering of Sichuan Province, Southwest Jiaotong University, Chengdu 611756, China
5 Dehua County Road and Bridge Construction Co., Ltd., Quanzhou 362599, China
6 Fujian Communications Planning & Design Institute Co., Ltd., Fuzhou 350004, China
7 Quanzhou Road and Bridge Engineering Construction Co., Ltd., Quanzhou 362002, China
* Correspondence: konglin@my.swjtu.edu.cn

Abstract: To enhance the adhesion between tuff and asphalt, this study investigates the efficacy of alkalinization treatment technology using a molecular self-assembly layer derived from the silane-coupling agent γ-aminopropyltriethoxysilane (APTES). APTES hydrolysis solutions at varying concentrations were prepared to assess their impact on the adhesive strength of the aggregate–asphalt interface and water damage resistance. Using surface energy theory, the interface adhesion work of tuff was analyzed, while SEM and EDS were employed to examine changes in surface morphology and composition after treatment. The results demonstrate that an APTES:water:ethanol mass ratio of 5:45:50, along with a curing temperature of 200 °C, significantly improves the bonding strength between tuff and asphalt. The silanol groups on APTES react with hydroxyl groups on the tuff surface to form siloxane bonds (Si-O-Si), anchoring APTES to the tuff. This study elucidates the self-assembly mechanisms of APTES on tuff aggregates and demonstrates the consequent enhancement of interfacial adhesion, providing valuable insights for the application of tuff as tunnel spoil in road engineering.

Keywords: adhesion enhancement; silane-coupling agents; asphalt–tuff interface; surface modification; bond strength analysis

1. Introduction

The construction of asphalt pavements necessitates substantial quantities of aggregates. Given that conventional aggregates are finite resources, they are unlikely to fulfill the long-term requirements of road construction [1–3]. Moreover, the extraction of these materials often involves extensive mining operations, which can lead to geological hazards and environmental degradation [4]. In regions like the Dehua section of the Yongxing Expressway in Fujian Province, China, characterized by mountainous terrain and numerous tunnels, the excavation process generates large quantities of tuff slag, which poses a potential environmental waste if not utilized effectively. Leveraging tuff aggregates from these tunnel excavations aligns with the principles of sustainable construction, reducing both material procurement costs and environmental impact [5–7]. This approach not only conserves land and stone resources but also lowers material procurement costs, minimizes mining impacts on vegetation, and yields notable economic and environmental advantages.

The primary tunnel excavation byproduct in this project section is tuff, a material currently underutilized in asphalt pavement construction. Tuff, predominantly formed from volcanic ash and debris through processes such as weathering, erosion, transportation, deposition, and consolidation [8–10]. It is rich in silicates and may possess an acidic surface texture [11,12]. Given that asphalt contains acidic components, the adhesion between tuff aggregates and asphalt is inherently poor, necessitating innovative strategies to enhance this interfacial bonding for effective use in road surfaces [13,14]. Improving this bond is crucial for enhancing the durability and performance of asphalt pavements, particularly in demanding environments.

Hot asphalt binders typically exhibit stronger initial adhesion due to the higher mixing and compaction temperatures, which facilitate better binder flow and aggregate wetting. Conversely, cold asphalt mixtures, valued for their ease of use and quick setting at ambient temperatures, often encounter adhesion and durability challenges. Research indicates that the adhesion between aggregate and asphalt in cement-emulsified asphalt mixtures is notably weaker than in hot-mix asphalt concrete, making them more susceptible to cracking, loosening, potholes, and premature rutting [15]. To enhance the adhesion and interface bonding between aggregates and asphalt, thereby extending the lifespan of asphalt pavements, researchers globally have explored various methods [16]. The incorporation of anti-stripping agents into asphalt mixtures has shown promise in improving aggregate–asphalt adhesion [17–19]. Hydrated lime, exemplifying the first generation of anti-stripping agents, offers the benefits of low cost and ease of use, enhancing aggregate surface roughness. However, its effectiveness is contingent upon the thoroughness of mixing with the aggregate [20–22]. Metal soaps, capable of forming adhesive films at the aggregate–asphalt interface, improve anti-stripping properties but suffer from compatibility issues due to density differences, limiting their application in engineering projects [23]. Surfactants, by improving aggregate surface tension, enhance interfacial bonding, but their thermal instability and susceptibility to decomposition during mixing necessitate further investigation regarding their durability [24,25].

The technology for modifying the surface of aggregates to enhance the surface energy represents an effective strategy for improving the interfacial performance of materials using polymer anti-flaking agents [26]. The silane-coupling agent, γ-aminopropyltriethoxysilane (H2NCH2CH2CH2Si(OC2H5)3, APTES), integrates both organic functional groups and alkoxy groups within a single molecule. Hydrolysis transforms alkoxy groups into silanol groups [27,28]. The organic functional groups engage with the polymer, while the silanol groups establish covalent bonds with the inorganic surface, thereby enhancing adhesion at the inorganic/polymer interface, as confirmed by studies showing improved adhesion in polymer matrices when using APTES [29]. Scholars globally have predominantly applied silane-coupling agents for the surface treatment of inorganic fillers [30,31]. Existing research has demonstrated the effectiveness of silane-coupling agents in enhancing the hydrophobicity and adhesion properties of various inorganic fillers [32–35]. However, the application of silane-coupling agents specifically to tuff aggregates and the mechanisms through which they improve interfacial adhesion with asphalt have not been systematically studied. This study addresses this gap by investigating the unique application of APTES to enhance the adhesion between tuff aggregates and asphalt.

Silane-coupling agents can create a chemically bonded film on aggregate surfaces, with one end forming a chemical bond with the aggregate and the other end achieving physical or chemical adsorption with asphalt, thus significantly enhancing adhesion [36,37]. Treatment with silane-coupling agents introduces additional basic functional groups, such as hydroxyl groups (-OH), on the aggregate surface. These groups can react with acidic components in asphalt to form stable chemical bonds, thereby boosting adhesion [38,39]. Furthermore, alkalinization treatments can increase the aggregate surface's microscopic roughness through etching, providing more points for physical interlocking and enhancing mechanical adhesion with asphalt [40]. By improving adhesion to asphalt, alkalinization contributes to the water stability of asphalt mixtures, reducing water damage risks and extending pavement lifespan. Research on the surface modification of tuff using APTES is currently in its preliminary stages, lacking definitive conclusions regarding enhancement effects or modification mechanisms. The self-assembly mechanism of APTES and its impact on improving interfacial adhesion on tuff aggregate surfaces have not been systematically elucidated.

In summary, this study utilized APTES solutions of varying concentrations to create self-assembled monolayers on the surface of tuff. The interaction between APTES and tuff, along with the mechanisms enhancing interfacial performance, was investigated across multiple scales. Initially, the optimal conditions for silane-coupling agents were identified using the binder bond strength test, which determined the most effective modification process. Using unmodified tuff as a control, this study quantified the enhancement effect of silane-coupling agents on the adhesion work at tuff interfaces based on surface energy theory. Lastly, the micromorphology and elemental changes in silane-modified tuff were analyzed at the microscopic level using SEM and EDS tests, elucidating the modification mechanism of the self-assembled silane layer in enhancing adhesion between tuff and asphalt. This study presents a novel application of APTES to enhance the interfacial adhesion of tuff aggregates in asphalt pavements, offering a new perspective on the sustainable use of tunnel spoil in road engineering.

2. Materials and Methods

2.1. Materials

The materials used in this study include tuff aggregates, the silane-coupling agent APTES for surface treatment, and bitumen as the binder for evaluating interfacial adhesion.

2.1.1. Tuff

The tuff used in this study was sourced from tunnel excavation projects managed by the Quanzhou Municipal Government in Fujian Province and the Dehua section of the Yangyuan-Yongding Expressway. Performance testing was conducted in accordance with the JTG E42-2005 "Experimental Procedures for Aggregate in Highway Engineering", with the results presented in Table 1.

Table 1. Basic physical characteristics of tuff.

Testing Indexes	Unit	Requirements	Tested Parameters	Test Methods
Crushing value	%	≤26%	14.4	T0316: Determines the crushing value to assess aggregate resistance to compressive loads.
Los Angeles abrasion value	%	≤28%	7.9	T0317: Measures the Los Angeles abrasion value to evaluate wear resistance.
Bulk volume relative density	g/cm^3	≥2.60	2.684	T0304: Tests bulk volume relative density, apparent relative density, and water absorption.
Apparent relative density	g/cm^3	≥2.70	2.749	
Water absorption	%	≤2.0	1.034	
Acicular content	%	≤12	8.5	T0312: Assesses the acicular content, indicating the percentage of elongated and flaky particles.

2.1.2. Silane-Coupling Agent

This research introduces γ-aminopropyltriethoxysilane (APTES) as a silane-coupling agent to modify the tuff surface, aiming to enhance interfacial bonding with asphalt through the formation of an APTES self-assembly layer. The APTES used was procured from Dongguan Kangmian New Materials Chemical Reagent Co., Ltd. (Dongguan, China), with its performance metrics detailed in Table 2.

Table 2. Basic physical characteristics of APTES.

Molecular Formula	Boiling Point (°C)	Refractive Index (nD25)	Flash Point (°C)	Purity (%)
C9H23O3NSi	214	1.415	86	>99

2.1.3. Bitumen

ES70# bitumen, supplied by ExxonMobil Corporation, is characterized within Table 3.

Table 3. Physical characteristics of ES70# bitumen.

PG Grade	Penetration (25 °C, 0.1 mm)	Ductility (10 °C, cm)	Softening Point (°C)	Dynamic Viscosity (60 °C, Pa·s)	Ignition Point (°C)
58-16	66	46	49.3	286	354

2.2. Tuff Aggregate Surface Modification Procedures

This study prepared APTES hydrolysates at varying concentrations to identify the optimal modification process and experimental conditions, setting unmodified tuff aggregate at different curing temperatures as the control group. The preparation of hydrolysate solutions ("m" refers to the mass ratio) is enumerated in Table 4.

Table 4. Numbering of APTES hydrolysis solutions with different concentrations.

Item	APTES (m)	Water (m)	Ethanol (m)	Curing Temperature
A1	0	0	0	150
A2	0	0	0	180
A3	0	0	0	200
B1	2	49	49	150
B2	2	49	49	180
B3	2	49	49	200
C1	5	45	50	150
C2	5	45	50	180
C3	5	45	50	200
D1	10	45	45	150
D2	10	45	45	180
D3	10	45	45	200
E1	15	45	40	150
E2	15	45	40	180
E3	15	45	40	200

The hydrolysis stage is crucial for activating the coupling agent, enabling effective bonding with various materials, including inorganic fillers and organic polymers. Cleaned and surface-dried tuff aggregates were immersed in the APTES solution for 60 min to ensure thorough interaction with the aggregate surface. After treatment, the tuff was oven-dried for 60 min at a temperature simulating actual production mixing conditions, followed by cooling to room temperature to complete the self-assembly of the APTES coating. The aggregate-processing workflow is illustrated in Figure 1.

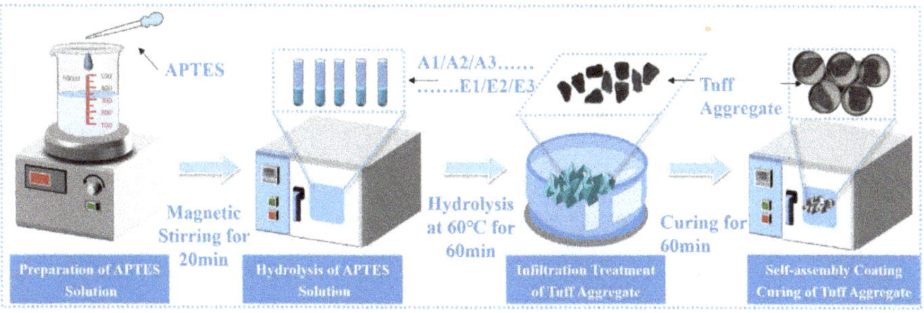

Figure 1. Tuff aggregate surface modification procedures.

2.3. *Experimental Plan Design*

The experimental plan is outlined in Figure 2. APTES hydrolysis solutions with varying concentrations were prepared, and their impact on the adhesive strength of the aggregate–asphalt interface and its resistance to water damage were assessed to identify optimal testing conditions. Using surface energy theory, the interface adhesion work of tuff was analyzed, while SEM and EDS were employed to examine changes in surface morphology and composition after treatment.

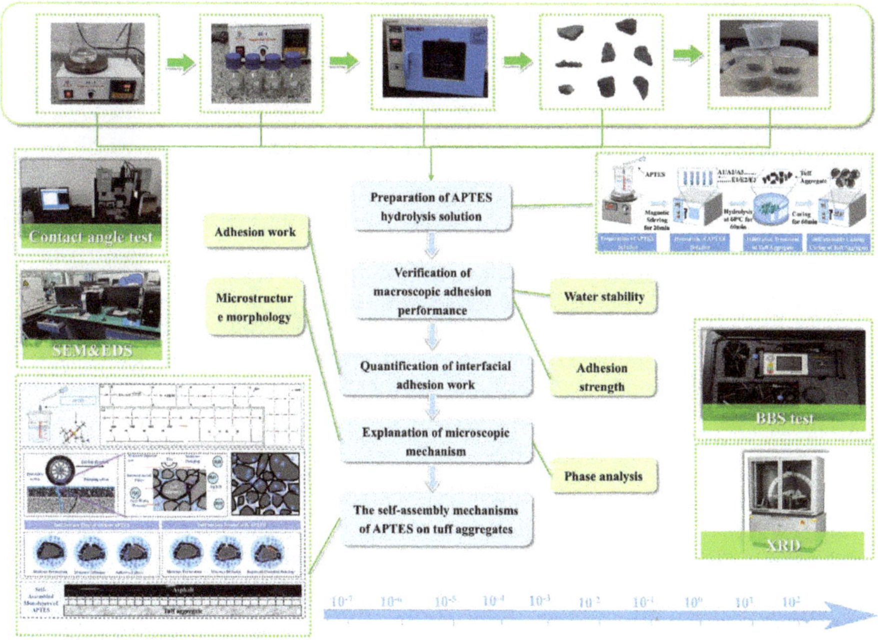

Figure 2. The experimental procedure.

2.4. *X-Ray Diffraction*

X-ray diffraction (XRD) was employed as the primary method for analyzing the physical phases and crystal structures of the materials, using a Bruker AXS D8 ADVANCE X-ray diffractometer. The apparatus operated with Cu Kα radiation (λ = 1.789 Å), a tube voltage of 40 kV, and a current of 30 mA, scanning from 3° to 80° at a speed of 10°/min.

2.5. Binder Bond Strength (BBS) Test

This study adheres to the methodologies outlined in AASHTO TP-91 (AASHTO 2004) and ASTM D4541 (ASTM 2017), employing an American PosiTest AT-M pull-out adhesion tester, which operates within a range of 0.4–3.5 MPa, to evaluate the adhesion performance of tuff aggregate samples. As illustrated in Figure 3, the testing setup includes a schematic of the binder bond strength (BBS) test procedure and the preparation of test samples. The device features a 20 mm supporting drawing head and a resolution of 0.01 MPa (1 psi) [41]. To maintain the consistency of test data, a controlled spreading amount method is applied, with the spreading amount for all samples uniformly maintained at 0.3 kg/m^2.

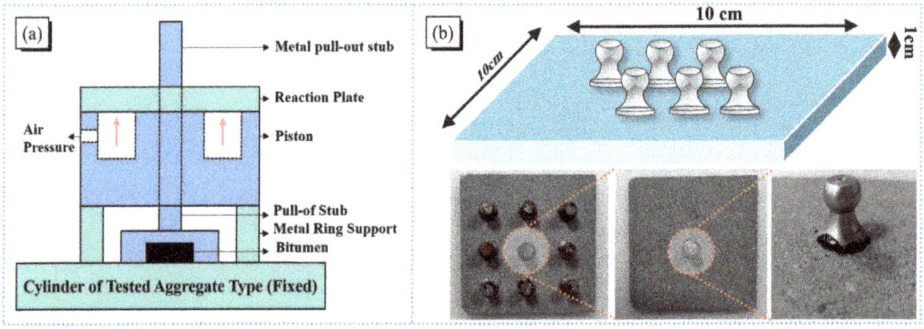

Figure 3. Binder bond strength (BBS) test: (**a**) schematic diagram BBS test; (**b**) test sample preparation.

2.6. Immersion Residue Stability Test

The water stability of an asphalt mixture is a critical metric for assessing its capacity to retain functionality under prolonged water immersion or through freeze–thaw cycles. In this research, a soaked Marshall residual stability test was employed to examine the resistance of tuff asphalt mixtures to water-induced damage, both pre- and post-modification. Figure 4 shows the AC-20 gradation curve and the Marshall stability test procedure. Following the "Testing Procedures for Asphalt and Asphalt Mixtures in Highway Engineering" T0719-2011, a tuff asphalt mixture with an oil-to-stone ratio of 4.3% was prepared, using AC-20 grading. One set of Marshall specimens underwent a conditioning process at 60 ± 1 °C for 30 min, with its stability thereafter denoted as MS_2; another set was subjected to the same temperature for 48 h, with its stability measured and labeled as MS_1. The residual stability post-water-immersion, MS_0, was then calculated utilizing formula 1.

$$MS_0 = \frac{MS_1}{MS_2} \times 100 \quad (1)$$

where MS_0 represents the residual stability after water immersion (%), MS_2 denotes the Marshall stability after a 30 min water bath (kN), and MS_1 signifies the Marshall stability following a 48 h water bath (kN). To minimize experimental errors, each mixture was subjected to three replicate tests. The average of these three trials was then computed to determine the test result.

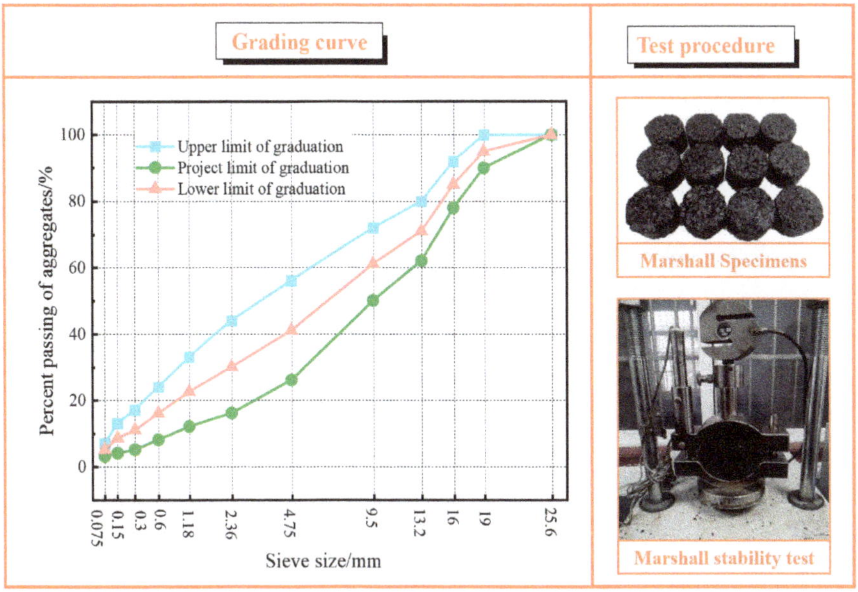

Figure 4. The Marshall stability test procedure.

2.7. Contact Angle Test

The performance of asphalt mixtures is significantly influenced by the adhesion between the aggregate and asphalt [42]. Evaluating the surface energy of aggregate and asphalt through testing can reveal the efficacy of surface treatment and the interfacial bonding capabilities of the APTES self-assembly layer. These assessments employ the German KRUSS DSA100 contact angle measuring instrument, which offers a contact angle measurement range of 0–180° and an accuracy of ±0.1°, with an image magnification capability of 0.7×.

To prepare the asphalt sample, the procedure involves heating the asphalt to 150 °C in an oven until it reaches a fluid state. The heated asphalt is then carefully titrated onto a transparent glass slide to create a smooth film. This film is allowed to cool to room temperature over a 12 h period. Once cooled, the hanging drop method is used to measure the contact angle.

The method for preparing aggregate samples involves initially using a high-precision cutting machine to achieve a relatively flat test surface on irregular aggregates. Subsequently, the aggregates are cleaned and oven-dried until completely moisture-free. To ensure data precision, testing is conducted on three parallel specimens within each group, with the results being averaged. Figure 5 shows the contact angle test procedure.

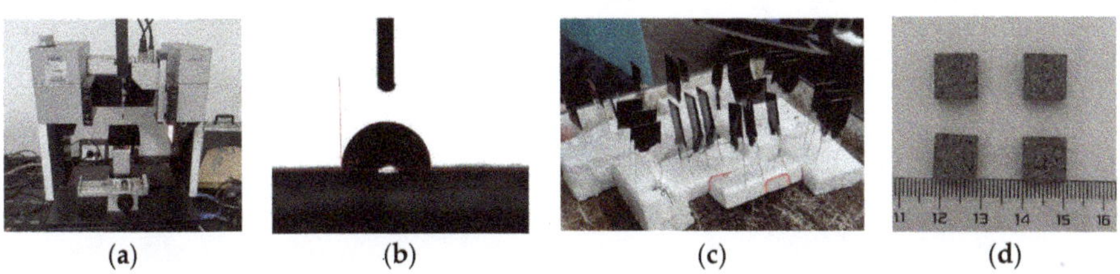

Figure 5. Contact angle test: (**a**) contact angle test; (**b**) contact angle diagram; (**c**) asphalt sample; (**d**) sliced aggregate pieces.

The relationship between the contact angle and interfacial tension in the equilibrium state of a solid surface is described via the Young–Dupre equation [43]. At standard atmospheric pressure, the work of interfacial adhesion between solid and liquid phases can be calculated using Equation (2) [44]:

$$W_a = \gamma_l(1 + \cos\theta) \tag{2}$$

The polar and non-polar components of the surface energy of tuff aggregate and asphalt are determined using Equation (3). This Equation facilitates the calculation of surface energy, distinguishing between its polar and non-polar constituents.

$$\gamma_l(1 + \cos\theta) = 2\left(\gamma_s^d \cdot \gamma_l^d\right)^{1/2} + 2\left(\gamma_s^p \cdot \gamma_l^p\right)^{1/2} \tag{3}$$

where γ_l—surface energy, mJ/m^2;

γ_s^d, γ_l^d—dispersive component of solid and liquid, mJ/m^2;

γ_s^p, γ_l^p—nonpolar and polar part of solid and liquid, mJ/m^2.

For the experimental investigation, distilled water, ethylene glycol, and glycerol were selected as liquid probes for surface energy analysis. Their surface energies are 72.1 mJ/m^2 for distilled water, 48.0 mJ/m^2 for ethylene glycol, and 64.6 mJ/m^2 for glycerol [45].

2.8. Scanning Electron Microscopy (SEM) and X-Ray Energy Dispersive Spectroscopy (EDS)

The impact of the APTES self-assembly layer on the micromorphology and elemental composition of the tuff aggregates was investigated using scanning-electron microscopy (SEM) and X-ray energy-dispersive spectroscopy (EDS). Comparing the elemental content on the tuff surface before and after treatment with solutions of varying concentrations facilitated the identification of the optimal treatment concentration. To ensure data accuracy, tests were conducted on three parallel specimens per group, with results averaged. The SEM employed was a Zeiss Sigma 300, with specifications including an accelerating voltage of 0.3–30 kV, a magnification range of 8–10 k, and a working distance of 10 mm. EDS data were captured over an area of 280 μm × 210 μm using Aztec Version 5.1, with a resolution of 1024 × 768 pixels.

3. Results and Discussions

3.1. X-Ray Diffraction Test Results

X-ray diffraction (XRD) analysis reveals the diffraction patterns of tuff and basalt, with each peak corresponding to diffraction from specific crystallographic planes within the minerals. This allows for the identification of minerals in the samples by analyzing the position and intensity of these peaks. Figure 6 shows that tuff, a volcanic rock primarily composed of volcanic ash, contains minerals such as volcanic glass, quartz, plagioclase feldspar, and minor clay minerals, depending on the source and weathering processes. A peak at 26.6° in the XRD spectrum suggests the presence of quartz (SiO_2) within the tuff. Basalt, an igneous rock rich in iron and magnesium, primarily consists of plagioclase, pyroxene, and olivine. In the XRD spectrum, pyroxene typically shows peaks between 35° and 40° (2θ values). The differences in adhesion performance between tuff and basalt are largely influenced by their mineralogical composition, texture, and surface energy.

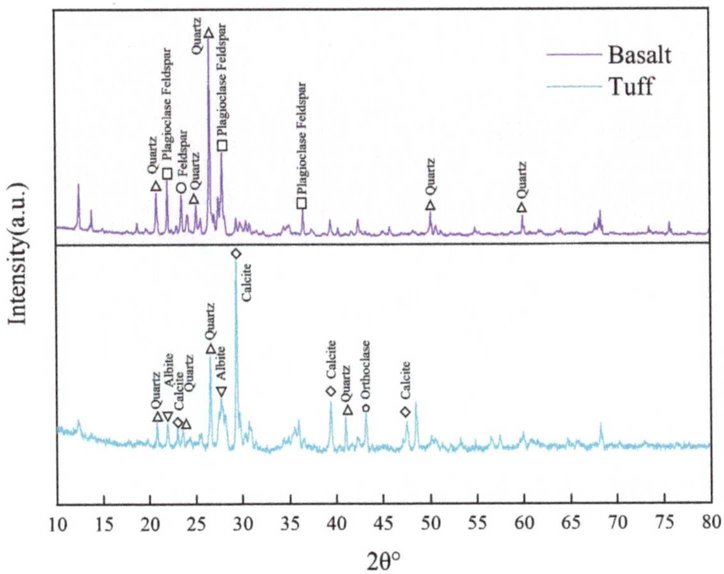

Figure 6. X-ray diffraction test results of basalt and tuff.

These experimental outcomes elucidate the disparities in surface energy between the two rocks from a phase composition perspective, highlighting basalt's superior adhesion to asphalt, rendering it more apt for use in asphalt mixtures. This underscores the need for surface treatment of tuff with silane-coupling agents to enhance its compatibility with asphalt, facilitating a 'bridging' effect that improves asphalt's wettability on tuff surfaces [46]. Future experiments will investigate the enhancement in adhesion and the modification mechanisms of silane-coupling agents on tuff aggregates at both macroscopic and microscopic levels.

3.2. Binder Bond Strength (BBS) Test Results

Figure 7 demonstrates that the bonding strength between tuff aggregate and asphalt is significantly influenced by both the curing temperature and the concentration of the silane-coupling agent. At a curing temperature of 180 °C, the bonding strengths for tuff and asphalt, with increasing silane-coupling-agent hydrolysate mass fractions from 0% to 2% and 5%, are 0.89 MPa, 0.93 MPa, and 1.11 MPa, respectively. Post hoc analysis revealed that the 5% solution group exhibits a significant increase in bonding strength, with a 24.7% improvement (0.22 MPa) compared to the control (0% solution), indicating optimal performance at this concentration ($p < 0.05$). However, further increases in the silane-coupling-agent concentration to 10% and 15% resulted in reduced bonding strengths of 0.94 MPa and 0.76 MPa, respectively, suggesting that overly high concentrations may negatively impact adhesion due to excessive film thickness or other limiting factors ($p < 0.05$). ANOVA results indicated that both curing temperature and silane concentration significantly affect bonding strength ($p < 0.01$). Additionally, correlation analysis showed a positive relationship between curing temperature and bonding strength, with an average increase of 0.08 MPa for every 10 °C rise in temperature across all concentrations, highlighting the importance of curing temperature in enhancing adhesion.

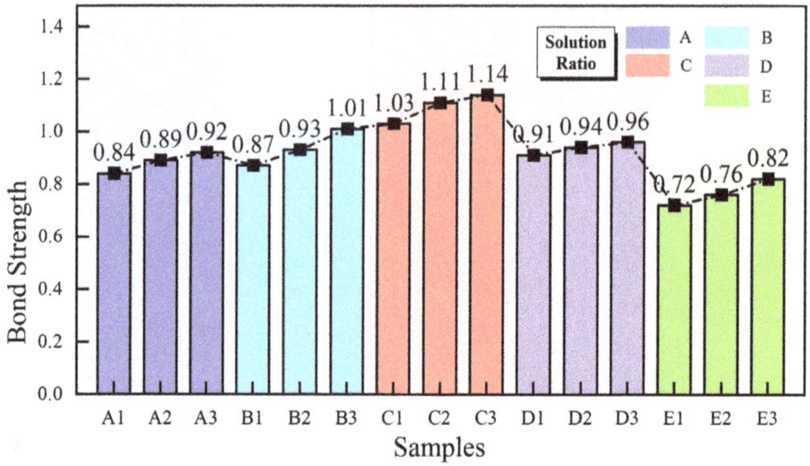

Figure 7. Binder bond strength (BBS) test results.

The analysis indicates that APTES primarily forms covalent bonds, specifically siloxane bonds (Si-O-Si), between the tuff and asphalt, thereby enhancing their interfacial compatibility and bonding properties. Increasing concentrations of the silane-coupling-agent hydrolysate expand the film coverage on tuff surfaces, establishing more chemical bonds and physical adsorption points, thereby augmenting the bond strength by increasing the actual contact area and mechanical engagement points with the asphalt. Post-treatment with the coupling agent modifies the surface energy of the tuff, making it more compatible with asphalt and facilitating stronger bonds. Additionally, the coupling agent forms new interfacial bonds through chemical reactions, further enhancing the bonding properties. Nonetheless, overly high concentrations of the silane-coupling agent may lead to surface saturation and excessive film thickness, impeding effective bonding due to the potential formation of chemical bonds or increased stress concentration at the interface, thus diminishing bond strength. Consequently, the recommended silane-coupling-agent hydrolysis concentration is m (silane-coupling agent)/m (water)/m (ethanol) = 5:45:50, with an optimal curing temperature of 200 °C. Future studies will focus on the effects of the silane-coupling agent's self-assembled film on the contact angle, surface elemental composition, functional group variations, and micromorphology of tuff.

3.3. Immersion Residue Stability Test Results

The water-immersion Marshall stability test serves as an instrumental method for assessing the stability of asphalt mixtures under conditions of water immersion, thereby indicating the mixture's capability to withstand water damage in real-world application settings. As depicted in Figure 8, the Marshall stability of untreated tuff changes from 9.14 kN before immersion to 7.13 kN after immersion, with the residual stability recorded at 78%. Treatment with APTES modestly enhances the Marshall stability of tuff from 9.14 kN to 9.46 kN. More notably, the stability after water immersion escalates from 7.13 kN to 8.25 kN, and the residual stability improves from 78% to 89%, indicating a significant enhancement. At an aggregate consolidation temperature of 200 °C, the residual stability of tuff treated with APTES approaches that of basalt under equivalent conditions. Post hoc testing (Tukey's HSD) showed that the increase in post-immersion stability between untreated tuff (A1) and the 200 °C treated sample (C3) was significant ($p < 0.05$), supporting the conclusion that higher curing temperatures enhance the efficacy of APTES treatment.

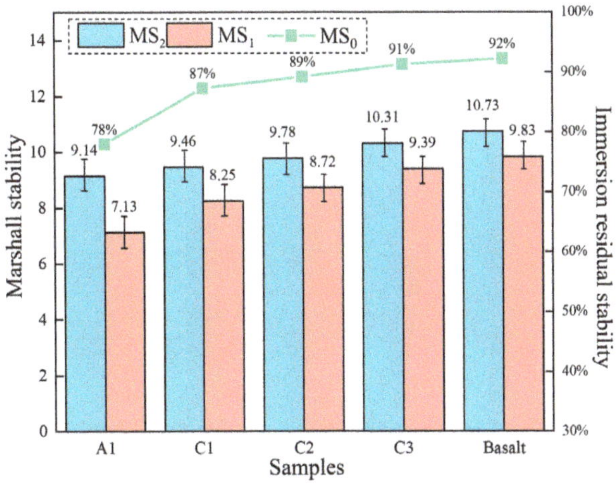

Figure 8. Immersion residue stability test results.

This analysis indicates that immersion in a 60 °C water bath accelerates aging and replicates the effects of hot and humid environments. Water penetration at the fragile bonding interface between asphalt and aggregate leads to the detachment of the asphalt film on the aggregate surface and a reduction in the adhesive force between asphalt and aggregate, consequently diminishing the internal cohesive strength of the asphalt mixture [47]. This improvement is supported by the BBS test results, which demonstrate increased bonding strength at the asphalt–tuff interface, directly correlating with the observed macro-level performance. Unprocessed tuff exhibits weak adhesion to asphalt, rendering the mixture susceptible to delamination in hot and humid conditions and thus decreasing its stability [48]. APTES enhances the bond between asphalt and aggregate by creating a self-assembled layer on the aggregate's surface, thereby increasing the affinity between the two materials. This self-assembled layer acts to inhibit water penetration at the asphalt-aggregate interface, bolster cohesion within the mixture, preserve structural integrity post-immersion, and mitigate water-induced damage.

3.4. Contact Angle Test Results

Ethylene glycol, glycerol, and distilled water were selected as liquid probes to mitigate experimental errors, with the average of three sets of parallel experiments calculated for each sample. The work of adhesion and peeling are delineated in Equations (4) and (5) [37,49], respectively, with the results illustrated in Figure 9. The modification effect can be more accurately assessed by using the energy ratio (ER), which provides a more comprehensive evaluation of the interaction between the asphalt and aggregate [50].

$$W_{\text{adhension}} = 2\sqrt{\gamma_a^d \gamma_s^d} + 2\gamma_a^p \gamma_s^p \quad (4)$$

$$W_{\text{tripping}} = 2\gamma_w - 2(\sqrt{\gamma_w^d \gamma_a^d} + \sqrt{\gamma_w^p \gamma_a^p}) + 2(\sqrt{\gamma_a^d} - \sqrt{\gamma_w^d}\sqrt{\gamma_s^d} + 2\sqrt{\gamma_a^p} - \sqrt{\gamma_w^p})\sqrt{\gamma_s^p} \quad (5)$$

$$ER = \frac{W_a}{|W_t|} \quad (6)$$

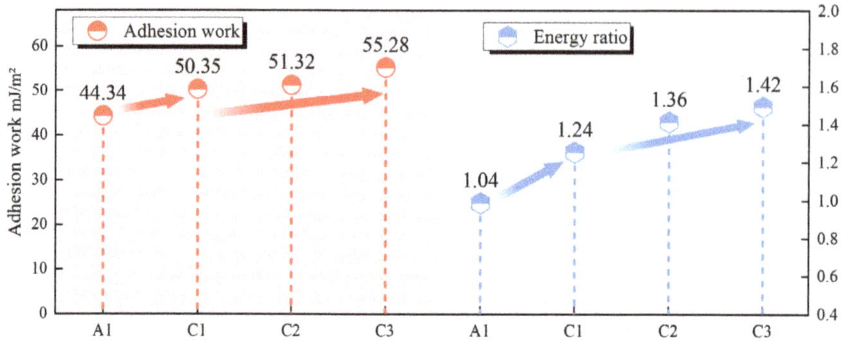

Figure 9. Adhesion work and energy ratio.

Figure 9 shows that the adhesion work for unmodified tuff, derived from surface energy components, is 44.34 mJ/m^2, with an energy ratio of 1.04. In contrast, for the C3 sample treated with the optimal concentration of silane-coupling agent, the adhesion work and energy ratio increase significantly to 55.28 mJ/m^2 and 1.42, respectively.

Post-treatment with an optimal silane-coupling agent concentration notably increases both adhesion works and energy ratio, indicating enhanced resistance of tuff to stripping. This improvement is attributed to the formation of a self-assembled silane film on the tuff surface, increasing its polar components and surface polarity, thereby improving the adhesion between the acidic tuff and asphalt due to the introduction of polar functional groups, such as amino groups.

3.5. SEM Test Results

Scanning-electron microscopy was employed to analyze the microscopic morphology of the tuff aggregate, both under optimal modification conditions and in its untreated state. Observations were conducted at magnifications of 500× and 2500×, with the findings presented in Figure 10.

Scanning-electron microscopy (SEM) analysis at 500× and 2500× magnification of the optimally modified and untreated tuff aggregates revealed significant differences in surface morphology, as shown in Figure 10. SEM images indicate that the self-assembled silane-coupling agent film partially fills or covers the tuff's scale structure, creating an uniform layer. The silanol groups in APTES react with surface hydroxyl groups on the tuff, forming stable siloxane (Si-O-Si) bonds. This modification enhances surface uniformity, potentially increasing erosion and chemical resistance. The resulting denser surface is crucial for improving the interface with materials like bitumen by filling micropores and strengthening the bond.

Analysis suggests that the silanol groups in APTES can react with hydroxyl groups on tuff surfaces, typically resulting in the formation of stable siloxane bonds (Si-O-Si). This reaction not only establishes stable chemical bonds but also significantly alters the surface properties of tuff. The formation of siloxane bonds alters the tuff surface, transforming it from smooth to a network-like structure of villi "membranes". This modification is crucial for integrating tuff with materials such as bitumen. Furthermore, the reaction products fill the microporous structure of the tuff surface, thereby strengthening the interface between tuff and asphalt. The SEM analysis further supports these findings by revealing the formation of a uniform and dense APTES layer on the tuff surface.

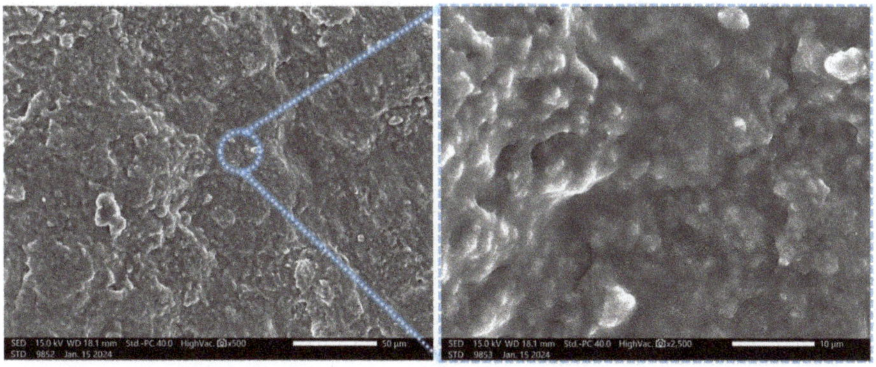

(a)

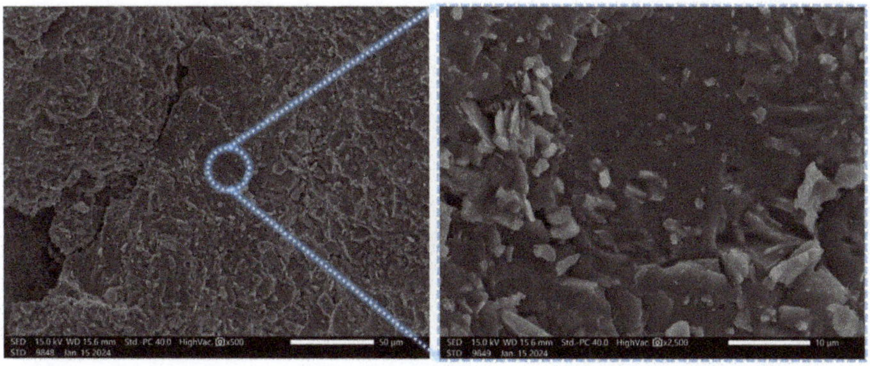

(b)

Figure 10. SEM micrographs: (**a**) SEM micrographs (500× magnifcation, 2500× magnifcation) of tuff surface treated with APTES. (**b**) SEM micrographs (500× magnifcation, 2500× magnifcation) of tuff surface treated without APTES.

3.6. EDS Test Results

Quantitative analysis of the EDS spectra (Figure 11) reveals significant changes in the surface elemental composition. The untreated tuff's EDS spectrum shows characteristic peaks for silicon (Si), oxygen (O), aluminum (Al), and trace elements such as calcium (Ca), sodium (Na), and potassium (K). After APTES modification, there is a noticeable increase in the intensities of the silicon and carbon peaks, with carbon content rising from 3.8% to 10.3%, as shown in Figure 12. These changes confirm the formation of Si-O-Si bonds and a successful silane-bonding process. The increased silicon content, along with shifts in elemental proportions, further validates this interaction. As the concentration of the silane-coupling agent increases, the formation of a self-assembled film on the tuff surface is evident. The EDS analysis supports the expected increase in silicon content, consistent with the chemical reaction where the silanol groups of the silane agent bond with hydroxyl groups on the tuff, leading to the formation of siloxane bonds and an enhanced silicon concentration at the surface.

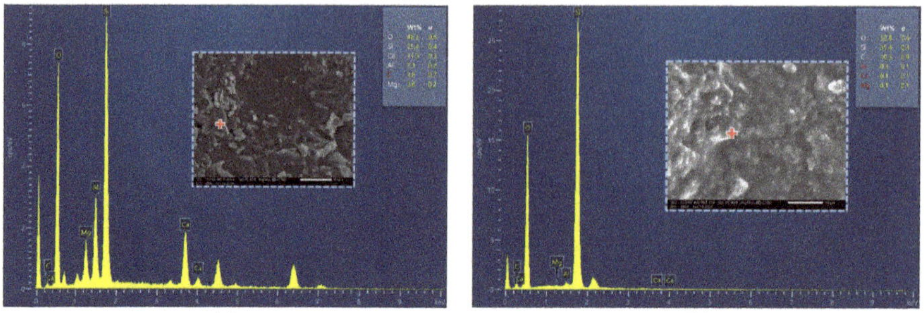

(a) (b)

Figure 11. EDS analysis results: (**a**)untreated specimen; (**b**)treated specimen.

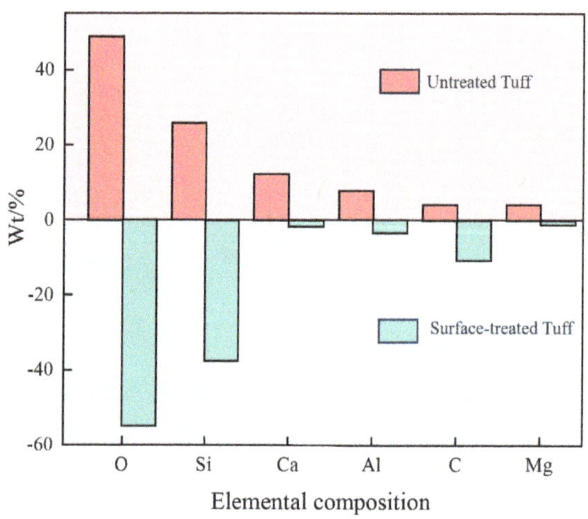

Figure 12. Elemental composition.

3.7. Mechanisms of APTES-Induced Interface Enhancement Between Tuff Aggregates and Asphalt

Figure 13 methodically depicts the modification and interface enhancement mechanisms between APTES and tuff aggregates through the formation of an APTES self-assembled molecular layer, involving three key steps: APTES hydrolysis, self-condensation, and condensation with inorganic materials.

Figure 13(1) illustrates the molecular structure of APTES ($NH_2(CH_2)_3Si(OC_2H_5)_3$) and the stages of its hydrolysis and self-condensation. During hydrolysis, the alkoxy groups on APTES are replaced by hydroxyl groups, producing ethanol as a by-product. These newly formed silanol groups can subsequently undergo self-condensation, forming Si-O-Si bonds that create a cohesive network. This foundational molecular process allows for APTES to chemically bond to the tuff surface in the subsequent stages. Figure 13(2) depicts the environmental stresses experienced by the aggregate–asphalt system, specifically water infiltration and dynamic wheel loading. The mechanism of adhesion failure between aggregate and asphalt involves the dynamic interaction of water molecules at the aggregate–asphalt interface due to the driving load's "squeezing" and "pumping" effects. This repeated interaction leads to bond failure at the interface, exacerbated by the penetration of water molecules into microscopic defects and voids on the aggregate surface.

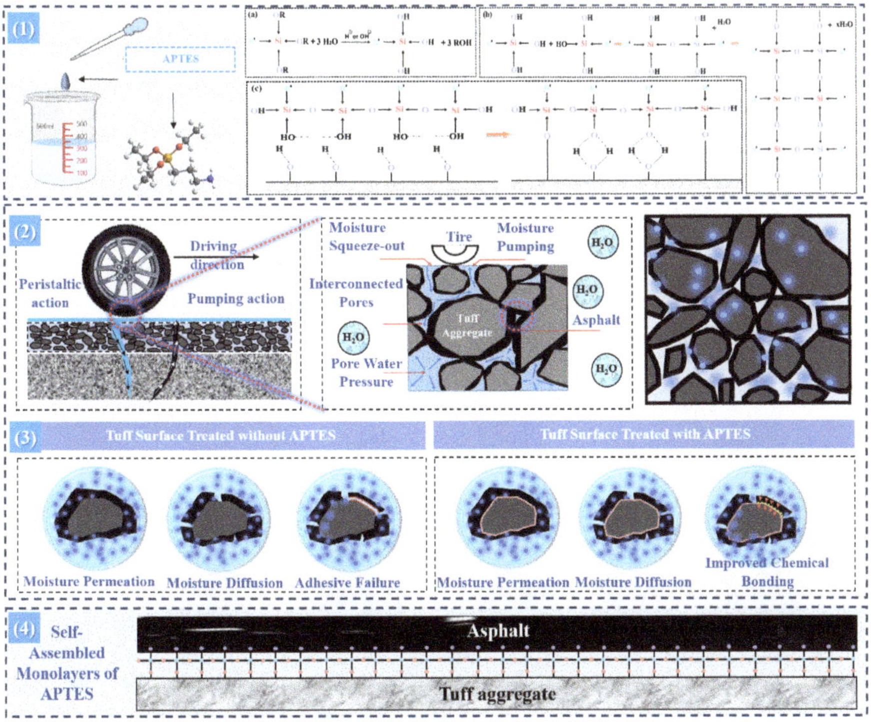

Figure 13. Schematic diagram of the enhancement mechanism.

Figure 13(3) compares the adhesion performance of untreated and APTES-treated aggregates. In untreated samples, water infiltration leads to adhesion failure due to insufficient bonding at the interface. When APTES bonds to the tuff, the γ-aminopropyl group protrudes from the surface, forming a stronger and more durable bond than physical interlocking alone. This bond reduces the risk of water infiltration at the asphalt–aggregate interface, thereby mitigating moisture-induced damage and enhancing long-term adhesion and pavement durability. The silane-treated tuff creates a hydrophobic barrier, repelling water and preserving the aggregate–asphalt bond even in wet conditions. The final part illustrates the aggregate–APTES–asphalt interface after APTES treatment. The γ-aminopropyl group of APTES is shown as a functional bridge between the inorganic tuff and organic asphalt, providing a robust, moisture-resistant bond. This enhanced interface prevents moisture-induced degradation, effectively preserving long-term adhesion and improving the durability of asphalt pavements.

In summary, while untreated tuff provides some adhesion through physical interlocking and limited chemical compatibility, treating tuff with a silane-coupling agent significantly enhances both chemical and physical bonding mechanisms.

4. Conclusions

This study addresses the challenge of enhancing adhesion between tuff and asphalt by evaluating the impact of an APTES self-assembly layer on tuff interface properties across various scales, encompassing macroscopic bonding, surface energy, and micromorphology. Utilizing the BBS test, Marshall stability test, contact angle measurements, SEM, and EDS analyses, this research explores the bond strength, water damage resistance, adhesion work, and elemental composition changes in tuff before and after surface treatment with the silane-coupling agent. The self-assembly mechanism of APTES and its effect on enhancing interfacial adhesion on tuff aggregate surfaces are systematically elucidated. The conclusions drawn from this work are as follows:

1. A mass ratio of water:ethanol:APTES solution of 5:45:50, combined with a curing temperature of 200 °C, significantly enhances the bond strength between tuff and asphalt.
2. Under optimal modification conditions, the adhesion work and energy ratio are improved to 55.28 mJ/m^2 and 1.42, respectively, indicating substantial enhancements in both adhesion strength and resistance to peeling.
3. The APTES treatment results in the formation of a denser silane-coupling agent layer on the tuff surface, which is confirmed by increased silicon content and Si-O-Si bonding, contributing to stronger chemical bonds at the asphalt–aggregate interface.
4. The chemical reaction between APTES and the tuff surface introduces γ-aminopropyl groups, significantly enhancing interfacial adhesion with asphalt. This stronger chemical bond reduces water infiltration and mitigates moisture-induced damage.

The results of this study demonstrate that APTES self-assembly on tuff aggregates significantly enhances interfacial adhesion with asphalt, suggesting that APTES-treated tuff has strong potential for durable applications in road engineering, particularly in utilizing tunnel spoil.

Author Contributions: Conceptualization, Y.L.; data curation, G.G. and J.Z.; formal analysis, H.L. and G.Z.; funding acquisition, X.Z.; investigation, G.G., J.Z. and H.Z.; methodology, M.L.; project administration, G.Z.; software, X.G. and H.Z.; supervision, L.K.; validation, L.C.; visualization, L.C. and Y.L.; writing—original draft, M.L.; writing—review and editing, X.G. All authors have read and agreed to the published version of the manuscript.

Funding: (1) Supported by the National Key R&D Program of China (No. 2022YFB2602602). (2) Supported by the Research on Application Technology of Asphalt Pavement Floor Materials Based on Tunnel Slag Tuff (No. KYL202307-0014). (3) Supported by the National Key R&D Program of China (No. 2022YFB2602604).

Institutional Review Board Statement: Not applicable.

Informed Consent Statement: Not applicable.

Data Availability Statement: Data are contained within the article.

Conflicts of Interest: Authors Mingxin Lai, Gen Zhang, Hongming Zhu and Xinping Zhang were employed by Fujian First Highway Engineering Group Co., Ltd. Author Xiaoying Gao was employed by Fujian Expressway Group Co., Ltd. Author Lizong Chen was employed by Dehua County Road and Bridge Construction Co., Ltd. Author Haixing Lin was employed by Fujian Communications Planning & Design Institute CO., Ltd. Author Yueling Lin was employed by Quanzhou Road and Bridge Engineering Construction Co., Ltd. The remaining authors declare that the research was conducted in the absence of any commercial or financial relationships that could be construed as a potential conflict of interest.

References

1. Mills-Beale, J.; You, Z. The mechanical properties of asphalt mixtures with recycled concrete aggregates. *Constr. Build. Mater.* **2010**, *24*, 230–235. [CrossRef]
2. Lewis, S.; Coleri, E.; Sukhija, M.; Sreedhar, S. Blending of Virgin and RAP Binder for Asphalt Mixes with High RAP Contents: A Pilot Study. *Int. J. Pavement Res. Technol.* **2024**, 1–15. [CrossRef]
3. Zhang, K.; Wei, G.; Luo, Y.; Zhao, Y.; Zhao, Y.; Zhang, J. Evaluation of aggregate distribution homogeneity for asphalt pavement based on the fractal characteristic of three-dimensional texture. *Int. J. Pavement Res. Technol.* **2024**, *17*, 577–594. [CrossRef]
4. Fang, M.; Park, D.; Singuranayo, J.L.; Chen, H.; Li, Y. Aggregate gradation theory, design and its impact on asphalt pavement performance: A review. *Int. J. Pavement Eng.* **2019**, *20*, 1408–1424. [CrossRef]
5. Luhar, S.; Cheng, T.-W.; Luhar, I. Incorporation of natural waste from agricultural and aquacultural farming as supplementary materials with green concrete: A review. *Compos. Part B Eng.* **2019**, *175*, 107076. [CrossRef]
6. Gautam, P.K.; Kalla, P.; Jethoo, A.S.; Agrawal, R.; Singh, H. Sustainable use of waste in flexible pavement: A review. *Constr. Build. Mater.* **2018**, *180*, 239–253. [CrossRef]

Fang, F.T.; Chong, Y.C.; Nyunt, T.T.; Loi, S.S. Development of environmentally sustainable pavement mix. *Int. J. Pavement Res. Technol.* **2013**, *6*, 440.

Zhang, X.; Zhang, Y.; Yang, H.; Lin, Q.; Tang, B. Laboratory Investigation of the Water Damage Resistance of Tuff Asphalt Mixture Modified with Additives. *Coatings* **2023**, *13*, 302. [CrossRef]

Al-Omari, A.A.; Al-Rousan, T.M.; Al-Facara, E.A.S. Laboratory Characterization of Pozzolan-Modified Asphalt Binder. *Int. J. Pavement Res. Technol.* **2022**, *15*, 1051–1061. [CrossRef]

0. Yang, T.; Zheng, J.; Xie, B. Experimental study on high and low temperature and water stability of tuff asphalt mixture. *J. Highw. Transp. Res. Dev.* **2016**, *33*, 1–6. (In Chinese)

1. Chen, C.; Xu, J.; Okubo, S.; Peng, S. Damage evolution of tuff under cyclic tension–compression loading based on 3D digital image correlation. *Eng. Geol.* **2020**, *275*, 105736. [CrossRef]

2. Alkofahi, N.; Khedaywi, T. Stripping potential of asphalt mixtures: State of the art. *Int. J. Pavement Res. Technol.* **2022**, *15*, 29–43. [CrossRef]

3. Abouelsaad, A.; White, G. Review of asphalt mixture ravelling mechanisms, causes and testing. *Int. J. Pavement Res. Technol.* **2021**, *15*, 1448–1462. [CrossRef]

4. Chakravarty, H.; Sinha, S.; Kumar, A. Influence of air voids and aggregates composition on moisture damage sensitivity of bituminous mixes. *Int. J. Pavement Res. Technol.* **2023**, *16*, 862–872. [CrossRef]

5. Wang, L.; Shen, A.; Yao, J. Effect of different coarse aggregate surface morphologies on cement emulsified asphalt adhesion. *Constr. Build. Mater.* **2020**, *262*, 120030. [CrossRef]

6. Wang, W.; Luo, R.; Yang, H.; Wang, L. Analysis of water erosion on asphalt binder using multi-scale experimental methods. *Int. J. Pavement Res. Technol.* **2022**, *15*, 485–495. [CrossRef]

7. Nazirizad, M.; Kavussi, A.; Abdi, A. Evaluation of the effects of anti-stripping agents on the performance of asphalt mixtures. *Constr. Build. Mater.* **2015**, *84*, 348–353. [CrossRef]

8. Alam, M.N.; Aggarwal, P. Effectiveness of anti stripping agents on moisture susceptibility of bituminous mix. *Constr. Build. Mater.* **2020**, *264*, 120274.

9. Ameli, A.; Norouzi, N.; Khabbaz, E.H.; Babagoli, R. Influence of anti stripping agents on performance of binders and asphalt mixtures containing Crumb Rubber and Styrene-Butadiene-Rubber. *Constr. Build. Mater.* **2020**, *261*, 119880. [CrossRef]

0. Wu, H.; Li, P.; Nian, T.; Zhang, G.; He, T.; Wei, X. Evaluation of asphalt and asphalt mixtures' water stability method under multiple freeze-thaw cycles. *Constr. Build. Mater.* **2019**, *228*, 117089. [CrossRef]

1. Sun, H.; Ding, Y.; Jiang, P.; Wang, B.; Zhang, A.; Wang, D. Study on the interaction mechanism in the hardening process of cement-asphalt mortar. *Constr. Build. Mater.* **2019**, *227*, 116663. [CrossRef]

2. Zhang, H.; Duan, H.; Tang, J. Effects of different anti-stripping agents on physical, rheological and aging properties of asphalt. *J. Highw. Transp. Res. Dev.* **2021**, *38*, 1–9+111+18. (In Chinese)

3. Peng, C.; Lu, L.; You, Z.; Xu, F.; You, L.; Miljković, M.; Guo, C.; Huang, S.; Ma, H.; Hu, Y. Influence of silane-hydrolysate coupling agents on bitumen–aggregate interfacial adhesion: An exploration from molecular dynamics simulation. *Int. J. Adhes. Adhes.* **2022**, *112*, 102993. [CrossRef]

4. Pérez-Martínez, M.; Moreno-Navarro, F.; Martín-Marín, J.; Ríos-Losada, C.; Rubio-Gámez, M.C. Analysis of cleaner technologies based on waxes and surfactant additives in road construction. *J. Clean. Prod.* **2014**, *65*, 374–379. [CrossRef]

5. Wang, L.; Liu, J.; Wang, H. Effects of surfactants on the performance of diluted asphalt and its mixtures. *J. Highw. Transp. Res. Dev.* **2023**, *40*, 1–8. (In Chinese)

6. Nejad, F.M.; Asadi, M.; Hamedi, G.H. Determination of moisture damage mechanism in asphalt mixtures using thermodynamic and mix design parameters. *Int. J. Pavement Res. Technol.* **2020**, *13*, 176–186. [CrossRef]

7. Han, X.; Cao, Z.; Wang, R.; Hė, P.; Zhang, Y.; Yu, J.; Ge, Y. Effect of silane coupling agent modified zeolite warm mix additives on properties of asphalt. *Constr. Build. Mater.* **2020**, *259*, 119713. [CrossRef]

8. Hao, H.; Zhang, A.; Cheng, Y.; Cong, P. The modification mechanisms of silane coupling agent (SCA) on the physical properties of thermosetting polyurethane asphalt binder (PUAB). *Constr. Build. Mater.* **2022**, *350*, 128836. [CrossRef]

9. Maslennikova, V.V.; Filatov, S.N.; Orlov, A.V.; Surin, N.M.; Svidchenko, E.A.; Chistyakov, E.M. Luminescent coatings based on (3-Aminopropyl) triethoxysilane and europium complex β-Diketophosphazene. *Polymers* **2022**, *14*, 728. [CrossRef]

0. Nurdina, A.; Mariatti, M.; Samayamutthirian, P. Effect of filler surface treatment on mechanical properties and thermal properties of single and hybrid filler–filled PP composites. *J. Appl. Polym. Sci.* **2011**, *120*, 857–865. [CrossRef]

1. Tee, D.; Mariatti, M.; Azizan, A.; See, C.; Chong, K. Effect of silane-based coupling agent on the properties of silver nanoparticles filled epoxy composites. *Compos. Sci. Technol.* **2007**, *67*, 2584–2591. [CrossRef]

2. Shang, X.; Zhu, Y.; Li, Z. Surface modification of silicon carbide with silane coupling agent and hexadecyl iodiele. *Appl. Surf. Sci.* **2017**, *394*, 169–177. [CrossRef]

3. Yang, R.; Liu, Y.; Wang, K.; Yu, J. Characterization of surface interaction of inorganic fillers with silane coupling agents. *J. Anal. Appl. Pyrolysis* **2003**, *70*, 413–425. [CrossRef]

4. Wang, L.; Tang, C.; Wang, X.; Zheng, W. Molecular dynamics simulation on the thermodynamic properties of insulating paper cellulose modified by silane coupling agent grafted nano-SiO_2. *AIP Adv.* **2019**, *9*, 125134. [CrossRef]

5. Min, Y.; Fang, Y.; Huang, X.; Zhu, Y.; Li, W.; Yuan, J.; Tan, L.; Wang, S.; Wu, Z. Surface modification of basalt with silane coupling agent on asphalt mixture moisture damage. *Appl. Surf. Sci.* **2015**, *346*, 497–502. [CrossRef]

36. Feng, X.; Chen, W.; Li, W. Effects of silane coupling agent modified coal waste powder on performance of asphalt mortar an asphalt mixture. *Int. J. Pavement Res. Technol.* **2020**, *13*, 383–391. [CrossRef]
37. Yang, Y.Z.; Pinxue; Zhang, Y. Effect of basalt fiber surface modification on asphalt mortar performance. *J. Highw. Transp. Res. De* **2023**, *40*, 9–15+25. (In Chinese)
38. Ahangaran, F.; Navarchian, A.H. Recent advances in chemical surface modification of metal oxide nanoparticles with silan coupling agents: A review. *Adv. Colloid Interface Sci.* **2020**, *286*, 102298. [CrossRef]
39. Han, X.; Yu, J.; Cao, Z.; Wang, R.; Du, W.; He, P.; Ge, Y. Preparation and properties of silane coupling agent modified zeolite a warm mix additive. *Constr. Build. Mater.* **2020**, *244*, 118408. [CrossRef]
40. Chmielewska, B.; Czarnecki, L.; Sustersic, J.; Zajc, A. The influence of silane coupling agents on the polymer mortar. *Cem. Conc Compos.* **2006**, *28*, 803–810. [CrossRef]
41. Magdaleno-López, C.; de Jesús Pérez-Bueno, J. Quantitative evaluation for the ASTM D4541-17/D7234 and ASTM D335 adhesion norms with digital optical microscopy for surface modifications with flame and APPJ. *Int. J. Adhes. Adhes.* **2020**, *9* 102551. [CrossRef]
42. Zhang, D.; Luo, R. Modifying the BET model for accurately determining specific surface area and surface energy components c aggregates. *Constr. Build. Mater.* **2018**, *175*, 653–663. [CrossRef]
43. Wayner, P.C., Jr. Spreading of a liquid film with a finite contact angle by the evaporation/condensation process. *Langmuir* **1993**, 294–299. [CrossRef]
44. Gouin, H. The wetting problem of fluids on solid surfaces. Part 2: The contact angle hysteresis. *Continuum Mech. Thermody* **2003**, *15*, 597–611. [CrossRef]
45. Lou, K.; Xiao, P.; Tang, Q.; Wu, Y.; Wu, Z.; Pan, X. Research on the micro-nano characteristic of basalt fiber and its impact on th performance of relevant asphalt mastic. *Constr. Build. Mater.* **2022**, *318*, 126048. [CrossRef]
46. Feng, P.; Wang, H.; Ding, H.; Xiao, J.; Hassan, M. Effects of surface texture and its mineral composition on interfacial behavio between asphalt binder and coarse aggregate. *Constr. Build. Mater.* **2020**, *262*, 120869. [CrossRef]
47. Gao, J.; Liu, P.; Wu, Y.; Xu, Y.; Lu, H. Moisture damage of asphalt mixture and its evaluation under the long-term soaked duratio *Int. J. Pavement Res. Technol.* **2021**, *14*, 607–614. [CrossRef]
48. Liu, K.; Deng, L.; Zheng, J. Nanoscale study on water damage for different warm mix asphalt binders. *Int. J. Pavement Res. Techno* **2016**, *9*, 405–413. [CrossRef]
49. Guo, F.; Pei, J.; Zhang, J.; Xue, B.; Sun, G.; Li, R. Study on the adhesion property between asphalt binder and aggregate: state-of-the-art review. *Constr. Build. Mater.* **2020**, *256*, 119474. [CrossRef]
50. Bhasin, A.; Little, D.N.; Vasconcelos, K.L.; Masad, E. Surface free energy to identify moisture sensitivity of materials for aspha mixes. *Transp. Res. Rec.* **2007**, *2001*, 37–45. [CrossRef]

coatings

MDPI

.rticle

Method of Forming Road Surface Replicas Using 3D Printing Technology

Vojciech Owczarzak *, Sławomir Sommer and Grzegorz Ronowski

Institute of Mechanics and Machine Design, Faculty of Mechanical Engineering and Ship Technology and EkoTech Center, Gdańsk University of Technology, ul. Narutowicza 11/12, 80-233 Gdańsk, Poland; slasomme@pg.edu.pl (S.S.); gronowsk@pg.edu.pl (G.R.)
* Correspondence: wojowcza@pg.edu.pl

Abstract: Rolling resistance is a critical factor that influences vehicle energy consumption, emissions, and overall performance. It directly impacts fuel efficiency, tire longevity, and driving dynamics. Traditional rolling resistance tests are conducted on smooth steel drums, which fail to replicate real-world road surface textures, potentially skewing results. This article presents the process of designing surface replicas using 3D printing technology, which consisted of selecting the internal structure, material, and print parameters of the surface sample. In order to verify the designed structures, an original mechanical strength test was performed. The test was based on pressing the tire onto the test sample with an appropriate force that corresponded to typical conditions during rolling resistance measurements. The test results included surface texture profiles before and after the application of load, which were then superimposed to detect any possible sample deformation. The obtained strength test results confirmed the validity of using 3D printing technology in the process of obtaining road surface replicas.

Keywords: tire; road pavement; texture; rolling resistance; 3D printing; measurements

1. Introduction

The rolling resistance of car tires is one of the most significant parameters describing the interaction between the tire and the road surface. It directly affects the energy consumption of the moving vehicle, the emission of toxic compounds into the atmosphere (such as NH_3, CO_2, or NO_2), as well as its performance, such as maximum speed, acceleration, and maximum range. What is more, a high rolling resistance value also increases the tire temperature, which accelerates the aging process and reduces its mechanical strength. Rolling resistance is one of the primary resistances to motion (along with air resistance, incline resistance, and inertia). Depending on the driving conditions, reducing rolling resistance by 10% may lead to a 2%–4% decrease in fuel consumption. The rolling resistance of car tires results directly from energy losses that occur during cyclic deformation of the tire structure during pressures occurring on the track of its contact with the road. Tire deflection requires some work that must be supplied to the system from the vehicle side. The materials used to produce tires are characterized by significant hysteresis, which means that the amount of energy used to deform the tire is greater than the amount of energy released when the tire returns to its original shape. The vast majority of the energy lost by the vehicle as a result of rolling resistance is converted into heat. The remaining, much smaller part of the energy is converted into acoustic energy, permanent deformation of the road surface, and tire abrasion [1–4].

The rolling resistance of a car tire is determined by factors such as [2]:

1. Tire construction (width, height, tread pattern and depth, number of cord layers, belt material, type of rubber compound).
2. Traffic conditions, i.e., wheel load, inflation, tire temperature, rolling direction, rolling speed.

3. Road surface, i.e., its type, texture, stiffness, and technical condition.

This article focuses on the last of these: the influence of texture. The figure below shows the texture sub-ranges along with their influence on the tire-road interaction (Figure 1).

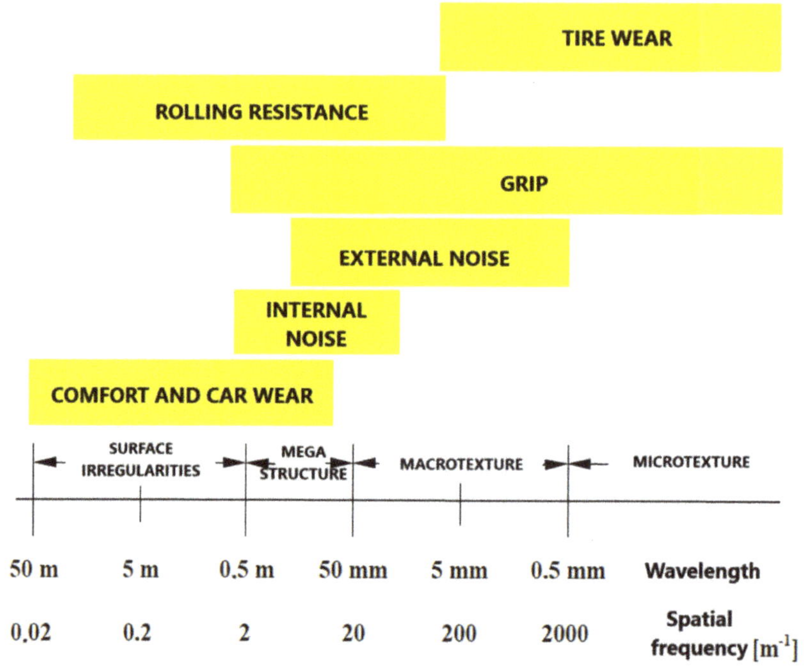

Figure 1. Texture ranges and their impact on tire-road interaction [5].

The texture of the road surface significantly affects the deformation of the tire tread elements, and therefore contributes to the increase in rolling resistance. Deformations also occur on the surface side, but due to the fact that the mechanical impedance of the surface is much greater than that of the tire, the deformations of the former are much smaller. Many previous studies have tried to describe the interactions between the tire and the road surface [6–8]. When talking about the influence of surface texture on rolling resistance, we cannot forget about the phenomenon of so-called enveloping. Too little elasticity of the tread elements does not allow them to be deformed to accurately reflect the shape of the surface. Contact with the surface is limited only to the highest parts of the surface (aggregate peaks). The tire tread is pressed and deflected by the highest points of the surface, but due to the stiffness of the rubber compound and the belt, its elements do not penetrate into the surface texture recesses but create "bridges" between adjacent peaks. As a result, air pockets and channels are formed on the contact patch of the tire with the surface. In the figure below we can see how two different texture profiles A and B deflect the tire in the same way, but the air pockets and channels differ to a large extent (Figure 2) [9].

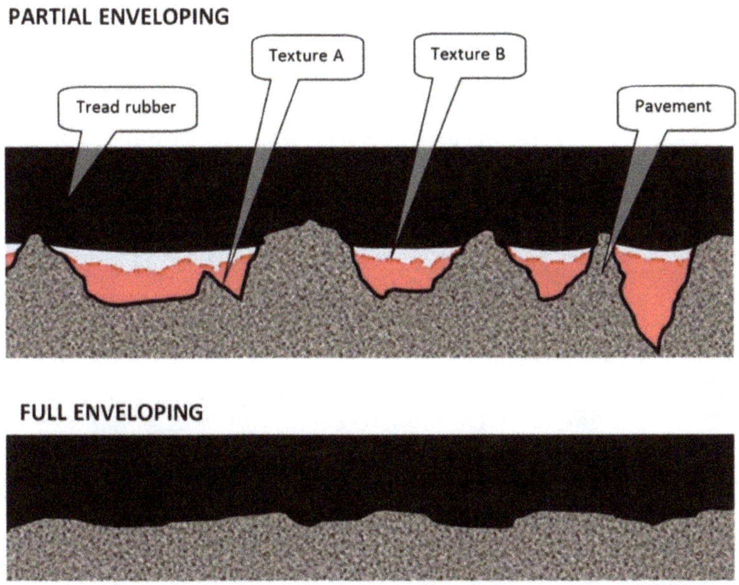

Figure 2. Interface between tire and road surface for pavements with different texture [9].

Rolling resistance can be reduced by improving tire production technologies and road surfaces. In order to achieve this, it is essential to conduct measurements that verify the effectiveness of the solutions applied in this regard. Rolling resistance is measured using both road and laboratory methods [10]. When reviewing contemporary literature, one can also find attempts to simulate the rolling resistance of car tires using computer simulations [11,12]. Tire manufacturers highly prefer laboratory testing methods, which allow tire testing regardless of weather conditions and enable greater accuracy and repeatability of results. Currently applicable standards, such as ISO 28580 (2018) and SAE J2452 (2017), require that standard tests be conducted on tire testing machines equipped with smooth steel drums, which is called the torque method (Figure 3) [13–16]. The main reasons for using steel drums are the difficulties in replicating the texture of typical road surfaces on the drums and the challenge of ensuring that all testing devices used for standardized tests have identically textured drums. It is well known that road surface texture strongly affects tire rolling resistance, this means that tests conducted on smooth steel drums do not yield results that accurately correspond to real road conditions. Rolling resistance coefficients (CRR) measured during laboratory tests are lower due to the lack of tread element deformation caused by the road surface texture [9]. In order to maintain the advantages of laboratory testing while ensuring a high level of representativeness in the results the drums are equipped with replicas of road surfaces that have the same texture as the surfaces on which vehicles actually drive.

Currently, replicas are produced using the casting method (replicas are made with the help of casting molds using materials such as epoxy resins and fiberglass). Based on our own experience and a review of the literature [6–8] on the influence of road surface texture on rolling resistance results, there is a strong need to improve the method of manufacturing road surface replicas. In contrast to the currently used technology, the use of 3D printing will allow for filling the research gap in the form of creating the possibility of designing original surface textures (current technology allows for the creation of surface replicas with existing textures). In addition to the possibility of rapid prototyping of surface textures in terms of their impact on rolling resistance, it will also allow for a more thorough exploration of the previously mentioned enveloping phenomenon (creating replicas with a similar texture, differing in profile depth). This will also allow for an extension of the range of measurement conditions. The smaller mass of the replicas will allow for the drum to accelerate to higher speeds (lower centrifugal force), or it will improve the fit of the replica segments to each other, which will allow them to be used for accurate measurements of car tire noise (no acoustic interference when the tire drives over the replica joints). The research presented in the article is an introduction to the development of a method for forming road surface replicas using 3D printing technology. The research objective was to determine the internal structure and printing parameters so that the replicas would meet the measurement conditions occurring during tests on drum stands.

Figure 3. A running machine, which belongs to the equipment of the Mechanical Vehicles and Military Techniques Department of the Gdańsk University of Technology, drum with a diameter of 2 m.

2. Materials and Methods

The Gdańsk University of Technology has the world's only laboratory that uses replicas of actual road surfaces, made from reinforced laminates, which are produced using epoxy resins while preserving the original surface texture (Figure 4). Tests conducted on these replicas are more reliable compared to those performed on steel surfaces, as they better reflect the real operating conditions of vehicles. The results of rolling resistance and noise measurements on these replicas are consistent with those obtained on actual road surfaces under real traffic conditions. Road surface replicas adapted for mounting on tire testing machine drums must meet several requirements. The first is mechanical strength, which ensures safe installation on rapidly rotating drums. The centrifugal force generated during drum rotation would destroy almost any road surface material used for wearing courses. The road surface replica and its mounting system must withstand both the centrifugal force and the localized pressure of the tires on the surface. The second requirement is that the replicas shall accurately reflect the texture of the road surface, which they imitate.

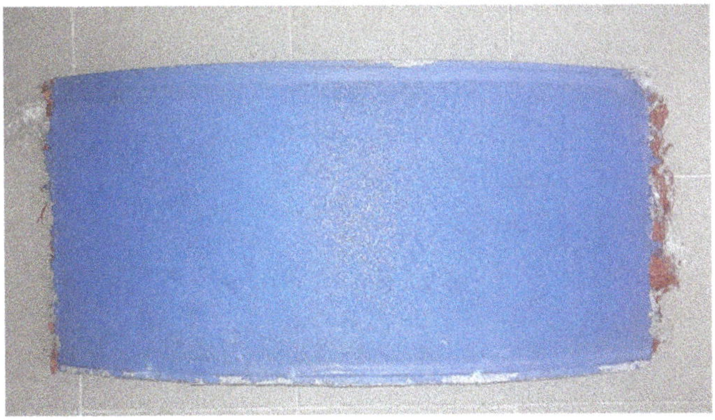

Figure 4. Road surface replica (PERS surface).

The simplified diagram of the production of this type of road surface replica is presented below:

1. First, a layer of silicone rubber is poured onto the original surface, which after hardening will constitute a negative representation of the road surface (Figure 5). This layer is additionally reinforced with a glass fiber mat.

Figure 5. The process of creating an elastic layer that is a negative representation of the road surface.

2. The obtained elastic coating is next placed in the concave half of the mold (Figure 4). In the same part of the mold, gelcoat layers (the outer layer of the replica) and epoxy resin equalization layers are applied successively.
3. In the second part of the mold, layers of glass fiber immersed in epoxy resin are applied.
4. In the final phase, both parts of the mold are joined and the empty space between them is filled with epoxy casting resin. It is worth noting that the finished replica in the form of a shell is only a part of the entire surface on which the tested car wheel rolls (Figure 6). Usually, three to eight shells are attached to the drum track.

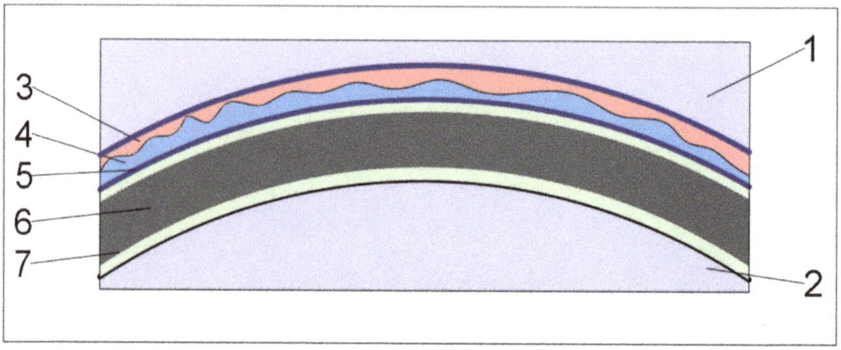

Figure 6. Diagram showing the method of forming a replica of a road surface for running machines with an external drum (1, 2—mold halves; 3—elastic coating constituting a negative reproduction of the road surface; 4—top layer of the replica (gelcoat layer mixed with finely cut glass fiber); 5—leveling layer of epoxy resin; 6—layer of epoxy casting resin; 7—layer consisting of glass fiber mats immersed in epoxy resin).

The results obtained on the replicas are fully satisfactory, but the process of their production has many drawbacks, hence the need to test a new technology for their production. The main disadvantages are high time consumption, laborious processing of replica segments in order to fit them to the drum surface, and the production process itself generates a lot of chemical waste requiring disposal. Due to the growing interest in tire and automotive industry concerns in research on surface replicas, there was a need to improve the method of their production. The research presented later in the article also aims to determine whether 3D printing technology will shorten the time needed to produce replicas, reduce costs, and reduce hazardous chemical waste. This method will also simplify the modification of existing road surface textures or enable the creation of new road surface textures. The process of developing a road surface replica using 3D printing technology is shown in the flowchart below (see Figure 7).

In order to develop a method for forming full-size replicas of road surfaces for covering the external tread of the wheel drum using 3D printing technology, it was necessary to select the internal structure and the appropriate material (filament).

For this purpose, predefined fill geometries from the library of the 3D printer software were used, and several types of fill were chosen.

Description of the different types of fill for the replicas:

- Geometry called "grid", which is based on a square filling with a filling level from 5 to 15% (see Figure 8). The internal structure of the "grid" type in 3D printing is one of the popular methods of filling the print. It consists of regular connections in the form of a grid, creating spatial cells. The supporting nodes are the points of intersection of the grid lines. In the classic "grid" filling, the nodes are evenly distributed, creating a symmetrical, mesh structure, resembling a rectangular or square system.
- Geometry called "concentric", with a filling level from 5 to 15% (see Figure 9). The "concentric" fill in 3D printing differs from the "grid" structure primarily in the way the fill lines are arranged. Instead of creating grid-like, rectangular patterns, the fill lines are arranged concentrically (around the object) in concentric layers that reproduce the external shape of the model. The "concentric" fill does not have the typical supporting nodes of the "grid" structure because the fill lines are arranged in layers in a continuous manner, without regular intersection points. In practice, supporting nodes could only appear in places where successive circles, ellipses, or other curves end.

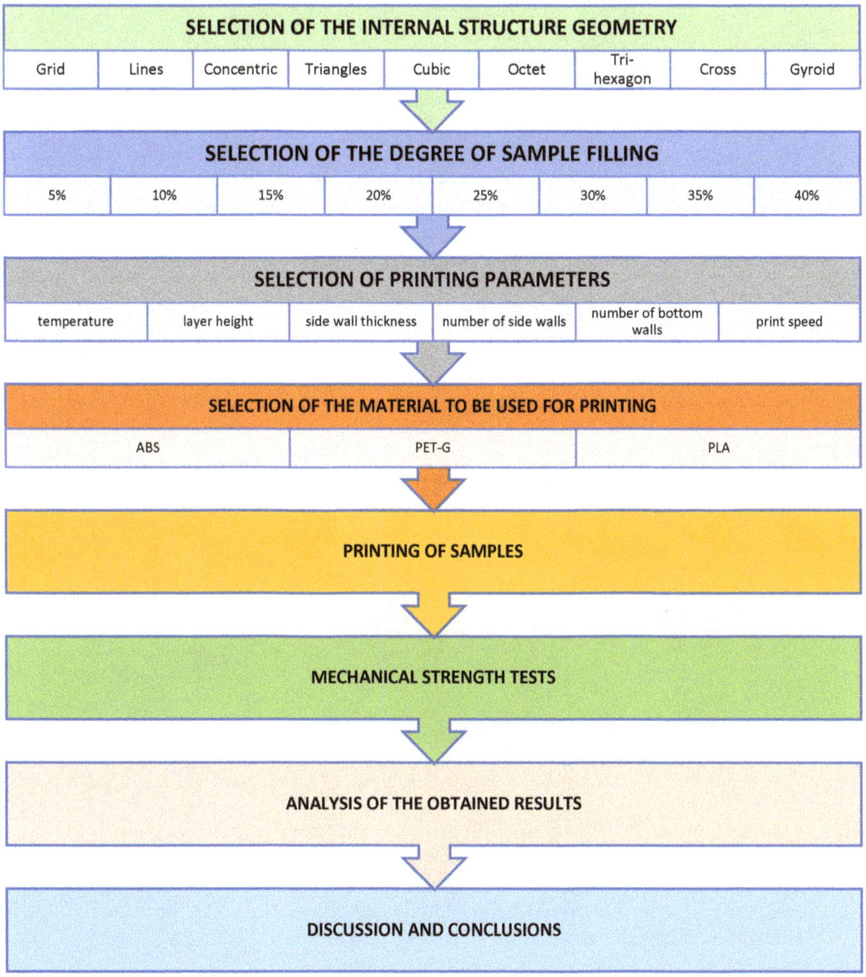

Figure 7. Flowchart showing the research stages.

Figure 8. "Grid" geometry of pavement replica fillings (5%, 10%, and 15% of infill density).

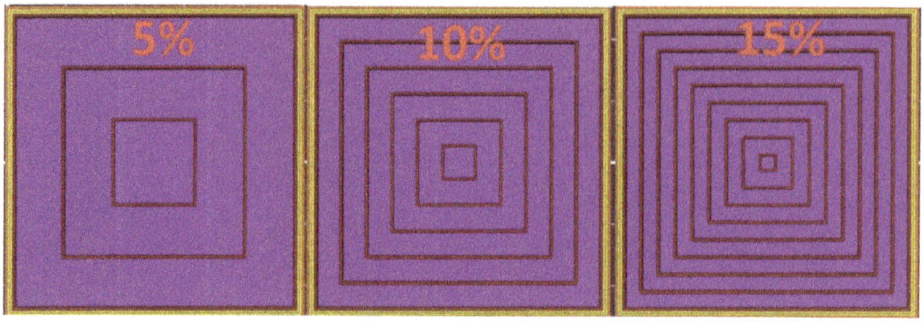

Figure 9. "Concentric" geometry of pavement replica fillings (5%, 10%, and 15% of infill density).

- Geometry called "triangles", with a filling level from 5 to 15% (see Figure 10). In a cross section, the "triangles" structure presents regular, equilateral triangles. In the "triangles" structure, the supporting nodes are located at the points of intersection of the lines forming the triangles.

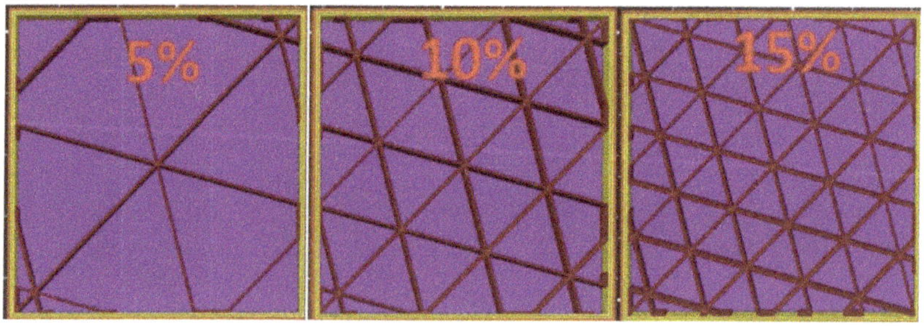

Figure 10. "Triangles" geometry of pavement replica fillings (5%, 10%, and 15% of infill density).

- Geometry called "tri-hexagon", with a filling level from 5 to 15% (see Figure 11). The "tri-hexagon" structure in 3D printing is more complex than simple infills because it combines two different shapes—triangles and hexagons. The supporting nodes in the "tri-hexagon" structure are located at the places where the lines that make up both triangles and hexagons meet.

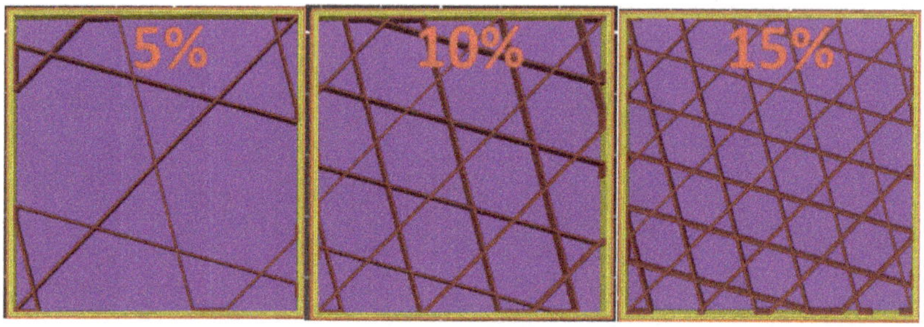

Figure 11. "Trihexagon" geometry of pavement replica fillings (5%, 10%, and 15% of infill density).

- Geometry called "cross", with a filling level from 5 to 15% (see Figure 12). The "cross" infill in 3D printing is based on a system of lines that create a cross pattern. The "cross" infill does not have typical load-bearing nodes. The lines are arranged perpendicularly, but do not create nodal points, as in the case of more complex structures, e.g., "triangles" or "tri-hexagon". The strength of the infill comes rather from the overall geometric stability of the lines, which distributes forces over the entire surface of the model, without concentrating them in single points.

Figure 12. "Cross" geometry of pavement replica fillings (5%, 10%, and 15% of infill density).

Below is a description of the common features of all models:

- Each replica was printed using layers with a height of 0.15 mm. The print layer height was selected based on previous experience. A thicker layer (0.20 mm) would create a "stepped" top layer, which would result in differences in the texture of the original and the replica. On the other hand, a thinner layer would significantly extend the printing time, which is one of the parameters that requires optimization in the currently used method of replica production.
- The bottom and top layers of the replicas consisted of a solid fill with a thickness of 1.05 mm.
- The peripheral layer consisted of a triple wall with a thickness of 1.05 mm.

The exterior of the replica is a road surface texture. To develop a method for forming such replicas, the decision was made to replicate the rough APS4 surface (see Figure 13). Due to the protruding aggregate elements, this surface is prone to mechanical damage [17]. Therefore, it was assumed that if the internal structure and suitable material could be developed for this type of road surface, the other types would also meet the required standards. The geometries of the APS4 surface texture were obtained using a 3D scanner.

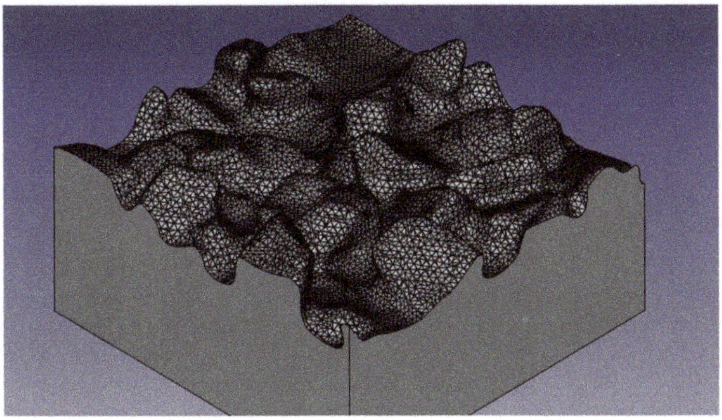

Figure 13. APS4 surface 3D model.

To create a replica it was decided to use the material PET-G [18]. This choice was dictated by high mechanical strength and a correspondingly high softening temperature of 82 °C [19]. Under standard test conditions, the surface temperature does not exceed 50 °C [4]. To verify the design assumptions, samples of dimensions 50 × 50 × 20 mm were printed. These samples were used for strength testing to determine if the designed replica met the requirements for rolling resistance testing on wheel drums machines. The road surface samples were subjected to a proprietary compression strength test. Due to the specific interaction between the tire and the road surface (the elastic deformation of tread elements over surface roughness), a traditional strength test using a testing machine was not chosen. Instead, it was decided to use an original method, which consisted of pressing a sample of the surface with a tire (smooth slick tread). The tests were performed starting from a load of 3 kg/cm^2 (standard load in passenger car tires), ending with 8 k/cm^2 (load in trucks) [2]. It was decided to measure only the compressive strength because during the actual measurement of rolling resistance the wheel rolls freely. During the measurement there is no braking or driving torque (it occurs only during the acceleration and braking of the drum, but the times of these processes have been selected so that the generated torques are small). The test was performed on a drum wheel machine, which is used in the torque method mentioned in the introduction (see Figure 14). To better understand the idea of carrying out rolling resistance tests using the torque method, its description is given below. It is based on the use of wheel drum machines, where rolling resistance is measured by measuring the torque on the drum drive shaft. When determining rolling resistance, the running machine's own resistances are taken into account, which are determined by measuring the torque driving the drum, on which the wheel rolls pressed with a force of only 100 N (the rolling resistance during such a test is negligible and is less than 0.5 N). The advantage of this method is its accuracy and low sensitivity to deviations or shifts in the line of action of the radial force loading the wheel from the straight line normal to the drum surface and passing through the center of the tested wheel. The rolling resistance force and the rolling resistance coefficient are determined based on the following Formula (1).

$$f_t = \frac{\frac{M - M_{SKIM}}{R_b} \cdot [1 + k(T - T_o)] \cdot \left(1 + \frac{r}{R_b}\right)}{N} \tag{1}$$

where

M—moment of force measured on the shaft of the running machine drum [Nm]
M_{SKIM}—moment of force of the running machine's own resistances [Nm]
R_b—radius of the running machine drum [m]
r—radius of the tested wheel [m]
k—empirical coefficient
T—ambient temperature during rolling resistance tests [°C]
T_o—reference temperature 25 °C
N—force normal to the drum surface [N]

In order to determine the deformation of the samples, a laser profilometer was used. The surface profile was measured before and after the compression, and the area between these profiles was compared (see Figure 15).

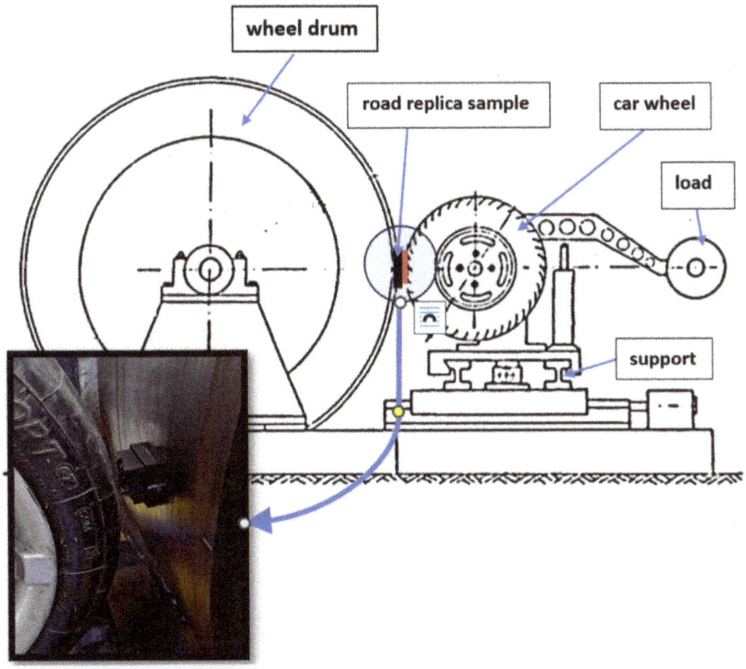

Figure 14. A stand for carrying out strength tests.

Figure 15. Road replica sample scanned by laser profilometer.

3. Results

The results obtained using the research method described above are presented in figures below in the form of superimposed surface texture profiles.

3.1. Replicas with Internal Structures of the "Grid", "Trihexagon" and "Cross" Types

Road replicas with internal structures of the "grid", "trihexagon" and "cross" types proved to be fully mechanically resistant to the loads occurring during the test. Even the lowest filling level of 5% was completely sufficient. As can be seen in the graphs below, the surface profiles obtained after each load match the profile obtained before the measurements (Figure 16). This proves that there was no mechanical damage to the tested replicas.

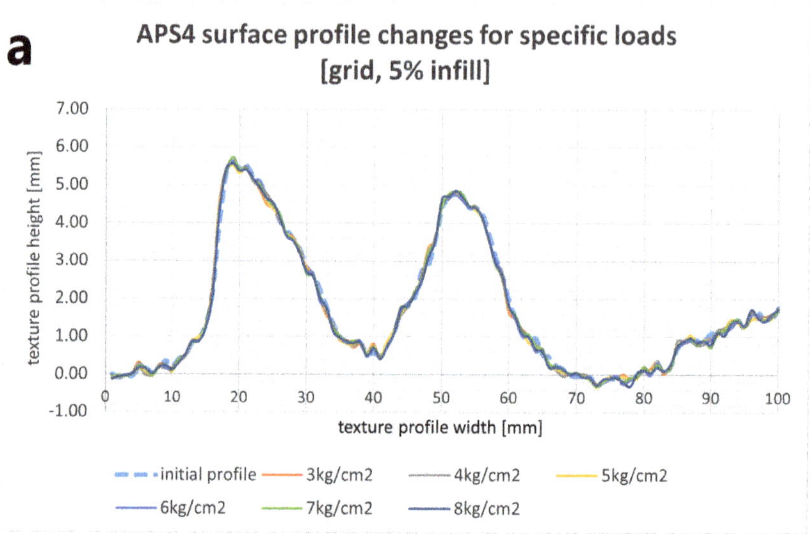

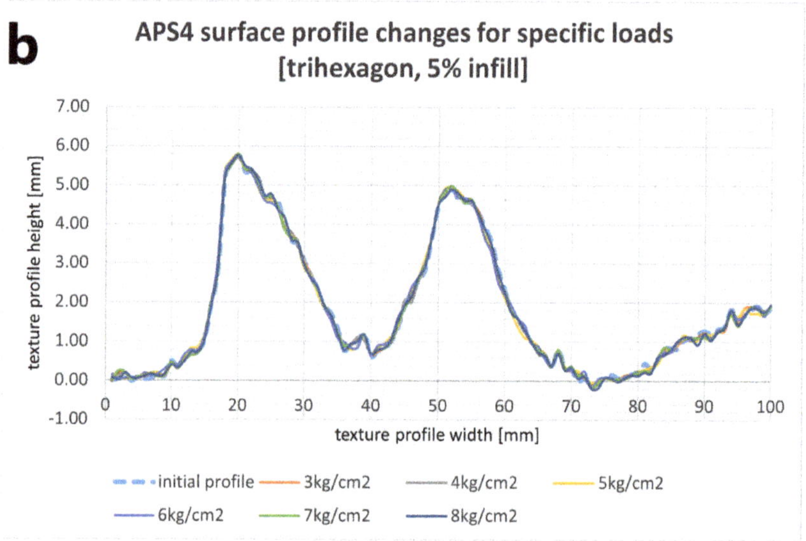

Figure 16. *Cont.*

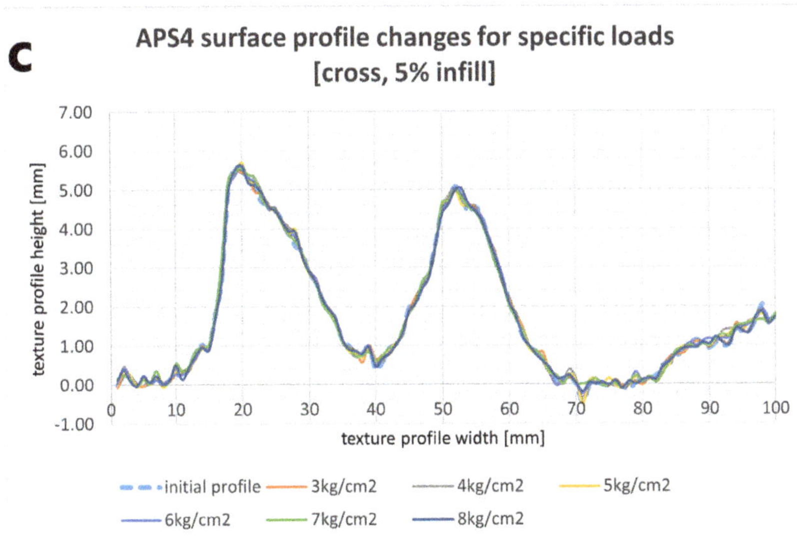

Figure 16. Graphs showing the strength test results for individual samples: (**a**) Grid; (**b**) Trihexagon; (**c**) Cross.

3.2. Replica with Internal Structure of the "Concentic" Type

For a 5% fill level, the profiles are superimposed as follows (Figure 17). The protruding "peaks" (texture profile width 22 mm and 55 mm) of the texture and "valleys" (texture profile width 0–15 mm) show material deformation. Material deformation had already occurred at a load of 3 kg/cm^2. The deformations were small (the largest reached 0.6 mm), but in the case of road surface replicas no deformations are not allowed. Replicas must maintain texture continuity and stiffness throughout their period of use.

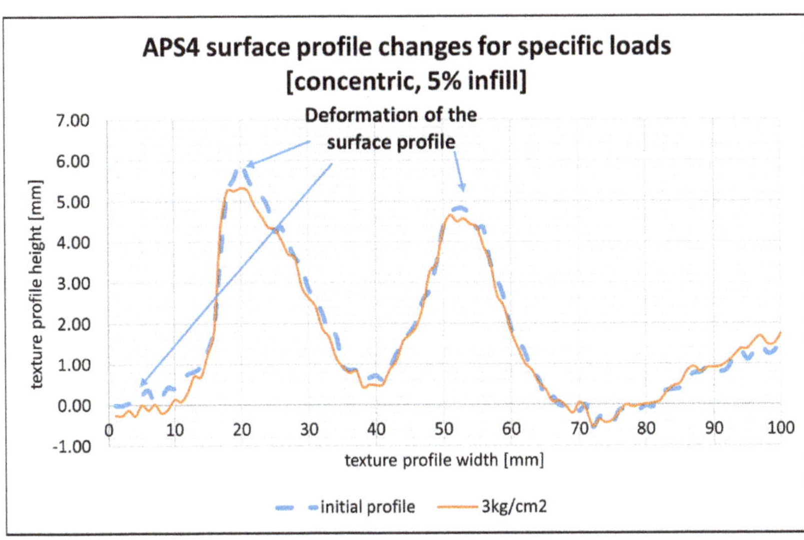

Figure 17. APS4 surface profile changes for specific loads (concentric, 5% infill).

Next, a sample with a 10% fill level was tested. As can be seen, the damage is much smaller, with only one peak deformation (texture profile width 50–55 mm, deformation reach 0.4 mm) at a load of 6 kg/cm^2 (Figure 18). As in the previous case, this is unacceptable.

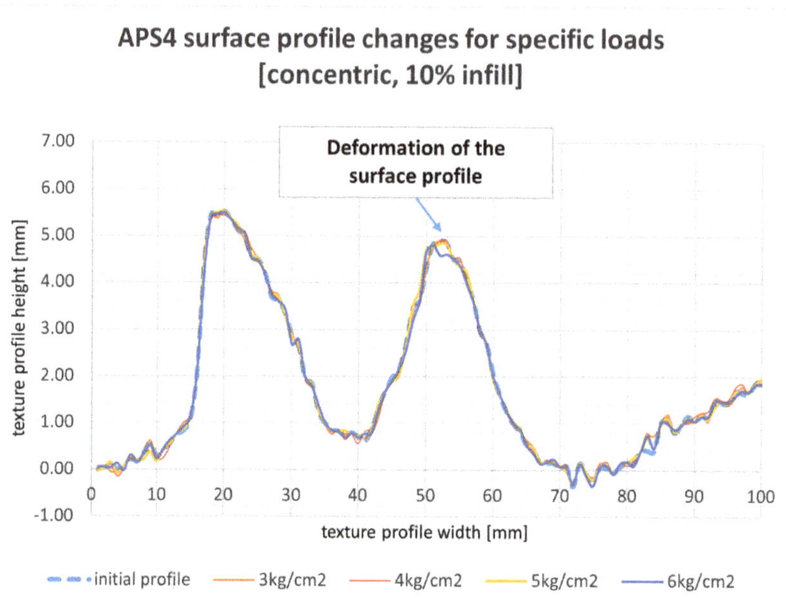

Figure 18. APS4 surface profile changes for specific loads (concentric, 10% infill).

Only a filling degree of 15% ensured the appropriate strength of the tested sample (Figure 19). For all loads, the surface texture profiles match the initial profile.

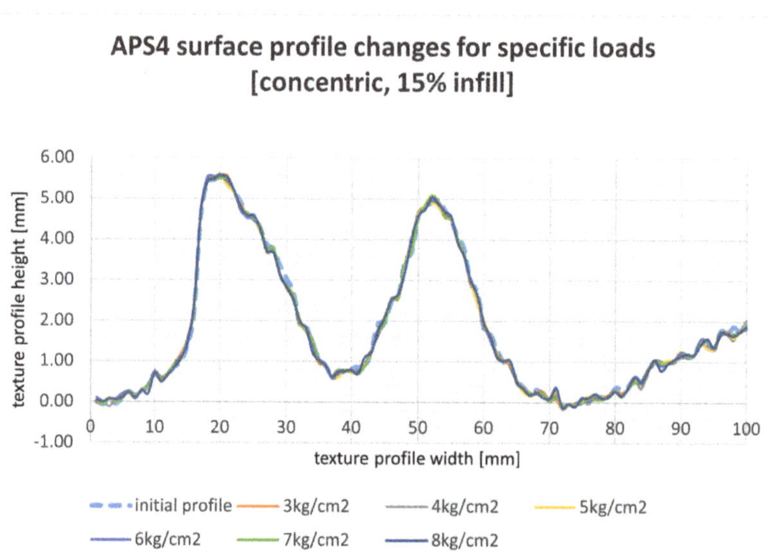

Figure 19. APS4 surface profile changes for specific loads (concentric, 15% infill).

3.3. Replica with Internal Structure of "Triangle" Type

In this type of internal structure, the filling level of 5% was insufficient. In the drawing below (Figure 20), when the tire was pressed with a weight of 6 kg/cm^2, the "peaks" of the replica were deformed, and the internal structure was crushed (deformlations visible at widths of 10 mm, 20 mm, and 84 mm). With a filling level of 10%, this problem no longer occurred (Figure 21).

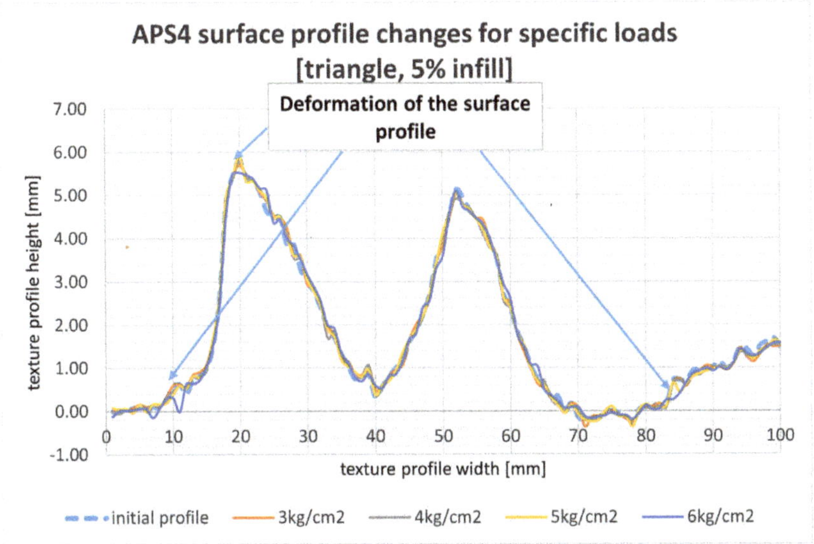

Figure 20. APS4 surface profile changes for specific loads (triangle, 5% infill).

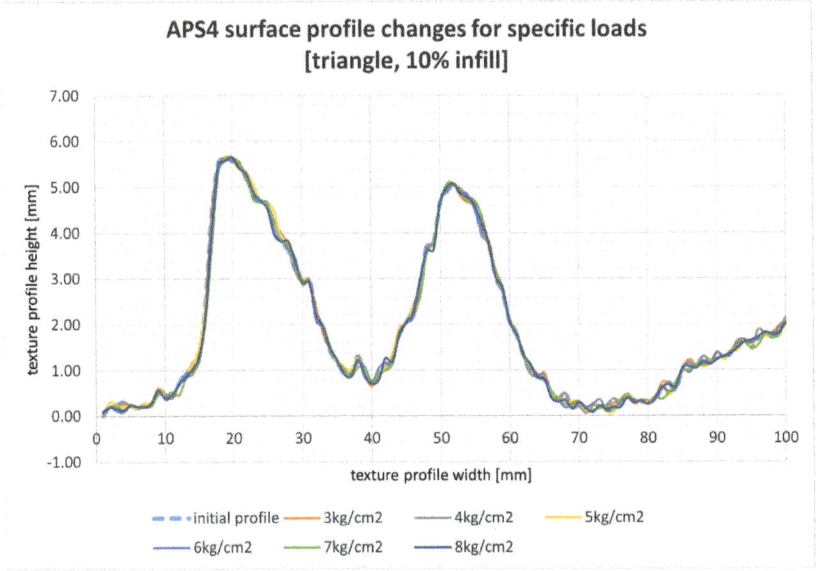

Figure 21. APS4 surface profile changes for specific loads (triangle, 10% infill).

3.4. Printing Time and Amount of Filament Used

Next, two other criteria were used to select the most advantageous internal structure: print time and amount of filament used. The table below presents parameters for samples that met the strength requirements (Table 1). The difference between the longest (Concentric) and shortest (Trihexagon) printing times was 44%. The difference in filament consumption was 25% (difference between Concentric and Grid).

Table 1. Summary of printing time and filament consumption for individual fill types.

Filling Type	Printing Time [Min]	Amount of Used Material [g]
Grid [5%]	78	17.83
Trihexagon [5%]	52	18.21
Cross [5%]	53	18.26
Concentric [15%]	92	23.55
Triangle [10%]	53	20.6

4. Discussion

At this stage of research, the aim was to select an internal structure that would provide satisfactory load-bearing capacity using the least amount of material and the shortest printing time. The results presented above indicate that the maximum infill level, which provides mechanical strength in the tire-to-road contact is 15% (taking into account all types of infill). However, for three types of infill geometry: "Grid", "Trihexagon", and "Cross", 5% infill was sufficient. Therefore, further analysis was limited to these three filling types (material consumption criterion).

When analyzing replicas for print time, the difference between the shortest ("Trihexagon") and longest ("Grid") time was 33%. When considering the same three types of filling in terms of material consumption, the differences between the lowest ("Grid") and the highest ("Cross") were only 2.35%. These were small differences; therefore it was decided that the parameter determining the choice of structure would be the printing time. It might seem that 26 min is not a big difference. However, it should be remembered that the target is to print a full-size replica of the surface to cover the outer track of the drum wheel machine. The estimated size of such a replica is 38 × 78 × 2 cm, and eight segments need to be printed to cover the entire circumference of the drum. Taking into account all of the abovementioned criteria, the optimal solution seems to be the use of the "Tri-hexagon" filling, due to its sufficient mechanical strength and the fastest printing time, as well as its material consumption.

To further verify the usefulness of the proposed method, it is planned to print larger samples of different types of road surfaces, characterized by a high and low rolling resistance coefficient. These samples would be used to perform tests using the so-called oscillation method. The method is based on indirect measurement of energy losses occurring as a result of cyclic pressing of the tire against the surface, forced by free fall causing oscillations of the system containing mass, elasticity, and damping [20]. This test will be used to determine the influence of the filament material used (its stiffness) and layer height (texture accuracy) on the obtained results of energy losses during oscillations, which are directly correlated with the rolling resistance of the tire on a given surface.

5. Conclusions

The research results presented in this article initially indicate the possibility of using 3D printing technology to produce surface replicas used to measure rolling resistance. Comparing the above research results with previous experiences on surface replicas obtained using the traditional method (casting from resin, reinforced with glass fiber), one can conclude that the use of 3D printing technology will allow for:

- Simplification and reduction in the time needed to obtain a road surface matrix by performing an accurate scan of its texture. Currently, the matrix is obtained by making a rubber cast of the actual road surface section, which is then glued to a casting mold. This operation is problematic due to the several-hour process of rubber gelation, and the entire procedure takes place in traffic conditions (it rarely happens that the road section is closed). The use of a high-resolution 3D handheld scanner would shorten the time needed to obtain a surface texture scan to several minutes, and thus increase the comfort and safety of people performing the entire procedure.

- Reduction in replica production time. Currently, using the casting method, the process of making a full replica takes approximately 100 days. This time consists of obtaining a rubber matrix of the road surface, preparing a casting mold, laminating and casting the replica, allowing the resins to harden, machining the front and side surfaces of the replica segments so that they will fit the drum circumference, and drilling holes for mounting screws. The method proposed in the application would enable printing of ready replica segments in an estimated time of 35 days (replicas could be printed around the clock). The appropriate design of the replica segments would eliminate the time-consuming processing in order to fit them to the drum surface.
- Reducing the costs of producing road surface replicas. Currently, the cost of producing a replica using the casting method is approximately $4000 (gross). This consists of making a casting mold, making a surface texture matrix cast using silicone rubber, and the purchase of resins, separator, gelcoat, glass fiber mats, mounting screws etc. The estimated cost of printing a replica is approximately $1000 (purchase of filament and electricity consumption). Of course, the proposed method requires the purchase of a 3D printer and a portable 3D scanner; however, this is a one-time purchase that would pay off in the case of producing a large number of replicas.
- Expanding the research area. Current replicas are based on textures of existing surfaces. Three-dimensional printing technology allows for modifying or designing completely new road surface textures using CAD software. This enables rapid prototyping of new types of road surfaces and testing them for rolling resistance and noise. Printed replicas would also be lighter, which allows for testing tires at higher speeds, fitting into the current measurement trend (the value of the centrifugal force acting on the segments attached to the drum surface would decrease). The replica segments would fit together better, and as a result smaller gaps would be achieved in the surface texture between adjacent segments, having a positive impact on the rolling resistance and noise of car tires (in the case of poorly fitted segments, the tire driving over the gaps between them generates noise in the form of a "knock").

Author Contributions: Conceptualization, W.O., S.S. and G.R.; measurements, S.S.; validation, G.R.; formal analysis, W.O.; writing, W.O.; visualization, W.O.; supervision, W.O.; funding acquisition, W.O. All authors have read and agreed to the published version of the manuscript.

Funding: Financial support of these studies from Gdańsk University of Technology by the DEC-037406 grant under the ARGENTUM TRIGGERING RESEARCH GRANTS—'Excellence Initiative—Research University' program is gratefully acknowledged.

Institutional Review Board Statement: Not applicable.

Informed Consent Statement: Not applicable.

Data Availability Statement: The data presented in this study are available on request from the corresponding author due to plans related to patenting the presented solution.

Acknowledgments: The authors would like to acknowledge Kornel Piłat for his contribution to measurements and evaluations.

Conflicts of Interest: The authors declare no conflicts of interest. The funders had no role in the design of the study; in the collection, analyses, or interpretation of data; in the writing of the manuscript, or in the decision to publish the results.

References

1. Reimpell, J.; Betzler, J. *Podwozia Samochodów Podstawy Konstrukcji*; WKŁ: Warszawa, Poland, 2001; s.456.
2. Taryma, S. *Opór Toczenia Opon Samochodowych*; Wydawnictwo PG: Gdańsk, Poland, 2007.
3. Clark, S.K. A brief history of tire rolling resistance. In Proceedings of the Symposium on Tire Rolling Resistance at 122nd Meetin of Rubber Division, American Chemical Society, Chicago, IL, USA, 5–7 October 1982.
4. Clark, S.K.; Dodge, R.N. *A Handbook for the Rolling Resistance of Pneumatic Tires*; Institute of Science and Technology, The Universit of Michigen: Ann Arbor, MI, USA, 1979.
5. Sandberg, U. *Influence of Road Surface Texture on Traffic Characteristics Related to Environment, Economy and Safety*; VTI Nota 53A-1997; Swedish National Road and Transport Research Institute: Linkoeping, Sweden, 1997.
6. Kane, M.; Riahi, E.; Do, M.-T. Tire/Road Rolling Resistance Modeling: Discussing the Surface Macrotexture Effect. *Coatings* **202** *11*, 538. [CrossRef]
7. Riahi, E.; Ropert, C.; Do, M.-T. Developing a laboratory test method for rolling resistance characterisation of road surface textur *Surf. Topogr. Metrol. Prop.* **2020**, *8*, 024006. [CrossRef]
8. Sandberg, U.; Bergiers, A.; Ejsmont, J.A.; Goubert, L.; Karlsson, R.; Zöller, M. Road Surface Influence on Tyre/Roa Rolling Resistance [Internet]. 2011. Available online: https://urn.kb.se/resolve?urn=urn:nbn:se:vti:diva-15699 (accessed o 12 November 2024).
9. Ejsmont, J.A.; Ronowski, G.; Świeczko-Żurek, B.; Sommer, S. Road texture influence on tyre rolling resistance. *Road Mate Pavement Des.* **2016**, *18*, 181–198. [CrossRef]
10. Ejsmont, J.; Świeczko-Żurek, B. Methods of Tire Rolling Resistance Measurements. In Proceedings of the COTUME'2014, Congrè Tunisien de Mècanique, Sousse, Tunisia, 24–26 March 2014.
11. Rafei, M.; Ghoreishy, M.H.R.; Naderi, G. Computer simulation of tire rolling resistance using finite element method: Effect o linear and nonlinear viscoelastic models. *Proc. Inst. Mech. Eng. Part D J. Automob. Eng.* **2019**, *233*, 2746–2760. [CrossRef]
12. Shida, Z.; Koishi, M.; Kogure, T.; Kabe, K. A Rolling Resistance Simulation of Tires Using Static Finite Element Analysis. *Tire Sc Technol. TSTCA* **1999**, *27*, 84–105. [CrossRef]
13. *ISO28580*; Passenger Car, Truck and Bus Tyres—Methods of Measuring Rolling Resistance—Single Point Test and Correlation o Measurement Results, 2nd ed. International Organization for Standardization: Genève, Switzerland, 2018.
14. *ISO18164*; Passenger Car, Truck, Bus and Motorcycle Tyres—Methods of Measuring Rolling Resistance, 2nd ed. Internationa Organization for Standardization: Genève, Switzerland, 2005.
15. *J2452_201707*; Society of Automotive Engineers. Stepwise Coastdown Methodology for Measuring Tire Rolling Resistanc Society of Automotive Engineers: Warrendale, PA, USA, 2017.
16. *J1269_201912*; Rolling Resistance Measurement Procedure for Passenger Car, Light Truck, and Highway Truck and Bus Tire Society of Automotive Engineers: Warrendale, PA, USA, 2019.
17. Ejsmont, J.; Sandberg, U.; Świeczko-Żurek, B.; Mioduszewski, P. Tyre/road noise reduction by a poroelastic road surface. I Proceedings of the 43rd International Congress on Noise Control Engineering Improving the World through Noise Contro Proceedings of Inter-Noise, Melbourne, Australia, 16–19 November 2014; pp. 1–12.
18. Szykiedans, K.; Credo, W.; Osiński, D.P. Selected mechanical properties of PetG 3-D prints. In Proceedings of the XXI Polish Slovak Scientific Conference Machine Modeling and Simulations (MMS 2016), Rydzyna, Poland, 4–7 September 2018; pp. 455–461 [CrossRef]
19. van der Vegt, A.K.; Govaert, L.E. *Polymeren: Van Keten tot Kunststof*, 5th ed.; VSSD: Delft, The Netherland, 2005.
20. Ejsmont, J.; Owczarzak, W. Engineering method of tire rolling resistance evaluation. *Measurement* **2019**, *145*, 144–149. [CrossRef

coatings

rticle

₹heological Performance and UV Aging Resistance of Buton ₹ock Asphalt and Styrene-Butadiene-Styrene Composite ɅModified Asphalt Under High Temperature

ian Li [1], Chunxiang Huang [2,3], Xu Zeng [4], Jinfeng Yang [5], Wei Gao [6], Yaping Chen [2] and Haibo Yang [7,*]

1 Department of Vehicle Engineering, Guizhou Technological College of Machinery and Electricity, Duyun 558000, China
2 Chongqing Architectural Design Institute Co., Ltd., Chongqing 400015, China
3 School of Civil Engineering, Chongqing Jiaotong University, Chongqing 400047, China
4 Department of Electrical Engineering, Guizhou Technological College of Machinery and Electricity, Duyun 558000, China
5 School of Transportation and Municipal Engineering, Chongqing Jianzhu College, Chongqing 400072, China
6 Guizhou Zhongsheng Highway Maintenance Technology Research Co., Ltd., Guiyang 550014, China
7 School of Civil Engineering, Southwest Jiaotong University, Chengdu 610031, China
* Correspondence: yhaibo2259389889@163.com

Abstract: This study aims to investigate the high-temperature rheological properties and ultraviolet (UV) aging resistance of Buton rock asphalt (BRA) and styrene-butadiene-styrene (SBS) composite modified asphalt. Initially, the surface morphology of BRA and SBS was examined using scanning electron microscopy. The high-temperature rheological properties and chemical characteristics of BRA modified asphalt, SBS modified asphalt, and BRA/SBS composite modified asphalt before and after UV aging were investigated. The results reveal that numerous pores and folds on the surface of BRA form a distinctive honeycomb-like structure that offers a large surface area, enhancing asphalt adsorption and thereby improving the high-temperature properties of asphalt. The addition of SBS enhances the elastic response of Buton rock asphalt modified asphalt (BRAMA). The Fourier transform infrared (FTIR) spectra indicate that BRA effectively reduces the carbonyl and sulfoxide index of asphalt, and the complex modulus aging index (*CMAI*) and phase angle aging index (*PAI*) suggest that the BRA/SBS composite modified asphalt exhibits better resistance to ultraviolet aging.

Keywords: BRA/SBS composite modified asphalt; rheological performance; ultraviolet aging; FTIR

1. Introduction

Buton rock asphalt (BRA), renowned for its excellent compatibility with asphalt and its economic and environmental benefits, is employed in asphalt modification [1,2]. As a derivative of asphalt, BRA has been extensively studied for its significant advantages in enhancing high-temperature rheology, fatigue resistance, and aging performance [3,4]. At the same time, BRA contains high-alkali minerals that have high ionic adsorption ability, which can be used as an excellent filler at the asphalt–aggregate interface to improve the bonding performance between asphalt and aggregate [5]. Researchers have utilized dynamic shear rheometers (DSRs) to assess the rheological properties of asphalt modified with varying amounts of BRA, revealing that with increasing BRA content, the complex shear modulus (G^*) and dynamic viscosity at 60 °C increase [6,7]. Compared to neat asphalt,

BRA modified asphalt exhibits superior high-temperature rheological and aging resistance properties, albeit poorer low-temperature cracking resistance due to stress concentration effects from BRA particles [8,9]. Consequently, some scholars have utilized composite modifiers combining other additives with BRA to modify asphalt, aiming to enhance its overall performance [7,10]. Songtao Lv et al. assessed the low-temperature performance of SBR/BRA and nano-$CaCO_3$/BRA composite modified asphalts, finding that the SBR/BRA composite modified asphalt had better low-temperature performance [11]. Yanbing He et al. compared rheological properties between 3.5% SBS and 15% rock compound additive (RCA) composite modified asphalt, with the latter showing superior performance; the RCA was made of 53% natural rock asphalt, nanopolymer materials, and stable adhesives [12]. Yankai Wen et al. explored the microstructure and rheological properties of rock asphalt/SBS modified asphalt before and after aging, revealing that there is a physical modification, with rock asphalt particles forming a denser, more stable network within the SBS matrix [13]. Composite modification ensures a balanced performance of asphalt in both high and low temperatures. Scholars have shown that SBS can effectively enhance the crack resistance of BRA modified asphalt at low temperatures [14]. However, comprehensive and systematic studies on the UV aging resistance and high-temperature rheological properties of BRA/SBS composite modified asphalt before and after aging have not been conducted.

Therefore, this study utilized a dynamic shear rheometer (DSR) with temperature sweep (TS) and multiple stress creep recovery (MSCR) tests to obtain various rheological indices of BRA/SBS composite modified asphalt, enabling a comprehensive analysis of its high-temperature rheological properties before and after UV aging. The surface morphology of the modifiers was analyzed using field emission scanning electron microscopy (FESEM) to elucidate the enhancement mechanisms at the microscopic level. Simultaneously, the UV aging resistance of the asphalts was also analyzed from both macroscopic and microscopic perspectives. This study provides theoretical support for the design and engineering applications of BRA/SBS composite modified asphalt, offering valuable insights for expanding the utilization of Buton rock asphalt and addressing its shortcomings.

The overall framework of this study is shown in Figure 1.

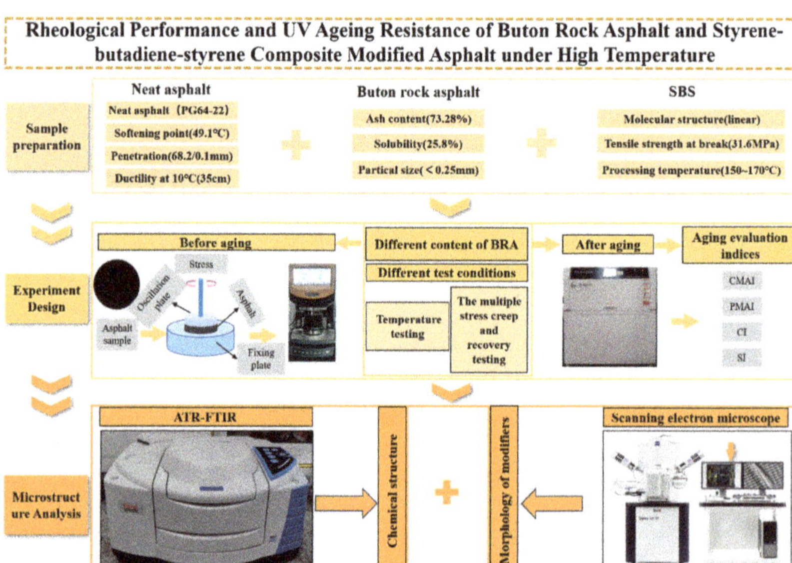

Figure 1. Framework of this study.

2. Experimental Design

2.1. Materials

A neat asphalt (SK70) with a PG 64-22 was selected to be tested in the laboratory, and its physical properties are listed in Table 1 according to the test methods in JTG E20-2011 "Standard Test Methods of Bitumen and Bituminous Mixtures for Highway Engineering" [15,16].

Table 1. The physical properties of neat asphalt.

Test Items		Test Results	Industry Standard	Test Methods
Softening point/°C		48.5	≥46	JTG E20-2011
Penetration (25 °C)/0.1 mm		67.7	60–80	
Ductility (10 °C, 5 cm/min)/cm		36.0	≥15	
RTFOT	Mass loss/%	0.15	≤±0.8	
	Residual penetration rate (25 °C)/%	73	≥61	
	Residual ductility (10 °C)/cm	8	≥6	

The modifier BRA, which is a type of rock asphalt produced in Indonesia, was used in this study. To ensure the BRA's uniformity in the modified asphalt, its particle size was less than 0.25 mm [4,17,18]. Table 2 presents the main technical indicators of BRA. For a representative study, SBS (YH-791) with a block ratio (S/B) of 30/70 purchased from Sinopec Baling Petrochemical Corp was used as another modifier. Table 3 presents the basic technical indicators of SBS.

Table 2. The technical indicators of BRA modifier.

Test Items		Industry Standard	Test Results	Test Methods
Appearance		Brown powder	Brown powder	AASHTO T315-09
Ash content/%		<75	72.32	
Solubility/%		>25	26.4	
Water content/%		<2	0.78	
Percentage passing/%	<4.75 mm	100	100	
	<2.36 mm	90~100	100	
	<0.6 mm	10~60	57	

Table 3. The basic index of SBS modifier.

Test Items	Unit	Test Results
Specific gravity	/	0.92
Molecular structure	/	Linear
Molecular weight	/	140,000
Processing temperature	°C	150–170

2.2. Preparation of Modified Asphalts

All modified asphalts were prepared in the laboratory by mixing neat asphalt with the BRA and SBS modifiers. To ensure the modified asphalts possessed satisfactory storage stability, the preparation parameters for modified asphalts were selected based on previous research [8,19]. The compositions and abbreviations of the modified asphalts used in this study are shown in Table 4.

Table 4. Compositions and abbreviations of the modified asphalts.

Sample	Composition	Abbreviation
SBSMA	3%SBS	3SMA
RAMA	10%RA	10RMA
	5%RA + 2%SBS	5R/2SMA
R/SMA	10%RA + 2%SBS	10R/2SMA
	15%RA + 2%SBS	15R/2SMA

2.2.1. Preparation of SBSMA

The preparation temperature was maintained at 170 °C to fully swell the SBS. The blend was sheared for 30 min (5000 rpm) and then stirred for 35 min (2000 rpm) [20].

2.2.2. Preparation of BRAMA

The preparation temperature was maintained at 170 °C in the process of preparation. The blend was stirred for 15 min (1500 rpm) and then sheared for 20 min (2000 rpm). Then, the blend was placed in an oven for 30 min at 150 °C [21,22].

2.2.3. Preparation of R/SMA

Initially, 2% SBS modified asphalt was prepared by the method described above, and different dosages of BRA were then mixed with it. The mixture was stirred for 30 min at 165 °C (2500 rpm) and sheared for an additional 30 min at 165 °C (2500 rpm). Finally, the mixture was placed in an oven for 30 min at 150 °C. Figure 2 presents a flow chart illustrating the R/SMA preparation process [23].

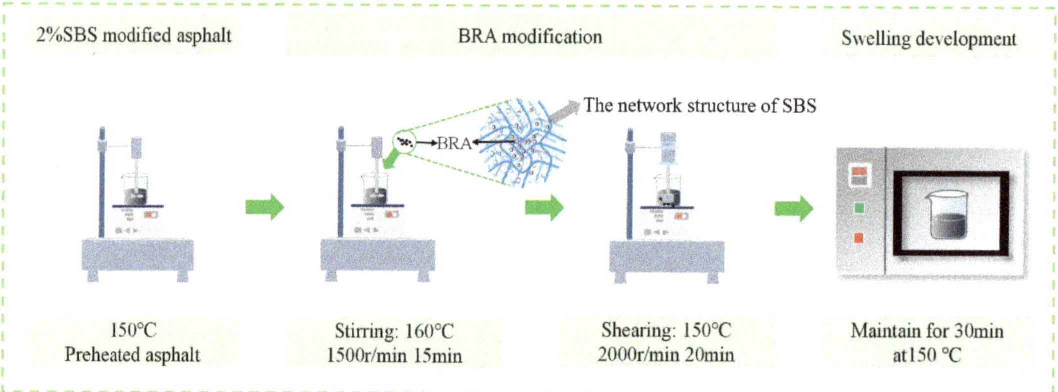

Figure 2. The preparation process of R/SMA.

2.3. *Test Methods*

2.3.1. FESEM Characterization

The morphology of BRA and SBS was examined using the FESEM equipment under a magnification of 1000×. Before testing, a gold film was sprayed onto the surface of samples to eliminate the effects electrical charging.

2.3.2. FTIR Test

Fourier transform infrared spectroscopy (Nicolet iS50,Thermo Fisher Scientific, Massachusetts, America) was employed to analyze the chemical structural changes of each modified asphalt before and after aging. The infrared testing spanned from 400 cm^{-1} to

4000 cm^{-1} [24]. The carbonyl index ($I_{C=O}$) and the sulfoxide index ($I_{S=O}$) were calculated based on the peak areas of the infrared spectra.

$$I_{C=O} = \frac{A_{1700}}{\sum A} \tag{1}$$

$$I_{S=O} = \frac{A_{1030}}{\sum A} \tag{2}$$

where $\sum A$ is the area of the absorption peaks between 2000 cm^{-1} and 600 cm^{-1}; A_{1030} represents the $S = O$ absorption peak area; and A_{1700} represents the $C = O$ peak area [14].

2.3.3. Extended Temperature Sweep Test

A traditional temperature sweep test is typically constrained to temperatures below 90 °C, which limits the assessment of the high-temperature rheological properties of asphalt. To overcome this constraint, the temperature range for testing was expanded to 40 °C to 140 °C, with intervals of 10 °C [25]. A dynamic loading frequency of 1.59 Hz was used [26].

2.3.4. The Multiple Stress Creep and Recovery(MSCR) Test

AASHTO T350 recommends employing a multiple stress creep and recovery (MSCR) test to evaluate the creep and elastic recovery behavior of asphalt. This experiment assessed the creep and elastic recovery behaviors of modified asphalt within the temperature range of 40 °C to 100 °C, with temperature intervals of 10 °C. The MSCR test can be divided into two stages during experimentation. The first consists of applying a constant 1 s long stress to the sample during the creep stage. The second step is the sample recovery phase without any external load, lasting for 9 s. Finally, the creep and elastic recovery performance of the sample is calculated based on the degree of deformation in the creep and recovery stages [27].

$$R = \frac{\varepsilon_r}{\varepsilon_p} \times 100\% \tag{3}$$

$$J_{nr} = \frac{\varepsilon_u}{\sigma} \times 100\% \tag{4}$$

where ε_r is the strain at 10 s in each cycle; ε_p is the peak strain at 1 s in each cycle; ε_u is the unrecovered strain at 10 s in each cycle; and σ is the creep stress.

The shear stress was fixed at 0.1 kPa to ensure the linearity of the strain–stress response. The nonlinear stress–strain response was tested by increasing the shear stress to 3.2 kPa [26].

2.4. *Aging Procedure and Aging Evaluation Indices*

The asphalt is subjected to a rolling thin-film oven test prior to UV aging to simulate the aging of the asphalt during construction. A standard-film oven-test sample tray of known quality is then placed on a balance, 50 g ± 0.5 g of asphalt is added to the tray and spread to a film thickness of approximately 3.2 mm, and the prepared asphalt sample is placed in a UV aging vessel for aging. The temperature of the UV aging vessel is set at 60 °C, with UV radiation intensity set at 8 W/m^2, and the aging time is 6 days [28]. UVA-340 was used as the light source because UVA-340 simulates the sunlight spectrum in the critical shortwave wavelength range (295–360 nm).

The complex modulus aging index (*CMAI*) and phase angle aging index (*PAI*) were employed to assess the changes in the high-temperature rheological properties of asphalt after aging. These indices are defined as follows [22].

$$CMAI = \frac{Age\ complex\ modulus}{Unaged\ complex\ modulus} \tag{5}$$

$$PAI = \frac{Age\ phase\ angle}{Unaged\ phase\ angle} \tag{6}$$

3. Results and Discussions

3.1. Surface Morphology of Modifiers

Figure 3 illustrates the microscopic morphology of the modifiers BRA and SBS. It is evident that SBS exhibits aggregated flocculent structures, whereas BRA appears as dispersed granular particles. Being an asphalt derivative, BRA demonstrates better compatibility with asphalt, facilitating its uniform and stable distribution within the asphalt [29]. Further magnification using FESEM reveals numerous pores and folds on the surface of BRA forming a distinctive honeycomb-like structure. This rough and irregular honeycomb structure offers a large surface area, enhancing asphalt adsorption and thereby improving the high-temperature rheological properties.

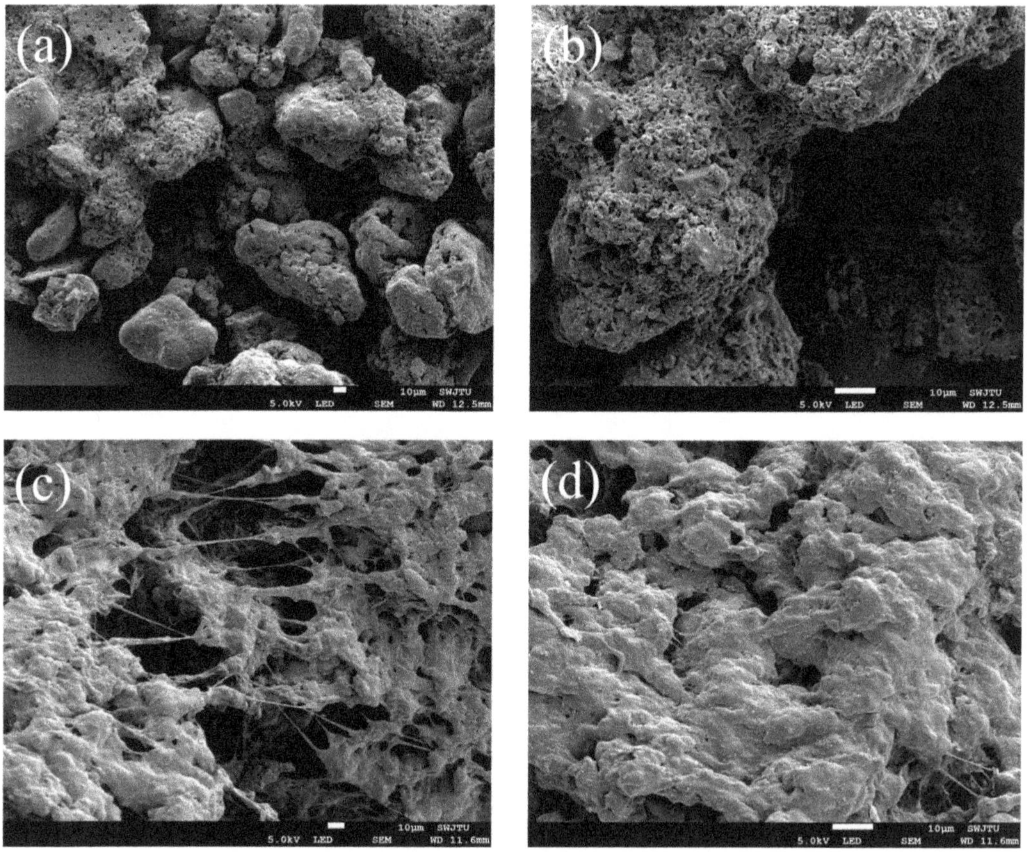

Figure 3. Surface morphology of modifiers: (**a**) BRA (×300), (**b**) BRA (×1000), (**c**) SBS (×300), (**d**) SBS (×1000).

3.2. High-Temperature Rheological Properties

3.2.1. The TS Test

The Superpave rutting parameter ($G^*/sin\delta$) of modified asphalt with temperature variation is depicted in Figure 4. Compared to SBS modified asphalt, the $G^*/sin\delta$ of BRA/SBS composite modified asphalt notably increases with the addition of BRA. This suggests that BRA improve asphalt's high-temperature rheological properties. Below 80 °C, the $G^*/sin\delta$ values are ranked as follows: 15R2S > 10R2S > 5R2S > 10R > 3S, indicating that 15R2S has the most significant impact on enhancing asphalt's resistance to high-temperature deformation. Additionally, it reduces the quantity of SBS modifier used, yielding economic benefits. Above 80 °C, the $G^*/sin\delta$ parameters of the five modified asphalts are nearly identical, suggesting that higher temperatures diminish the beneficial effects of BRA.

To enable us to comprehensively understand the high-temperature rheological characteristics of the five modified asphalts, an analysis was conducted on the complex shear modulus (G^*) and phase angle (δ) over the temperature range of 80 °C to 140 °C.

The variation in the G^* of modified asphalt according to temperature in various temperature ranges (40–140 °C, 80–140 °C, 80–120 °C) is depicted in Figure 5. Below 80 °C, the order of G^* for different modified asphalts is as follows: R/SMA > RAMA > SBSMA. It is also observed that all modified asphalts exhibit a similar trend in G^* reduction with increasing temperature, with nearly identical reduction rates. Above 80 °C, due to the network structure formed by SBS, the rate of G^* reduction for 3SMA decreases as temperature rises, showing a pronounced platform region in the 80–120 °C range [20]. In contrast to SBS modified asphalt, RA modified asphalt does not exhibit a platform region. At temperatures above 120 °C, the G^* curves of composite modified asphalts align with those of SBS modified asphalt.

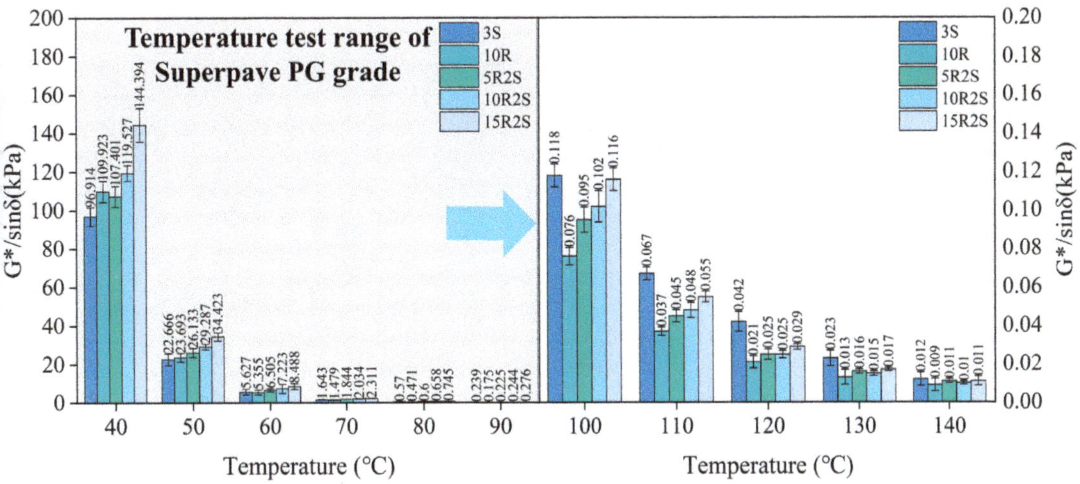

Figure 4. Effects of BRA and SBS on rheological properties of modified asphalts under different temperatures.

Figure 6 presents the phase angle test results for the five types of asphalt evaluated in this study. The curves illustrate that δ more distinctly reveals the complex rheological behaviors of various modified asphalts compared to the G^*. Consistent with existing research, SBS modified asphalt (SBSMA) exhibits a plateau region and valley region due to the network structure formed by SBS, presenting an S-shaped curve in the phase angle plot [20]. In contrast, RAMA, categorized as non-elastomeric modified asphalt (non-EMA), shows a C-shaped curve influenced by the inertia effect of the DSR instrument, with δ initially rising and then declining [30]. However, the phase angle curve of R/SMA differs from the aforementioned two types of modified asphalts. Compared to SBSMA, R/SMA shows no distinct valley region (80–120 °C). Compared to RAMA, although δ also decreases across the range of 80–140 °C, the reduction rate significantly diminishes, approaching zero. Under the combined influence of RA and SBS, the curve shape resembles a semi-C shape, indicating that SBS mitigates the inertial effect of the instrument on RAMA, thereby making the phase angle of RAMA closer to its true value.

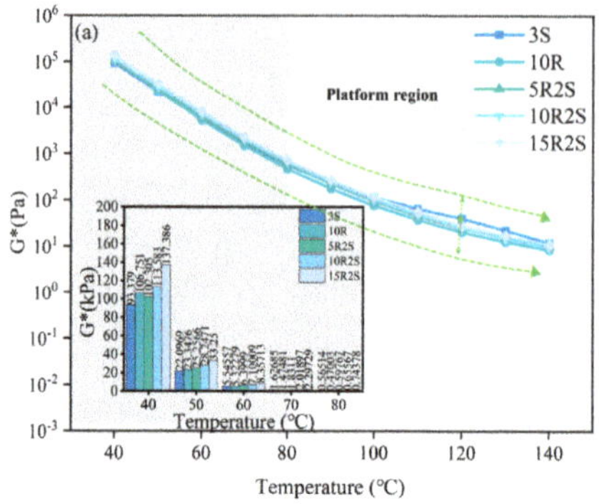

Figure 5. *Cont.*

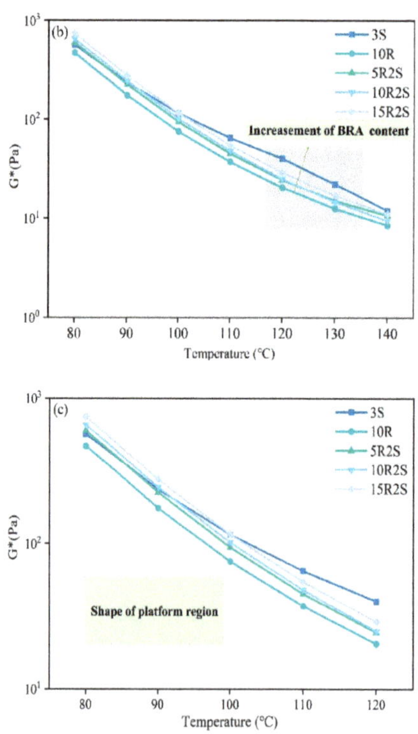

Figure 5. G* of different temperature ranges: (**a**) 40–140 °C, (**b**) 80–140 °C, (**c**) 80–120 °C.

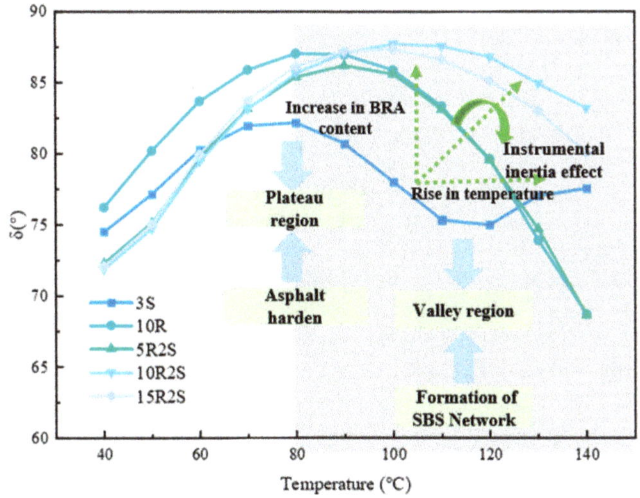

Figure 6. δ of modified asphalts.

The elastic storage modulus (G′) represents the recoverable elastic portion of an asphalt's response under cyclic loading, while the viscous loss modulus (G″) reflects the irrecoverable viscous portion. Figure 7 presents the test results for both G′ and G″, showing that both moduli decrease with increasing temperature. As illustrated in Figure 7a, the G′ values of the five types of asphalt exhibit varying plateau regions. At the same temperature, a higher BRA content results in a lower G′ for the modified asphalt. Conversely, the corresponding G″ values increase progressively with higher BRA content.

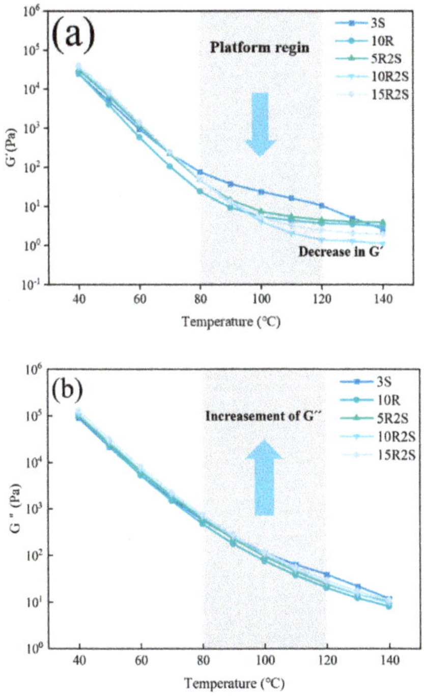

Figure 7. G′ and G″ of modified asphalts: (**a**) G′, (**b**) G″.

3.2.2. The MSCR Test

According to the findings in Section 3.2.1, changes in the elastic components of modified asphalt are the primary factors influencing its performance. To further evaluate the anti-rutting ability of the five modified asphalts under repeated loading, the MSCR test was conducted. The viscoelastic properties were assessed at two stress levels across a temperature range of 40–100 °C. The measured parameters included creep compliance (J_{nr}) and percent recovery (R).

R and J_{nr} at a stress level of 0.1 kPa are presented in Figures 8 and 9. At 70 °C, for example, the R of 10R2SMA is 90.8% higher than that of 10RMA, while the J_{nr} is 92.3% lower. These results demonstrate that the addition of SBS significantly enhances the elastic component and anti-rutting performance of modified asphalt. Under medium temperature conditions (40–70 °C), 3SMA exhibits an overall increasing trend in $R_{0.1}$, attributed to its network structure [20]. In contrast, the R of RMA decreases linearly with increasing temperature. While R for R/SMA also decreases with temperature, an interesting phenomenon emerges: at constant temperatures, the addition of rock asphalt not only avoids a reduction in R but also enhances it, improving the delayed elasticity of the asphalt. Additionally, with the incorporation of BRA, the shaded area between the R curves of R/SMA and 3SMA gradually narrows, approaching the curve of SBSMA and forming a pattern reminiscent of a tilted vase, as shown in Figure 8 BRA influences the balance between elastic and viscous responses in modified asphalt, increasing R under consistent conditions. Analysis reveals that rock asphalt integrates within the SBS network structure, further improving the elastic recovery behavior of asphalt [31].

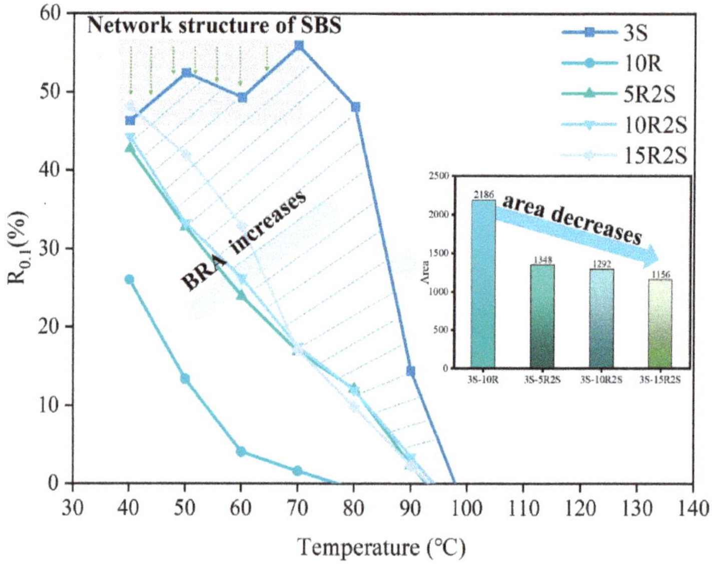

Figure 8. R of modified asphalts at 0.1 kPa.

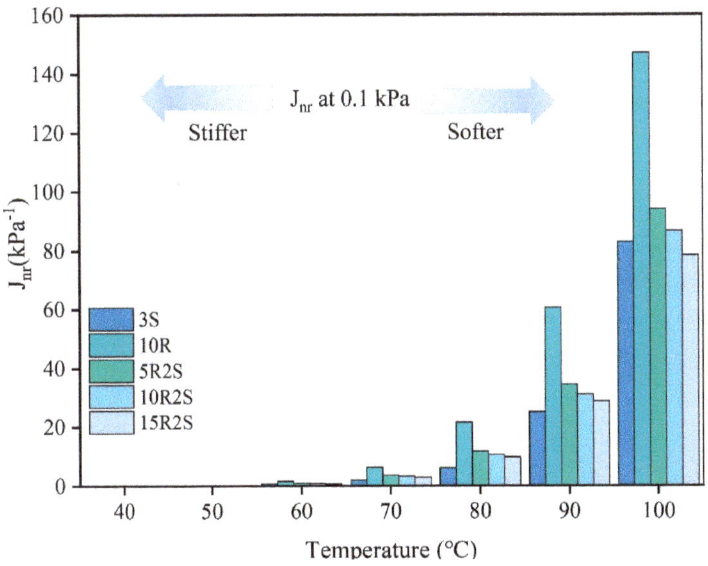

Figure 9. J_{nr} of modified asphalts at 0.1 kPa.

The R and J_{nr} at a stress level of 3.2 kPa are presented in Figure 10.

Figure 10 depicts the variation in R and J_{nr} of modified asphalt under a stress level of 3.2 kPa (non-LVE range) with temperature. Figure 10 illustrates that changes in stress level and temperature exhibit similar effects on asphalt performance. As the stress level increases, R decreases and J_{nr} increases, indicating a weakening of the elastic response. At a stress level of 3.2 kPa, the area enclosed by the R curves of R/SMA and SBSMA decreases to approximately 1/4 to 1/5 of the area observed at 0.1 kPa, with the difference in R between the two significantly reduced. Similarly, as shown in Figure 10, the J_{nr} of R/SMA initially exceeds that of SBSMA but transitions to being lower as the stress level increases. This shift indicates that higher stress levels disrupt the network structure of SBS, diminishing its ability to maintain elastic recovery.

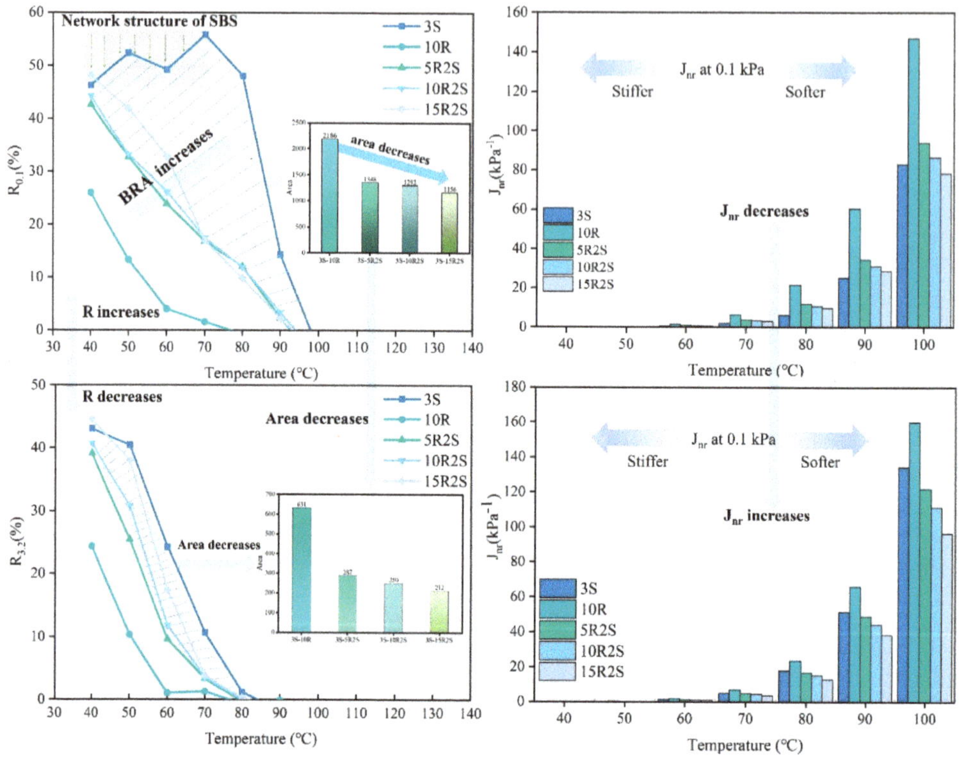

Figure 10. R and J_{nr} of modified asphalts at 3.2 kPa.

3.3. Rheological Performance of RA/SBS Modified Asphalts After Aging

3.3.1. The Results of TS After Aging

Figure 11 illustrates the variation in $G^*/sin\delta$ of five types of asphalt after aging. Aging transforms the light aromatic phenols in asphalt into higher-molecular-weight components, thereby increasing the $G^*/sin\delta$ [24,26]. Post-aging, the $G^*/sin\delta$ values rank as follows: 10R > 5R2S > 10R2S > 15R2S > 3S. At the same temperature, the $G^*/sin\delta$ value of SBS modified asphalt with BRA is higher than that of SBS modified asphalt without BRA. This indicates that even under aging conditions, the addition of BRA further enhances the anti-rutting performance of SBS modified asphalt.

As shown in Figure 12, under temperature conditions (40–100°C), the *CMAI* values of the modified asphalt are ranked as follows: 10R > 5R2S > 3S > 10R2S > 15R2S. Similarly, the PAI values follow this order: 10R < 3S < 5R2S < 15R2S < 10R2S. The CMAI of asphalt modified with both rock asphalt and SBS is lower, while the *PAI* is higher compared to single modified asphalt. This demonstrates that the synergistic effect of these two modifiers enhances the asphalt's resistance to ultraviolet aging.

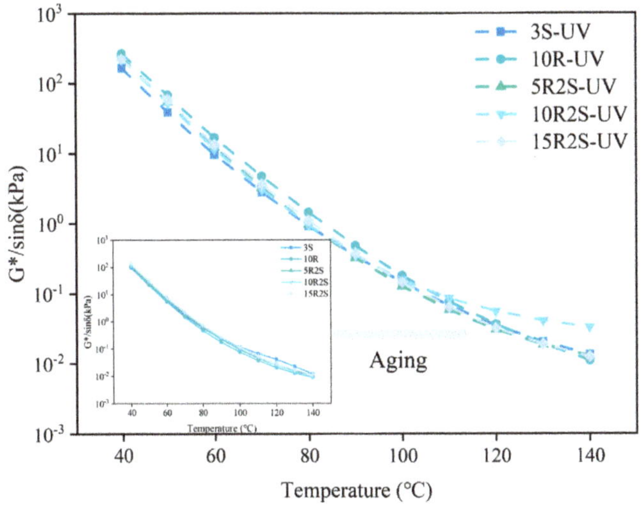

Figure 11. *G*/sinδ* test results after aging.

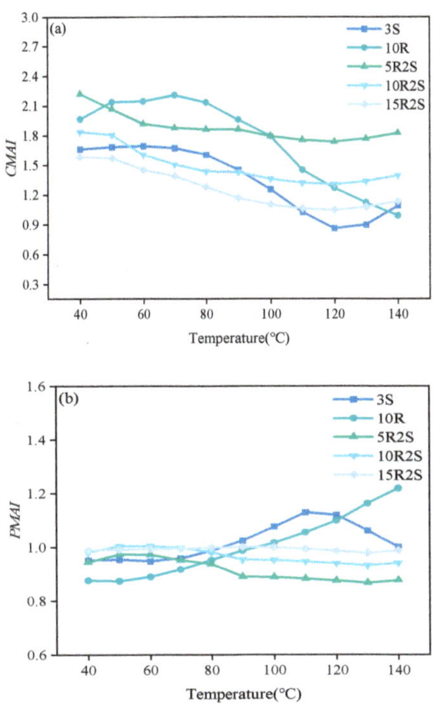

Figure 12. *CMAI* and *PAI* of modified asphalts after aging: (**a**) *CMAI*, (**b**) *PAI*.

3.3.2. The Results of MSCR After Aging

Figure 13 illustrates the variation in *R* with temperature for different-aged modified asphalts. The *R* value of all five asphalts declined following UV aging. SBS enhances the recovery ratio of BRA modified asphalt, thereby strengthening its elastic response. For instance, at 70 °C and 3.2 kPa, the recovery ratio (*R*) of 10R2S-UV was 110% higher than that of 10R-UV. Concurrently, the synergistic effects of light, oxygen, and temperature degrade SBS, leading to a narrowing difference in R between 10R, 5R2S, 10R2S, and 15R2S compared to 3S, as depicted by the diminishing area enclosed by the curves in the graph.

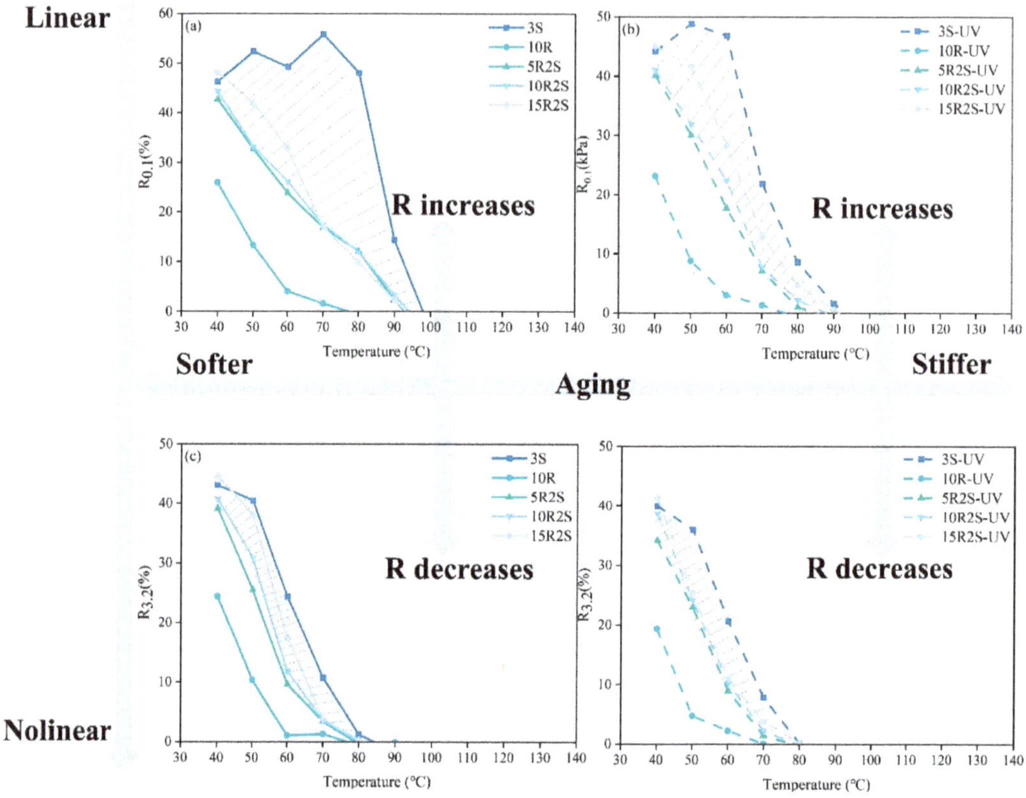

Figure 13. R of modified asphalts after aging.

3.4. FTIR Spectra

UV aging induces physical and chemical changes in modified asphalt, altering its ability to absorb and transmit infrared radiation. These changes result in noticeable shifts in the peaks and valleys of the asphalt's FTIR spectrum, as shown in Figure 14. The peak at 1030 cm^{-1} is ascribed to the $S = O$ stretching vibration. The peak at 1700 cm^{-1} corresponds to the conjugated $C = O$ [32,33]. The presence of carbonyl and sulfoxide groups serves as an indicator of the extent of UV-induced oxidation.

Figure 15 highlights the significant changes in the carbonyl and sulfoxide indices of modified asphalt under UV aging conditions. Both BRA and the BRA/SBS composite modified asphalt exhibit lower $I_{C=O}$ and $I_{S=O}$ compared to SBS modified asphalt. An increased proportion of BRA further reduces these indices, indicating enhanced UV aging resistance. Compared to the single SBS modifier, the addition of BRA significantly improves the asphalt's resistance to UV aging.

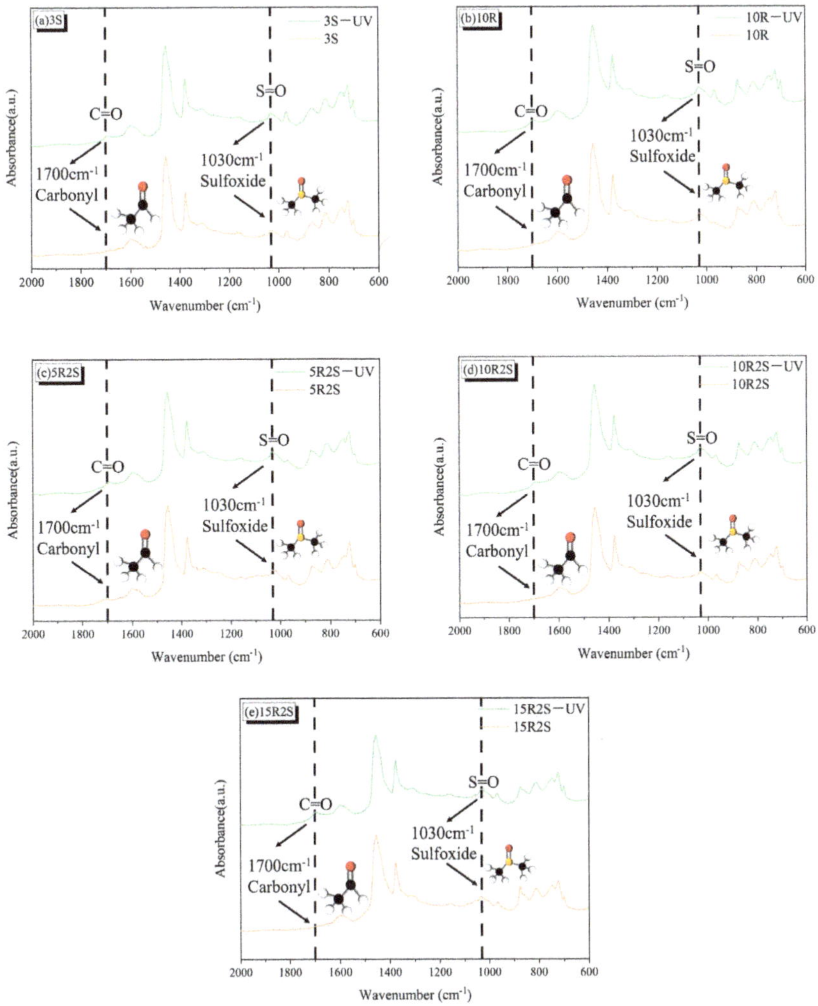

Figure 14. FTIR spectra of modified asphalts under aging conditions: (**a**) 3S, (**b**) 10R, (**c**) 5R2S, (**d**) 10R2S, (**e**) 15R2S.

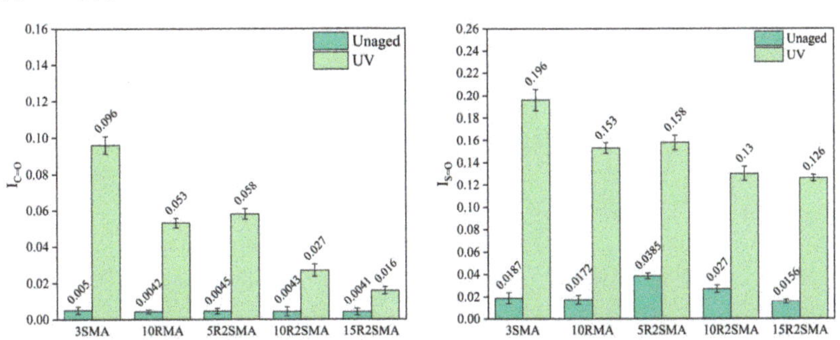

(**a**) Carbonyl index $I_{C=O}$ (**b**) Sulfoxide index $I_{S=O}$

Figure 15. Carbonyl index ($I_{C=O}$) and sulfoxide index ($I_{S=O}$) of modified asphalts.

4. Conclusions

To investigate the effects of BRA on the rheological properties and UV aging resistance of asphalt, five different modified asphalts were prepared, using SBS modified asphalt as the control group. This study employed FESEM, DSR, and FTIR to examine the mechanisms by which BRA enhances these properties, focusing on micromorphology, anti-rutting performance, and functional groups. The key findings are summarized as follows:

- BRA exhibits a large number of irregular surface features, such as pores and folds, forming a distinctive honeycomb structure, which offers a larger specific surface area (1.5–6.8 m^3/g). This structure improves asphalt adsorption and enhances high-temperature rheological performance.
- The high-temperature rheological properties of asphalt modified with different dosages of BRA were compared. An increase in BRA content reduces the elastic modulus (G') of the modified asphalt, while the viscous modulus (G'') progressively rises with higher BRA content.
- At 70 °C, compared to 10RMA, 10R2SMA shows an increase in R of 90.8% and a decrease in J_{nr} of 92.3%. The addition of SBS enhances the elastic response of BRAMA. However, both aging and increased stress levels can weaken the enhancing effect of SBS.
- The *CMAI* and *PAI* suggest that the composite modified asphalt exhibits better resistance to UV aging. Compared with 3SMA, the $I_{C=O}$ and $I_{S=O}$ of 10R2SMA decreased by 71.9% and 33.7%, respectively, after aging, and the composite modified asphalt had a lower $I_{C=O}$ and $I_{S=O}$ under the same aging conditions, indicating that BRA has advantages in enhancing the asphalt's resistance to UV aging.

This study analyzed the effect of UV aging on the high-temperature rheological properties of BRA/SBS composite modified asphalt. However, fatigue cracking resistance, viscosity, and low-temperature cracking resistance are equally critical parameters that require further investigation. Future research should focus on evaluating these properties to establish a comprehensive and systematic performance-evaluation framework.

Author Contributions: L.L.: methodology, data curation, writing—original draft preparation; C.H.: software; X.Z.: conceptualization, validation; J.Y.: methodology, software; W.G.: visualization; Y.C.: investigation; H.Y.: conceptualization, methodology, resources, supervision, writing—review and editing. All authors have read and agreed to the published version of the manuscript.

Funding: This research was funded by the National Natural Science Foundation of China, grant number 50603452.

Institutional Review Board Statement: Not applicable.

Informed Consent Statement: Not applicable.

Data Availability Statement: Data are contained within the article.

Acknowledgments: The authors are grateful for the support given by the National Natural Science Foundation of China (No. 50603452).

Conflicts of Interest: Author Chunxiang Huang and Yaping Chen were employed by the company Chongqing Architectural Design Institute Co., Ltd. The remaining authors declare that the research was conducted in the absence of any commercial or financial relationships that could be construed as a potential conflict of interest.

References

1. Anupam, K.; Akinmade, D.; Kasbergen, C.; Erkens, S.; Adebiyi, F. A state-of-the-art review of Natural bitumen in pavemen Underlining challenges and the way forward. *J. Clean. Prod.* **2023**, *382*, 134957. [CrossRef]

Zhang, X.; Zeng, Y.; Feng, Y.; Zhang, C.; Zhang, L. Carbon emissions analysis of producing modified asphalt with natural asphalt. *Green Process. Synth.* **2023**, *12*, 20228146. [CrossRef]

Lu, W.; Peng, X.; Lv, S.; Yang, Y.; Wang, J.; Wang, Z.; Xie, N. High-temperature properties and aging resistance of rock asphalt ash modified asphalt based on rutting index. *Constr. Build. Mater.* **2023**, *363*, 129774. [CrossRef]

Lv, S.; Peng, X.; Liu, C.; Qu, F.; Zhu, X.; Tian, W.; Zheng, J. Aging resistance evaluation of asphalt modified by Buton-rock asphalt and bio-oil based on the rheological and microscopic characteristics. *J. Clean. Prod.* **2020**, *257*, 120589. [CrossRef]

Lima, M.S.S.; Thives, L.P.; Haritonovs, V.; Gschösser, F. The Influence of Alternative Fillers on the Adhesive Properties of Mastics Fabricated with Red Mud. *Materials* **2020**, *13*, 484. [CrossRef]

Cai, L.; Shi, X.; Xue, J. Laboratory evaluation of composed modified asphalt binder and mixture containing nano-silica/rock asphalt/SBS. *Constr. Build. Mater.* **2018**, *172*, 204–211. [CrossRef]

Hadiwardoyo, S.P.; Sumabrata, R.J.; Nissa, A.C.; Muhammad, F.A.; Hia, M.; Iskandar, D.; Lumingkewas, R.H. Improvement of Buton Rock Asphalt Performance by Adding Nano-Crumb Rubber and Waste Engine Oil. *Int. J. Pavement Res. Technol.* **2023**, *16*, 1181–1195. [CrossRef]

Li, Y.; Yan, X.; Guo, J.; Wu, W.; Shi, W.; Xu, Q.; Ji, Z. Performance and Verification of High-Modulus Asphalt Modified by Styrene-Butadiene-Styrene Block Copolymer (SBS) and Rock Asphalt. *Coatings* **2022**, *13*, 38. [CrossRef]

Liu, H.; Zhang, Z.; Li, Z.; Li, N. Effects of polyphosphoric acid (PPA), styrene-butadiene-styrene (SBS), or rock asphalt on the performance of desulfurized rubber modified asphalt. *J. Appl. Polym. Sci.* **2021**, *138*, 50621. [CrossRef]

Jin, X.; Guo, N.; You, Z.; Wang, L.; Wen, Y.; Tan, Y. Rheological properties and micro-characteristics of polyurethane composite modified asphalt. *Constr. Build. Mater.* **2020**, *234*, 117395. [CrossRef]

Lv, S.; Wang, S.; Guo, T.; Xia, C.; Li, J.; Hou, G. Laboratory evaluation on performance of compound-modified asphalt for rock asphalt/styrene–butadiene rubber (sbr) and rock asphalt/nano-$CaCO_3$. *Appl. Sci.* **2018**, *8*, 1009. [CrossRef]

He, Y.; Yi, J.; Huang, T. Study on Rheological and Mechanical Properties of Rock-Compound-Additive-Modified Asphalt and Its Mixture. *Materials* **2023**, *16*, 3771. [CrossRef]

Wen, Y.; Guo, N.; Wang, L.; Jiao, B.; Li, W. Rheological properties and microscopic mechanism of rock asphalt composite modified asphalts. *Constr. Build. Mater.* **2021**, *281*, 122543. [CrossRef]

Ren, D.; Yang, H.; Wu, P.; Xu, J.; Kong, L.; Wang, Z.; Ai, C. Green and synergistic effects of buton rock asphalt and styrene-butadiene-styrene on asphalt performance: Focus on resistance to thermal oxygen aging. *Constr. Build. Mater.* **2024**, *453*, 139027. [CrossRef]

Zhou, S.; Shi, B.; Kong, L.; Mao, C.; Yan, C.; Ai, C. Rheological and adhesive improvements of terminal blend rubberized asphalt via fumed silica nanoparticle modification. *Constr. Build. Mater.* **2024**, *449*, 138543. [CrossRef]

Lv, S.; Xia, C.; Yang, Q.; Guo, S.; You, L.; Guo, Y.; Zheng, J. Improvements on high-temperature stability; rheology, and stiffness of asphalt binder modified with waste crayfish shell powder. *J. Clean. Prod.* **2020**, *264*, 121745. [CrossRef]

Lv, S. Laboratory experiments of various bio-asphalt on rheological and microscopic properties. *J. Clean. Prod.* **2021**, *320*, 128770. [CrossRef]

Ren, D.; Yang, H.; Xu, J.; Wu, P.; Huang, Y.; Kong, L.; Wang, Z.; Ai, C. Investigation on fatigue damage of buton rock asphalt mixtures using semi-circular bending (SCB) and digital image correlation (DIC) techniques. *Constr. Build. Mater.* **2024**, *451*, 138797. [CrossRef]

Wen, Y.; Guo, N.; Wang, L.; Jin, X.; Li, W.; Wen, H. Assessment of various fatigue life indicators and fatigue properties of rock asphalt composite. *Constr. Build. Mater.* **2021**, *289*, 123147. [CrossRef]

Zhou, S.; Wang, J.; Li, S.; Yan, C. Investigating the rheological behaviour of styrene–butadiene–styrene modified asphalt. *Road Mater. Pavement Des.* **2024**, *25*, 1964–1979. [CrossRef]

Hao, Y.; Ye, Y.; Zhuang, C.; Hou, F. Research on High-temperature evaluation indexes and performance of Qingchuan rock-SBS composite modified asphalt. *Materials* **2022**, *15*, 7455. [CrossRef] [PubMed]

Cheng, L.; Yu, J.; Zhao, Q.; Wu, J.; Zhang, L. Chemical, rheological and aging characteristic properties of Xinjiang rock asphalt-modified bitumen. *Constr. Build. Mater.* **2020**, *240*, 117908. [CrossRef]

Yan, K.; Li, Y.; Long, Z.; You, L.; Wang, M.; Zhang, M.; Diab, A. Mechanical behaviors of asphalt mixtures modified with European rock bitumen and waste cooking oil. *Constr. Build. Mater.* **2022**, *319*, 125909. [CrossRef]

Zhou, S.; Li, S.; Yan, C. Influence of fumed silica nanoparticles on the rheological and anti-aging properties of bitumen. *Constr. Build. Mater.* **2023**, *397*, 132388. [CrossRef]

Zhou, S.; Yan, J.; Shi, B.; Li, S.; Ai, C. Cost-effective enhancement of high viscosity modified bitumen anti-aging properties using organic layered double hydroxide/fume silica nanoparticles composite nanomaterials. *J. Clean. Prod.* **2024**, *459*, 142538. [CrossRef]

Zhou, S.; Wang, J.; Li, S.; Yan, C. Investigating the Effects of Antioxidants on the Aging Characteristics of Crumb Rubber Modified Asphalt. *Mater. Today Commun.* **2023**, *35*, 106040. [CrossRef]

27. Zeng, G.; Zhang, J.; Huang, H.; Xiao, X.; Yan, C. A Comparative Study for Creep and Recovery Behavior Characterization c Modified Bitumens Using the MSCR Test. *Coatings* **2023**, *13*, 1445. [CrossRef]
28. Zhou, S.; Yan, J.; Shi, B.; Li, S.; Ai, C.; Yan, C. Green synthesis of a broad-spectrum UV-blocking bitumen modifier: Investigatio of anti-aging performance and mechanism in bitumen. *J. Clean. Prod.* **2023**, *425*, 139064. [CrossRef]
29. Arisona, A.; Nawawi, M.; Nuraddeen, U.K.; Hamzah, M. A preliminary mineralogical evaluation study of natural asphalt roc characterization, southeast Sulawesi, Indonesia. *Arab. J. Geosci.* **2016**, *9*, 272. [CrossRef]
30. Yan, C.; Yu, X.; Tao, Y.; Zhou, S.; Zhan, Y. Influence of the Instrumental Inertia Effect on the Asphalt Binder Oscillatio Measurement. *J. Mater. Civ. Eng.* **2023**, *35*, 04023207. [CrossRef]
31. Liu, W.; Wu, S.; Jiang, Q.; Zhao, Z.; Yang, X.; Xie, J. Investigation on Rheological Properties and Microscopic Mechanisms c Sasobit/Buton Rock Asphalt Modified Asphalt. *Sustainability* **2023**, *15*, 10224. [CrossRef]
32. Yadykova, A.Y.; Strelets, L.A.; Ilyin, S.O. Infrared Spectral Classification of Natural Bitumens for Their Rheological and Therm physical Characterization. *Molecules* **2023**, *28*, 2065. [CrossRef] [PubMed]
33. Zhang, H.; Zheng, F.; Zhang, J.; Hu, D.; Minelli, M.; Pei, J.; Sangiorgi, C. Contributions of fourier-transform infrared spectroscop technologies to the research of asphalt materials: A comprehensive review. *Fuel* **2024**, *371*, 132078. [CrossRef]

coatings

MDPI

Article

Experimental and Numerical Investigation on the Mechanical Properties of Concrete with High Volumes of Modified Phosphogypsum

Xiang Cheng [1,2], Qizhi Li [3], Peng Liu [1,2,4,*], Jingxiang Huang [1,2], Lingling Wang [5], Ying Chen [1,6], Feng Zhang [7], Wei Li [7], Zhiwu Yu [2,4], Lei Liu [1,2], Guangqiang Shao [3] and Shuaifeng Wang [1,2]

1 School of Civil Engineering, Central South University, Changsha 410000, China; chengxiang0458@163.com (X.C.); 18090271678@163.com (J.H.); cheny83@csu.edu.cn (Y.C.); sanlingzhang30@163.com (L.L.); wangshuaifeng1119@163.com (S.W.)
2 National Engineering Research Center for High Speed Railway Construction, Changsha 410000, China; zhwyu0512@163.com
3 China Construction Second Engineering Bureau Ltd., Beijing 100160, China; liqizhi998@163.com (Q.L.); 234801029@csu.edu.cn (G.S.)
4 China Railway Group Ltd., Beijing 100039, China
5 School of Civil Engineering, Guizhou University, Guiyang 550025, China; czt1035994754@126.com
6 School of Civil Engineering, Central South University of Forestry and Technology, Changsha 410000, China
7 China Railway 20th Bureau Group Sixth Engineering Ltd., Xi'an 710000, China; 224812282@csu.edu.cn (F.Z.); 224801007@csu.edu.cn (W.L.)
* Correspondence: lop868@163.com

Abstract: The effects of high-temperature modified phosphogypsum (HPG), incorporated at contents of 40%, 50%, and 60%, on the compressive strength and elastic modulus of mortar and concrete were investigated. Additionally, the influence of graded granulated blast furnace slag powder (GGBS), quicklime, and silica fume on the mechanical properties of HPG-based mortar (HPGM) and HPG-based concrete (HPGC) was discussed. Moreover, the microstructure of HPGM was analyzed using scanning electron microscopy (SEM). A two-dimensional mesoscale model of HPGC was developed to predict how variations in HPG content, coarse aggregate characteristics, and interfacial transition zone (ITZ) characteristics influence the compressive strength and elastic modulus of HPGC. The experimental results showed that high volumes of HPG weakened the mechanical properties of HPGM and HPGC, while appropriate amounts of mineral admixtures offset the negative effects caused by calcium hydroxide ($Ca(OH)_2$) crystals and impurities within the system. The simulation results indicated that the maximum deviation between the mesoscale model prediction and experimental data was only 8.38%, which verified the accuracy of the mesoscale model prediction. The compressive strength of HPGC initially decreased and subsequently increased with the rise in the modulus and content of coarse aggregate, whereas it declined with higher HPG dosage and increased ITZ thickness. In contrast, the elastic modulus of HPGC showed a gradual increase with rising coarse aggregate content and improved ITZ mechanical properties, while it decreased as HPG content and ITZ thickness increased.

Keywords: high-temperature modified phosphogypsum-based concrete; mechanical properties; mesoscale model; microstructure

1. Introduction

In the face of global carbon reduction targets, the traditional cement industry urgently needs to transform to reduce its environmental impact and dependence on non-renewable

https://doi.org/10.3390/coatings15010065

resources. Phosphogypsum, a by-product of phosphate fertilizer production, is characterized by high production volumes and low utilization rates. It primarily consists of calcium sulfate dihydrate ($CaSO_4 \cdot 2H_2O$) and has potential as a novel cementitious material. Research on the application of phosphogypsum in building materials can not only promote the green transformation of the cement industry, but also facilitate the resourceful utilization of industrial waste, which is of great significance to the construction of an environmentally friendly building materials industry [1,2].

In recent years, researchers have intensified their efforts to improve the utilization rate of phosphogypsum by advancing studies on its fundamental properties, modification techniques, and reaction mechanisms. As a result, phosphogypsum has been widely applied in building materials, including cementitious materials, mortar, concrete, and filling materials [3–10]. Regarding the research on the basic properties of phosphogypsum, Zhang et al. [11] and Gao et al. [12] conducted experiments to investigate the mixing ratios, compressive properties, elastic modulus, and water resistance of phosphogypsum, confirming its potential as a non-load-bearing material in engineering applications. Based on the above research, Ma et al. [13] successfully employed phosphogypsum as a filler material for partition walls. In addition, phosphogypsum can serve as a partial replacement for cement in concrete, enabling the adjustment of concrete strength through the optimization of its proportion. Bagade et al. [14] studied the impact of phosphogypsum substitution rates on concrete compressive strength using 5%, 10%, and 15% phosphogypsum replacements. The results showed that the compressive strength of concrete increased with 5% and 10% phosphogypsum replacements but significantly decreased at a 15% phosphogypsum substitution rate. Similarly, Wang et al. [15] tested concrete samples with 20%, 25%, and 30% phosphogypsum replacements. Their results indicated a decrease in compressive strength with increasing phosphogypsum content, suggesting that the phosphogypsum dosage should not exceed 25%.

The above research findings reveal that an appropriate amount of phosphogypsum enhances concrete strength, while an excessive amount weakens it. This is mainly attributed to the lower strength of phosphogypsum compared to cement, which reduces the matrix strength when used in large quantities. To enhance the utilization of phosphogypsum, modified technology has emerged. Huang et al. [16] proposed a novel cementitious material composed primarily of phosphogypsum, supplemented with grade granulated blast furnace slag powder (GGBS), steel slag, cement, and other admixtures. Through carefully designing and optimizing the mix ratio, they successfully developed a water-hardened material with a 28-day compressive strength exceeding 40 MPa. Tian et al. [17] utilized phosphogypsum, combined with cement, GGBS, and other admixtures, to create foam concrete. The experimental results indicated that the strength of the phosphogypsum foam concrete reached its peak when the phosphogypsum content ranged from 45% to 55%, and the ratio of cement to mineral powder was 1:1. Additionally, the research results of Li et al. [18] and An et al. [19] confirmed the strengthening effect of mechanically and thermally modified phosphogypsum on phosphogypsum-based cementitious materials. To elucidate the enhancement mechanism, Liu et al. [20] investigated the phase composition and microstructure of thermally modified phosphogypsum-based cement pastes using XRD and SEM. The results showed that the modified samples exhibited increased hydration products and denser microstructures. Jiang et al. [21] suggested that quicklime could neutralize phosphorus and fluorine impurities in phosphogypsum, increasing the number of contact points of $CaSO_4 \cdot 2H_2O$, thus enhancing the properties of phosphogypsum-based building materials.

On the mesoscale, concrete is recognized as a composite material consisting of coarse aggregate, mortar, and the interfacial transition zone (ITZ). Consequently, employing mesoscale models to reveal the intricate connections between the mesoscale component characteristics and macroscale mechanical properties of concrete represents an efficient and viable approach [22,23]. Wu et al. [24] and Zheng et al. [25] established a two-dimensional (2D) concrete mesoscopic model, revealing the influence of factors, such as coarse aggregate content, the elastic modulus of coarse aggregate, and mortar material, on the elastic modulus of concrete. Zhou et al. [23] and Qu et al. [26], respectively, constructed concrete mesoscopic models of different dimensions to predict the influence of coarse aggregate shape, particle size, and porosity on the compressive strength of concrete. The above research results confirm that the mesoscale model can effectively reflect the effect of mesoscopic components on the compressive strength and elastic modulus of concrete.

In summary, the existing research on phosphogypsum-based materials has predominantly focused on low content levels, with limited studies on high-volume phosphogypsum materials, and performance enhancement technologies remain immature. Additionally, there is a relative lack of combined experimental and simulation studies for phosphogypsum-based concrete at the microscale, mesoscale, and macroscale. Therefore, this study proposed to prepare HPG-based mortar (HPGM) and HPG-based concrete (HPGC) specimens with HPG replacement contents of 40%, 50%, and 60%. The impact of HPG content on the compressive strength and elastic modulus of HPGM and HPGC were investigated. Additionally, the improvements in the macro-mechanical properties of HPGM and HPGC, achieved by incorporating GGBS, quicklime, and silica fume, were systematically analyzed. A microstructural analysis of HPGM was also conducted using SEM. To predict the macroscopic mechanical properties of HPGC under varying conditions, a 2D mesoscale model of HPGC was developed. Based on this, the influence of HPG content, coarse aggregate characteristics (shape, content, elastic modulus) and ITZ characteristics (thickness and mechanical properties) on the compressive strength and elastic modulus of HPGC was quantified. This provides important theoretical support and practical guidance for the engineering application of phosphogypsum at large dosages.

2. Materials and Method

2.1. Raw Materials

Phosphogypsum (PG), primarily consisting of $CaSO_4 \cdot 2H_2O$, inherently lacks cementitious properties. However, some research has shown that high-temperature modification can induce partial dehydration, converting phosphogypsum into α- or β-hemihydrate gypsum ($CaSO_4 \cdot 0.5H_2O$) or even anhydrous gypsum ($CaSO_4$) [18–20]. Specifically, high-temperature modification at 160 °C for 2 h effectively promotes the transformation of $CaSO_4 \cdot 2H_2O$ into $CaSO_4 \cdot 0.5H_2O$, while requiring a relatively short processing time. The resulting phosphogypsum-based materials exhibit excellent mechanical properties. Accordingly, this study employs high-temperature modification at 160 °C for 2 h to process PG with a particle size of 0.15 mm, aiming to produce excellent performance high-temperature modified phosphogypsum (HPG).

The P·O 42.5 Portland cement was provided by Hailuo Cement Co., Ltd., Guiyang, China, and its specific surface area was 330 m^2/kg. The S95 grade granulated blast furnace slag powder (GGBS) was bought from Gongyi Longze Water Purification Materials Co., Ltd., Zhengzhou, China, and has a specific surface area of 450 m^2/kg. The ultrafine quicklime with 98.3% effective CaO content, produced by Yibin Sichuan Ash Biotechnology Co., Ltd., Yibin, China, was selected for the test. The silica fume with SiO_2 content of 96% or more was selected for the test, which was bought from Henan Province Gongyi Qingyang Water Treatment Materials Co., Ltd., Zhengzhou, China. The chemical compositions of the materials, including cement, PG, HPG, GGBS, quicklime, and silica fume, were shown in Table 1. The experimental photography and microstructures of PG and HPG are shown in Figure 1. Manufactured sand from Guizhou Province, with a fineness modulus of 2.8, was used as fine aggregate. Continuous grading basalt gravels from Guizhou Province with a grain size of 5–25 mm were collected. The retarders belonging to the protein class of retarders were gained from Shanghai Chenqi Chemical Technology Co., Ltd., Shanghai, China. The polycarboxylate superplasticizer with a water reduction rate of 25–27% was obtained from Guizhou Zhucheng Hengchuang Construction Engineering Co., Ltd., Guiyang, China.

Table 1. Main chemical compositions of raw materials (wt%).

Materials	SiO_2	Al_2O_3	CaO	Fe_2O_3	SO_3	P_2O_5	K_2O	Others
Cement	19.90	5.15	61.71	4.46	3.96	0.17	1.19	3.45
PG	2.68	0.30	39.52	0.37	55.28	0.89	0.07	0.89
HPG	2.71	0.29	41.84	0.38	53.60	0.86	0.07	0.25
GGBS	34.99	15.70	35.18	0.84	2.61	0.03	7.58	3.07
Quicklime	0.60	0.14	98.29	0.11	0.24	0.01	0.58	0.03
Silica fume	96.74	0.32	0.11	0.08	-	-	-	2.75

Note: PG refers to phosphogypsum; HPG is high-temperature modification phosphogypsum.

Figure 1. Powdered forms and microscopic morphology of PG and HPG. (**a**) PG powder. (**b**) Microscopic PG. (**c**) HPG powder. (**d**) Microscopic HPG.

2.2. Preparation of Specimens

Referring to the Chinese standard GB/T 50081-2019 [27] and JCJ/T 70-2009 [28], the HPGM and HPGC specimens were prepared and tested during the experiment. The size of HPGM specimens for the compression and modulus of elasticity test was 70.7 mm × 70.7 mm × 70.7 mm, 70.7 mm × 70.7 mm × 220 mm, respectively. The size of HPGC specimens for the compression test was 100 mm × 100 mm × 100 mm, and the size for modulus of elasticity was 150 mm × 150 mm × 300 mm. In addition, there were three specimens for each group of performance tests. All the HPGM and HPGC specimens used in this experiment were prepared according to the mixing proportion shown in Tables 2 and 3. Firstly, the specimens were de-molded 24 h after casting. Then, they were cured at the required age in a standard curing room, maintained at a temperature of (20 ± 2) °C and a relative humidity no less than 95%.

Table 2. Mixture proportion of HPGM.

	HPG (%)	Cement (%)	GGBS (%)	Limestone (%)	Silica Fume (%)	Water (kg/m^3)	Sand (kg/m^3)	Water–Cement Ratio	Water Reducing (%)	Retarder (%)
HPGM	0	100	0	-	-	170	832	0.45	1	-
	40	50	10	-	-	170	832	0.45	1.8	0.3
	40	45	15	-	-	170	832	0.45	1.8	0.3
	40	40	20	-	-	170	832	0.45	1.8	0.3
	40	35	25	-	-	170	832	0.45	1.8	0.3
	50	40	10	-	-	170	832	0.45	1.8	0.3
	50	35	15	-	-	170	832	0.45	1.8	0.3
	50	30	20	-	-	170	832	0.45	1.8	0.3
	50	25	25	-	-	170	832	0.45	1.8	0.3
	60	30	10	-	-	170	832	0.45	1.8	0.3
	60	25	15	-	-	170	832	0.45	1.8	0.3
	60	20	20	-	-	170	832	0.45	1.8	0.3
	60	15	25	-	-	170	832	0.45	1.8	0.3
	40	40	20	7	5	170	832	0.45	1.8	0.3
	50	30	20	7	5	170	832	0.45	1.8	0.3
	60	20	20	7	5	170	832	0.45	1.8	0.3

Table 3. Mixture proportion of HPGC.

	HPG (%)	Cement (%)	GGBS (%)	Limestone (%)	Silica Fume (%)	Water (kg/m^3)	Sand (kg/m^3)	Water–Cement Ratio	Aggregate (kg/m^3)	Water Reducing (%)	Retarder (%)
HPGC	0	100	-	-	-	170	832	0.45	1017	1	-
	40	60	-	-	-	170	832	0.45	1017	2	0.3
	40	40	20	7	5	170	832	0.45	1017	2	0.3
	50	50	-	-	-	170	832	0.45	1017	2	0.3
	50	30	20	7	5	170	832	0.45	1017	2	0.3
	60	40	-	-	-	170	832	0.45	1017	2	0.3
	60	20	20	7	5	170	832	0.45	1017	2	0.3

2.3. Experimental Process

In this research, the PG with a particle size of 0.15 mm was dried in an electric blast drying oven for 2 h at 160 °C. The testing machine model is 101-4B. The microstructural morphology of PG, HPG, and HPGM was observed using SEM, ZEISS GeminiSEM 300, Jena, Germany. Concerning the Chinese standard JGJ/T 70-2009 [29], the uniaxial compression and modulus of elasticity tests of the HPGM specimens were carried out. Meanwhile, the uniaxial compression, splitting tensile, and modulus of elasticity tests of the HPGC specimens were conducted with reference to the Chinese standard GB/T 50081-2019 [28]. The rock and concrete test system (RMT), with the RMT-301 model, was used to evaluate the mechanical properties of HPGM and HPGC.

3. Experimental Results

3.1. Effect of HPG Content on Mechanical Properties of HPGM

To investigate the effect of a high volume of HPG incorporation on the mechanical properties of HPGM, a series of specimens were prepared with HPG substituting cement at ratios of 40%, 50%, and 60%. And the mortar without HPG was the reference group mortar. The outcomes of the 28-day compressive strength and elastic modulus of HPGM are depicted in Figure 2.

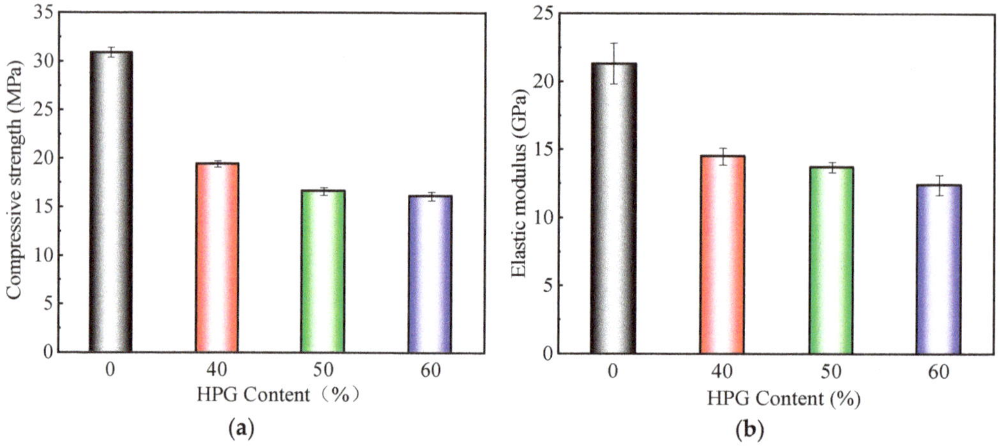

Figure 2. Effect of HPG content on mechanical properties of HPGM. (**a**) Compressive strength. (**b**) Elastic modulus.

As observed from Figure 2, the incorporation of a high volume of HPG significantly affected both the compressive strength and elastic modulus of HPGM. Compared to the reference mortar, which exhibited a compressive strength of 30.9 MPa and an elastic modulus of 31.3 GPa, the HPGM with 40%, 50%, and 60% HPG content showed compressive strengths of 19.4, 16.6, and 16.1 MPa, respectively, with the maximum reduction reaching 47.89%. Similarly, the elastic modulus of HPGM decreased to 14.5, 14.3, and 12.1 GPa, respectively, with a maximum decrease of 43.19%. The experimental findings demonstrate that substituting cement with a high volume of HPG significantly degrades the mechanical properties of the mortar. Notably, both compressive strength and elastic modulus exhibit a gradual decline as the HPG content increases. This is mainly due to the high volume of HPG and its rapid coagulation when in contact with water, which leads to a poor mixing uniformity of the HPGM and increases internal structural defects. These defects weaken the overall integrity of the material, thereby reducing both its compressive strength and elastic modulus. Additionally, the primary hydration product of HPG, $CaSO_4{\cdot}2H_2O$, exhibits inferior strength and wear resistance compared to the hydration products of cement. This disparity further exacerbates the decline in mechanical properties. As HPG content increases, the reduction in compressive strength and elastic modulus becomes more pronounced. Moreover, when excessive HPG is incorporated, the reaction between $CaSO_4{\cdot}2H_2O$ and the aluminum phase in the system occurs more readily, generating a significant number of ettringite with expansion. This exacerbates internal stresses and defects, further degrading the mechanical properties of HPGM.

3.2. Effect of Mineral Admixtures on the Mechanical Properties of HPGM

3.2.1. Effect of GGBS on the Compressive Strength of HPGM

Some studies have shown that common mineral admixtures, such as GGBS, silica fume, and quicklime, exhibit excellent pozzolanic or micro-aggregate effects, which can significantly enhance the mechanical properties of HPGM when incorporated in appropriate amounts [29,30]. In this section, HPG was used to replace cement at replacement rates of 40%, 50%, and 60%. Additionally, varying amounts of GGBS (10%, 15%, 20%, 25%) were used to replace part of the remaining cement used to prepare HPGM, with the mortar without GGBS serving as the reference group mortar. The influence of GGBS content on the 28-day compressive strength of HPGM was investigated, and the results are presented in Figure 3.

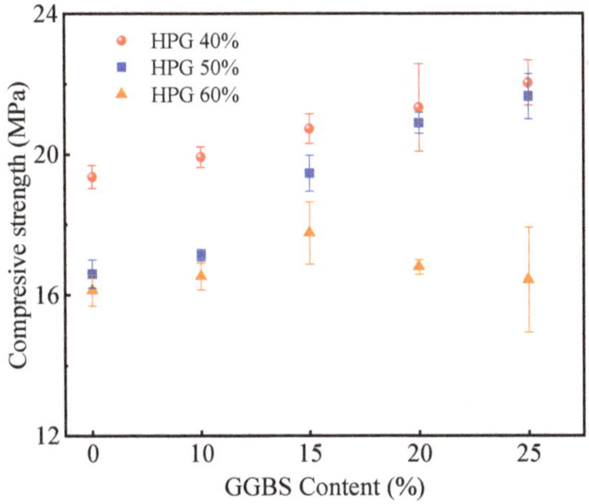

Figure 3. Effect of GGBS on the compressive strength of HPGM.

Figure 3 demonstrated that GGBS, within a range of 10% to 25%, had a notable impact on the 28-day compressive strength of HPGM. For HPGM with 60% HPG, the compressive strength initially increased and reached a maximum value of 17.8 MPa at 15% GGBS, representing a 10.2% improvement over the reference mortar (16.1 MPa). However, further increases in GGBS content led to a subsequent decrease in compressive strength. For HPGM containing 40% or 50% HPG, the compressive strength gradually increased with the increase in GGBS content. However, when the GGBS content was increased from 20% to 25%, the compressive strength of HPGM increased only slightly. These findings suggested that moderate amounts of GGBS could effectively improve the compressive strength of HPGM, while excessive amounts of GGBS had diminishing returns or even negative effects. The observed phenomenon is primarily attributed to the hydration of GGBS, which consumes $Ca(OH)_2$ crystals within the system, thereby mitigating their adverse effects in the ITZ. Additionally, under the double activation of $Ca(OH)_2$ and sulfate ions, the hydration of GGBS generates a flocculent calcium silicate hydrate (C-S-H) gel, calcium aluminate hydrated (C-A-H) gel, and acicular ettringite. These products interlock and densify the HPGM, thereby further enhancing their compressive strength. However, excessive GGBS substitution reduces cement content, limiting the formation of C-S-H and C-A-H gels from cement hydration and subsequently, weakening the HPGM. Considering the effect of GGBS on the compressive strength of HPGM with 40%, 50% and 60% HPG content, the optimal GGBS substitution rate is 20%.

3.2.2. Effect of Quicklime and Silica Fume on the Mechanical Properties of HPGM

To further enhance the mechanical properties of HPGM, this study prepared control groups of HPGM by using 40%, 50%, and 60% HPG to substitute cement. Meanwhile, based on the previous experimental results and relevant literature, a fixed replacement rate of 20% GGBS was used, with an additional 7% quicklime and 5% silica fume incorporated to prepare HPGM [10]. Using this approach, the effects of these mineral admixtures on the 28-day compressive strength and elastic modulus of HPGM were investigated. The results are shown in Figure 4.

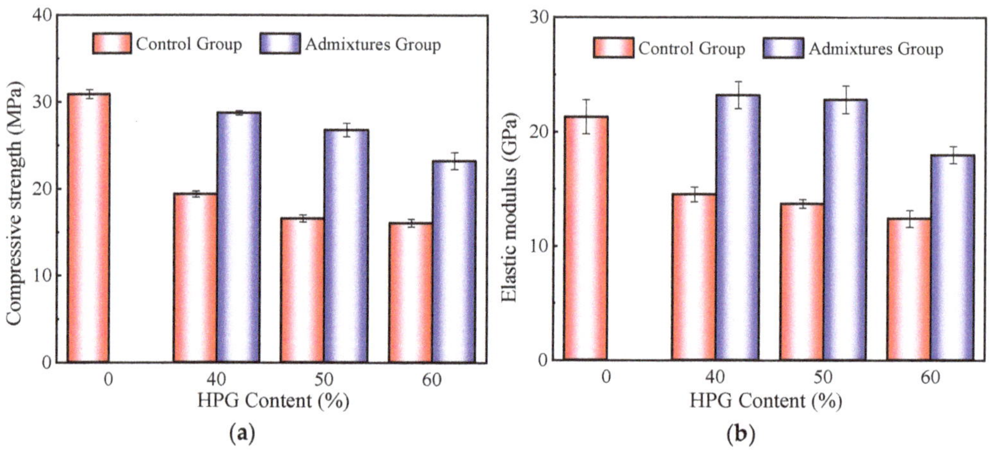

Figure 4. Effect of mineral admixtures on mechanical properties of HPGM. (**a**) Compressive strength. (**b**) Elastic modulus.

As seen from Figure 4, the combined addition of GGBS, quicklime, and silica fume significantly enhances the 28-day compressive strength and elastic modulus of HPGM. Taking HPGM with 50% HPG as an example, the compressive strength and elastic modulus of the HPGM with mineral admixtures were 26.8 MPa and 22.8 GPa, respectively. This represented a remarkable increase of 61.4% and 66.4% compared to the control group HPGM, which had a compressive strength of 16.6 MPa and an elastic modulus of 13.7 GPa. In addition, the compressive strength of HPGM with multiphase admixtures was further improved, compared with HPGM with only 20% GGBS. These findings suggested that the mechanical properties of HPGM can be further bolstered by the proper addition of quicklime and silica fume. The main reason is that quicklime and silica fume possess fine particles, which effectively improve the particle gradation of the HPGM cementitious system and fill the internal pores. Furthermore, the superposition effect is easily produced between multiphase admixtures. The addition of an appropriate amount of quicklime not only eliminates the retardation and deterioration effects produced by the impurities contained in HPG, but also generates a large amount of $Ca(OH)_2$, which activates the activity of GGBS and silica fume materials. During this process, the hydration of GGBS and silica fume consumes $Ca(OH)_2$, mitigating its adverse effects in the ITZ. Moreover, the hydration products, such as C-S-H and C-A-H gels, refine the pore size of the HPGM, improving the bonding between the components, and enhancing the densification of the HPGM. In summary, the appropriate combination of GGBS, quicklime, and silica fume effectively offsets the negative impact of HPG as a cement replacement, significantly enhancing the mechanical properties of HPGM.

3.3. Effect of HPG Content on Mechanical Properties of HPGC

For a systematic investigation of the influence of a high volume of HPG on the mechanical properties of HPGC, this study prepared HPGC specimens containing 40%, 50%, and 60% HPG, respectively. Additionally, there was a reference group concrete without HPG. Subsequently, the 28-day compressive strength and elastic modulus of HPGC were measured. The test results are presented in Figure 5.

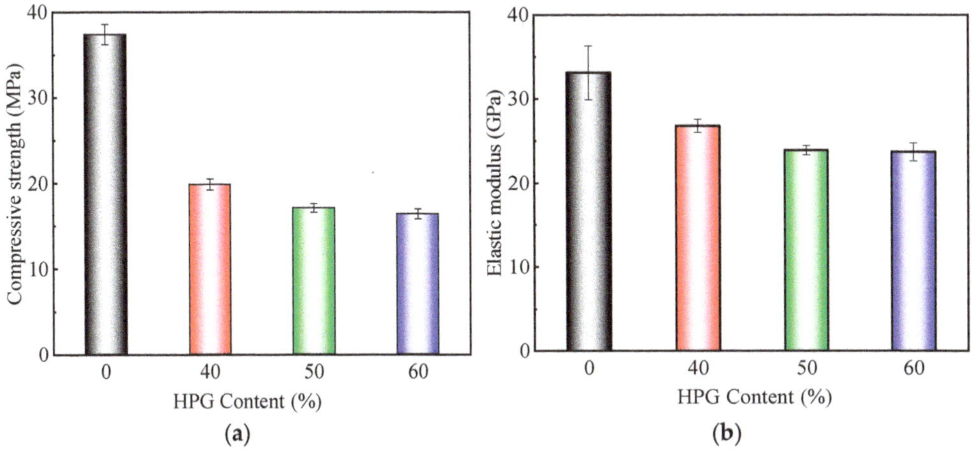

Figure 5. Effect of HPG content on mechanical properties of HPGC. (**a**) Compressive strength. (**b**) Elastic modulus.

Figure 5 illustrated that the compressive strength and elastic modulus of HPGC were significantly reduced after the incorporation of a high volume of HPG. Moreover, as the HPG content increased, the compressive strength and elastic modulus of HPGC decreased gradually. When the HPG content was 40%, the HPGC specimens exhibited a compressive strength of 19.9 MPa and an elastic modulus of 23.7 GPa, representing decreases of 46.8% and 19.0%, respectively, compared to the reference group. As the HPG content increased to 60%, the reductions in compressive strength and elastic modulus of the HPGC escalated to 56.1% and 28.4%, respectively. The reason for the observed phenomenon is that when HPGC is prepared by replacing cement with HPG in high volumes, the amount of C-S-H gel formed by cement hydration in the system is significantly reduced. Correspondingly, the content of $CaSO_4{\cdot}2H_2O$, which exhibits low strength and stability, increases due to the hydration of HPG. This leads to the formation of massive ettringite when $CaSO_4{\cdot}2H_2O$ reacts with the aluminum phase in the system. With the substantial accumulation of ettringite, microscopic cracks are easily induced inside the HPGC, resulting in a significant decrease in the compressive strength and elastic modulus. In addition, HPG contains abundant soluble phosphorus, fluorides, and other impurities, which hinder the hydration reaction of the system. Consequently, the internal structure of HPGC becomes more porous, and its compactness is reduced.

3.4. Effect of Mineral Admixtures on Mechanical Properties of HPGC

To eliminate the adverse effects of HPG incorporation on the mechanical properties of HPGC, this study investigated the improvement of compressive strength and elastic modulus of HPGC through the incorporation of mineral admixtures. The results are shown in Figure 6. The control groups were the HPGC specimens with HPG institution contents of 0, 40%, 50%, and 60%. The admixture groups were based on the control groups, with the further addition of 20% GGBS, 7% quicklime, and 5% silica fume.

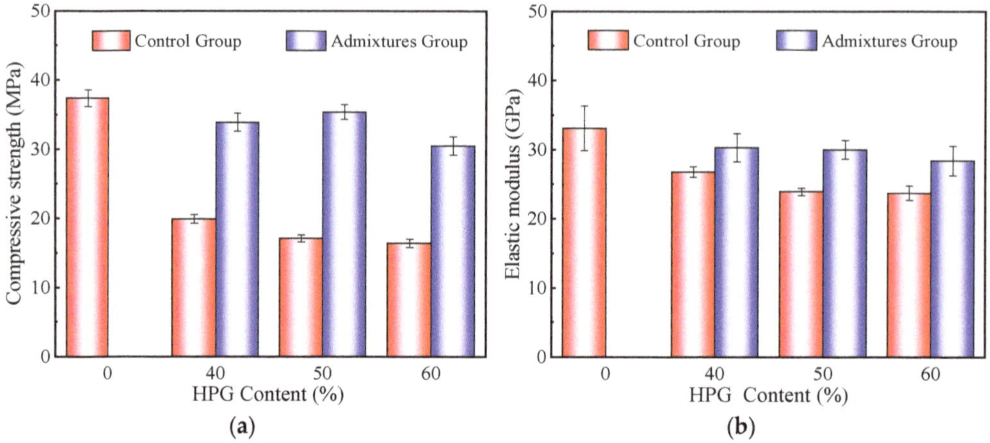

Figure 6. Effect of mineral admixtures on mechanical properties of HPGC. (**a**) Compressive strength. (**b**) Elastic modulus.

Figure 6 indicated that the compressive strength and elastic modulus of HPGC could be effectively improved by adding appropriate amounts of mineral admixtures. Compared with that of 40%, 50% and 60% HPGC (without mineral admixtures), the compressive strength and elastic modulus of HPGC with mineral admixtures increased by 77.9%, 98.2% and 86.0%, respectively, while the elastic modulus increased by 13.1%, 25.5% and 19.8%, respectively. The results showed that the mechanical properties of HPGC can be effectively optimized through the proper proportioning of mineral admixtures, achieving a strength grade of C30. The main reason is the occurrence of a gradient hydration phenomenon when multiple mineral admixtures are mixed. The addition of an appropriate amount of quicklime not only weakens the negative impact of impurities in HPG, but also strengthens the alkalinity of the system. With the hydration reaction, on the one hand, the hydration of GGBS and silica fume consumes the enriched $Ca(OH)_2$ crystals in the system, improving the microstructural morphology of ITZ. On the other hand, the hydration of GGBS and silica fume generates abundant C-S-H and C-A-H gels under the activation of $Ca(OH)_2$. Through that, the internal microstructural pores of HPGC are effectively filled, and the bonding between the components is enhanced. Thus, the compactness of HPGC is improved. In summary, the multiphase admixtures tend to produce advantageous superposition effects among them, which effectively improve the mechanical properties of HPGC.

3.5. Microstructure

To reveal the improvement mechanism of HPGM performance by the above admixtures, the microstructure and product morphology of three groups of HPGM specimens at 28 days were analyzed using SEM in this study. The three groups of HPGM specimens consisted of 50% HPG, 50% HPG-20% GGBS, and 50% HPG-20% GGBS-7% quicklime-5% silica fume, respectively. Among them, the last group of HPGM samples were added externally with quicklime and silica fume. The results are shown in Figure 7.

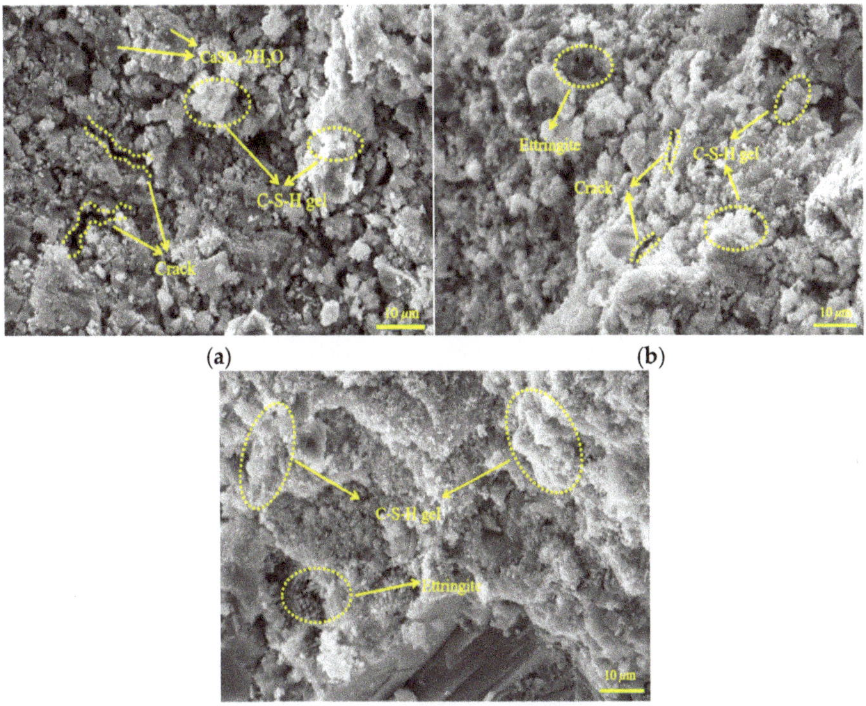

(**a**) (**b**) (**c**)

Figure 7. Microstructure of HPGM. (**a**) 50% HPG. (**b**) 50% HPG-20% GGBS. (**c**) 50% HPG-20% GGBS-7% quicklime-5% silica fume.

As seen from Figure 7a, the hydration products of HPGM with a 50% HPG substitution content were predominantly lumpy $CaSO_4·2H_2O$ and flocculent C-S-H gel. However, these hydration products exhibited inadequate interconnectivity among each other, characterized by high porosity and obvious microcracks, resulting in the inferior macroscopic mechanical properties of HPGM. In contrast, when 20% GGBS was doped (e.g., Figure 7b), the internal microstructure of HPGM became significantly denser. The content of pores and microcracks was reduced, and the sizes of microcracks significantly decreased, compared with the HPGM without GGBS substitution. This is because the incorporation of a moderate amount of GGBS reduces the generation of $Ca(OH)_2$ crystals in the system, thereby mitigating the adverse effects stemming from the excessive accumulation of $Ca(OH)_2$ at the ITZ. Furthermore, the synergistic effect of $Ca(OH)_2$ and sulfate ions in the system activates the latent reactivity of the GGBS, fostering the formation of C-S-H and C-A-H gels, as well as ettringite. Based on the above study, when 7% quicklime and 5% silica fume were externally added (e.g., Figure 7c), the hydration products within the HPGM exhibited a tight interlocking arrangement. Moreover, the pores were partially filled with ettringite, and no obvious microcracks were observed. The primary reason is the gradient hydration process between multiphase admixtures, which leads to a superposition effect. This effect not only mitigates the negative impact of impurities in HPG on the hardened matrix, but also enhances the alkaline environment within the system. Subsequently, the alkaline conditions stimulate the Pozzolanic effect of GGBS and silica fume, generating a greater quantity of C-S-H and C-A-H gels, as well as ettringite. This optimization effectively refines the ITZ, fills internal pores, and enhances the interfacial bonding between components, ultimately promoting the mechanical properties of the HPGM.

4. Numerical Simulation

4.1. Mesoscale Model

On the mesoscale, concrete is regarded as a multiphase inhomogeneous composite material. The material properties of the internal components of concrete, e.g., coarse aggregate, mortar matrix, and ITZ, have a significant influence on its macroscopic mechanical properties [31–33]. Through constructing concrete mesoscale models, the shape, position, and grading distribution of coarse aggregates in concrete can be accurately portrayed. Moreover, these models can also describe the damage of ITZ and the expansion of cracks, thereby realizing the cross-scale prediction of the mechanical properties of concrete [34,35]. Among them, 2D circular and polygonal aggregate models are the mainstream forms of existing research [32,36]. Therefore, this study employed Python language for the secondary development of the finite element software ABAQUS 2022, constructing the 2D circular and polygonal HPGC model.

4.1.1. Coarse Aggregate Grading and Particle Size Distribution

The gradation and particle size of coarse aggregates are important factors affecting the compactness of concrete. Reasonable coarse aggregate gradation design not only helps to reduce the amounts of cementitious materials, but also effectively fills the internal voids, thereby improving the overall compactness of concrete. Therefore, based on the Fuller grading theory and Walraven formula, this section initially determined the content and particle size range of the coarse aggregates [37,38]. Meanwhile, considering that the range of the coarse aggregate particle sizes used in the actual tests is 5–25 mm, this study divided the range into three intervals, 5–10 mm, 10–15 mm, and 15–25 mm, and assigned each particle size interval in the ratio of approximately 3:3:4.

The specific expression for the Fuller grading theory is given below:

$$p = 100\sqrt{\frac{D_0}{D_{\max}}} \tag{1}$$

where D_0 and $D_{\max}$ are the sieve diameter and maximum coarse aggregate size, respectively; and p is the percentage of coarse aggregate mass.

The specific expression of the Walraven formula is as follows:

$$p_c(D < D_0) = p_k \begin{bmatrix} 1.065\left(\frac{D_0}{D_{\max}}\right)^{0.5} - 0.053\left(\frac{D_0}{D_{\max}}\right)^4 - 0.012\left(\frac{D_0}{D_{\max}}\right)^6 \\ -0.0045\left(\frac{D_0}{D_{\max}}\right)^8 + 0.0025\left(\frac{D_0}{D_{\max}}\right)^{10} \end{bmatrix} \tag{2}$$

where D refers to the particle size in the concrete section; $p_c(D < D_0)$ is the probability that the particle size D in the concrete section is smaller than the sieve aperture D_0; andp_k refers to aggregate percentage, where p_k = 0.75 is generally taken.

4.1.2. Coarse Aggregate Generation and Placement

During the actual mixing, pouring, and molding of concrete, the position of coarse aggregates in concrete presents randomness. In addition, there is no cross-over between coarse aggregate particles. Therefore, the generation, placement, and interference test of coarse aggregates were an important part of the establishment of the mesoscale model. The specific steps for the model generation are as follows:

(1) The volume fraction of coarse aggregates cannot exceed the volume of the mesoscale model, and the location of coarse aggregates cannot exceed the boundary constraints of the mesoscale model. (2) To improve the generation efficiency of the model, priority is given to the generation and placement of coarse aggregates with a larger particle

size. (3) The newly generated coarse aggregates must not overlap or intersect with those that are already placed. (4) A sufficient distance must be maintained between coarse aggregates to ensure that a certain thickness of the ITZ surrounds each aggregate. This distance prevents interference between the ITZs.

Based on the above coarse aggregate gradation theory, combined with the steps for coarse aggregate generation and placement, this paper constructed circular aggregate and polygonal aggregate mesoscale models, respectively. Figure 8 gives the schematic diagram of the single coarse aggregate generation and the schematic diagram of the coarse aggregate model under three working conditions.

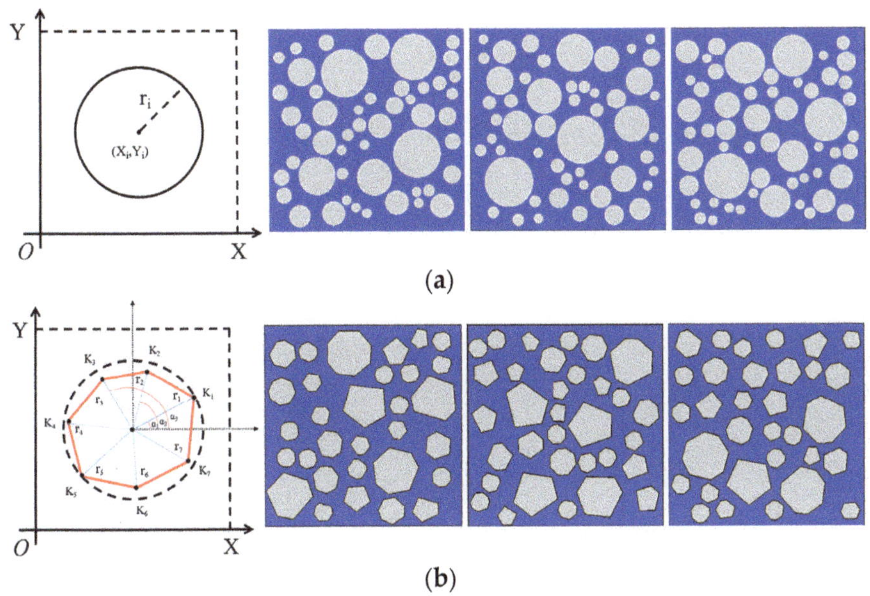

Figure 8. Diagram of circular and polygonal aggregate model generation. (**a**) Model with circular aggregate shape. (**b**) Model with polygonal aggregate shape.

4.2. Constitutive Model

4.2.1. Concrete Plastic Damage Model

Concrete, as a typical brittle material, produces different degrees of damage and stiffness degradation when subjected to external loads. To accurately portray the cumulative effect of damage, stiffness degradation, and plastic deformation of concrete under external loading, the Concrete Damage Plastic (CDP) model has been employed. This model incorporates damage variables, enabling a sophisticated simulation of the nonlinear response of concrete-like brittle materials to applied forces. Recognized for its efficacy in modeling the fracture and damage mechanisms in concrete structures, the CDP model has been widely used for mechanical analyses and safety assessments of concrete structures [39,40].

The CDP model attributes the fracture damage of concrete materials to compression damage or tensile cracking and characterizes the stiffness degradation of the material by introducing the damage variables d_c and d_t. Meanwhile, the model controls the damage evolution process of concrete materials through tensile and compressive equivalent plastic strain parameters ($\tilde{\varepsilon}_c^{pl}$, $\tilde{\varepsilon}_t^{pl}$), as shown in Figure 9. This not only accurately portrays the degradation characteristics of material stiffness, but also effectively controls the whole process of concrete damage evolution. Considering this, this paper adopted the concrete constitutive relationship provided by the Chinese standard GB 50010-2010 [41] to determine the parameters in the CDP model. The equations are as follows:

$$\sigma_c = (1 - d_c)E_0\left(\varepsilon_c - \widetilde{\varepsilon}_c^{pl}\right) \tag{3}$$

$$\widetilde{\varepsilon}_c^{pl} = \widetilde{\varepsilon}_c^{in} - \frac{d_c}{1 - d_c} \cdot \frac{\sigma_c}{E_0} \tag{4}$$

$$\sigma_t = (1 - d_t)E_0(\varepsilon_t - \widetilde{\varepsilon}_t^{pl}) \tag{5}$$

$$\widetilde{\varepsilon}_t^{pl} = \widetilde{\varepsilon}_t^{ck} - \frac{d_t}{1 - d_t} \cdot \frac{\sigma_t}{E_0} \tag{6}$$

where σ_c and σ_t denote the concrete compressive and tensile stresses, respectively. d_t is the tensile damage factor. d_c refers to the compressive damage factor. E_0 refers to the initial modulus of elasticity. $\widetilde{\varepsilon}_c^{pl}$ and $\widetilde{\varepsilon}_t^{pl}$ denote the compressive and tensile equivalent plastic strain, respectively. $\widetilde{\varepsilon}_c^{in}$ is the inelastic strain. $\widetilde{\varepsilon}_t^{ck}$ refers to the cracking strain.

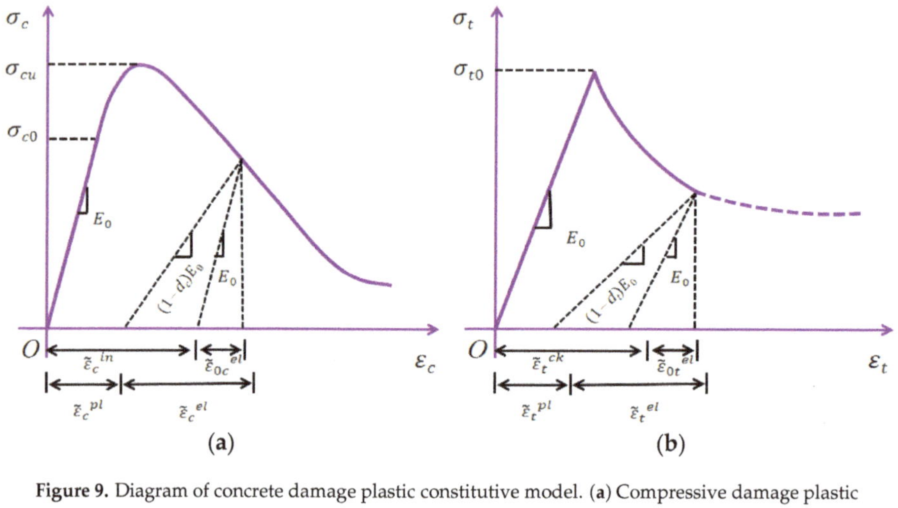

Figure 9. Diagram of concrete damage plastic constitutive model. (**a**) Compressive damage plastic constitutive. (**b**) Tensile damage plastic constitutive.

4.2.2. Parameters of Each Component Material

In the mesoscopic simulation of concrete, coarse aggregates were considered linearly elastic materials due to their strength characteristics, and the numerical simulation results agree with the experimental data [42]. For the complex mechanical behavior of mortar and ITZ, the CDP model was adopted. This kind of treatment was consistent with the research ideas of Jin [43] and Xiao [44]. The mechanical parameters for the mortar were set as follows: the compressive strength and elastic modulus were directly taken from the experimental data shown in Figure 2. The tensile strength was determined as one-tenth of the compressive strength [45,46]. The Poisson ratio was established based on the data provided in the literature [47]. Regarding the mechanical properties of the ITZ, some studies [48,49] suggested that the ratio of ITZ to mortar mechanical properties typically ranges from 0.75 to 0.90. To quantify this characteristic, a value of 0.8 was adopted in this study. The specific material parameters for HPGC are detailed in Table 4.

Table 4. Mechanical property parameters of materials.

Items	Elastic Modulus (GPa)	Poisson Ratio (-)	Compressive Strength (MPa)	Tensile Strength (MPa)
Aggregate	73.00	0.16	-	-
Cement mortar	21.30	0.20	30.90	3.09
Cement ITZ	17.00	0.16	24.70	2.47
40% HPGM	14.50	0.20	19.40	1.94
40% HPGM ITZ	11.60	0.16	15.50	1.55
50% HPGM	14.30	0.20	16.60	1.66
50% HPGM ITZ	11.40	0.16	13.30	1.33
60% HPGM	12.10	0.20	16.10	1.61
60% HPGM ITZ	9.70	0.16	12.90	1.29

4.3. Validation of the Mesoscale Model

Based on the mechanical property test parameters of HPGM and HPGC, a 2D mesoscale concrete model was constructed in this study, aiming to reveal the mechanical response of HPGC under uniaxial compression conditions using numerical simulation. Considering the randomness of coarse aggregate distribution in actual concrete, the mesoscale model simulated three circular aggregate specimens with a size of 100 mm × 100 mm and a coarse aggregate content of approximately 40%. The ITZ thickness of the mesoscale model was determined to be 0.2 mm, according to Kim et al. [48]. To balance the computational efficiency and simulation accuracy, the model mesh size was set to 1 mm and divided using a quadrilateral-dominated mesh in ABAQUS software.

4.3.1. Verification of Damage Modes of HPGC

Some experimental studies have shown that the compressive damage patterns of concrete specimens in high- and low-friction environments exhibit two typical patterns: "X" and "V". This phenomenon was also confirmed by numerical simulation results [50,51]. Given the results presented above, this study simulated the failure of HPGC under the conditions of high friction and low friction, respectively, using fixed hinges to support the whole bottom or support only the midpoint of the bottom. The schematic diagrams are shown in Figure 10. Additionally, the detailed material parameters of coarse aggregates, 60% HPGM, and the corresponding ITZ in Table 4 were embedded in the model. As seen in Figure 11, HPGC specimens in the field test exhibited a bi-directional "X" damage pattern, which is consistent with the damage characteristics observed in the mesoscale model under high friction constraints. Therefore, the fixed hinge supporting all the bottoms was chosen for the subsequent studies.

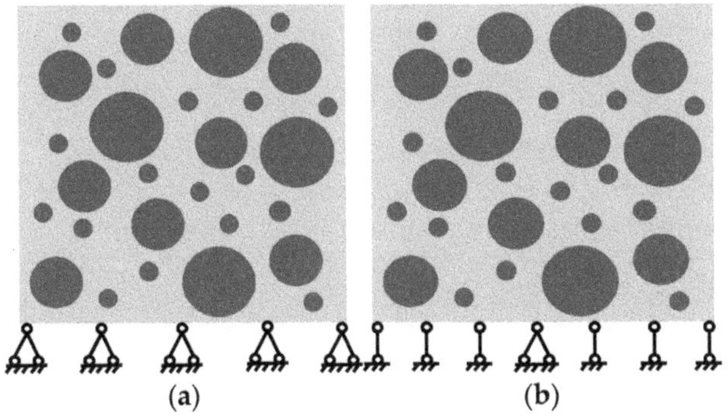

Figure 10. Bottom boundary conditions. (**a**) High friction. (**b**) Low friction.

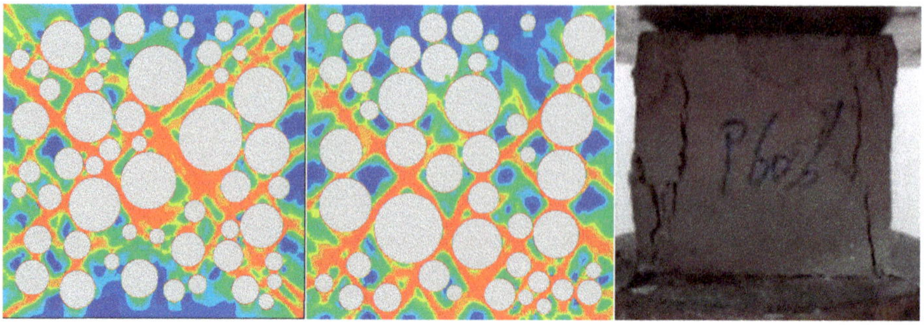

(a) (b) (c)

Figure 11. Comparison of failure modes of different bottom boundary conditions. (**a**) High friction. (**b**) Low friction. (**c**) Experimental shooting.

4.3.2. Verification of Mechanical Properties of HPGC

To further verify the effectiveness of the HPGC mesoscale model, the detailed material parameters of coarse aggregates, different components of HPGM, and the corresponding ITZ in Table 4 were embedded in the model. Subsequently, a compression test simulation study was carried out for HPGC. The comparison of deviation between the simulation results and the experimental results is summarized in Table 5. Through an in-depth analysis of Table 5, the uniaxial compressive strength and elastic modulus predicted by the HPGC model agreed well with the experimental results. Taking the HPG content as a variable, the maximum deviation of the simulated compressive strength from the measured value was only 5.74% when the HPG dosage was 40%, while the maximum difference in the elastic modulus was 8.38% when the HPG dosage was 60%. It is noteworthy that the maximum deviation between model predictions and experimental data never exceeded the 10% threshold, regardless of the variation in HPG content. This verified the high accuracy and reliability of the model in predicting the compressive strength and elastic modulus of HPGC.

Table 5. Deviations in mechanical properties of HPG concrete for each group.

	HPG (%)	Compressive Strength Experiment (MPa)	Compressive Strength Simulation (MPa)	Deviation (%)	Elastic Modulus Experiment (GPa)	Elastic Modulus Simulation (GPa)	Deviation (%)
HPGC	0	37.37	37.33	0.11	33.07	32.30	2.33
	40	19.87	18.73	5.74	26.83	24.95	7.12
	50	17.10	16.42	3.98	23.85	23.89	−0.17
	60	16.43	15.62	4.93	22.90	20.98	8.38

4.4. Analysis of Simulation Results

4.4.1. Effect of HPG Content on Compressive Mechanical Properties of HPGC

To investigate the effect of HPG content on the compressive mechanical properties of HPGC, this study employed a circular aggregate model to predict the compressive strength and elastic modulus of HPGC containing 40%, 50%, and 60% HPG content. Additionally, to minimize the effect of random aggregate distribution on the model outcomes, three working conditions were set for each group of HPGC to conduct the simulations. The results are shown in Figures 12 and 13.

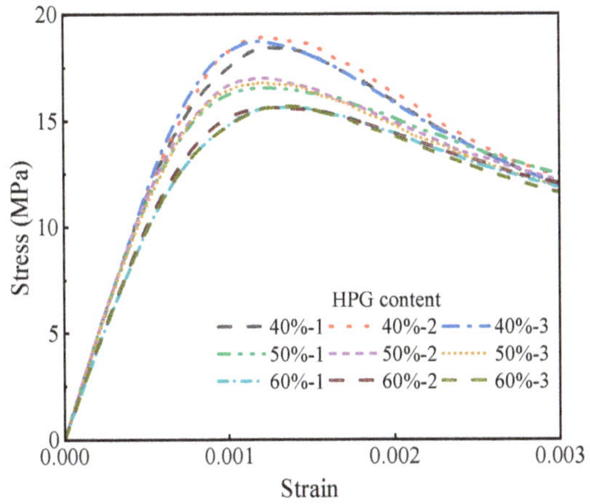

Figure 12. Simulation results of stress-strain curves for HPGC with varying HPG content.

Figure 12 revealed that the HPG content exerted a significant influence on the mechanical properties of HPGC, manifesting in alterations to the stress-strain curve morphology, compressive strength, and elastic modulus. When the HPG content increased, the peak stress and elastic modulus of HPGC showed a decreasing trend, while the peak strain shifted to the right. Under a fixed HPG dosage, the stress-strain curves of each group of HPGC specimens under the three working conditions showed the same overall direction, but the curves fluctuated slightly near the peak stress and in the softening stage. Specifically, these curves can be categorized into three stages: (a) The initial loading stage, where the load is small and there is no significant change in the specimens. During this stage, all specimens exhibit approximately linear behavior. (b) With the increase in load, damage and cracking begin to occur at the ITZ. Subsequently, the cracks gradually propagate into the mortar, resulting in a nonlinear behavior characterized by a reduction in stiffness in the ascending portion of the curve. (c) After the peak load, the cracks rapidly propagate and penetrate through the specimens, causing a decrease in stress until failure occurs. As seen in Figure 13, the increase in HPG substitution rate significantly reduced the compressive strength and elastic modulus of HPGC. The simulated trends were consistent with the experimental observations. Compared to the experimental reference concrete (37.37 MPa compressive strength, 33.07 GPa elastic modulus), HPGC simulations with 40%, 50%, and 60% HPG content exhibited decreases in compressive strength of 49.9%, 56.1%, and 58.1%, respectively, and in elastic moduli of 24.6%, 27.8%, and 36.6%, respectively. These findings indicated that with the HPG content above 40%, both the compressive strength and elastic modulus of HPGC decreased significantly. However, within the 40–60% HPG substitution range, the relative differences between the compressive strength and elastic modulus of HPGC remained narrow, with just 16.4% and 15.9%, respectively. This phenomenon is attributed to the fact that under compressive loading, the coarse aggregate remains largely intact, with the strength of HPGC primarily depending on the mechanical properties of the mortar matrix and ITZ. When the HPG substitution rate ranges from 40% to 60%, the mechanical properties of both the mortar matrix and the ITZ are significantly weakened, resulting in a marked decrease in the mechanical performance of HPGC. However, within this range, the differences in the mechanical properties of the mortar matrix and ITZ become insignificant, leading to minor variations in the results of the mechanical properties predicted by the HPGC mesoscale model.

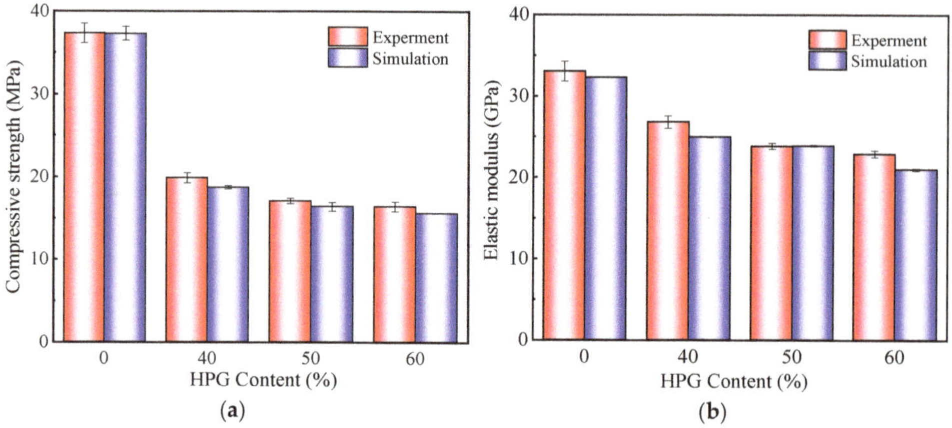

Figure 13. Comparison between experimental and simulation results of HPGC. (**a**) Compressive strength. (**b**) Elastic modulus.

4.4.2. Effect of Coarse Aggregate Shape on Compressive Mechanical Properties of HPGC

Due to the difference in the shape of coarse aggregates, concrete can be categorized into two main types: pebble-type concrete and gravel-type concrete. Accordingly, in this study, two types of HPGC mesoscale models with circular and polygonal aggregate shapes were constructed, with the other conditions remaining unchanged, aiming to investigate the effect of coarse aggregate shape on the mechanical properties and damage modes of HPGC. The results are, respectively, shown in Figures 14 and 15.

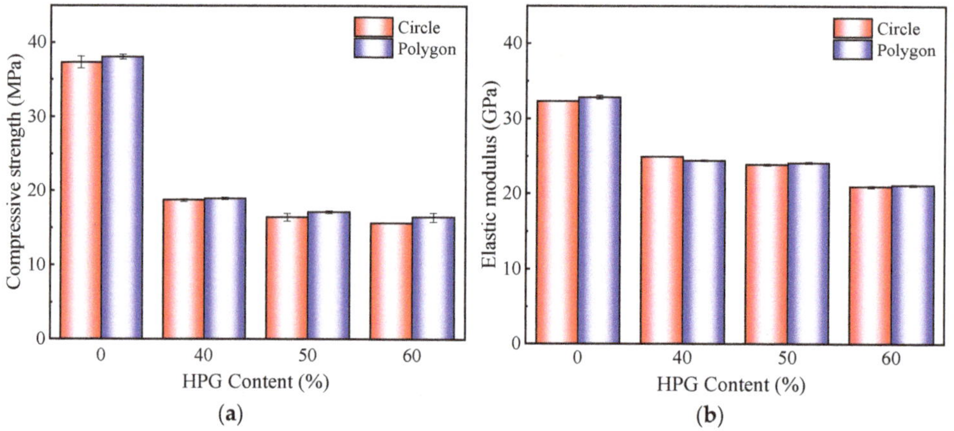

Figure 14. Comparison of simulation results of mechanical properties of HPGC with different aggregate shapes. (**a**) Compressive strength. (**b**) Elastic modulus.

As shown in Figure 14, the compressive strength and elastic modulus predictions of the polygonal aggregate model were slightly higher than those of the circular aggregate model. However, the discrepancies between the predictions of the two models were minimal, with none exceeding 1%. The differences in compressive strength were 1.0%, 0.9%, and 0.5%, respectively, and the differences in elastic moduli were 0.8%, 0.8%, and 0.5%, respectively. This comparison indicated that the shape of coarse aggregates had a negligible effect on the compressive strength and elastic modulus of HPGC. Although polygonal aggregates, being more angular than circular ones, enhance interlocking with the mortar matrix and strengthen HPGC integrity, they also lead to a significant increase in the ITZ content. Since the ITZ is the vulnerable section with inferior mechanical properties in HPGC, excessive ITZ negatively impacts its overall performance. In summary, the polygonal aggregate model shows a slight but insignificant advantage over the circular aggregate model. Figure 15 revealed the damage pattern of HPGC when subjected to external loading. In the initial loading stage, the damage first occurred at the ITZ. Then, as the internal stress of HPGC continuously increased, the damaged cracks at the ITZ gradually connected. Subsequently, the connected damage cracks extended to the adjacent mortar matrix region, leading to the redistribution of internal stresses in the HPGC. When the compressive stress in the HPGC reached its peak, the internal cracks in the mortar rapidly expanded and connected. Ultimately, this led to the formation of a penetrating main crack, which triggered the fracture damage of the HPGC. Comparing the damage modes of the circular and polygonal aggregate models, the damage process and final morphology of the two models were similar. In brief, the two models exhibited their characteristics in performance prediction for HPGC, and the numerical differences were minimal. To simplify the simulation process, the coarse aggregates in the subsequent models were uniformly adopted as circles.

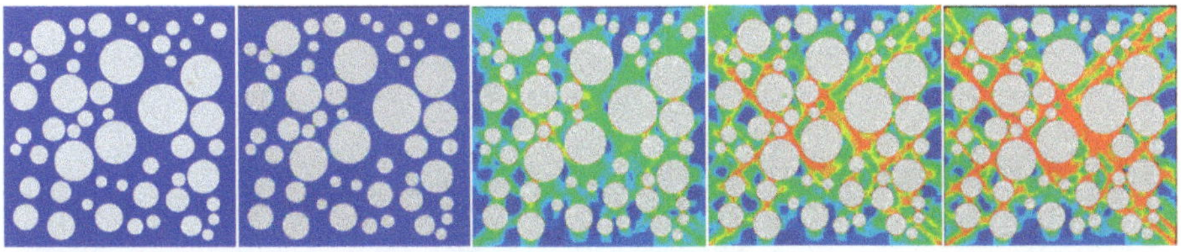

(a)

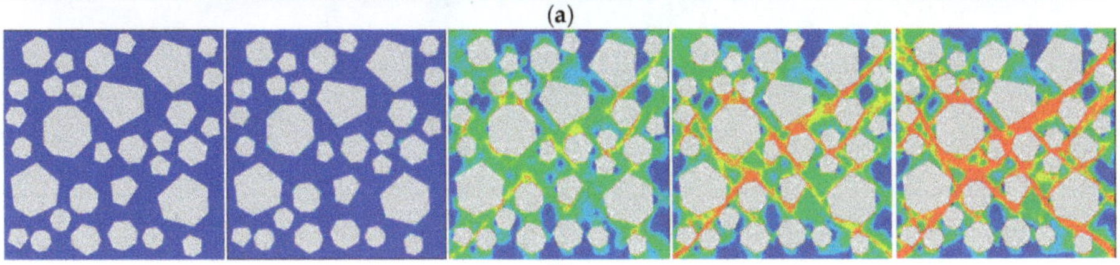

(b)

Figure 15. Comparison of failure mode of HPGC with different aggregate shapes. (**a**) The failure mode of HPGC with circular aggregate shape. (**b**) The failure mode of HPGC with polygonal aggregate shape.

4.4.3. Effect of Coarse Aggregate Content on Compressive Mechanical Properties of HPGC

Aggregates typically serve as the skeleton of concrete, and their content significantly influences the mechanical properties of the composite material. Therefore, the present study, varying the volume fractions of coarse aggregate at 20%, 25%, 30%, 35%, 40%, and 45%, established mesoscale models to assess the influence of aggregate content on the stress-strain behavior, compressive strength, and elastic modulus of HPGC. The simulation results are depicted in Figures 16 and 17.

As seen in Figure 16, the stress-strain curves for HPGC with 40%, 50%, and 60% HPG content exhibited similar overall trends upon comparison. However, slight fluctuations in the curves were noted near the peak stress and during the softening stage for each group of HPGC. Taking the HPGC with 40% HPG as an example, the peak strain of HPGC gradually shifted to the left with the increase in coarse aggregate content. The softening part of the HPGC stress-strain curve gradually became steeper, indicating that HPGC exhibited greater brittleness as the aggregate content increased. Specifically, as the coarse aggregate content increased, the compressive strength of HPGC with 40%, 50%, and 60%

HPG content showed a tendency to first decrease and then increase, while the elastic moduli gradually increased with the increase in coarse aggregate content, which was depicted in Figure 17. Fixing the coarse aggregate content at 25%, the compressive strengths of the three groups of HPGC all decreased to their lowest levels, which were 0.8%, 1.3%, and 1.5% lower than those at a coarse aggregate content of 20%, respectively. However, when the coarse aggregate content was increased to 45%, the compressive strength of the three groups of HPGC improved by 3.4%, 4.8%, and 6.7%, respectively, compared to the coarse aggregate content of 25%. The main reason is that, in the mortar matrix, coarse aggregates can be regarded as impurities. As the coarse aggregate content increases from 20% to 25%, the impurity content in the mortar matrix also rises, leading to an increase in internal defects, such as ITZ, pore, and crack content. This, in turn, reduces the overall mechanical properties of the HPGC. However, coarse aggregates generally exhibit better strength than the mortar matrix and can enhance the interlocking between coarse aggregates and the mortar matrix. With the further increase in coarse aggregate content, its improvement effect on the compressive strength of HPGC offsets its negative impact. In contrast, the elastic moduli of the three groups of HPGC show a positive correlation with the increases in coarse aggregate content. When the coarse aggregate content increases from 20% to 45%, the elastic modulus enhancement is 46.3%, 39.3%, and 50.7% for HPGC containing 40%, 50%, and 60% HPG, respectively. Thus, it is evident that the effect of coarse aggregate content on the elastic modulus of HPGC is more significant than its effect on compressive strength.

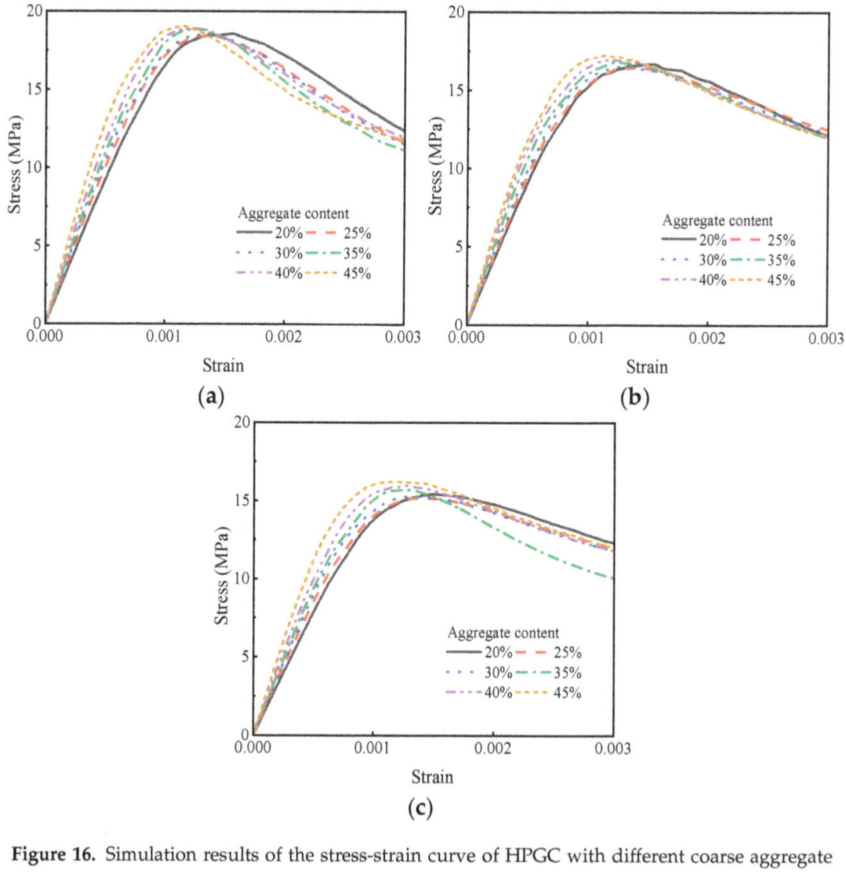

Figure 16. Simulation results of the stress-strain curve of HPGC with different coarse aggregate content. (**a**) HPGC with 40% HPG. (**b**) HPGC with 50% HPG. (**c**) HPGC with 60% HPG.

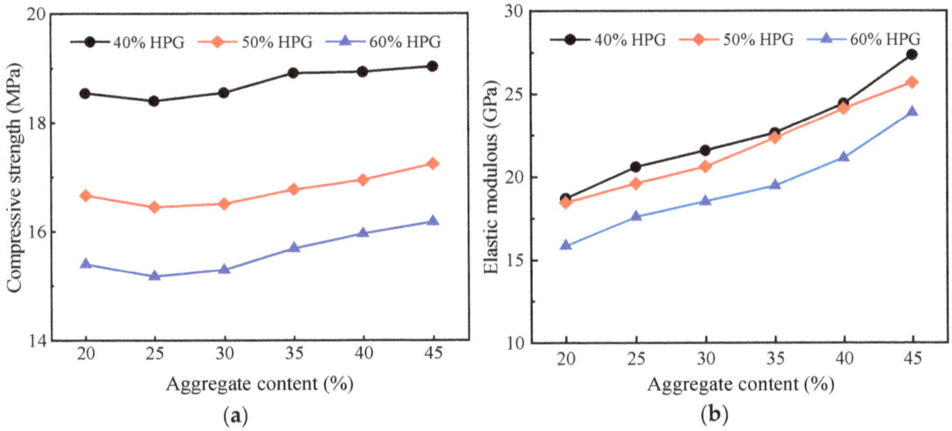

Figure 17. Simulation results of mechanical properties of HPGC with different aggregate content. (**a**) Compressive strength. (**b**) Elastic modulus.

4.4.4. Effect of Coarse Aggregate Elastic Modulus on Compressive Mechanical Properties of HPGC

Due to the variation in geographic locations, environments, and economic conditions, there are significant differences in the properties of coarse aggregates selected from various regions. Therefore, this study fixed both HPG and coarse aggregate content at 40%, varying the elastic modulus of coarse aggregates at 53 GPa, 63 GPa, 73 GPa, 83 GPa, and 93 GPa, aiming to investigate the impact of the elastic modulus of aggregates on the mechanical properties of HPGC. The resulting simulations are presented in Figure 18.

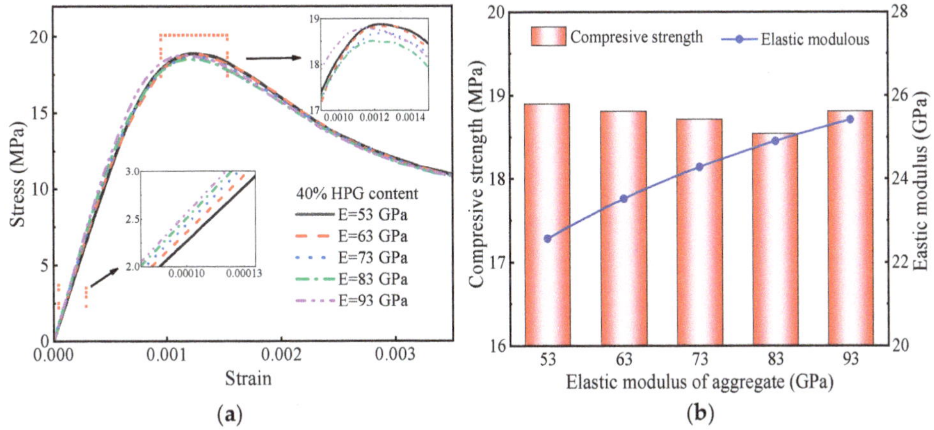

Figure 18. Simulation results of mechanical properties of HPGC with different elastic moduli of coarse aggregate. (**a**) Stress-strain curve. (**b**) Compressive strength and elastic modulus.

As shown in Figure 18, changes in the elastic modulus of the coarse aggregate had minimal impact on the overall trend of the HPGC stress-strain curve. However, this significantly influenced the compressive strength and elastic modulus of HPGC, which was particularly evident in the linear stage and near the peak stress, where noticeable fluctuations occurred. Specifically, as the elastic modulus of the coarse aggregate increased, the compressive strength of HPGC displayed an initial decrease, followed by an increase. When the elastic modulus of the coarse aggregate was set at 83 GPa, the HPGC exhibited its lowest compressive strength, at 18.54 MPa, which was 1.9% lower than the compressive strength of HPGC with a coarse aggregate elastic modulus of 53 GPa. The elastic modulus of HPGC gradually increased as the elastic modulus of the coarse aggregate rose. Notably, as the elastic modulus of the coarse aggregate elevated to 93 GPa, the elastic modulus of HPGC increased to 25.42 GPa, marking a 12.6% improvement compared to HPGC with a 53 GPa coarse aggregate elastic modulus. The above results revealed that the fluctuations in the elastic modulus of coarse aggregates, ranging from 53 to 93 GPa, obviously affected the compressive strength and elastic modulus of HPGC. However, the magnitude of the impact appeared to be slight, especially in the compressive strength. Additionally, the high elastic modulus of coarse aggregates did not necessarily improve the performance of HPGM. Similarly, other studies in the literature [52] also observed that concrete achieves greater strength when the elastic modulus of the aggregate closely matches that of the mortar. Conversely, the greater the difference between them, the lower the strength of the concrete. Therefore, in the subsequent preparation and applications of HPGC, it is advisable to choose coarse aggregates with an elastic modulus similar to the mortar, based on specific requirements, to achieve a higher compressive strength.

4.4.5. Effect of ITZ Thickness on Compressive Mechanical Properties of HPGC

At the mesoscale, ITZ is one of the critical components of concrete. Some studies indicated that the actual thickness of the ITZ in concrete is approximately 10–50 μm. However, when the ITZ thickness is set too small in the mesoscale model, mesh deformations may occur, making model convergence difficult. If the ITZ thickness is set between 100 and 800 μm, and appropriate mechanical parameters are selected, the numerical simulation results align well with experimental data [48,49]. Therefore, to investigate the effect of ITZ thickness on the mechanical properties of HPGC, the ITZ thicknesses were set to 0, 0.02, 0.06, 0.2, 0.6, and 1 mm in this study. The contents of coarse aggregates and HPG in the mesoscale model were fixed at 40%, and the elastic modulus of the coarse aggregates was set at 73 GPa. Based on these parameters, mechanical simulations were performed to analyze the concrete behavior at different ITZ thicknesses. The results are depicted in Figure 19.

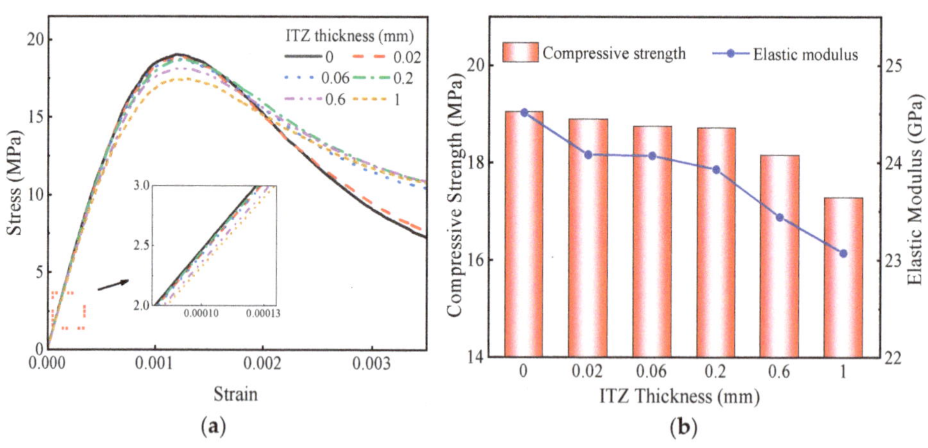

Figure 19. Simulation results of mechanical properties of HPGC with different ITZ thicknesses. (**a**) Stress-strain curve. (**b**) Compressive strength and elastic modulus.

As shown in Figure 19, the overall trend of HPGC stress-strain curves of different ITZ thicknesses appeared to be similar. Notably, for HPGC without ITZ or with an ITZ thickness in the range of 0 to 0.01 mm, the slope of the descending portion of the stress-strain curve exhibited a greater magnitude. As the ITZ thickness further increased, the slope tended to flatten. For the compressive strength and elastic modulus of HPGC, they showed a gradual decrease with the increase in ITZ thickness. When the ITZ thickness increased to 0.2 mm, the compressive strength and elastic modulus of the HPGC decreased to 18.89 MPa and 24.08 GPa, respectively. Compared to HPGC without an ITZ or with an ITZ thickness smaller than 10^{-2} mm, the differences were within 2% and could be negligible. As the ITZ thickness further increased to 1 mm, the HPGC exhibited its lowest compressive strength and elastic modulus at 17.29 MPa and 23.07 GPa, respectively. Compared to HPGC without an ITZ, the maximum reduction in compressive strength and elastic modulus was 9.2% and 5.9%, respectively. This phenomenon is attributed to the complex structure and inferior mechanical properties of the ITZ, which make it more susceptible to damage under external loads. Specifically, an increase in ITZ thickness results in a relative decrease in the effective mortar matrix content within the model. Consequently, compression damage and crack propagation occur more easily under compressive loads, reducing the mechanical performance of HPGM. In conclusion, the ITZ thickness significantly influences the mechanical properties and stress-strain behavior of concrete. Moreover, the simulation results indicate that an ITZ thickness of 0.2 mm is considered reasonable in this simulation and should not be ignored or taken to be too large or too small.

4.4.6. Effect of ITZ Mechanical Parameters (ITZMP) on Compressive Mechanical Properties of HPGC

Research has shown that coarse aggregates are considered linearly elastic materials, which have little effect on the mechanical properties of concrete [42]. However, the ITZ in concrete is a fragile component that directly and significantly affects the overall mechanical performance of the material [38,40]. Hence, in the mesoscale model, the ITZMP were set at ratios of 0.45, 0.6, 0.75, 0.8, 0.85, and 0.9 relative to the mechanical parameters of the mortar. The ITZ thickness was maintained at 0.2 mm in the mesoscale model, with 40% content of both coarse aggregates and HPG, and the elastic modulus of the coarse aggregate was fixed at 73 GPa. This approach was used to quantify the influence of ITZMP on the stress-strain curve, compressive strength, and elastic modulus of HPGC. The results are depicted in Figures 20 and 21.

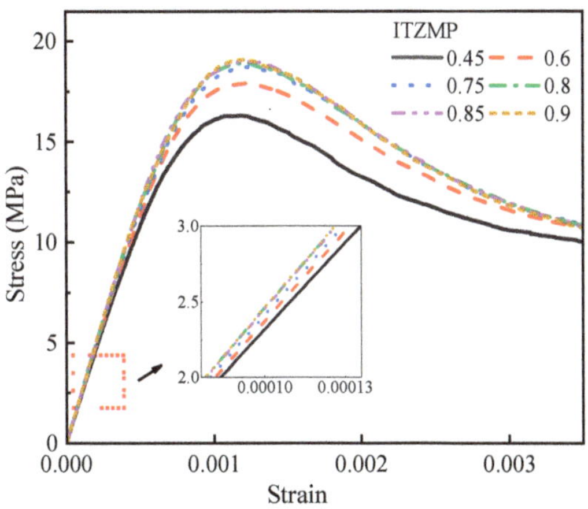

Figure 20. Simulation results of stress-strain curves of HPGC with different ITZMP.

As illustrated in Figure 20, with the increase in the ITZMP, both the compressive strength and elastic modulus of HPGC exhibited a gradual increasing tendency. Notably, as the ITZMP increased from 0.45 to 0.75, the softening stage of the stress-strain curve experienced a more pronounced change, characterized by a significant reduction in residual stress. As the ITZMP continued to increase, the overall change pattern of the stress-strain curve tended to moderate, with the variation being relatively limited. Specifically, from Figure 21, when the ITZMP increased from 0.6 to 0.75, the HPGC compressive strength rose by 4.6%. However, as the ITZMP increases from 0.75 to 0.9, the enhancement in HPGC compressive strength was only 1.7%. The findings indicate that although a moderate increase in the ITZMP positively influences the compressive strength of HPGC, this effect diminishes once the ITZMP exceeds a certain threshold. Data analysis reveals a significant quadratic relationship between the compressive strength of HPGC and ITZMP, with a correlation coefficient of 0.9991. In contrast, the correlation between the elastic modulus and ITZMP follows a linear trend, with a correlation coefficient of 0.9850. This is attributed to the fact that the ITZ, being the weakest section of HPGC, significantly affects its performance. An increase in ITZMP improves the mechanical properties of the ITZ, thereby enhancing the overall mechanical properties of HPGC.

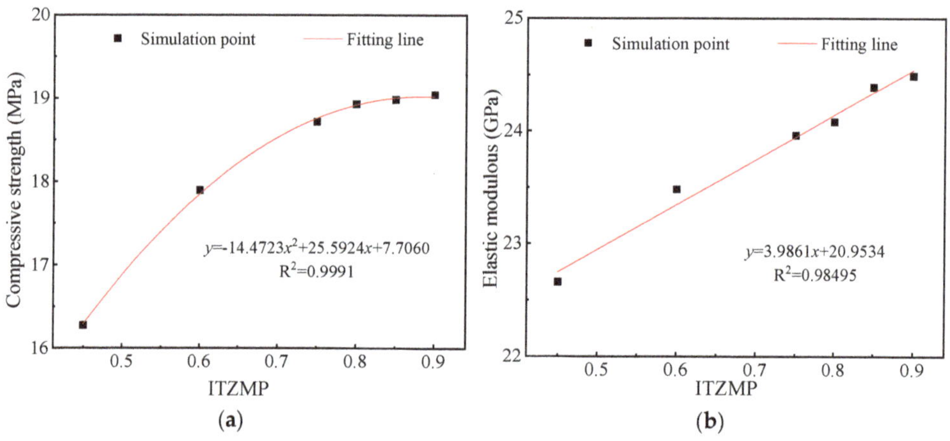

Figure 21. Simulation results of mechanical properties of HPGC with different ITZMP. (**a**) Compressive strength. (**b**) Elastic modulus.

5. Conclusions

In this study, a comprehensive investigation into the mechanical properties of HPGM and HPGC was conducted. Initially, experiments were conducted to explore the impact of HPG content and the addition of mineral admixtures on the mechanical properties of these materials. Following these experiments, a mesoscale model for HPGC was constructed to predict the effects of HPG content, coarse aggregate, and ITZ on the compressive strength and elastic modulus of HPGC. The main conclusions are as follows:

(1) The replacement cement with a high volume of HPG resulted in a significant decrease in the mechanical properties of HPGM and HPGC. Moreover, as the HPG content increased, the compressive strength and elastic modulus of HPGM and HPGC gradually decreased. Notably, the influence of HPG content on the compressive strength of HPGM and HPGC was higher than that on the elastic modulus.

(2) The incorporation of GGBS could significantly improve the mechanical properties of HPGM and HPGC, with 20% GGBS content being the optimal ratio for HPGM with a high volume of HPG. Additionally, adding an appropriate amount of quicklime and silica fume further enhanced the compressive strength and elastic moduli of HPGM and HPGC. The results demonstrate that the appropriate incorporation of mineral

admixtures can significantly enhance the mechanical properties of HPGM and HPGC at 40%, 50%, and 60% dosages. This improvement brings their strength levels in line with, or even exceeding, the requirements for the corresponding grades of mortar and concrete. This approach not only offers innovative strategies for the development and application of green building materials in practical engineering, but also provides an effective solution for mitigating the environmental impact of phosphogypsum.

(3) A 2D mesoscale model for HPGC was established, and the compressive strength and elastic moduli of HPGC with 40%, 50%, and 60% HPG content were predicted. Upon analysis, the simulation results agreed well with the experimental results. Furthermore, the maximum deviations of compressive strength and elastic modulus were 5.74% and 8.38%, respectively. Both of them were less than 10%, which verified the accuracy of the mesoscale model. Mesoscopic numerical simulation not only facilitates an in-depth analysis of the mechanical properties and failure modes of concrete in practical engineering applications, but also accurately predicts the effects of each component on the mechanical properties of concrete. Furthermore, it provides essential theoretical support for engineering design, reduces the trial-and-deviation process, and significantly improves engineering efficiency.

(4) The shape of the coarse aggregate had a minimal effect on the compressive strength and elastic modulus of HPGC, as predicted by the mesoscale model. As the elastic modulus and content of coarse aggregate increased, the compressive strength of HPGC initially decreased and then increased, while the elastic modulus of HPGC gradually increased. When the ITZ thickness increased, both the compressive strength and elastic modulus of HPGC decreased gradually. Conversely, as the ITZMP increased, the compressive strength and elastic modulus of HPGC gradually enhanced, although the rate of change diminished once a certain threshold was surpassed.

Author Contributions: Methodology, Y.C. and S.W.; Validation, F.Z., W.L. and G.S.; Formal analysis, Q.L., J.H. and Z.Y.; Investigation, L.W.; Resources, L.L.; Writing—original draft, X.C.; Writing—review & editing, P.L.; Funding acquisition, P.L. All authors have read and agreed to the published version of the manuscript.

Funding: This study was funded by the National Natural Science Foundation of China (grant numbers 52178182, 52108262 and U1934217), the Science and Technology Research and Development Program Project of China Railway Group Limited (Major special project, No.: 2020-Special-02, 2021-Special-08, 2022-Special-09; Major projects, No.: 2023-Major projects-07, 2023-Major projects-08, 2021-Special-02; Key project, No.: 2022-Key-46), and the Science and Technology Research and Development Subject of China Construction Second Engineering Bureau Ltd. (No.: 2024ZC120012 and 2023ZX12001). The authors have also received research grants from the Natural Science Foundation for Distinguished Young Scholars of Hunan Province (2022JJ10075), and the Hunan Science and Technology Plan Project (2023SK2014).

Institutional Review Board Statement: Not applicable.

Informed Consent Statement: Not applicable.

Data Availability Statement: The original contributions presented in the study are included in the article; further inquiries can be directed to the corresponding authors.

Conflicts of Interest: Authors Qizhi Li and Guangqiang Shao were employed by the China Construction Second Engineering Bureau Ltd. Authors Peng Liu and Zhiwu Yu were employed by the China Railway Group Ltd. Authors Feng Zhang and Wei Li were employed by the China Railway 20th Bureau Group Sixth Engineering Ltd. The remaining authors declare that the research was conducted in the absence of any commercial or financial relationships that could be construed as a potential conflict of interest.

References

1. Zhang, J.; Xie, W.; Dong, X.; Yang, H. Research progress on comprehensive utilization of phosphogypsum for materials: A revie *Mater. Rep.* **2023**, *37*, 22010110.
2. Wei, Z.Q.; Deng, Z.B. Research hotpots and trends of comprehensive utilization of phosphogypsum: Bibliometric analysis. *Environ. Radioact.* **2022**, *242*, 106778. [CrossRef]
3. Qin, X.; Cao, Y.; Guan, H.; Hu, Q.; Liu, Z.; Xu, J.; Hu, B.; Zhang, Z.; Luo, R. Resource utilization and development phosphogypsum-based materials in civil engineering. *J. Clean. Prod.* **2023**, *387*, 135858. [CrossRef]
4. Calderón-Morales, B.R.; García-Martínez, A.; Pineda, P.; García-Tenório, R. Valorization of phosphogypsum in cement-base materials: Limits and potential in eco-efficient construction. *J. Build. Eng.* **2021**, *44*, 102506. [CrossRef]
5. Zhang, T.Y.; Xie, F.; Guo, J.Y. Study on performance optimization and mechanism of phosphogypsumbased composite cemen tious materials. *Inorg. Chem. Ind.* **2022**, *54*, 136–145.
6. Shen, Y.; Qian, J.; Huang, Y.; Yang, D. Synthesis of belite sulfoaluminate-ternesite cements with phosphogypsum. *Cem. Con Compos.* **2015**, *63*, 67–75. [CrossRef]
7. Yang, L.; Zhang, Y.S.; Yan, Y. Utilization of original phosphogypsum as raw material for the preparation of self-leveling mortar. *Clean. Prod.* **2016**, *127*, 204–213. [CrossRef]
8. Huang, Y.; Lu, J.; Chen, F.; Shui, Z. The chloride permeability of persulphated phosphogypsum-slag cement concrete. *J. Wuha Univ. Technol.-Mater.* **2016**, *31*, 1031–1037. [CrossRef]
9. Jiang, G.; Wu, A.; Wang, Y.; Lan, W. Low cost and high efficiency utilization of hemihydrate phosphogypsum: Used as binder t prepare filling material. *Constr. Build. Mater.* **2018**, *167*, 263–270. [CrossRef]
10. Wu, H.K.; Cheng, X.; Wang, L.L. Research on the performances of high-temperature modified phosphogypsum-mineral powde cement mortar. *Non-Met. Mines* **2022**, *45*, 30–33.
11. Zhang, H.G.; Wu, Q.; Jia, X.F.; Chen, H.N.; Xie, G.Y. Experimental investigation on compressive strength of cast-in-sit phosphogypsum. *J. Hunan Univ.* **2016**, *43*, 127–134.
12. Gao, F.; Zhao, Z.M.; Quan, S.C.; Yao, Y.; Yang, L. Experimental study on elastic modulus and poisson ration of the buildin phosphogypsum. *Low Temp. Archit. Technol.* **2017**, *39*, 1–4.
13. Ma, K.J.; Gao, F.G.; Zhang, H.G.; Luo, Y.Q.; Duan, Y.Z.; Jiang, Y.W. Energy-saving and structure integration of new reinforce concrete structure system. *J. Guizhou Univ. Technol.* **2008**, *37*, 34–43.
14. Bagade, M.A.; Satone, S.R. An experimental investigation of partial replacement of cement by various percentage of phosphogy sum in cement concrete. *Int. J. Eng. Res. Appl.* **2012**, *2*, 785–787.
15. Wang, Y.Y.; Wang, J.R.; Liu, J.X.; Ren, J. Research on preparation of binding material and concrete with phosphogypsun *Multipurp. Util. Miner. Resour.* **2014**, *5*, 70–73.
16. Huang, Y.; Lin, Z.S. Investigation on phosphogypsum-steel slag-granulated blast-furnace slag-limestone cement. *Constr. Buil Mater.* **2010**, *24*, 1296–1301. [CrossRef]
17. Tian, T.; Yan, Y.; Hu, Z.; Xu, Y.; Chen, Y.; Shi, J. Utilization of original phosphogypsum for the preparation of foam concret *Constr. Build. Mater.* **2016**, *115*, 143–152. [CrossRef]
18. Li, B.; Li, L.; Chen, X.; Ma, Y.; Zhou, M. Modification of phosphogypsum using circulating fluidized bed fly ash and carbide sla for use as cement retarder. *Constr. Build. Mater.* **2022**, *338*, 127630. [CrossRef]
19. An, H.; Fu, R.; Lu, Y.; Kong, D.; Luo, S.; Wang, L.; Lv, F. Effect of heat treatment time on splitting tensile properties o phosphogypsum-based composite cementitious material. *Bull. Chin. Ceram. Soc.* **2022**, *41*, 545–552.
20. Liu, S.; Ouyang, J.; Ren, J. Mechanism of calcination modification of phosphogypsum and its effect on the hydration properties c phosphogypsum-based supersulfated cement. *Constr. Build. Mater.* **2020**, *243*, 118226. [CrossRef]
21. Jiang, G.Z.; Wu, A.X.; Wang, Y.M.; Li, J.Q. Effect of lime on properties of filling cementitious material prepared by hemihydrat phosphogypsum. *J. Chin. Ceram. Soc.* **2020**, *48*, 86–93.
22. Zhou, C.S.; Li, K.F.; Ma, F. Numerical and statistical analysis of elastic modulus of concrete as a three-phase heterogeneou composite. *Comput. Struct.* **2014**, *139*, 33–42. [CrossRef]
23. Zhou, Y.L.; Jin, H.; Wang, B.L. Modeling and mechanical influence of meso-scale concrete considering actual aggregate shape *Constr. Build. Mater.* **2019**, *228*, 116785. [CrossRef]
24. Wu, B.J.; Li, Z.X.; Tang, K.K. Numerical modeling on micro-to-macro evolution of crack network for concrete materials. *Theo Appl. Fract. Mech.* **2020**, *107*, 102525. [CrossRef]
25. Zheng, J.; Zhou, X.; Wu, Z.; Jin, X. Numerical method for predicting young's modulus of concrete with aggregate shape effect. *Mater. Civ. Eng.* **2011**, *23*, 1609–1615. [CrossRef]
26. Qu, H.C.; Chen, G.R. Numerical simulation of different-shaped random aggregates' influence on concrete's compression strengt *Adv. Mater. Res.* **2008**, *33–37*, 623–630. [CrossRef]
27. *GB/T 50081-2019*; Standard for Test Methods of Concrete Physical and Mechanical Properties. Ministry of Housing and Urba Rural Development of the People's Republic of China: Beijing, China, 2019.

- *JGJ/T 70-2009*; Standard for Test Method of Basic Properties of Construction Mortar. Ministry of Housing and Urban-Rural Development of the People's Republic of China: Beijing, China, 2009.
- Cheng, X.; Chen, S.H.; Hu, Q.; Liu, C.; Wang, L.L.; Kong, D.W. Study on the strength performance of modified phosphogypsum-based cement mortar. *Non-Met. Mines* **2022**, *45*, 37–40+45.
- Buhari, B.M.; Raju, R. Flexural behavior of ternary blended concrete containing silica fume and phosphogypsum. *Int. J. Sci. Technol. Eng.* **2016**, *3*, 9–14.
- Li, K.-Q.; Li, D.-Q.; Li, P.-T.; Liu, Y. Meso-mechanical investigations on the overall elastic properties of multi-phase construction materials using finite element method. *Constr. Build. Mater.* **2019**, *228*, 116727. [CrossRef]
- Zhou, X.Q.; Hao, H. Mesoscale modelling of concrete tensile failure mechanism at high strain rates. *Comput. Struct.* **2008**, *86*, 2013–2026. [CrossRef]
- Liu, Z.; Peng, H.; Cai, S.C. Mesoscale analysis of stress distribution along ITZs in recycled concrete with variously shaped aggregates under uniaxial compression. *J. Mater. Civ. Eng.* **2015**, *27*, 04015024. [CrossRef]
- Jin, L.; Wan, S.; Li, D.; Miao, L.; Du, X. Influence of concrete strength and aggregate size on the size effect of shear capacity of RC beam-column joints under monotonic loading. *Mesoscale Numer. Tests* **2022**, *37*, 671–685. [CrossRef]
- Cheng, X.; Cao, J.X.; Wang, L.L.; Zhao, Y.S.; Kong, D.W. Research progress in mechanical properties of concrete based on the mesoscopic model. *China Concr. Cem. Prod.* **2022**, *1*, 27–32+37. [CrossRef]
- Zheng, Z.; Zeng, C.; Tian, C.; Wei, X. Mesoscale numerical investigation on the size effect of concrete uniaxial compressive strength under different contact friction. *Constr. Build. Mater.* **2022**, *346*, 128416. [CrossRef]
- Wittmann, F.H.; Roelfstra, P.E.; Sadouki, H. Simulation and analysis of composite structures. *Mater. Sci. Eng.* **1985**, *68*, 239–248. [CrossRef]
- Walraven, J.C.; Reinhardt, H.W. Theory and experiments on the mechanical behavior of cracks in plain and reinforced concrete subjected to shear loading. *Heron* **1981**, *26*, 26–35.
- Jin, L.; Yu, W.; Du, X.; Yang, W. Meso-scale simulations of size effect on concrete dynamic splitting tensile strength: Influence of aggregate content and maximum aggregate size. *Eng. Fract. Mech.* **2020**, *230*, 106979. [CrossRef]
- Zheng, Y.X.; Zhang, Y.; Zhuo, J.B. Mesoscale synergistic effect mechanism of aggregate grading and specimen size on compressive strength of concrete with large aggregate size. *Constr. Build. Mater.* **2023**, *367*, 130346. [CrossRef]
- *GB 50010-2010*; Code for Design of Concrete Structures. Ministry of Housing and Urban-Rural Development of the People's Republic of China: Beijing, China, 2010.
- Xu, L.; Jin, Y.; Jing, S.; Liu, J.; Huang, Y.; Zhou, C. Comparisons of tensile fracturing behaviors of hydraulic fully graded and wet-screened concretes: A mesoscale study. *Adv. Mater. Sci. Eng.* **2018**, *2018*, 6965723. [CrossRef]
- Jin, L.; Yu, W.X.; Du, X.L. Size effect on static splitting tensile strength of concrete: Experimental and numerical studies. *J. Mater. Civ. Eng.* **2020**, *32*, 04020308. [CrossRef]
- Xiao, J.; Li, W.; Corr, D.J.; Shah, S.P. Effects of interfacial transition zones on the stress-strain behavior of modeled recycled aggregate concrete. *Cem. Concr. Res.* **2013**, *52*, 82–99. [CrossRef]
- Peng, Y.L.; Chu, H.; Pu, J.W. Numerical simulation of recycled concrete using convex aggregate model and base force element method. *Adv. Mater. Sci. Eng.* **2016**, *2016*, 5075109. [CrossRef]
- Jayasuriya, A.; Adams, M.P.; Bandelt, M.J. Understanding variability in recycled aggregate concrete mechanical properties through numerical simulation and statistical evaluation. *Constr. Build. Mater.* **2018**, *178*, 301–312. [CrossRef]
- Ma, H.F.; Chen, H.X.; Li, B.K. Meso-structure numerical simulation of concrete specimens. *J. Hydraul. Eng.* **2004**, *35*, 27–35.
- Kim, S.M.; Al-Rub, R.K.A. Meso-scale computational modeling of the plastic-damage response of cementitious composites. *Cem. Concr. Res.* **2011**, *41*, 339–358. [CrossRef]
- Pedersen, R.R.; Simone, A.; Sluys, L.J. Mesoscopic modeling and simulation of the dynamic tensile behavior of concrete. *Cem. Concr. Res.* **2013**, *50*, 74–87. [CrossRef]
- Chen, H.B.; Xu, B.; Wang, J. Parametric analysis on compressive strain rate effect of concrete using mesoscale modeling approach. *Constr. Build. Mater.* **2020**, *246*, 118375. [CrossRef]
- Xiao, J.; Li, W.; Liu, Q. Meso-level numerical simulation on mechanical properties of modeled recycled concrete under uniaxial compression. *J. Tongji Univ.* **2011**, *39*, 791–797.
- Kuang, G.P. *Study on Strain Rate Effect of Concrete Based on Three-Dimensional Mesomodel*; Harbin Institute of Technology: Harbin, China, 2017.

coatings

Article

Damage Detection of Beam Structures Using Displacement Differences and an Artificial Neural Network

Xudi Huang [1,2], Xi Peng [1,2,3,*], Fengjiang Qin [4,*], Qiuwei Yang [1,2,*] and Bin Xu [3]

1 School of Civil and Transportation Engineering, Ningbo University of Technology, Ningbo 315211, China; huangxudi0116@nbut.edu.cn
2 Engineering Research Center of Industrial Construction in Civil Engineering of Zhejiang, Ningbo University of Technology, Ningbo 315211, China
3 Ningbo Roaby Technology Industrial Group Co., Ltd., Ningbo 315800, China; 5632374376@aliyun.com
4 Key Laboratory of New Technology for Construction of Cities in Mountain Area, School of Civil Engineering, Chongqing University, Chongqing 400045, China
* Correspondence: pengxi@nbut.edu.cn (X.P.); qinfengjiang@cqu.edu.cn (F.Q.); yangqiuwei@nbut.edu.cn (Q.Y.)

Abstract: The beam structure constitutes a vital element in construction and bridge engineering. Static damage detection technology provides a method for identifying potential damage by measuring static displacements, with the advantage of being easy to implement. In this work, a two-stage damage detection method is proposed to determine the location and severity of damage in beam structures. The first stage identifies the damage location based on the displacement difference curves of the beam structure under static loading before and after the damage occurs. The second stage employs an artificial neural network to determine the severity of the damage. The proposed two-stage damage detection method has been validated in both a numerical model and an experimental model of beam structures. The following conclusions can be drawn from both numerical simulations and experimental studies. Regardless of the loading position, the turning points in the displacement difference curves always occur in the damaged regions, indicating that the damage locations in the beam structure can be determined by the turning points of the displacement difference curves. A single inflection point in the displacement difference curve indicates the presence of a single damage, while multiple inflection points indicate the existence of multiple damaged elements, with each inflection point corresponding to a damaged location. Furthermore, the severity of the damage can be accurately calculated using an artificial neural network. For experimental example 1, the damage locations identified by the proposed method all fall within the actual damage area, and the average error between the obtained damage severity and the true value is approximately 3.8%. For experimental example 2, the distance error between the damage location identified by the method and the actual damage location is approximately 1.4%, and the error between the obtained damage severity and the true value is approximately 2.8%. This two-stage damage detection method is more convenient to implement than traditional detection methods because it can precisely identify damage in beam structures using only a small amount of displacement data, providing a simple and highly practical solution for detecting defects in beam structures.

Keywords: artificial neural network; beam structures; damage detection; displacement difference

1. Introduction

Damage identification in beam structures is crucial for maintaining the safety, reliability, and efficiency of numerous infrastructures, including bridges, buildings, and various industrial applications. Beam structures, as fundamental load-bearing elements, are constantly exposed to a variety of stressors and loads that can lead to the gradual onset of damage over time. Identifying damage in beam structures is vital for ensuring public safety. Structures that support significant loads or house critical functionalities cannot afford to have undetected structural weaknesses. By regularly inspecting and identifying damages, engineers can predict potential failure points and implement necessary repairs to prevent catastrophic collapses, thereby safeguarding human life and property. Early detection of damage can significantly extend the service life of beam structures and reduce maintenance costs. Minor cracks or corrosion, if identified promptly, can often be repaired with minimal intervention and cost. However, if left undetected, these minor issues can evolve into major structural problems, necessitating extensive and costly repairs or even complete replacement of components.

Various technologies and software have been used for static testing and assessing damage/deformation in beam structures. Wen et al. [1] proposed a baseline-free damage localization method for statically indeterminate beam structures using dual responses (strain and displacement) excited by quasi-static moving loads. They found that combining strain and displacement responses in damaged states avoids the need for baselines. Wen et al. [2] developed a two-stage method that quantifies structural damage using displacement responses from quasi-static moving loads. They discovered this method utilizes rich test data from moving load tests better and quickly locates and quantifies damage with fewer sensors. Greco et al. [3] proposed a program for identifying and reconstructing damage in beam structures using genetic algorithms in specialized software. Oudah et al. [4] studied damage evolution and deformation in SSB and DSB connections in large-scale tests, using digital image techniques to evaluate deformation contributions. They found that moving vertical grooves on columns reduced SSB ductility, with negligible effects in DSB within a certain distance.

Ma et al. [5] focused on using wavelet analysis to detect and identify multiple beam damages via static deflection. Experiments were conducted to pinpoint damage using the wavelet maximum line and assess severity based on coefficients. Jahangir et al. [6] applied wavelet transform to analyze curvature in reinforced concrete beams, accurately detecting damage locations and evaluating severity trends. Wu et al. [7] experimentally studied crack detection in beam structures under static displacement using spatial wavelet transform, finding it effective in identifying damaged areas. Pooya et al. [8] introduced a damage detection method using only dynamic data, which proved effective through finite element models of beams with different shapes and boundary conditions. Yang et al. [9] proposed a static algorithm for locating damage in beams under moving loads, verified through simulations. The algorithm identifies damage by analyzing deflection curve curvature spikes and works without baseline data. Yazdanpanah et al. [10] introduced a crack location method in beam–column structures considering axial loads, proven accurate in low-load conditions. Shi et al. [11] derived a sensitivity matrix linking strain mode to stiffness, treating noise as probabilistic, and found the combination index useful in damage identification. Peng et al. [12] proposed a static shear energy algorithm for beam-like structures, using energy changes to locate damage. Hashemi et al. [13] proposed a crack diagnosis method for beam–column structures considering axial loads and random noise, effectively identifying single and multiple damages.

Ma et al. [14] proposed a method to identify damage from static measurements by analyzing structural deflection slopes and comparing bending stiffness. This method can detect and quantify small damages. Le et al. [15] enhanced damage location and quantification using modal flexibility matrix-estimated deflection changes, achieving accurate results with multiple vibration modes, often better than traditional static methods. Zhang et al. [16] introduced a method using quasi-static displacement measurements for beam bridge damage detection. Yang et al. [17] introduced a combined static and dynamic method for structural damage detection, which mixes static and dynamic data in the equilibrium equation solved by Moore Penrose inverse. This method considers the displacement difference and modal force vector variation between damaged and undamaged states. Our static and dynamic methods were tested on shear-type frames and effectively identified single and multiple damage locations and severity levels. For structures with multiple damages, this combination method is superior to individual static or dynamic methods. Chen et al. [18] developed a bridge structure damage identification model. This model effectively solves the problems of low accuracy and high positioning error. The experimental results show that the use of static test data and BP neural network significantly improves the accuracy of bridge structure damage identification, reducing identification and positioning errors. Wu et al. [19] introduced a new damage identification method for beam structures with uncertain static data, considering both measurement and modeling errors. Results showed high accuracy and efficiency, especially with large errors, and successfully identified beam damage in tests. Shen et al. [20] proposed a damage identification method using force residual vectors defined by static displacement data and finite element model stiffness matrix. The non-zero elements in this vector represent the structural elements of the damage. The severity of the damage is calculated using the equilibrium equation based on the overall stiffness matrix of the damaged element. Alkhabbaz, Yang, and Daabo, among others, have investigated the issue of damage identification in offshore wind turbine structures [21–23].

Overall, existing damage identification methods often require the placement of a large number of sensors, resulting in relatively high detection costs. In this study, a novel two-phase approach is introduced to assess the location and extent of damage in beam structures. The initial phase pinpoints damage locations by analyzing displacement deviation profiles under static loading conditions, comparing pre- and post-damage states. The subsequent phase leverages an artificial neural network to quantify the damage severity. This two-phase methodology has undergone rigorous validation through both numerical simulations and experimental setups involving beam structures. Key observations from both avenues of study are as follows: Irrespective of the loading point, the displacement deviation profiles consistently exhibit inflection points within damaged areas, thus serving as reliable indicators of damage locations. A solitary inflection point signifies a single damage instance, whereas multiple inflection points denote the presence of several damaged components, each corresponding to a specific location. Additionally, the severity of damage can be precisely estimated using the artificial neural network. The main advantages of the proposed method are as follows: (1) Only a small number of displacement meters are required for damage localization in beam structures, leading to lower detection costs; (2) After determining the damage location, the damage severity is calculated using an artificial neural network (ANN). This significantly reduces the scale of the required ANN, allowing for the construction of a simpler neural network model to solve for the damage severity with low computational costs. This two-phase approach boasts ease of implementation, necessitating only minimal displacement data to accurately detect damage in beam structures, thereby furnishing a straightforward and highly practical strategy for identifying structural flaws. The structure of this article is outlined below: Section 2 elaborates on the computational formula and operational steps of the introduced two-phase damage detection technique, which integrates displacement difference analysis with an artificial neural network (ANN). Section 3 substantiates the proposed methodology using a numerical simulation of a 32-element simply supported beam structure. Section 4 reinforces the method's effectiveness through experimental beam testing. Finally, Section 5 summarizes the primary research findings.

2. Theoretical Development

2.1. Damage Localization

Assume that K_u and d_u are the stiffness matrix and displacement before damage, and K_d and d_d represent the stiffness matrix and displacement after damage, respectively. The damage can be the result of various factors, such as material fatigue, external impacts, or environmental conditions, and these changes are reflected in the altered stiffness matrix and displacement vector. For a complete structure, the static analysis model can be represented as [24]

$$K_u \cdot d_u = f \tag{1}$$

$$K_d \cdot d_d = f \tag{2}$$

Suppose Δd is the displacement before and after damage, f is the load vector, and ΔK is the change in the stiffness matrix caused by damage. Therefore, the change in static displacement Δd and stiffness matrix ΔK before and after structural damage can be calculated by the following formula:

$$\Delta d = d_d - d_u \tag{3}$$

$$\Delta K = K_u - K_d \tag{4}$$

Substitute Equations (3) and (4) into (2) to obtain:

$$K_u \cdot d_u + K_u \cdot \Delta d \ - \Delta K \cdot d_u - \Delta K \cdot \Delta d = f \tag{5}$$

Substituting Equation (1) into Equation (5) yields

$$K_u \cdot \Delta d = \Delta K(\Delta d + d_u) = \Delta K \cdot d_d \tag{6}$$

Therefore, the displacement variation (DV) Δd can be obtained as

$$\Delta d = K_u^{-1} \cdot \Delta K \cdot d_d \tag{7}$$

In the structural finite element model, the global stiffness matrix K of the intact structure is a sum of the elemental stiffness matrices, i.e.,

$$\Delta K = \sum_{i=1}^{N} \alpha_i K_i \tag{8}$$

In Equation (8), α_i is the damage extent of the i-th element, and K_i is the i-th elemental stiffness matrix. For a single damage, without loss of generality, the i-th unit damage is generally taken, and Formula (7) is simplified as follows.

$$\Delta d = K_u^{-1} \cdot K_i \cdot \alpha_i \cdot d_d \tag{9}$$

The elemental stiffness matrix K_i is a sparse matrix. Let $\eta_1, \cdots, \eta_r$ represent the non-zero column vectors in matrix K_i, i.e., $K_i = [0, \cdots, \eta_1, \cdots, \eta_r, \cdots, 0]$, then Equation (9) can be simplified as

$$\Delta d = \sum_{j=1}^{r} \beta_j \cdot \xi_j \tag{10}$$

$$\beta_j = \alpha_i \cdot d_{d,j} \tag{11}$$

$$\xi_j = K_u^{-1} \cdot \eta_j \tag{12}$$

where $d_{d,j}$ is the j-th coefficient of the vector d_d, ξ_j is referred to as the characteristic displacement vector, and η_j is referred to as the characteristic force vector.

The Equation (10) holds significant physical meaning, as it indicates that the displacement before and after structural damage is a linear combination of the characteristic displacements generated under the action of the characteristic force vectors corresponding to the damaged elements. The characteristic forces and characteristic displacements possess specificity. Next, the beam structure is used as an example to discuss some patterns or laws related to characteristic forces and characteristic displacements.

Let us consider a double nodal Bernoulli Euler plane beam element, which is characterized by its simplicity and effectiveness in simulating structural bending deformation. This special element embodies four degrees of freedom, as shown in Figure 1, where each node has two degrees of freedom: one for translation in a direction perpendicular to the beam axis and the other for rotation around the beam axis. The vector denoting the displacement of the nodes and the stiffness matrix of the element in local coordinates can be described as follows.

$$u^e = [v_1, \theta_1, v_2, \theta_2]^T \tag{13}$$

$$K^e = \frac{EI}{L^3}\begin{bmatrix} 12 & 6L & -12 & 6L \\ 6L & 4L^2 & -6L & 2L^2 \\ -12 & -6L & 12 & -6L \\ 6L & 2L^2 & -6L & 4L^2 \end{bmatrix} \tag{14}$$

where E is the elastic modulus, I is the moment of inertia, and L is the length of the beam element. From Equation (14), the four characteristic forces of the beam element are given as follows:

$$[\eta_1, \eta_2, \eta_3, \eta_4] = H \cdot K^e \tag{15}$$

where H is the transformation matrix in the finite element model, which converts coordinates from the local coordinate system to the global coordinate system. From Equations (14) and (15), it can be observed that the non-zero elements of the characteristic force vector constitute a self-balancing force system. When this inherent self-equilibrating force is applied to the structure, as illustrated in Figure 2, a remarkable phenomenon occurs: the majority of internal forces within the structure remain null, with the exception of those elements directly associated with this unique force. This indicates that the force is highly localized, and its influence is confined to specific components of the structure. From the observations made in Figure 2, we can draw several important conclusions: (1) The characteristic force is an independent, self-equilibrating force that does not rely on external forces for its existence or balance. This makes it a distinct and isolated force within the structural system. (2) The characteristic force exerts its influence solely on its own constituents, leaving the rest of the structure unaffected. This means that the force does not propagate through the structure but rather remains confined to the specific elements upon which it acts upon.

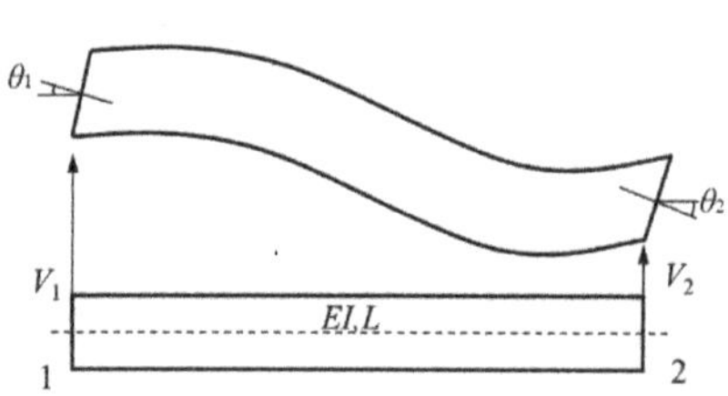

Figure 1. The two-node Bernoulli-Euler plane beam element with four DOFs [24].

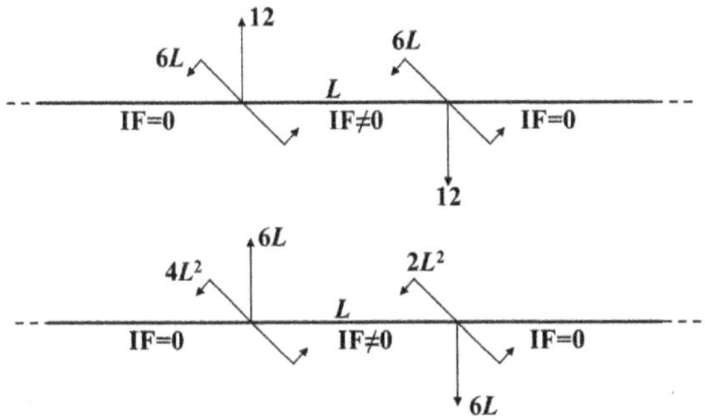

Figure 2. The internal force (IF) distribution in the structure under the action of the characteristic force [24].

In the context of a statically determinate structure, these conclusions have significant implications. Since only the element corresponding to the characteristic force undergoes deformation, the rest of the structure remains rigid and exhibits only rigid body motion. This means that the overall displacement of the structure under the influence of the characteristic force is composed of several components of rigid displacement, excluding the segment directly related to the force. Consequently, in the case of a statically determinate structure, the deformation is confined solely to the element affiliated with the characteristic force. The rest of the structure maintains its original shape and exhibits only rigid body motion, meaning it moves without any internal deformation. Essentially, when the characteristic force is applied, the overall displacement of the statically determinate structure is composed of multiple components of rigid displacement, except for the segment directly influenced by this distinctive force, which undergoes deformation. As previously discussed, any structural displacement change caused by damage can be understood as a linear combination of the characteristic displacements associated with the damaged units. This means that, under an arbitrary load, the displacement variation of the statically determinate structure will consist of several components of rigid displacement, excluding the segments where damage has occurred. It is worth noting that the last two columns of the unit stiffness matrix are the opposite of the first two columns, and the expression is similar to the one illustrated in the figure mentioned earlier. This symmetry in the stiffness matrix reflects the balanced nature of the forces and displacements within the structure.

To illustrate this point, consider the example of a simply supported beam under point load. In this case, the displacement change at the support is zero and increases linearly with the mid-span movement towards the beam. For each damaged location, the displacement change reaches its peak at the damaged component. This indicates that for simply supported beams, the area with the greatest displacement change is the damaged area.

For statically indeterminate structures, internal forces in some parts under characteristic force are slightly positive due to redundant constraints. This results in characteristic and arbitrary load displacements comprising deformational and rigid displacements. Figure 3 illustrates this for a fixed-fixed beam, showing characteristic and damage-induced displacement variations with curved segments due to fixed-end constraints. These curves gradually straighten as they move away from the fixed end towards the damage. Except for damaged areas and near restrained ends, most displacement variations can be approximated as rigid. Therefore, most of the displacement difference segments are approximately straight lines.

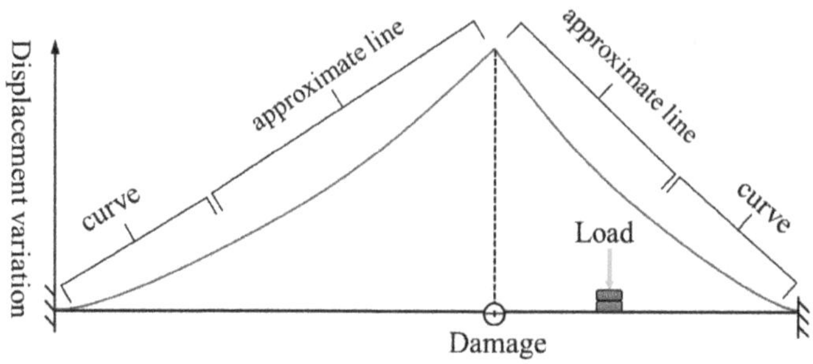

Figure 3. The displacement variation (DV) due to damage for the fixed-fixed beam under a point load [24].

To summarize, we can establish the following theorem: (1) For a statically determinate structure under any load, the displacement change caused by damage will be composed of several components of rigid displacement, excluding the damage position. This highlights the localized nature of the deformation in such structures. (2) For a statically indeterminate structure under an arbitrary load, the displacement variation due to damage may consist of both deformational displacements and rigid displacements. However, most of the deformation displacement changes can be approximated as rigid displacements, except for the damage position and the area near the restraint end. This indicates that, although the structure is more complex and the displacements are not entirely rigid, the majority of the displacement changes can still be understood in terms of rigid body motion. Based on the aforementioned principles, the criterion for locating damage in beam structures is derived from the fact that the inflection points in the displacement difference curves correspond to the damage locations in the structure. In cases of multiple damages, multiple inflection points will appear in the displacement difference curves.

2.2. Damage Quantification

On the basis that the damage locations have been identified, the artificial neural network (ANN) is further utilized to calculate the extent of damage in the beam structure. This approach of first locating the damage and then using an ANN to determine the damage severity can significantly reduce both the number of samples required for training the ANN and the number of neurons needed for the ANN model, as many undamaged units have already been excluded. The configuration of the ANN model is illustrated in Figure 4. The input layer's neuron count, labeled as n_i, and the output layer's neuron count, labeled as n_o, are determined by the quantity of measurement points and the quantity of damaged units, respectively. The hidden layer's neuron count, labeled as n_h, is established through empirical means. The number of neurons is crucial as it affects the model's ability to learn and generalize from the training data. A balance must be struck: too few neurons may lead to underfitting, where the model cannot capture potential patterns in the data, while too many neurons may lead to overfitting, where the model begins to model noise in the training data rather than actual relationships. In the following examples, the number of neurons used in each example is provided. This neural network model is trained using the backpropagation algorithm and is, therefore, also known as the BP neural network model. In the numerical example and experimental case presented later in this paper, the number of neurons in the input layer, hidden layer, and output layer are 4, 6, and 1, respectively.

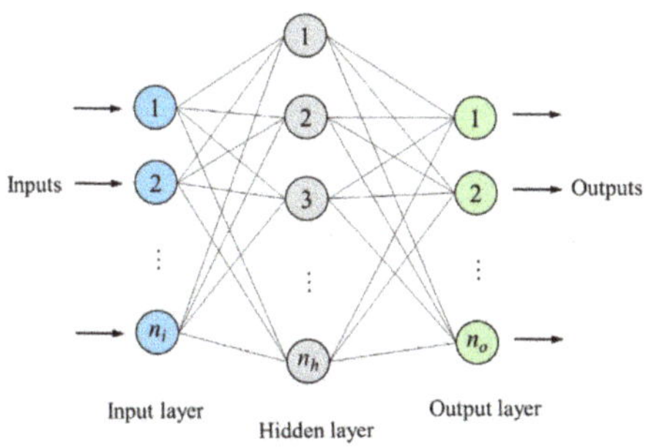

Figure 4. The ANN model.

In the end, the flow chart of the whole method is shown as Figure 5.

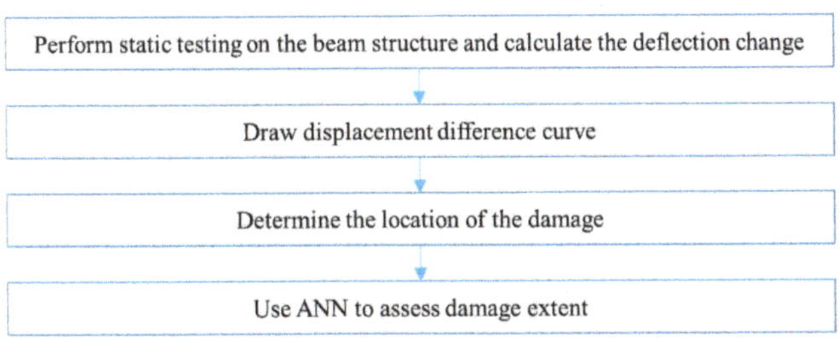

Figure 5. The flowchart of the two-stage damage detection method.

3. Numerical Example

The example of the numerical verification is a simply supported beam with 32 elements, as shown in Figure 6. The basic parameters of the structure are as follows: Young's modulus $E = 200$ GPa, density $\rho = 7.8 \times 10^3$ kg/m^3, moment of inertia $I = 1.04 \times 10^{-6}$ m^4, and cross-sectional area $A = 0.0025$ m^2. The cross-sectional width of the beam is about 3.5 cm, and the height is about 7.1 cm. Note that this example serves merely as a hypothetical model to illustrate the applicability of the method. This method is also applicable to beam structures with other geometric dimensions. The beam is modeled using 32 elements, and the length of each element is $L = 0.1$ m. The damage in the beam is simulated as a decrease in the Young's modulus of a single component.

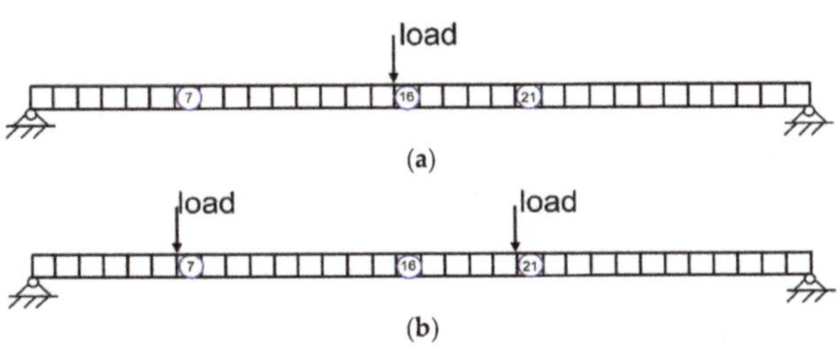

Figure 6. (**a**) Load condition 1; (**b**) Load condition 2. (The number in the circle represents the element number of this simply supported beam).

This example studied two types of damage scenarios. Damage scenario 1 involves a single damage to only unit 16, with damage severity values of 10%, 15%, and 20%, respectively. Damage scenario 2 involves units 7 and 21 sustaining the same degree of damage simultaneously, with damage severity values of 10%, 15%, and 20%, respectively.

Asymmetric loading increases the complexity of structural responses, which aids in identifying damage locations. Furthermore, asymmetric loading may be more conducive to revealing asymmetric damages within the structure compared to symmetric loading. It is noteworthy that the method presented in this paper has no special requirements for the loading mode and is applicable to both symmetric and asymmetric loading. Both types of loading are involved in the examples used in this paper. Figure 6a and 6b, respectively, represent two different static loading schemes. For load condition (LC) 1, Figure 7a shows the displacement differences before and after damage when element 16 is damaged, and Figure 7b shows the displacement differences before and after damage when elements 7 and 21 are simultaneously damaged. For load condition 2, Figure 7c shows the displacement differences before and after damage when element 16 is damaged, and Figure 7d shows the displacement differences before and after damage when elements 7 and 21 are simultaneously damaged. As can be seen from Figure 8, regardless of the variation in the position of the static load, the turning points on the displacement difference curves before and after damage consistently indicate the location of the damage. For a single damage scenario, the ANN has four neurons in the input layer, six neurons in the hidden layer, and one neuron in the output layer. For the scenario with two damage locations, the ANN has four neurons in the input layer, six neurons in the hidden layer, and two neurons in the output layer. By utilizing the trained ANN for this example based on simulated data, it can be determined that the damage degrees of the damaged elements in both damage scenarios are equal to the assumed values. These results demonstrate the feasibility of the method proposed in this paper for determining the damage location based on the inflection points of the displacement difference curves and further show the accuracy of using the trained ANN to calculate the damage severity.

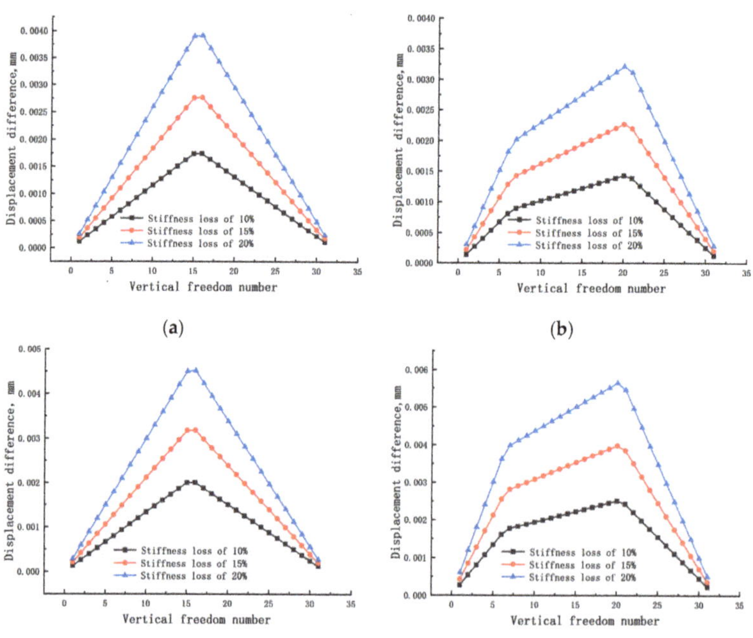

Figure 7. Displacement differences before and after damage. (**a**) Element 16 is damaged in LC1; (**b**) Elements 7 and 21 are damaged in LC1; (**c**) Element 16 is damaged in LC2; (**d**) Elements 7 and 21 are damaged in LC2. (unit: mm).

(**a**) Experimental steel beam under undamaged working conditions

(**b**) Experimental steel beam under damage conditions

(**c**) The damaged beams

Figure 8. Static test of beam structure.

4. Experimental Verification

4.1. Rectangular Section Beam

Due to the simulation of a simply supported beam in the above numerical example, the fixed beam at both ends is discussed in the experiment. This allows for a discussion on the applicability of the proposed method in various types of beam structures. As depicted in Figure 8, the methodology presented in this article has undergone further experimental validation on the fixed-end steel beam. The essential physical attributes of the structure encompassed an elastic modulus of 193 GPa, a density of 7850 kg/m^3, a beam length of 0.8 m, a cross-sectional width of 35 mm, and a cross-sectional height of 3 mm. Using a weight with a gravitational force of 10N, Figure 8a,b illustrate the static tests conducted under intact and damaged conditions, respectively. Figure 8c presents three test beam samples, each featuring a notch of the same dimensions, where the cut width (approximately 5 + 5 = 10 mm) accounted for 28.6% of the entire 35 mm cross-sectional width. Consequently, the actual extent of damage due to the cutting was approximately 28.6%. The geometric characteristics of the beam and the layout of measurement points are detailed in Figure 9.

During the static testing phase, four dial indicators (Ningbo, China) were employed to monitor the vertical displacements of nodes 3 to 6. The choice of these static measurement points was predicated on their location in the mid-span region, where the corresponding vertical displacement is notably large under the specified gravity load, facilitating accurate and straightforward measurements. As previously mentioned, asymmetric gravity loads can address the challenge of overlooked diagnoses in the detection of symmetrical structural flaws. Before conducting the experiment, a finite element model of this fixed-ended steel beam was established using finite element software, and the damage scenario was simulated to obtain numerical simulation results for the displacement differences at different loading points, as illustrated in Figure 10. In the figure, the solid line represents the actual measured displacement difference obtained from the measurement points, while the dashed line indicates the extension of the solid line. The hollow circle marks the intersection of the dashed lines, and its position on the beam is indicated by a dashed line with an arrow. The lines and points in the following figures have the same meaning with Figure 10. Regardless of the loading condition, the inflection points on the numerically simulated displacement difference curves in Figure 10 can clearly indicate the location of structural damage.

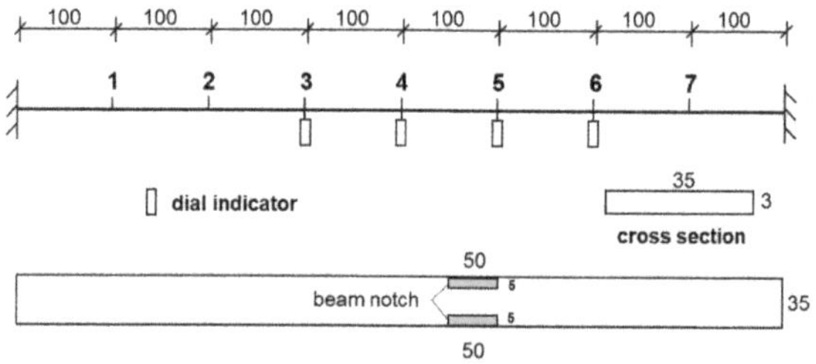

Figure 9. Geometric parameters, measurement locations, and size of the notch (unit: mm).

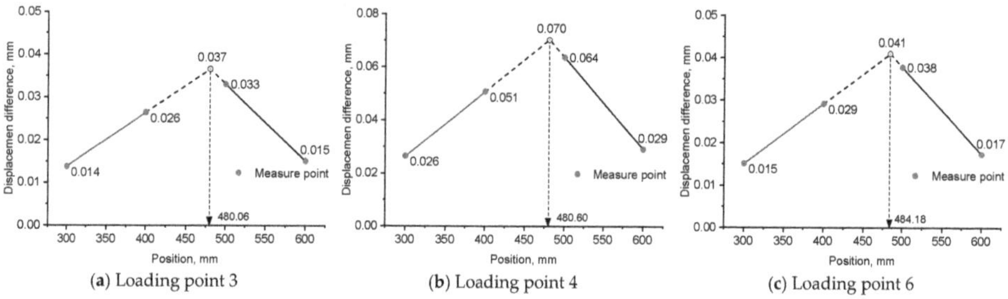

Figure 10. Numerical simulation results of the tested steel beam.

Subsequently, static loading tests were conducted on the three beam samples. The mean and standard deviation of the displacement difference data at various measurement points under three loading conditions are listed in Tables 1–3. Using the average displacement difference data from Tables 1–3, Figures 11b, 12b, and 13b present the displacement difference curves for the three loading conditions and the indicated damage locations. Under the three loading conditions, the identified damage positions from the left support were 465.17 mm, 463.64 mm, and 484.09 mm, respectively. All these positions fell within the actual range of the damage location, which spans between 450 mm and 500 mm. For this example, the ANN has four neurons in the input layer, six neurons in the hidden layer, and one neuron in the output layer. Using the trained ANN, the damage levels of the steel beam under three different loading methods were calculated as 27.63%, 29.11%, and 26.84%, which were close to the actual damage level (28.6%) of the steel beam due to cutting. The calculation errors for the damage levels under the three loading conditions were 3.40%, 1.78%, and 6.16%, respectively. Based on the above results, the proposed method accurately determined the location of damage defect in the steel beam structure. Furthermore, by utilizing an ANN, the severity of the damage can be precisely calculated.

Table 1. The mean and standard deviation of the displacement difference corresponding to loading point 3 (unit: mm).

Measure point	3	4	5	6
Mean value	0.013	0.027	0.030	0.014
Standard deviation	0.0020	0.0012	0.0020	0.0010

Table 2. The mean and standard deviation of the displacement difference corresponding to loading point 4 (unit: mm).

Measure point	3	4	5	6
Mean value	0.028	0.054	0.060	0.031
Standard deviation	0.0020	0.0026	0.0053	0.0026

Table 3. The mean and standard deviation of the displacement difference corresponding to loading point 6 (unit: mm).

Measure point	3	4	5	6
Mean value	0.015	0.027	0.035	0.018
Standard deviation	0.0044	0.0006	0.0026	0.0095

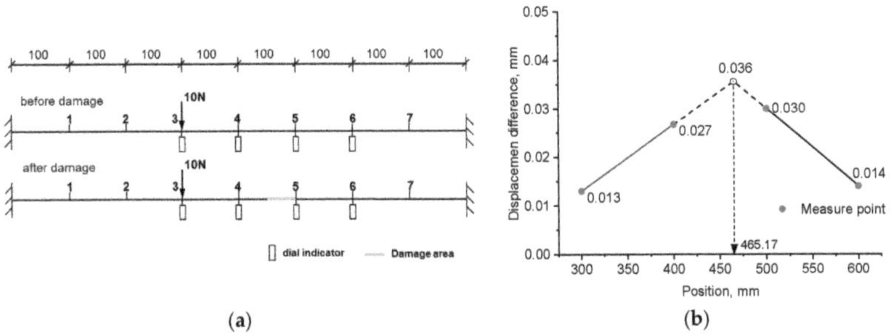

Figure 11. (**a**) Gravity load acts on node 6; (**b**) Displacement difference before and after damage obtained from the experiment (unit: mm).

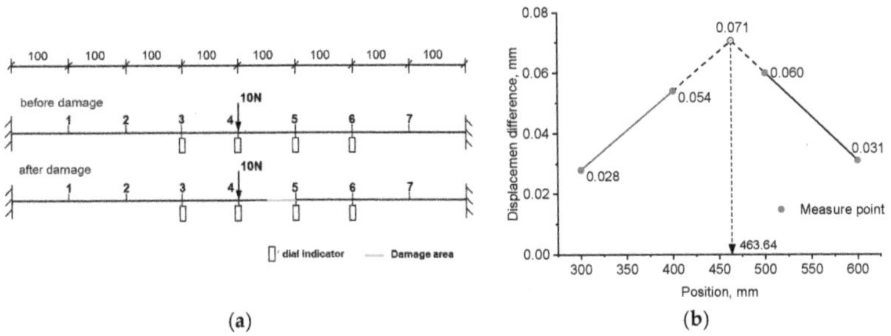

Figure 12. (**a**) Gravity load acts on node 4; (**b**) Displacement difference before and after damage obtained from the experiment (unit: mm).

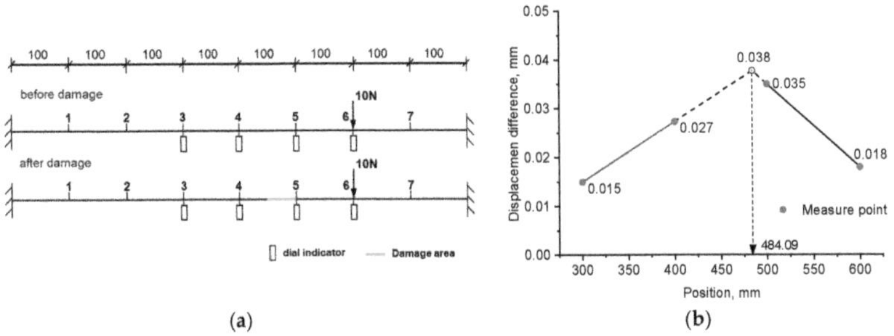

Figure 13. (**a**) Gravity load acts on node 3; (**b**) Displacement difference before and after damage obtained from the experiment (unit: mm).

4.2. Slotted Section Beam

An experimental steel beam conducted by Le et al. [25] was used to further verify the proposed method. The geometric dimensions and damage location of the beam structure are shown in Figure 14a. The specific physical parameters and process of the static test can be found in reference [25]. In the static test stage, several dial indicators were used to monitor the vertical displacement of several nodes (numbered from left to right as 1 to 6) shown in Figure 14a. The damage was simulated by cutting a part of the steel beam, and the cutting position was 1175 mm away from the left support. As shown in Figure 14b, the depths of the cuts were both 18 mm. The actual damage level of the beam, as shown in reference [25], was approximately 41.51%. Using the experimental data from reference [25], Figure 14c shows the line graph depicting the displacement difference before and after damage at measurement points 2–5 under gravity load, as well as the damage location indicated by the sharp corner of the curve. The identified damage location was 1158.3 mm away from the left support. Obviously, the identified damage location was very close to the actual damage location, with an error of approximately 1.4% when measured in terms of the distance from the left support. Similar to the previous two examples, since the location of the damage had already been identified, the constructed ANN model required only a small number of neurons. In this case, the ANN had four neurons in the input layer, six neurons in the hidden layer, and one neuron in the output layer. Using the trained ANN, the damage level of the steel beam was obtained as 42.67%, which was close to the actual damage level. These results demonstrate that the method proposed in this paper can accurately identify both the location and the extent of damage in this experimental steel beam.

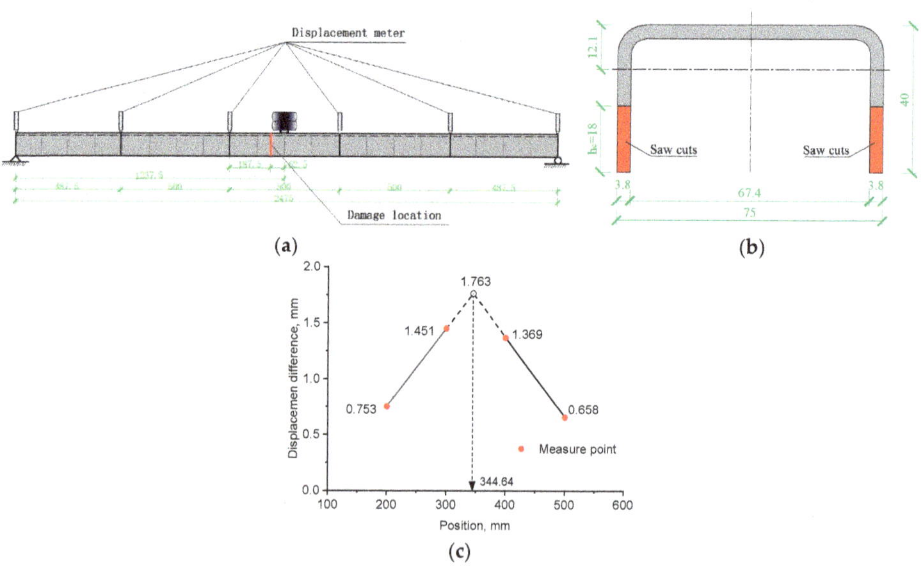

Figure 14. (**a**) Static test of beam structure; (**b**) Cross-sectional diagram (**c**) Displacement difference before and after damage (unit: μm).

5. Conclusions

This article introduced a two-phase approach for detecting damage in beam structures, aiming to pinpoint both the location and severity of the damage. In the first phase, the location of damage is identified by analyzing the displacement difference curve of the beam structure under static load. The second phase involves utilizing artificial neural networks to assess the severity of the damage. This two-phase damage detection method underwent validation through both numerical simulations and experimental models of beam structures. The results suggested that the algorithm has broad applicability in principle and significant value for structural damage detection. Based on the results presented in this work, the following main conclusions can be summarized:

As the severity of damage increases, the displacement difference before and after damage in the beam structure will also increase. However, the inflection points of the displacement difference curve always indicate the location of damage in the structure, regardless of the changing position of the static load.

1. When there is a single damage in the beam structure, the displacement difference curve will exhibit one inflection point. When multiple damages exist in the beam structure, the displacement difference curve will show multiple inflection points;
2. On the basis of the damage location already being determined, further adoption of a trained ANN can more accurately calculate the severity of structural damage. Pre-locating the damage can significantly reduce the number of samples required for training the ANN model, as well as decrease the number of neurons in the ANN model.
3. For experiment example 1, under three loading modes, the identified damage positions from the left support were 465.17 mm, 463.64 mm, and 484.09 mm, respectively. They all fell within the range of the actual damage location, which was between 450 mm and 500 mm. The trained neural network accurately calculated the severity of the damage, with errors of 3.40%, 1.78%, and 6.16% for the three loading modes, respectively.
4. For experiment example 2, the actual damage was simulated by cutting a portion of the steel beam at a distance of 1175 mm from the left support. The identified damage location was 1158.3 mm away from the left support. The distance error between the identified damage location and the actual damage location was 1.4%. The calculated severity of the damage (42.67%) was also very close to the actual degree of damage (41.51%), with an error of 2.8%.
5. The implementation of this two-phase damage detection approach is highly convenient, as it facilitates precise identification of damage in beam structures with minimal displacement data, offering a straightforward and immensely practical means for defect detection in such structures.

Author Contributions: Conceptualization, X.H., X.P., Q.Y. and F.Q.; methodology, X.P.; investigation, Q.Y., X.P. and F.Q.; writing—original draft preparation, X.H., X.P., Q.Y. and F.Q.; project administration, X.P. and B.X.; funding acquisition, B.X. All authors have read and agreed to the published version of the manuscript.

Funding: This work is supported by the Key Projects of Yongjiang Science and Technology Innovation 2035 in Ningbo (No. 2024Z090) and the Major Science and Technology Project of Ningbo High-Tech Zone in 2023 (No. 2023CX050001).

Institutional Review Board Statement: Not applicable.

Informed Consent Statement: Not applicable.

Data Availability Statement: The original contributions presented in this study are included in the article. Further inquiries can be directed to the corresponding authors.

Conflicts of Interest: Xi Peng and Bin Xu were employed by the company Ningbo Roaby Technology Industrial Group Co., Ltd. The remaining authors declare that the research was conducted in the absence of any commercial or financial relationships that could be construed as a potential conflict of interest.

Glossary

SSB	Single slotted beam system
DSB	Double slotted beam system
DC	Deflection change
BP	Backpropagation
ANN	Artificial neural network
IF	Internal force
diff.	Difference
MW	Megawatts
CFD	Computational fluid dynamics
DFBI	Dynamic fluid body interaction
VOF	Volume of fluid
LC	Load condition
Δd	Displacement before and after damage
K_u	Stiffness matrix before damage
d_u	Displacement before damage
K_d	Stiffness matrix after damage
d_d	Displacement after damage
f	Load vector
ΔK	The change in the stiffness matrix caused by damage
α_i	The damage extent of the i-th element
K_i	The i-th elemental stiffness matrix
$d_{d,j}$	The $j-$th coefficient of the vector d_d
ξ_j	Characteristic displacement vector
η_j	The characteristic force vector
K^e	The stiffness matrix of a single beam element
n_i	Input layer's neuron count
n_o	Output layer's neuron count
n_h	Hidden layer's neuron count
E	Young's modulus
ρ	Density
I	Moment of inertia
A	Cross-sectional area

References

1. He, W.Y.; Ren, W.X.; Zhu, S. Baseline-free damage localization method for statically determinate beam structures using dual-typ response induced by quasi-static moving load. *J. Sound Vib.* **2017**, *400*, 58–70. [CrossRef]
2. He, W.Y.; Ren, W.X.; Zhu, S. Damage detection of beam structures using quasi-static moving load induced displacement respons *Eng. Struct.* **2017**, *145*, 70–82. [CrossRef]
3. Greco, A.; Pluchino, A.; Cannizzaro, F. Closed-form solution based genetic algorithm software: Application to multiple crack detection on beam structures by static tests. *Appl. Soft Comput.* **2018**, *64*, 35–48. [CrossRef]
4. Oudah, F.; El-Hacha, R. Damage and deformation assessment of earthquake-resistant RC slotted beam-column connections usin digital image correlation technique. *Eng. Struct.* **2020**, *215*, 110442. [CrossRef]
5. Ma, Q.; Solís, M.; Galvín, P. Wavelet analysis of static deflections for multiple damage identification in beams. *Mech. Syst. Signa Process.* **2021**, *147*, 107103. [CrossRef]
6. Jahangir, H.; Hasani, H.; Esfahani, M.R. Wavelet-based damage localization and severity estimation of experimental RC beam subjected to gradual static bending tests. In *Structures*; Elsevier: Amsterdam, The Netherlands, 2021; Volume 34, pp. 3055–306
7. Wu, N.; Wang, Q. Experimental studies on damage detection of beam structures with wavelet transform. *Int. J. Eng. Sci.* **2011**, *4* 253–261. [CrossRef]
8. Pooya, S.M.H.; Massumi, A. A novel damage detection method in beam-like structures based on the relation between moda kinetic energy and modal strain energy and using only damaged structure data. *J. Sound Vib.* **2022**, *530*, 116943. [CrossRef]
9. Yang, Q.; Liu, J.; Sun, B. Damage localization for beam structure by moving load. *Adv. Mech. Eng.* **2017**, *9*, 91–105. [CrossRef]

Omid, Y.; Ali, R.I.; Mehdi, D. Static data based damage localization of beam-column structures considering axial load. *Mech. Adv. Mater. Struct.* **2020**, *27*, 1433–1450.

Shi, Q.; Hu, K.; Wang, L.; Wang, X. Uncertain identification method of structural damage for beam-like structures based on strain modes with noises. *Appl. Math. Comput.* **2021**, *390*, 125682. [CrossRef]

Peng, X.; Yang, Q.W. Damage detection in beam-like structures using static shear energy redistribution. *Front. Struct. Civ. Eng.* **2023**, *16*, 1552–1564. [CrossRef]

Hashemi, M.S.; Izadifard, R.A.; Yazdanpanah, O. Experimental static data-based damage localization of beam-like structures considering axial load. *Inverse Probl. Sci. Eng.* **2021**, *29*, 1729–1745. [CrossRef]

Ma, Q.Y.; Mario, S. Damage localization and quantification in beams from slope discontinuities in static deflections. *Smart Struct. Syst.* **2018**, *22*, 291–302.

Le, T.N.; Nguyen, A.; Thambiratnam, P.D.; Chan, T.H.T.; Khuc, T. Locating and Quantifying Damage in Beam-like Structures Using Modal Flexibility-based Deflection Changes. *Int. J. Struct. Stab. Dyn.* **2020**, *20*, 20. [CrossRef]

Zhang, S.; Liu, Y. Damage Detection in Beam Bridges Using Quasi-static Displacement Influence Lines. *Appl. Sci.* **2019**, *9*, 1805. [CrossRef]

Yang, J.; Li, P.; Yang, Y.; Xu, D. An improved EMD method for modal identification and a combined static-dynamic method for damage detection. *J. Sound Vib.* **2018**, *420*, 242–260. [CrossRef]

Chen, Y.; Liu, R.; Zheng, S. Identification and Diagnosis of Bridge Structural Damage Based on Static Test Data. *Iran. J. Sci. Technol. Trans. Civ. Eng.* **2024**, *48*, 2981–2996. [CrossRef]

Wang, Z.F.; Huang, B.; Tee, K.F.; Zhang, W.D. A Novel Stochastic Approach for Static Damage Identification of Beam Structures Using Homotopy Analysis Algorithm. *Sensors* **2021**, *21*, 2366. [CrossRef]

Shen, J.B.; Li, Z.K.; Luo, S.; Wang, W. A Structural Damage Identification Method Based on Arrangement of the Static Force Residual Vector. *Front. Mater.* **2022**, *9*, 918069. [CrossRef]

Alkhabbaz, A.; Hamza, H.; Daabo, M.A.; Yang, H.S.; Yoon, M.; Koprulu, A.; Lee, Y.H. The aero-hydrodynamic interference impact on the NREL 5-MW floating wind turbine experiencing surge motion. *Ocean Eng.* **2024**, *295*, 116970. [CrossRef]

Yang, S.H.; Alkhabbaz, A.; Tongphong, W.; Lee, H.Y. Cross-comparison analysis of environmental load components in extreme conditions for pontoon-connected semi-submersible FOWT using CFD and potential-based tools. *Ocean Eng.* **2024**, *304*, 117248. [CrossRef]

Daabo, M.A.; Alkhabbaz, A.; Ibrahim, S.S.; Hamzah, H.; Hassan, H.; Basem, A.; Easa, H.; Pavlovic, S. Thirteen vital factors for micro-scale radial turbine vane's design of geo-solar-powered Brayton cycle applications. *Energy Convers. Manag.* **2024**, *315*, 118774. [CrossRef]

Yang, Q.W.; Du, S.G.; Liang, C.F.; Yang, L.J. A Universal Model-Independent Algorithm for Structural Damage Localization. *CMES Comput. Model. Eng. Sci.* **2014**, *100*, 223–248.

Le, N.T.; Thambiratnam, D.P.; Nguyen, A.; Chan, T.H.T. A new method for locating and quantifying damage in beams from static deflection changes. *Eng. Struct.* **2019**, *180*, 779–792. [CrossRef]

MDPI AG
Grosspeteranlage 5
4052 Basel
Switzerland
Tel.: +41 61 683 77 34

Coatings Editorial Office
E-mail: coatings@mdpi.com
www.mdpi.com/journal/coatings

MDPI

www.ingramcontent.com/pod-product-compliance
Lightning Source LLC
Chambersburg PA
CBHW041016270926
51850CB00024B/372
* 9 7 8 3 7 2 5 8 5 6 9 9 2 *